Using Decision Support Systems for Transportation Planning Efficiency

Ebru V. Ocalir-Akunal
Gazi University, Turkey

A volume in the Advances in Civil and Industrial Engineering (ACIE) Book Series

ENGINEERING SCIENCE REFERENCE
An Imprint of IGI Global

Managing Director:	Lindsay Johnston
Managing Editor:	Keith Greenberg
Director of Intellectual Property & Contracts:	Jan Travers
Acquisitions Editor:	Kayla Wolfe
Production Editor:	Christina Henning
Development Editor:	Brandon Carbaugh
Cover Design:	Jason Mull

Published in the United States of America by
Engineering Science Reference (an imprint of IGI Global)
701 E. Chocolate Avenue
Hershey PA, USA 17033
Tel: 717-533-8845
Fax: 717-533-8661
E-mail: cust@igi-global.com
Web site: http://www.igi-global.com

Library of Congress Cataloging-in-Publication Data

Using decision support systems for transportation planning efficiency / Ebru Ocalir-Akunal, editor.
pages cm
Includes bibliographical references and index.
ISBN 978-1-4666-8648-9 (hardcover) -- ISBN 978-1-4666-8649-6 (ebook) 1. Transportation--Planning. 2. Transportation--Decision making. I. Ocalir-Akunal, Ebru, 1972-
HE147.5.U85 2015
388.068'4--dc23
2015015094

This book is published in the IGI Global book series Advances in Civil and Industrial Engineering (ACIE) (ISSN: 2326-6139; eISSN: 2326-6155)

British Cataloguing in Publication Data
A Cataloguing in Publication record for this book is available from the British Library.

Advances in Civil and Industrial Engineering (ACIE) Book Series

ISSN: 2326-6139
EISSN: 2326-6155

Mission

Private and public sector infrastructures begin to age, or require change in the face of developing technologies, the fields of civil and industrial engineering have become increasingly important as a method to mitigate and manage these changes. As governments and the public at large begin to grapple with climate change and growing populations, civil engineering has become more interdisciplinary and the need for publications that discuss the rapid changes and advancements in the field have become more in-demand. Additionally, private corporations and companies are facing similar changes and challenges, with the pressure for new and innovative methods being placed on those involved in industrial engineering.

The **Advances in Civil and Industrial Engineering (ACIE) Book Series** aims to present research and methodology that will provide solutions and discussions to meet such needs. The latest methodologies, applications, tools, and analysis will be published through the books included in **ACIE** in order to keep the available research in civil and industrial engineering as current and timely as possible.

Coverage

- Hydraulic Engineering
- Production Planning and Control
- Materials Management
- Coastal Engineering
- Ergonomics
- Productivity
- Operations Research
- Optimization Techniques
- Engineering Economics
- Structural Engineering

IGI Global is currently accepting manuscripts for publication within this series. To submit a proposal for a volume in this series, please contact our Acquisition Editors at Acquisitions@igi-global.com or visit: http://www.igi-global.com/publish/.

The Advances in Civil and Industrial Engineering (ACIE) Book Series (ISSN 2326-6139) is published by IGI Global, 701 E. Chocolate Avenue, Hershey, PA 17033-1240, USA, www.igi-global.com. This series is composed of titles available for purchase individually; each title is edited to be contextually exclusive from any other title within the series. For pricing and ordering information please visit http://www.igi-global.com/book-series/advances-civil-industrial-engineering/73673. Postmaster: Send all address changes to above address.

Titles in this Series

For a list of additional titles in this series, please visit: www.igi-global.com

Contemporary Ethical Issues in Engineering
Satya Sundar Sethy (Indian Institute of Technology Madras, India)
Engineering Science Reference • copyright 2015 • 343pp • H/C (ISBN: 9781466681309) • US $215.00 (our price)

Emerging Issues, Challenges, and Opportunities in Urban E-Planning
Carlos Nunes Silva (University of Lisbon, Portugal)
Engineering Science Reference • copyright 2015 • 380pp • H/C (ISBN: 9781466681507) • US $205.00 (our price)

Technology and Practice in Geotechnical Engineering
Joseph B. Adeyeri (Federal University Oye-Ekiti, Nigeria)
Engineering Science Reference • copyright 2015 • 836pp • H/C (ISBN: 9781466665057) • US $225.00 (our price)

Fracture and Damage Mechanics for Structural Engineering of Frames State-of-the-Art Industrial Applications
Julio Flórez-López (University of Los Andes, Venezuela) María Eugenia Marante (Lisandro Alvarado University, Venezuela) and Ricardo Picón (Lisandro Alvarado University, Venezuela)
Engineering Science Reference • copyright 2015 • 602pp • H/C (ISBN: 9781466663794) • US $225.00 (our price)

Computer-Mediated Briefing for Architects
Alexander Koutamanis (Delft University of Technology, The Netherlands)
Engineering Science Reference • copyright 2014 • 321pp • H/C (ISBN: 9781466646476) • US $180.00 (our price)

Technologies for Urban and Spatial Planning Virtual Cities and Territories
Nuno Norte Pinto (The University of Manchester, UK) José António Tenedório (Universidade NOVA de Lisboa, Portugal) António Pais Antunes (University of Coimbra, Portugal) and Josep Roca Cladera (Technical University of Catalonia, BarcelonaTech, Spain)
Information Science Reference • copyright 2014 • 349pp • H/C (ISBN: 9781466643499) • US $200.00 (our price)

Formal Methods in Manufacturing Systems Recent Advances
Zhiwu Li (Xidian University, People's Republic of China) and Abdulrahman M. Al-Ahmari (King Saud University, Saudi Arabia)
Engineering Science Reference • copyright 2013 • 531pp • H/C (ISBN: 9781466640344) • US $195.00 (our price)

Production and Manufacturing System Management Coordination Approaches and Multi-Site Planning
Paolo Renna (University of Basilicata, Italy)
Engineering Science Reference • copyright 2013 • 377pp • H/C (ISBN: 9781466620988) • US $195.00 (our price)

IGI GLOBAL
DISSEMINATOR OF KNOWLEDGE
www.igi-global.com

701 E. Chocolate Ave., Hershey, PA 17033
Order online at www.igi-global.com or call 717-533-8845 x100
To place a standing order for titles released in this series, contact: cust@igi-global.com
Mon-Fri 8:00 am - 5:00 pm (est) or fax 24 hours a day 717-533-8661

Editorial Advisory Board

Table of Contents

Detailed Table of Contents

rates of pier were predicted by ANN's. The superiorities of this proposed simulation-ANN model to other classical investment planning methods were the inclusion of uncertainties in the investment parameters like the change of cargo and costs variables in time, and the determination of project benefit/cost with an improved accuracy when compared to classical decision support models.

Since the European Union has certain targets and criteria that must be fulfilled by its members, Turkey's situation in transportation compared to the EU member countries (23 countries whose data were accessed) was examined through the approaches of decision support models (Data Envelopment Analysis and Malmquist Index). This study investigates whether Turkey utilizes its road, airway and railway transportation indicators efficiently or not within the framework of the European Union (EU) accession process. In addition, it aims to demonstrate Turkey's current position compared to other EU countries by performing relative efficiency analysis on road, airway and railway transportation indicators. The conclusions of this study will help policymakers to determine Turkey's policy objectives for its integration to the EU in terms of transportation indicators.

This article examines the capability of Extreme Learning Machine (ELM), Minimax Probability Machine Regression (MPMR) and Gaussian Process Regression (GPR) for determination of Work Zone Capacity. Number of lanes, number of open lanes, work zone layout, length, lane width, percentage trucks, grade, speed, work intensity, darkness factor, and proximity of ramps have been adopted as inputs of ELM, MPMR and GPR. ELM has excellent generalization performance, rapid training speed and little human intervention. MPMR is developed based on the concept of minimax probability machine classification. It does not assume any data distribution. GPR is a probabilistic, and non-parametric model. In GPR, different kinds of prior knowledge can be applied. This article describes a comparative study between the ELM, MPMR and GPR models.

Hubs are special centers that used for switching, transferring and handling freight or passengers in transportation systems. The hub location problem deals with locating hub facilities and allocating demand nodes to hubs in order to route the traffic between origin–destination pairs. Aiming at minimizing the

In transportation planning, design of intersections is required some concrete decisions. It is mostly related to capacity, safety and economy. In many applications, capacity is considered as the main factor for design purpose and type of intersection is determined regarding the capacity. But it is known that safety and economy are as important as the capacity. In order to handle this problem, a multi-criteria decision making approach (AHP) is used. Four intersection types are used. Fuel consumption, emission rates, delay, cost of intersection type and safety parameters are taken into account in the decision process. These intersection types are tested by SIDRA Intersection software using the parameters listed above. The data obtained by SIDRA software is used in the AHP model. Weights of the factors used in the AHP model are varied considering the views of capacity and economy. And results and applicability of the model are discussed.

Accessibility measures are generally concerned with equity and a better distribution of services in a territory and can be accepted as key variables for supporting supply/demand, location/allocation and service/catchment area related planning policies and strategies at national, regional, and local levels. Since accessibility measures need organization of huge and complex spatial data sets, accessibility modeling often lends itself to Geographical Information Systems (GIS) for analysis and presentation. Since numerous accessibility measures and modeling techniques ranging from simple to sophisticated can be found in the literature, this work aim to provide an overview of the theoretical framework and relevant background about GIS-based accessibility modeling process. The results could provide a significant support for the decision makers who are supposed to deal with transportation planning, accessibility modeling, location/allocation and service/catchment area related issues.

Transport systems are major emitters of greenhouse gases, which makes environmental sustainability of any transport a crucial issue. Another issue is the lack of a systematic approach to the modeling and implementation of public transport systems. Finally, there are problems with the human interfaces to public transport systems, which do not encourage, and many do not allow, comfortable and simple interaction with the system. In this chapter, the authors discuss their solutions for these problems, explaining how to cover the existing gaps in a methodological and systematic way. The main contribution of this chapter is a model of an on-demand transport system that covers all the points mentioned above and focuses on spatial planning and optimizations including environmental issues in transport planning.

The emergence of easy mapping tools, widespread cellular network, declining costs of smartphones and increasing internet use by public agencies provide new opportunities towards the betterment of public transport management. Applying decision support systems such as Geographic Information Systems (GIS) and Information and Communication Technologies (ICT) to public transportation tailors transit services to urban dwellers' needs, reducing eco-footprint while making them more accessible. This chapter aims to explore how GIS and ICT can be used for more eco-friendly public transport and improved sense of community. The chapter posits the positive impact of mobile phones and social media towards sustainable mobility. Multi-modal journey planners, mobile bus ticketing, demand responsive transit systems, eco-taxis, electric car and bicycle sharing are explained with examples from the world. The chapter concludes with a discussion of challenges and future options for using smartphones, social networking and the position of disadvantaged groups in sustainable mobility.

Our study focuses on the position of protected areas in relation to the transport infrastructure planning and their specific requirements (fragmentation, artificial edge effects, buffer zones, etc.). The presence of transport infrastructure in protected areas is discouraged by the European Union legislation and although their avoidance should be easy, there are numerous cases in which knowledge of the protected areas lacked from the transport infrastructure planning. Our chapter presents the theoretical aspects of integrating protected areas with transportation infrastructures and the main software and methodologies which can be used in the planning process exemplified with case studies. The multitude of challenges found in transportation and conservation planning requires complex decision support systems (DSS), such are the tools based on Geographical Information Systems (GIS) we showcased in this chapter.

to each of the evaluation criteria for a possible route. The Analytic Hierarchy Process (AHP) is used to assign weights to factors. The priority weights of each environmental criteria is calculated by using Analytical Hierarchy Process (AHP) before risk models are implemented in the road network of Izmir-Manisa-Aydın-Mugla-Denizli (Aegean Region of Turkey). The most convenient route according to the criteria set is determined. The results give decision maker the possibility to choose the best alternative among possible routes according to certain criteria.

Foreword

The man-made artificial world around us, to create a safer and more convenient life for the majority of people, is the result of human brains, of engineers, economists, politicians – and the society as well. Technology in the transport and information sector was developed exponentially over the last two centuries – and implemented with enthusiasm. We have interfered into the complex system of nature and society without knowing the system reaction and system effect. Transport planning, as well as urban- and land-use planning, is mainly based on plausible assumptions, extrapolation of personal experience and standards based on general agreement. Unwanted effects, accidents, air and noise pollution, environmental degradation, economic problems to finance and operate transport and urban systems, social problems, loss of local opportunities for jobs are indicators that this man-made environment has become complex in itself and even more complex in its interaction with the social and natural environment. These systems have not only overgrown our societies but can also endanger human future.

The infrastructure of the transport system was based on simple economic, strategic and/or political decisions. The success of railways, cars, airplanes tells its own tale. The flipside of this coin was not visible soon – one of the problems of dynamic complex systems – time delay. Exponential growth is an indicator that the eigendynamic of the system is out of control. Making the system bigger and bigger, will and cannot solve the inherent problems. The "lab" in urban and transport planning is the reality. And most of the transport and urban structures have long lasting effects. It is therefore important to find methods and tools to support decisions based on objective basic principles. One of the tools for a better understanding of the system behavior is models, since there is no laboratory available, where we can test the behavior of a city, a region, a country, or even of a harbor or airport.

Theories, mathematical-statistical models, dealing with multicriteria problems, in different disciplines like psychology were available, but not common in practice. First of all, there was a lack of data and even more important, there were no capable and easy to handle computers. But it is not enough to develop/build and test models, it is of equal importance to publish them. During the last 40 years a vast number of publications about modelling appeared in the literature. Very few of them have been applied in practice. Desk work is easier than the bloody time-consuming and expensive data collection. The combination of both is a piece of luck, as it happened in some chapters of these book.

There are professional books, you read with some concentrated attention, there are books you have to read, textbooks for example and there are books you read and you use it, at least from time to time. It depends on the content. 15 chapters seem to be difficult to accomplish. But the grouping into "Transport Mode Specific Decision Support Systems", "Decisions in Urban Transportation" and "Sustainability/ Environmental Risks of Transport," as well as the content of the single chapters make the book good readable and the reader curious. Even if you are building models since decades and using them in different field of professional practice it is a pleasure to find some well known models used in another environment and several interesting others. The elaborated references of each of the topics are a good source for researchers, practitioners and students. The book belong to the group of books you will pick it up from your book-shelf more often than once.

Hermann Knoflacher
Vienna University of Technology Research Center of Transport Planning and Traffic Engineering, Austria

Preface

During the last three decades, the use of computers has contributed a lot to the practice of transport planning. In addition to gaining the ability to build up more complex models in order to analyze collected data, some new approaches, which take human behavior into consideration, have been added to transport planners' tools.

Increasing use of computers gives transport planners possibilities to handle much more data with increasing precision in very limited timeframes. This modern perspective in decision-making in transportation planning is mainly fed by scientific approaches of systems thinking, GIS, information systems and artificial intelligence. The results of any analysis regarding any transport decision can easily be visualized and presented to decision maker. A single solution is usually replaced by a set of solutions with a comparative list of advantages and disadvantages.

Use of decision support systems help mainly to professionals of transport planning, urban planning, public-policy planning, public transport operating, service industry and logistics but also individual users of transport systems. Monitoring of traffic conditions, listing all possible travel alternatives or calculating fares are now a part of transport users' daily life. User friendly interfaces in web applications and smart devices in transport decisions not only help individuals but also pave the way for formation of smart societies. The decision support systems for transportation planners cover a wide perspective, ranging from traffic control centers, passenger movements, public transport management for the scheduling and routing of cargo, automated transport systems, etc.

This publication addresses transportation planning problem cases which are solved by decision support systems. The book is prepared for transportation planning professionals to extend their expert information and domain knowledge to drive decision support systems (DSSs). The principal audiences are transportation planning professionals and academicians who need expert information to handle their specific transportation-related problems. The book offers a broad range of subjects addressed in specific areas such as spatial planning, developing policies for PT, traffic information systems, handling of cargo, logistics, goods transport, and environmental issues in transport planning.

The submitted case studies, developed models and researched literature survey in 15 chapters show that modern transport planning professionals are no longer capable of solving ever developing problems by using traditional methods. Instead, contributions from other fields of science, such as artificial intelligence, optimization algorithms and some advanced statistical analyses, in addition to modern technologies, such as, GIS, ITS and smart solutions are the tools in modern transport planners' hands.

The content of the book is organized into three headings: transport mode specific decision support systems, decisions in urban transportation and sustainability/ environmental risks of transport.

Under the first heading (transport mode specific decision support systems) case studies related to different transport modes are presented in five chapters. A brief description of each of the chapters follows:

Chapter 1 presents an analysis of air transport mode. The addressed problem is airline crew scheduling, which is a part of airline schedule planning. Robust airline crew pairing process refers to the procedure of constructing anonymous crew pairings to cover the given flight schedule in near-optimal way that is less prone to disruptions (or at least easier to recover once disrupted). This process has to deal with complicated specifications of inputs and satisfy a number of different and often conflicting constraints. Also user inputs are often needed in several points in the process to obtain high quality and timely solutions. Bülent Soykan, and Serpil Erol present a model-driven Decision Support Framework (DSF) which combines an optimization algorithm and a simulation-based schedule evaluation scheme for automating real-world practical robust ACP problem. A Branch-and-Price (B & P) approach is utilized as a solution method in which column generation method is applied at each node of the B & B tree. The presented DSF can primarily assist airline crew schedule planners, who have to spend a lot of time and effort creating crew schedules taking into account non-linear cost structure, various amounts of constraints that affect cost and flight safety.

Chapter 2 presents an analysis of maritime transport. It proposes a typical investment planning problem with uncertainties and risk parameters that decrease the reliability of cost and income predictions, which coastal engineers are used to face. Can Elmar Balas presents a case study from the Iskenderun Pier of Turkey, which is under construction. The superiorities of this proposed simulation-ANN model to other classical investment planning methods were the inclusion of uncertainties in the investment parameters like the change of cargo and costs variables in time, and the determination of project benefit/cost with an improved accuracy. A feasibility analysis model was proposed in this chapter for ports. Three methods were utilized: 1) Artificial Neural Network (ANN) to determine the rates and capacity of cargo by considering the economic development of hinterland, 2) Queuing model to determine the waiting to service time and the berth occupancy ratios, 3) Importance Sampling Monte Carlo (ISMC) to simulate ship arrivals/departures from the quays. The required time to reach the operational capacity in various economic scenarios is determined by the cargo rate predictions of ANN, which is interrelated with ISMC simulation. The artificial intelligence methodology takes the economic developments of the region into account by defining the GNP as one of the input parameters of ANN structures.

Chapter 3 presents a comparison of three transport modes, road, airway and railway in a case study of Turkey's transportation system's efficiency. Adnan Sozen, and Fatih Cipil present Turkey's situation in transportation compared to the EU member countries (23 countries whose data were accessed) through the approaches of Data Envelopment Analysis (DEA) and Malmquist Index (MI) as decision support models. The study determines Turkey's position among EU member countries and defines its pros and cons in terms of road and railway transportation (safety) indicators. Within the scope of this study, relative performance evaluation was first performed by applying DEA approach so as to determine if 12 regions of Turkey were using their airline transportation effectively or not. Meanwhile, to monitor the variation of the performance with time, total factor change of the airline transportation was examined by MI approach. Using subjective calculation results, information on the strategies and targets to be followed are given so that the performances of the transportation indicators were maintained in their highest level.

Chapter 4 presents an important maintenance problem in road transport. Sangeeta Roy, Jagan J, and Pijush Samui review work zones. A work zone is the area of a highway where construction, maintenance, or utility work activities that can be identified by warning signs/signals/indicators, including those on transport devices. Work zones also include roadway sections where there is on-going, moving (mobile) work activity such as lane line painting or roadside mowing only if the beginning of the on-going, moving (mobile) work activity is designated by warning signs or signals. The various problems which presaged the increased number of work zones in today's date also created a myriad of complicated challenges for the various traffic agencies involved in work zone management and estimation. The creation of a work zone changes the road user conditions which then lead to increased number of accidents. The placement, duration and maintenance of work zones are a multi-faceted problem. The highway is not closed thus work zones can be defined as spatial and temporal restrictions on the roadway which negatively impacts traffic flow conditions. The major concern of most traffic management and planning agencies is the placement of long term. This article describes Extreme Learning Machine (ELM), Minimax Probability Machine Regression (MPMR) and Gaussian Process Regression (GPR) for prediction of work zone capacity.

Chapter 5 discusses logistic hub optimization in a transportation network. Serpil Erol, and Aykut Sureyya Duyguvar review a background which classifies the considered literature using three criteria relevant for the aforementioned topic: a literature review, the list of mathematical models/algorithms and the solution approaches which are optimizing logistics services. Some applicable algorithms, namely greedy heuristic algorithm, genetic algorithm, and benders decomposition method, are presented as decision support systems for hub location problems (HLP). The submitted future research directions also give good hints for researchers who have their own HLPs of people, commodities and data.

The second heading is specially been defined for decisions in urban transportation. Transport planning is basically studied in varying scales. The largest scales cover global, intercontinental, international and national transport of people, goods, information and energy. The smallest scales deepen in intraurban relationships, among districts, dwelling units, working places, point destinations, etc). Urban transport planning requires a specialization different from other scales of transport. By planning for urban transport, in addition to aggregates, special arrangements are done for any individual demand. Thus, four chapters related directly with urban transportation are grouped separately. A brief description of each of the chapters follows:

In Chapter 6, I, as author, present a case study example to a web based Decision Support System (DSS) for individuals' urban travel alternatives. The case study is from a project prepared for Ankara (Turkey) Municipality, which aims to give individuals the opportunity of choice among all travel alternatives under different criteria, both for public transport riders and car users. A literature survey gives background information from fields of applications of Intelligent Transport Systems (ITS), shortest path problems and web based DSSs. Special emphasis is given to Floyd-Warshall algorithm which is used to solve the shortest travel time between all pairs of points in the case specific problem. In the case study part, the design of the proposed system, its capabilities and examples with screenshots are submitted, to help other researchers design their projects according to the requirements specific for their own cities. The proposed ideas can be applied to the transport net of another city.

Chapter 7, written by Darcin Akin and Serdar Alasalvar, presents a study on modeling the change of urban spatial structure. The purpose of this chapter is to model urban growth and expansion based on observed and modeled travel patterns in a study area. In the analysis of interzonal trips using hierarchical cluster analysis (HCA) for 2012 and 2023 as a case study of the province of Sakarya (Turkey). The modeling in this study made possible to analyze the urban structure for different time horizons when O/D travel data are available to understand the interactions over the spatial structure of an urbanized area. The achieved results are useful tools for planners, administrators and policy makers as decision maker to understand the interactions over a spatial structure, and to plan for creating efficient subcenters with more balanced distribution of travel patterns over urban agglomerations. The approach also helps to chose potential areas for future developments.

Chapter 8 presents an example to use of analytical hierarchy process (AHP) as a decision support system in the case of Denizli (Turkey) for intersection type selection in urban areas. Yetis Sazi Murat, Turan Arslan, Ziya Cakici, and Cengiz Akcam review also many other applications related to the given one. This chapter is related to higher construction cost of at-grade intersections. The authors gave emphasis on "not signalized intersections", as the minimum interruptions in traffic flows, average delays, fuel consumption, operating cost and CO emission rates are lower than the others. The analyses that consider operating cost show that preference of at-grade intersections is in the second order. Results of the AHP analysis support also the findings obtained for signalized intersections. Signalized intersections with left turn bay and signalized intersections are also discussed in the chapter. Left turn bays provide an extra space, thus possibility of an accident occurrence can be reduced and also capacity of the adjacent lane can be improved by this way.

Chapter 9 presents a detailed review of the literature about definitions of accessibility, usage areas of accessibility, components of accessibility, GIS based accessibility modeling techniques, and role of Geographical Information Systems (GIS) in accessibility modeling. Accessibility measures are generally concerned with equity and a better distribution of services and help to evaluate the proximity/availability of several public, commercial and emergency services by considering one or more modes of transportation. By referring numerous accessibility measures, Kivanc Ertugay, and Sebnem Duzgun illustrate the need of GIS-based analyses in assessing the justice in the distribution of public services for plan evaluations.

The third heading of the book is devoted to decision support systems related with sustainability and environmental risks of transport. While approaching to three decades after Brundtland Commission's Report of the United Nations, the increasing awareness and concern on environmental matters today better evaluate the human interventions on earth's resources. Transport, despite its all positive effects, e.g. combining role in society, increasing mobility, supporting economic activity, can also be a major threat against social, environmental and economic sustainability if not examined and managed well. Thanks to decision support systems, environmental risks of transport are better evaluated and reflected to decisions in transport planning now. The six chapters in this section provide information on sustainability, dangerous aspects of transport and suggestions for solutions, including smart applications. A brief description of each of the chapters follows:

Chapter 10 presents a model of a sustainable transport system, the transport on-demand model, for making public transport sustainable. After reviewing the use of various definitions, examples and scenarios, Maria Spichkova, and Margaret Hamilton develop a methodology to provide smart solutions for sustainable outcomes, so that a transport system becomes a part of a 'smart city'. In this chapter DSSs are viewed as a special way of partially designing humans out of the main system actions, particularly if it is assumed that the human will follow the decision recommended by the support system. The proposed web-based DSS, which can be a part of a smart city, is designed for the public transport driver, where the request information for the route is collected from the passengers dynamically in real time. The model can help enabling the reduction of greenhouse gas emission and noise pollution, as well as the saving of fuel costs for the public transport service company. Having a more flexible/dense timetable and possible longer routes as well as subjective shorter travelling times are other possible outcomes of the model.

Chapter 11 presents a discussion of challenges and future options for using smartphones, social networking and the position of disadvantaged groups in sustainable mobility. Ozge Yalciner Ercoskun presents a comprehensive literature survey on the emergence of easy mapping tools, widespread cellular network, declining costs of smartphones and increasing internet use by public agencies, which provide new opportunities towards the betterment of public transport management. The chapter, focusing mainly on spatial DSSs and communications-driven DSSs, aims to explore how GIS and ICT can be used for more eco-friendly public transport and improved sense of community. Some success stories from different countries are also given as examples to the use of mobile systems in public transport.

Chapter 12 presents a case study from Romania for integrating spatial planning of protected areas and transportation infrastructures. For the case study, Mihai Răzvan Niţă and Mihăiă Iulian Niculae analyse the connectivity of the Natura 2000 protected areas network for terrestrial mammals in the Alpine Biogeographic Region from Romania by evaluating the spatial distribution of Sites of Community Importance in the Region and assessing the functional connectivity of the protected areas network on two scenarios: with and without the impact induced by the roads network. The analysis includes a number of 125 sites and the distribution of 17 species of terrestrial mammals, protected under the Habitats Directive, each having different dispersion distances and different home-range sizes. The protected areas crossed by the transportation infrastructure are considered to be disconnected. The achieved results for the case study area opens a horizon in discussions about the adverse effects of transport on environment.

Chapter 13 presents a multi-objective optimizer for multimodal distribution networks. Mauro Gamberi, Marco Bortolini, Francesco Pilati, and Alberto Regattieri discuss the effective design of multimodal distribution networks (DNs) from a multi-objective perspective. Three of the most relevant optimization drivers are jointly considered looking for the multi-objective optimal network configuration. The operating cost function stresses the importance of effective DNs to save money making the retailer/end-user full price convenient, the carbon footprint looks for environmentally sustainable DNs able to mitigate the impact on the climate change and to reduce emissions due to production, storage and shipment activities. Finally, the delivery time function forces to speed the distribution process to promptly supply the market demand. The tri-objective model is behind an optimizer decision support system (DSS) supporting the planner in the steps to design an effective supply chain network. The application is exemplified through a case study taken from the fresh food industry. Particularly, the peculiarities of food produces are considered to update the general multi-objective model, e.g. produce perishability. The data are from Italian producers distributing products to multiple European countries. Short and long shelf life products are studied to highlight differences in the most effective shipping strategies. The key outcomes stress opposite trends between operating cost and environmental impact on one side and the time objective function on the other.

Chapter 14 presents a literature survey, which is a collection of many DSS applications, for hazardous materials (hazmat), also known as dangerous goods transportation. Serpil Erol, and Zafer Yilmaz draw our attention to issues which are well-studied by many researchers and the ones underestimated so far. The researchers give the clues for developing new DSSs in transportation of dangerous goods on road by underlying the underestimated issues and possible research areas.

In Chapter 15 Serhan Karabulut, and I present a tool for GIS based risk analysis for transportation of dangerous goods on road (the RAGISADR). The proposed model has been developed to evaluate the environmental risks of transportation of dangerous goods on road. In this chapter a risk based decision support model is used as a tool for route decision making of fuel products. The relevance of the model as a decision support tool is demonstrated through a case application to a road network in a region in western Turkey. The priority weights of each environmental criterion are calculated by using Analytical Hierarchy Process (AHP). The most convenient route according to the criteria set is determined after assessing the outcomes of the risk analyses models.

Ebru V. Ocalir-Akunal
Gazi University, Turkey

Section 1
Transport Mode Specific Decision Support Systems

Chapter 1
An Optimization-Based Decision Support Framework for Robust Airline Crew Pairing Process

Bülent Soykan
Old Dominion University, USA

Serpil Erol
Gazi University, Turkey

ABSTRACT

Robust airline crew pairing process refers to the procedure of constructing anonymous crew pairings to cover the given flight schedule in near-optimal way that is less prone to disruptions or easier to recover once disrupted. This process has to deal with complicated specifications of inputs and satisfy a number of different and often conflicting constraints. Also user inputs are often needed in several points in the process to obtain high quality and timely solutions. This chapter presents a model-driven Decision Support Framework (DSF) which utilizes an optimization algorithm that generates robust crew schedules and a simulation model that evaluates the schedules in terms of their robustness quality. The presented DSF can primarily assist airline crew schedule planners, who have to spend a lot of time and effort creating crew schedules taking into account non-linear cost structure, various amounts of constraints that affect cost and flight safety. The proposed DSF can be used either as a standalone tool or as part of an integrated or enterprise-level DSS.

INTRODUCTION

In order to be competitive in the market, airlines need to solve a large number of decision problems, both in airline schedule planning and operations (schedule execution) stages. Airline schedule planning stage for a scheduled passenger airline is too large and complex to be dealt with as a whole. It is usually divided into four sequential sub stages: *flight schedule design, fleet assignment, aircraft routing,* and

DOI: 10.4018/978-1-4666-8648-9.ch001

crew scheduling. Due to the fact that crew costs are second only to fuel costs among all airline direct operating costs, crew scheduling constitutes a crucial part of the whole planning process. Above and beyond all other considerations, crew costs are manageable by airlines compared to fuel costs. Hence, it implies potential cost savings by establishing efficient crew schedule systems, and even small reductions in crew cost can lead to large savings. Moreover, crew schedules' performance contributes to the flight safety, service quality and as a result airline's goodwill as well. Therefore, decision support tools associated with crew scheduling are indispensably and highly utilized by airlines.

Airline crew scheduling mainly deals with finding the minimum cost set of legal crew schedules that cover all the flight legs in a given flight schedule while considering a large number of restrictive rules, which is done at the tactical planning level. This problem is traditionally divided into two stages: *crew pairing* (*tour-of-duty planning*) and *crew rostering*, and solved with a sequential approach. Airline crew pairing, denoted as ACP, mainly deals with finding the minimum cost set of legal crew pairings that cover all the flight legs in a given flight schedule. On the other hand, airline crew rostering focuses on assigning individual crew members to the fixed pairings. The leave times, training periods etc. are considered in order to construct work schedules usually for a month. Most of the cost improvements can be achieved by optimizing pairings. For this reason ACP is generally accepted as the most important stage of crew scheduling, (Barnhart et al., 2003). In this chapter, only the crew pairing problem, which is the frequently encountered real-life decision problem, is taken into account.

The prominent traditional approach to overcome the combinatorial nature of the ACP problem is to formulate it as a set partitioning or set covering problem and to deal with all nonlinearities by using a network-based pairing generation scheme. The resulting integer program is inherently difficult to solve, because exact illustration of the feasibility set is either very time consuming or even impossible. In order to avoid this difficulty, first Linear Programming (LP) relaxation of the problem is solved and then a Branch-and-Bound (B & B) scheme is applied to obtain an integer solution.

Nevertheless, the solution obtained from traditional ACP problem assumes that the schedule will be carried out as planned with complete information and in a deterministic environment. However, in practice, airline operations are prone to suffer disruptions caused by inclement weather, airport congestion, equipment failure, crew sickness, late passengers, etc. In such cases optimal crew pairings easily become sub-optimal. Traditional approaches have proved to be not capable to address such disruptions adequately. Furthermore, tight schedules created by deterministic approaches lead to propagation of the delays across the network, because globally optimized deterministic solutions have short buffer times between flights and also smooth recovery is not taken into account during schedule planning (Soykan & Erol, 2014).

In operations stage, the effect of an individual delay may be negligible, but the ripple effect (also referred as snowball effect or knock-on effect), which is the propagation of delays over the whole network, may lead significant negative complications. Actually, in case of disruptions the quasi-optimal schedules generated in the planning stage become far from optimal in practice. Usually external factors, such as inclement weather or airport congestion, are held responsible for delays. However, certain delays especially propagated delays, are predictable and avoidable with incorporating robustness to the schedules at the planning stage.

Due to the fact that airline industry's operational environment is complex and dynamic; disturbances or disruptions can easily render schedules impractical, and most probably result in high extra costs. Soykan & Erol (2014) define two general approaches to deal with the uncertainties related to flight disruptions in order to improve punctuality and reduce operational costs. First one is the *reactive approach* in which schedules are re-optimized after the disruptions occur by utilizing real-time Decision Support Systems (DSSs). Main drawback of this approach is the fact that a solution is expected immediately. In addition, real-time recovery requires considering several complex constraints. These aspects lead to intractability issues. Because of the need of immediate response, airlines currently employ manual or heuristic methods that yield sub-optimal solutions. The other approach is the *proactive approach* in which robustness is integrated into the DSS in the planning stage so that schedule adjustments can be made more easily and more conveniently. While reactive approach tries to find the most appropriate way to react to disruptions on the day of operations, proactive approach seeks to prevent disruptions in advance. Thus, crew schedules that are less susceptible to disturbances or easier to repair once disrupted, could be created.

Unfortunately, existing automated transportation DSSs for airline crew schedule planning do not reflect complex and stochastic nature of the airline operations which is associated with high degree of uncertainty. Currently, decisions related to robust crew scheduling are typically made in a heuristic manner based on limited information and support. Therefore, a DSS, which can generate robust solutions that perform well under various scenarios, needs to be deployed to provide decision support to airline crew schedule planners and suggest valid solutions to them.

The main objective of this chapter is to present a model-driven Decision Support Framework (DSF) that combines an optimization algorithm and a simulation-based schedule evaluation scheme for automating real-world practical robust ACP problem. As far as the optimization method is concerned, the problem is modeled as a Bi - Objective General Set Partitioning Problem to minimize the estimated propagated delay while at the same time maintaining a cost effective solution, and the price of robustness is limited by using a ε-constraint method. Then, a Branch-and-Price (B & P) approach is utilized as a solution method in which column generation method is applied at each node of the B & B tree.

Owing to the complex nature of airline operations, it is not practical to verify optimization results, and assess the robustness of the obtained solutions and the performance of the proposed approach in an analytical way. Thus, a discrete-event simulation model is incorporated into the framework in order to simulate airline operations and various unexpected disruptions that are faced during daily airline operations. The simulation model is designed as a trace-driven simulation, where the historical data is used as an input and this historical information on disruptions is also exploited in optimization stage to conduct several different scenarios for a final decision. The presented framework designed to automate crew pairing proactively by focusing on to generate robust schedules. Also, main users, airline crew schedule planners, can exploit model parameters in order to apply both sensitivity and "what if?" analysis. Due to the fact that it is quite a challenging task to construct a DSF from the published literature, design level technicalities are described in sufficient detail. However, in no way does this chapter intends to be an exhaustive account of the implementation and application details of the presented DSF.

The chapter is organized as follows. In "Literature Review" section, the detailed description and definition of the problems that constitutes the process is given. Also previous contributions on these problems and state-of-the-art techniques related to these problems are reviewed. In "ACP Optimization" section, a more detail description of the ACP problem, and mathematical formulation to solve the problem is explained. Also an overview of the solution approaches is given from both theoretical and practical perspectives. In "Robust ACP, an Optimization Algorithm and a Simulation Model" section, a mathematical model and a solution algorithm for generating robust schedules and a simulation model for evaluation of the schedules are given along with a review on some recent methods for incorporating a measure of robustness into the robust ACP problem. In "DSF" section, a dedicated decision support framework and its major components are described. Finally, the chapter presents a summary of main conclusions.

LITERATURE REVIEW

A high number of articles, research papers and technical reports have been published over past several decades to address the efficiency problems which arise in transportation planning, especially in the airline context. The underlying mathematical models, solution methodologies, and the DSSs developed to address these problems also contribute toward the concepts and approaches for providing support to other transportation related schedule planning decision problems.

Airline scheduling decision process includes various decisions not only before actual operating time of a flight, such as when and where to fly, with which aircraft and crew members etc. but also during the actual day of operations. Typically, flight network structure contributes to the characteristics of the entire process, so before going any further, the flight network structure needs to be explained.

Flight Network Structure

After the U.S. federal government deregulated the American airline industry in 1978, airlines were allowed to choose their flight network, which also affected the global airline industry. As a result, most airlines adopt to operate on a *hub-and-spoke* flight network in order to improve efficiency.

In hub-and-spoke flight network, *hub* is an airport with higher rate of daily departure/arrival of flights; on the other hand *spoke* represents an airport that has limited amount of flight activities which is connected to rest of the network via one or more hubs. Departures from several different destinations are fed into a single hub airport, from which connecting flights are operated to other destinations. The hub-and-spoke flight network enables the airline to connect passengers between two distant airports through its hub/hubs. On the contrary, in a *point-to-point* flight network, each airport is connected via a non-stop flight. In this type of network, air traffic is dispersed in terms of both space and time. Figure 1 illustrates small examples for a fully connected point-to-point flight network and a typical hub-and-spoke flight network.

Hub-and-spoke flight networks, usually allow many more passenger connections compared to point-to-point networks. Hence, hub-and-spoke network has proven advantages from both revenue and operational considerations point of view. However, any disruption at the hub, for instance extreme weather conditions or a security problem, can create ripple effects throughout the system. Therefore, the overall operating efficiency of a hub-and-spoke network is limited by the capacity of the hub and highly depends on proper scheduling of the resources.

Airline scheduling decision process can be divided into three level planning problems: *strategic, tactical and operational.* Strategic level planning problems entail long-term decisions about flight network type, airline's fleet size and composition, origin-destination airport connections, and location of the crew and maintenance bases, and typically account for more than 2 years. Tactical level planning process mainly involves allocation and scheduling of airline's resources, and starts at least 18 months before the day of operations. Operational level problems focus on controlling and ensuring that the operations are executed as possible as planned, which is hard to accomplish because of the stochastic nature of the

Figure 1. Flight network structures

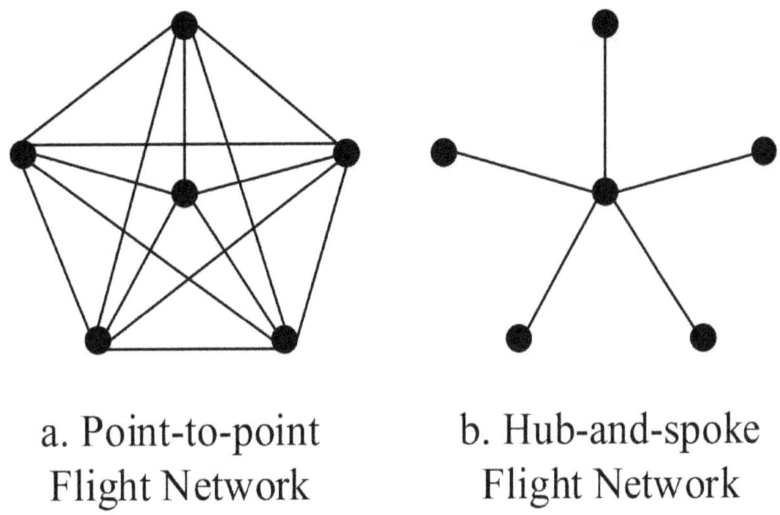

airline operations. Operational level problems can be divided into two parts: rescheduling and recovery. Rescheduling includes changes to the schedules due to a major change in conditions before the day of operations. On the contrary, recovery takes place on the actual day of operations in real-time by virtue of disruption.

Tactical level planning process encompasses high level of complexity with numerous resources and related decisions spanning temporally and spatially. Hence, the process is typically divided into four sequential steps in order to make the whole process computationally tractable. Tactical problems along with operational level problems are shown on Figure 2. Even though these problems are smaller in size and simpler than the whole problem, they are still large-scale and complex. Numerous heuristics and exact algorithms have been proposed to solve these problems over several past decades including integrated approaches comprise of combining two or more of the steps together.

Flight Schedule Design

Flight schedule design, also known as flight scheduling, is the initial problem addressed in tactical level planning process and it constitutes the foundation for the subsequent stages. In this stage, a flight schedule is generated to determine where and when to fly, which is the crucial product for the rest of the process consisting of flight legs between origin-destination airport and a given flight number for each flight leg, and in practice, it has substantial impact on the airline's profitability.

The factors that need to be considered when constructing a flight schedule are; demand forecasts and related revenue, available resources, airline's fleet size and composition, air traffic associated international agreements, gate availability, landing rights, route characteristics, and network constraints. Major part of the costs and revenues are determined after a flight schedule is fixed. Schedule design is usually done between 18 months to 3 months before the day of operations.

Recent publications are usually not deal with the flight schedule design as a separate optimization problem; instead it is integrated with other airline schedule planning problems, especially with fleet assignment. Lohatepanont and Barnhart (2004) proposed several integrated models and algorithms that

Figure 2. Tactical and operational level problems

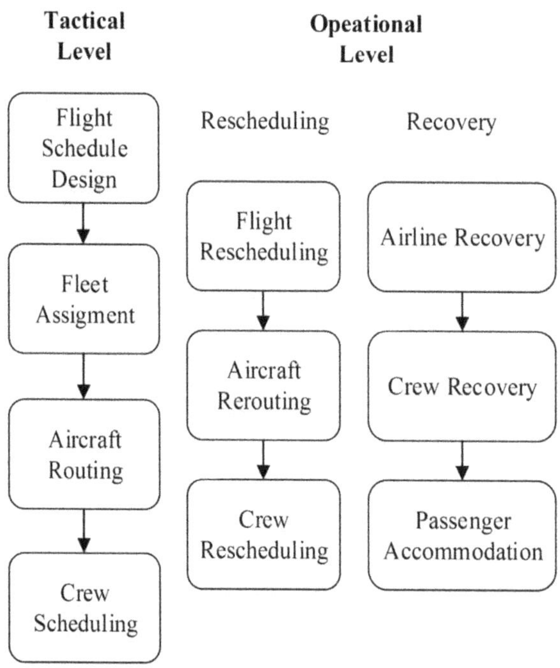

maximize potential revenue by concurrently adding or deleting a set of flights from the flight schedule and assigning fleet types to the flights. Mercer Mercier and Soumis (2007) presented an integrated aircraft routing, crew scheduling and flight schedule design model which allows the departure time of the flights to vary within a given time window.

Fleet Assignment

After the construction of a flight schedule, the fleet assignment stage is the next. This stage of the planning process mainly deals with assigning an aircraft type to each flight leg in a most cost-efficient way and also satisfying revenue, operating costs, operational constraints, maintenance rules, and crew restrictions. The main objective in fleet assignment is to maximize profit or minimize the total cost while considering only aircraft types instead of each individual aircraft. The reason why the planning process is partitioned into aircraft types is the fact that the capacities and operational characteristics of the each aircraft are considerably different. The inputs for this problem include the flight schedule, a set of aircraft type and the number of aircraft available for each type.

The requirements that must be considered to efficiently assign aircrafts to each fleet include: passenger demand, seating capacity, operational costs, and availability of aircraft maintenance at arrival and departure airports. Some of the apparent operational constraints in fleet assignment are flying distance, aircraft fuel and airport runway capacities. In order to schedule an aircraft type into a certain flight leg, aircraft needs to have enough fuel capacity to get to the destination and to be able to use the runways. Fleet assignments need to be determined usually at least 3 months in advance of actual day of operations due to governmental authority regulations and union contracts.

Barnhart, Farahat, and Lohatepanont (2009) proposed a model and solution method for fleet assignment problem that employs composite decision variables representing the simultaneous assignment of fleet types to sub networks of one or more flights. Recently, Li and Tan (2013) suggested an optimization model in which main objective is to maximize revenue, and considered comprehensively the difference of scheduled flights and aircraft models in flight areas and mean passenger flows.

Aircraft Routing

After each flight leg in the given flight schedule has an aircraft type, the next stage is the aircraft routing problem, which is also referred to as aircraft maintenance routing problem. In this problem each individual aircraft has to be routed to cover all these flight legs while satisfying the maintenance requirements of each aircraft and aircraft utilization restrictions. For each fleet type there will be a separate aircraft routing problem, because the solution to the previous stage (fleet assignment) partitions the given flight schedule into each fleet.

The aircraft routing problem is usually solved in a two-step approach, in which the first step is to generate a set of daily routes that covers all flights assigned to the given fleet and then, second step is to generate periodic schedules for individual aircraft that meet all maintenance requirements. For instance; each aircraft must undergo a safety check in every 3 to 4 days, which takes place only at a small number of hub airports that have maintenance facilities and technical personnel to provide periodic mechanical checks. The problem is typically modeled either as a set partitioning/covering problem, or as a multi-commodity flow problem in which aircraft types are viewed as commodities.

Liang and Chaovalitwongse (2009) provided a thorough discussion on the aircraft routing problem and reviewed the advances in optimization methods applied to airline schedule planning problems in general, and presented several mathematical formulations and solution methods for the problem. Liang, Chaovalitwongse, Huang, and Johnson (2011) proposed a rotation tour network model and a mixed integer linear programming formulation for the solution of the aircraft routing problem, and assessed the solution quality on real test instances from a major U.S. airline.

Crew Scheduling

The last stage of the airline planning process is crew scheduling, in which crew members are assigned to flights in the least expensive way possible, assuring that each flight is covered and that government/ safety regulations regarding crew scheduling are met (Soykan & Erol, 2014). The crew schedules are constructed typically at least one month in advance of operations.

Not only in industry but also in the academic literature, crew scheduling is commonly divided into two sequential sub problems, namely *crew pairing* and *crew rostering (assignment)*, mainly due to its large scale, complex work rules and safety regulations. Recently, some researches committed to integrate these two problems in order to generate better quality and lower cost crew schedules. However, in practice, this integration often leads to a disproportionate increase in complexity and renders the problem intractable.

Barnhart et al. (2003b), and Gopalakrishnan and Johnson (2005) comprehensively surveyed the models and solution methods for crew scheduling within the airline industry.

Crew Pairing

Crew pairing problem, also referred to as tour-of-duty planning, is the essential part of the crew scheduling and is done with the aim of constructing a set of legal and cost effective anonymous pairings for a given crew category and fleet type with minimal cost that covers all the flight legs in the given flight schedule. Similar to aircraft routings, pairings are consecutive set of flights that can be operated in a sequence, which usually span one to five days. Nevertheless, crew pairings usually involve more constraints due to governmental regulations and union restrictions.

ACP problem has got a lot of attention in the operations research literature and dates back to 1960s. This interest is particularly because of the complexity of the problem involving a huge number of feasible pairings, such that it is often impossible to enumerate all of them. Until the 1990s many heuristic methods were proposed to solve the problem, but recently the problem is usually solved with mathematical optimization methods.

Hoffman and Padberg (1993) proposed a pure set partitioning and a set partitioning with side constraints formulation to ACP problem, and presented a Branch–and-Cut (B & C) solution approach to optimality. Because of the difficulty in dealing with complex and various constraints, set partitioning/ covering models are much preferred by the researchers. Andersson, Housos, Kohl, and Wedelin (1998) reviewed the crew pairing problem and the formulation, and discussed previously applied mathematical optimization methods.

Desaulniers et al. (1997), and Barnhart and Shenoi (1998) formulated the ACP problem as a minimum cost multi-commodity flow problem with additional resource variables, and pairing generation is performed by a resource constrained shortest path approach. In the presented model, crews are modelled as commodities, airports are represented by nodes at different time points, and various crew activities, such as operating a flight, deadheading, resting and layover, are represented by arcs. Additional resource constraints are used to model different crew pairing constraints and restrictions.

Rasmussen, Lusby, Ryan, and Larsen (2011) introduced a solution approach based on generation of subsequences of flights and benchmarked this approach against column generation. In subsequence generation approach LP relaxation solution is kept as close to integral as possible, thereby providing a good starting point for B&B. The benchmarking has shown that the subsequence generation approach indeed is less fractional and faster, but the solution quality is not high compare to column generation. The results showed that subsequence is a worthwhile alternative to column generation.

Saddoune, Desaulniers, and Soumis (2013) proposed a rolling-horizon approach and formulated the problem as a set partitioning problem to find the minimal-cost pairings. Monthly problem is solved directly with the integrated rolling horizon/column generation approach.

The literature related to ACP problem is indeed abundant of underlying mathematical models and solution algorithms, but still there is an apparent shortage of study that fully describes a DSS and its principal components.

Crew Rostering

Once crew pairings are built, crew rostering problem needs to be solved. In this problem each individual crew member is assigned to these given pairings and additional reserve/standby crew duties are constructed for the entire planning time period, which is usually a month. The resulting monthly personalized work schedule is typically called *bidline* or *roster*, which includes line-of-works for each individual crew member. A *line-of-work* is basically a pattern of pairings with periods of time off in between. A legal line-of-work has to satisfy not only individual crew member needs, but also governmental regulations and union agreements, such as minimum consecutive number of days off per month, minimum total number of days off per month, maximum monthly total flight time and duty time, the number of days off between pairings, and flight experience requirements etc. It is worth mentioning that cockpit crews usually have more complex flight experience requirements than cabin crews. Typically, the cost of a line-of-work is typically calculated as a function of crew satisfaction and workload balance.

There are three basic approaches to the crew rostering problem: *bidline*, *preferential bidding* and *personalized rostering*. In bidline approach, first, lines-of-work are constructed without taking into consideration the individual crew members' preferences or constraints enforced by crews, and a generic set of monthly crew schedule is built and published. After then, individual crew members bid for particular line-of-work, usually based on seniority. Unlike bidline, in preferential bidding, each line-of-work is built consecutively (usually starting from most senior crew member) based on choices of individual crew member. While building lines-of-work, it is aimed to satisfy as many bids as possible given that crew members are considered in strict seniority order. In personalized rostering, rosters are built for individual crew members while considering peculiarities related to each crew such as vacation schedules, training activities etc. This approach can be based on fair share or individual preferences. Although most North American airlines commonly use bidline or preferential bidding approach with seniority principle, most European airlines prefer personalized rostering based on fair share without considering seniority at all. Both approaches can be formulated as a set partitioning/covering type model. The typical objective for this kind of model is to minimize overtime payment, which is an extra payment if the crews' total block hours exceed a certain threshold.

El Moudani, Cosenza, de Coligny, and Mora-Camino (2001) suggested a heuristic approach to solve the bi-criteria rostering problem in which objective function includes minimizing costs of the rosters and maximizing crew satisfaction, and tested on the medium-sized problems. Kohl and Karisch (2004) presented a comprehensive description of the airline rostering problem, reviewed different types of mathematical models associated with the problem and also a solution approach is proposed.

All these aforementioned airline schedule planning problems are conventionally solved in a deterministic way assuming that the schedules will be operated as planned.

Operational Recovery

Roughly two or three weeks before the actual day of operations all the schedules are passed to the airline's Operations Control Center (OCC) or Flight Dispatch Office, which is generally the department responsible for monitoring the actual day of operations, comprise of executing the approved schedules and handling all disruptions that may happen. In the course of disruption management (irregular operations), the OCC of the airline conducts coordination with the other internal/external entities and performs recovery actions to ensure continuous execution of the planned schedules. In order to resolve the disruptions during the operations, OCC implements essential changes to the planned schedules, so as to return to original schedule as soon as possible. Because of convoluted and tight crew schedules, and the fact that most crew members are operating at near-capacity, when disruptions occur in airline operations crew recovery typically functions as a bottleneck for the whole operational recovery process. Most of the time, these tight connections lead to fragile schedules that can be easily destroyed even by a minor delay.

The resource recovery activities which are carried out during the disruptions is often called *operational recovery* and primary goal of this process is usually to return the entire airline operation to its original schedule as soon as possible in a most cost-effective way. Operational recovery usually comprised of three sequential steps, namely *airline recovery, crew recovery* and *passenger accommodation*. First, in airline recovery phase, aircraft are rerouted and additionally delay or cancellation of the flight legs is decided. The main output of this step is the new aircraft routings along with new departure times and flight cancellations if any. Then, in crew recovery phase, crews are reassigned to flights and standby/reserve crews are called for uncovered flights. The objective might be to return back to the original crew schedule in a timely manner. Finally, in passenger accommodation phase, passenger itineraries are updated and they are directed to newly established flight schedules. Notably, crew recovery constitutes the bottleneck of the operational recovery problem because of the complex rules related to crews and hub-and-spoke flight network implemented by most of the airlines.

Once disruptions happen, the typical operational recovery strategies are delaying a set of flights and/or even canceling flights, changing the type of aircraft that is assigned to a flight leg or swapping aircraft and crews, where these measures can be interrelated. For example; in recovering a disrupted crew pairing, it may be indispensable to delay or cancel more flights or swap with another crew pairing, causing a disruption in additional aircraft rotations. Also re-accommodation of passengers, aircraft maintenance requirements and other factors must be considered, and a modified flight schedule for the recovery plan must be created.

This interrelated nature of the problem and multifaceted constraints considerably complicates the recovery problem. Therefore, it is barely possible to obtain optimal solutions after the practical disruptions occurs. As a matter of fact, it is usually not clear what goal needs to be taken into account. But in any case, it is critical to determine a proper recovery plan to make sure that the formulated problem is computationally tractable in real-time. In practice, typical goal is to minimize the additional costs of implementing the original schedule to cover the remaining flights. In practice, most of the time manual or heuristic methods are employed to reach a solution immediately, which is far from optimal.

Disruption Management

In airline operations a *disruption* is an unexpected event that cause or may cause either some or all operational schedules to become inoperable as planned. The main reasons for these disruptions may be inclement weather, equipment failure, crew sickness, late passengers, airport congestion etc. (Soykan & Erol, 2014). Such disruptions most probably may result with flight delays and cancellations that could not only considerably increase the operational crew costs, but they may also impair airline's goodwill. *Disruption management* (irregular operations) refers to the process of adjusting the flight schedule and reallocating the resources during and after the disruption occurs. A ripple effect occurs when a small disruption result in a major one. Effective disruption management is required in order to cope with this ripple effect and thus, to confine the negative effect of a single disruption.

In practice, during disruptions the likelihood of flight schedule change is not very common; usually one or more flight leg is delayed or cancelled. However, delays and cancellations are also critical because of the strong interconnectivity arising from hub-and-spoke network structure. The negative effects of these delays or cancellations on the schedules differ, depending on the schedule's vulnerability to delay propagation. It is essential to distinguish between delays that do not cause ripple effect in the airline network and delays that do cause propagation of delay, and require complicated and timely recovery procedures.

Robustness and Robust Airline Schedule Planning

Airline crew schedules are often subject to various sources of uncertainties that affect negatively the flight duration and costs. In order to reduce the undesirable impact of disruptions, and to ensure some level of stability and an organized recovery after a disruption, it is crucial to develop effective approaches to generate robust schedules that are less vulnerable to disruptions. Actually, robust scheduling is a multi-objective optimization problem that generates schedules with a balance between the individual robustness objective and a cost minimization one. However, robustness is not an end unto itself and trying to achieve maximum robustness is often not desirable.

In airline schedule planning context, main challenge of constructing robust schedules is to define and find abstract measures of robustness that can make possible to formulize mathematically the scheduling problems. There are various ways to incorporate robustness into the airline schedules, for instance, adding slack time to absorb disruptions, employing standby or reserve crews at certain airports to recover easily when recovered etc. The published work emerging in robust scheduling area appears to be theoretically diverse and there is no methodological way in the literature to define and incorporate robustness. Nevertheless, there is a general agreement that robustness encompasses two key capabilities: *stability* and *resilience* of the schedules. Stability is the ability to maintain near cost-optimal or at least feasible after disruptions occur. Resilience is the ability to recover from disruptions in a cost-efficiently and timely manner. While in the literature these attributes of robustness are often treated as analytically distinct, in practice they are generally interdependent.

It is noteworthy to mention that the term "robust scheduling" is not meant to refer to obtaining schedules from *robust optimization*. It only refers to generating schedules that are less prone to disruptions caused by uncontrollable factors and/or easier to reschedule once disrupted. Robust optimization is an alternative approach to handle cases where uncertainty cannot be described by any probability distribution due to the fact that the distributions are hard to justify or estimate. Therefore, it yields a solution that is resistant to any realization of uncertainty (Bertsimas & Sim, 2004). However, the solution attained from robust optimization approach is characteristically very conservative and mainly not well-suited for large-scale combinatorial problems, and not very commonly used in practice.

Even though most literature focuses on deterministic airline schedule planning problems and assumes that scheduled flights depart and arrive on time, there is a growing body of literature that models the robustness of the airline schedules. Smith and Johnson (2006) proposed a robust scheme for fleet assignment problem in which opportunities for aircraft and crew swaps are tried to be increased. The main advantage of this approach is that delays and cancellations might be eliminated if aircraft or crew can swap rotations or pairings, by simply restricting the number of fleet types that serve an airport. The same idea is extended by Gao, Johnson, and Smith (2009) to limit not only the number fleet types but also crew bases that serve an airport in order to construct integrated fleet assignment and robust crew planning. Weide, Ryan, and Ehrgott (2010) introduced an iterative approach to robust and integrated aircraft routing and crew pairing problem where the number of times crews change aircraft is penalized in the objective function of the crew pairing phase. The objective function of the aircraft routing phase is to maximize the total weights of short connections used by both crew pairing solution and aircraft routing solution. A detailed literature review on robust ACP is given under "Robust ACP Models" subsection.

One of the main challenges related to robustness of a schedule is to measure it in advance of operations. The reason is the difficulty to estimate what kind of disruptions may happen and when will happen. This also depends on the airline's disruption recovery strategy. Ionescu & Kliewer (2011) deafine a realistic way to measure the robustness of schedules is the On-Time Performance (OTP) of operated flights in a real or simulation environment. OTP is the most practically valid and well-known indicator of punctuality of the schedules, which is the fraction of flights arriving on-time at their destination airport. OTP is a key measure for all stakeholders especially for airlines, because it is related to direct and indirect costs due to the loss of efficiency. Also it needs be noted that propagated delays have a great impact on OTP.

Propagated Delay

Flight delays can be separated into two components: *primary delay* which is the delay that is not affected by any earlier or accumulated delay, and *propagated delay* which is the delay that is caused by ripple effect from a primary delay and generally referred to be caused by late arriving flight. In case of delays it is crucial to identify whether it is a primary delay or a propagated delay, because propagated delay is identified as the largest delay cause (Guest, 2007). Due to higher additional operational costs incurred by delay propagation, nowadays airlines are interested beyond just crew schedules with optimal planned crew cost, but robust schedules with low operational crew cost, in which optimal planned cost is not considered as necessary. Crew schedules that have reduced level of propagated delay appear more robust when confronted with unforeseen volatility in the operations, largely due to their capability to adapt better and more quickly.

Formerly, it was limited to better realize and handle propagated delay in practice and usually primary delays were taken into account as the main factor for better OTP (Guest, 2007). However, propagated delays create significantly more impact than the primary delays and an individual primary delay can create ripple effect through the entire schedule, affecting a large number of subsequent flights (AhmadBeygi, Cohn, Guan, & Belobaba, 2008).

AIRLINE CREW PAIRING OPTIMIZATION

Similar to airline scheduling decision process, airline crew management is also included in all planning levels: *strategic*, *tactical* and *operational*. Strategic level focuses on long term capacity planning of crew members, such as location of the crew bases. Tactical level deals with allocation of existing crew members to flight legs in the planned flight schedule. And operational level concentrate on rescheduling the planned crew schedules to make some necessary adjustments and recovery of the crew schedules to react on unexpected changes in the flight schedule as a result of disruptions. The main objective in operational level planning is to restore the crew schedules to become feasible rather than cost minimization. In general rescheduling is completed before the day operation and recovery is performed on the day of actual operations.

Problem Characteristics

There are several characteristics define the ACP problem: sequential solution stages, distinct crew categories, and the route type which the airline operates.

ACP problem is computationally difficult because of its nonlinear cost structure, complex feasibility rules and combinatorial nature. Even small amount of flight legs lead to millions of pairings. In order to overcome this burden, the problem is usually solved in three stages: *daily*, *weekly exceptions* and *monthly* (*dated*). In the daily problem, it is assumed that all flights operate on a daily basis. Therefore, all flight legs are included while the feasible pairings are constructed. Since most of the flights operate every day, the solution for the daily problem can be accepted as a rough estimate. Considering the fact that the daily problem appears most similar to flight operations in North American airlines compare to European airlines, most of the North American airlines can obtain an acceptable solution from the daily problem. Most published research on ACP problem deals with daily problem, because it is the most challenging part of the entire process. In weekly exceptions problem, similar to daily problem, the assumption is that flight schedule is fixed for a week period. In this stage, basically, the daily problem solution is adjusted to fit the weekly exceptions. In monthly problem, the differences between weeks such as holidays, flight schedule change etc. is considered. Actually, the daily and weekly exception problems are approximations of the monthly problem. Notably, a recent research attempted to solve the monthly problem directly with a rolling horizon/column generation approach and reported good results from experiments conducted with the data obtained from a major short and medium haul U.S. airline carrier (Saddoune et al., 2013).

Airlines usually separate the ACP problem for each crew type: *cockpit crew* and *cabin crew*. Unlike cabin crews, cockpit crews are usually qualified to operate only limited types of aircraft and hold a particular role in the cockpit. Hence, cockpit crews are not interchangeable among flights. Cabin crews, however, can generally serve on any type of aircraft. Also, cockpit crew scheduling is subject to more strict legal regulations and union agreements, while cabin crew scheduling has to deal with relatively fewer constraints. Apart from these differences, similar methodologies are typically applied to solve the problem for both types of crew.

There are two additional groups of crews besides that are in operation: *standby crew* and *reserve crew*. Standby crews are stationed at crew base airports and they are ready to substitute any other crew member who is not able to fly assigned flights. Reserve crews usually wait out of airport and a predefined period of time is given to allow them to get to the airport, and be ready to fly. If there is no crew available at an airport, crews from other airports need to be relocated.

Also there are considerable differences between the way that long-haul and short-haul route types are scheduled. Long-haul route type, which is usually assumed for international flights, is relatively sparse and generally operates on a weekly basis. Long-haul problem often has special modeling and solution features, because the number of feasible pairings is relatively small compare to short-haul one. On the contrary, short-haul route type is characterized by frequent flights that repeat on a daily basis and typically related to domestic flights.

Problem Definition

The ACP problem has a set of specific terms (Barnhart et al., 2003). A *flight* or a *flight leg* is a non-sop direct flight from an origin airport to a destination airport with specified departure and arrival times. *Duty period* is a legal workday comprise of at least one flight leg or deadhead that can be executed without the necessity for a rest period. A series of one or more duty periods is called a *pairing*, and typically lasts one to five days depending on the route type, the work rules of the airline and the fleet type. A pairing must start and end at an airport where crews are stationed, called a *crew base*. Airlines often have several crew bases. Airlines normally prefer short pairings, because the longer ones may cause greater difficulty in recovery when unexpected disruptions occur. Diagram of a pairing is illustrated in Figure 3.

The time between flight legs in a duty period is called the *connection* or *rest time*. Within a pairing, duty periods are separated by *overnight rest* periods, also called *layovers*. *Block time* is the scheduled duration of a flying leg from gate to gate. *Deadheading* is the activity that a crew travels as a passenger to get to a specific airport for a following flight or back to the crew base. The total duration of a pairing, that is the total time from departure to return back to crew base is called *Time Away From Base (TAFB)*. *Flying time* is the total duration of a pairing spent in flying.

The input for the ACP problem includes the flight schedule, aircraft rotations, cost structure, regulations and rules concerning the pairings. Only output is the set of anonymous crew pairings which usually covers a day, a week or a month.

A demanding set of restrictions affects ACP problem, resulting from governmental regulations determined to ensure the safety of airline operations, labor contracts, and airline's own rules. These constraints typically restrict the length of the working periods as well as the rest periods. The most important and practical rules are described below briefly;

- Maximum flying time in any given 24-hour period, which is generally 8 hours,
- Maximum duty time in any given 24-hour period, which is usually 13 hours,
- Minimum rest time between two legs that can be connected together in a duty, which usually cannot exceed 4 hours,
- Minimum layover time between two duty periods, which is usually 10 hours,

Figure 3. Diagram of a pairing

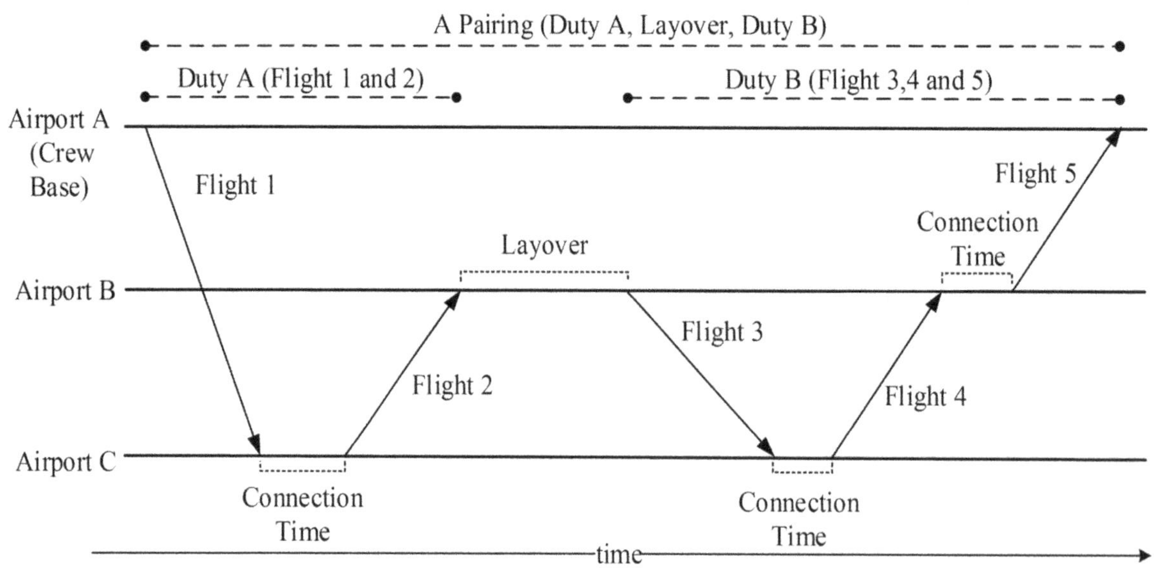

- Minimum layover time between two duty periods when scheduled for more than 8 hours flying in a 24-hour period, which is usually 12 hours,
- Maximum layover time between two duty periods, which is usually 36 hours,
- Maximum Time Away From Base (TAFB) which restricts the time that a crew will be away from crew base,
- Maximum number of duty periods in a pairing, which is usually 1 to 5 duties,
- Maximum number of daily landings, which is usually 4.

In ACP problem, usually the cost of a paring is not measured in monetary terms, because the pairings are not yet assigned to individual crews. Crews are often paid for their flying time along with the compensation for excessive ground times and TAFB, and also they may be eligible for per diem if they stay the night away from the crew base.

The crew cost associated with a duty period is usually considered as the maximum of three quantities: a fraction of the elapsed time of the duty period, minimum guaranteed number of hours, and total flying time in the duty period. Minimum guaranteed number of hours is a certain minimum threshold amount of time paid whether the crew flies or not.

The cost function of a duty period *d*, c_d, can then be expressed as follows:

$$c_d = max\ (f_d \times elapse,\ mgh_d,\ ft_d) \tag{1}$$

In Equation 1, f_d is a fractional value (such as 5/8), *elapse* is the elapsed time of the duty period *d*, mgh_d is the minimum guaranteed number of hours per duty period, and ft_d is the total flying time in the duty period *d*. And similar to the cost of a duty period, cost of a pairing is the maximum of aggregate three quantities: a fraction of the elapsed time of the pairing (TAFB), of the pairing, and total cost of the duties contained in the pairing. The cost of a pairing *p*, c_p, can be expressed in a mathematical term as follows:

$$c_p = max(\sum_{d=1}^{nd_p} c_d, f_p \times TAFB_p, mgh) \tag{2}$$

In Equation 2, nd_p is the number of duty periods in the pairing *p*, $TAFB_p$ is the elapsed time (TAFB) of the pairing *p*, f_p is a fractional value, and *mgh* is the minimum guaranteed number of hours per pairing.

North American airlines determine the cost of a crew schedule in terms of minutes of pay-and-credit and crews are paid proportionately to the number of pay-and-credit minutes they accumulate. The *pay-and-credit* time is the difference between the time actually needed to cover the subset of flights and the guaranteed hours of pay. On the contrary, in Europe, crews are often paid a fixed monthly payment, but still crew utilization is a significant component of the cost structure.

Mathematical Formulation

ACP is usually formulated as either Set Partitioning/Covering Problem (SPP/SCP) or network flow problem, but it is more often formulated as SPP/SCP to overcome the complexity of the constraints inherent in crew pairings. When using SPP/SCP, each legal pairing is modelled as a subset of flight legs along with its cost.

The SPP/SCP is one of the most studied mathematical programming models in combinatorial optimization with numerous real-world applications in various fields. The problem has been demonstrated to be *NP*-Hard (Garey & Johnson, 1979) which means when the size of the problem increases it becomes more complex and difficult to solve optimally in polynomial time.

SPP/SCP is the problem of partitioning/ covering the rows of an m-row, n-column zero-one matrix by a subset of the columns at a minimal cost. In simple terms, given a m-row, n-column, zero-one matrix $\boldsymbol{A}$, the main objective is to select minimum cost subset of columns while ensuring that every row has equally/at least one non-zero entry in the matrix covered by the columns. Each entry of the given zero-one matrix $\boldsymbol{A}$ is represented by a_{ij}, where $a_{ij} = 1$ if column i covers row j, $a_{ij} = 0$ otherwise. The cost of pairing j is given by c_j and decision variables x_{ij} indicate whether column i is selected as a part of the solution (x_{ij}=1) or not (x_{ij}=0). SPP with additional constraints is referred to as Generalized Set Partitioning Problem (GSPP) which can be formulated as a constrained optimization problem as follows;

$$min.c^T x \tag{3}$$

$$s.t.\boldsymbol{A}x = 1 \tag{4}$$

$$\boldsymbol{M}x \begin{Bmatrix} \geq \\ = \\ \leq \end{Bmatrix} b \tag{5}$$

$$x \in \{0,1\}^n \tag{6}$$

The objective function (3) minimizes the cost. The constraint set in Equation (4) guarantees that each row must be covered exactly one column (no over-covering). In ACP problem context, each column of $\boldsymbol{A}$ corresponds to a feasible pairing and each row denotes a flight that has to be operated.

Actually, the SPP is a special case of the Set Covering Problem (SCP), where the equalities in the constraint set (4) are replaced by greater-than or equal ($\geq$) type inequalities, which ensure that each column can cover at least one row that means deadheading is allowed. Deadheading incurs extra costs and for that reason it is often not desirable by airlines. However, totally ignoring deadheading sometimes may result in infeasible solutions.

The ACP problem is usually augmented with additional constraints (5), such as *crew base balancing constraints*, which ensure the evenly distribution of the workload across the crew bases in such a way as to comply with the actual allocation of the crews to crew bases, and confirm that crews contained in the solution that start/end at a crew base must be between given limits. Crew base constraints are in place as a result of union agreements and crew distributions. And last constraint (6) confirms that the decision variables are constrained to be binary values.

In order to generate pairings, two main types of network structures are used in the literature, *duty period network* and *flight (time-space) network*. In the flight network, nodes represent the departure or arrival of flight legs with "flight arcs" and "connection arcs" linking the nodes; whereas in the duty period network, nodes represent the start or the end of duty periods with "duty period arcs" and "layover arcs" to link the nodes. Duty period arcs represent the legal duty period that fulfill all constraints. In the flight network, the weight of the flight arcs represents the flight leg cost and the weight of the connection arcs may include additional costs. Similarly, in the duty period network, the weight of the duty period arcs represents the duty period cost and the weight of the layover arcs represents the rest costs.

The structure of the particular problem determines the network type for pairing generation. International flights are typically modelled as a duty period network, because duty periods rarely consist of more than one flight. On the contrary, domestic flights are typically modelled as a flight network, because duty periods can contain several flights and the total number of feasible duty periods is very large.

Solution Approaches

The main difficulty with the SPP model is that the number of feasible pairings (columns) grows exponentially with the number of flight legs (rows) and even for moderately sized instances, the computational CPU time and the required amount of memory is very large. Due to this combinatorial complexity of the model, only small sized instances can be solved directly with using standard commercial optimization software. For other than small size problem instances, the model is usually decomposed into master and sub problem based on the column generation framework in order to make the problem practically solvable.

Column generation (CG) is a widely used technique to solve linear optimization problems, particularly scheduling problems which have huge number of variables. In CG approach only a subset of columns, which have the potential to improve the objective function, needs to be dealt with in order to solve the problem, because in the optimal solution most of the columns will be non-basic with a value of zero. When implementing CG, first, the problem is divided into two sub problems, the Restricted Master Problem (RMP) and the Pricing Sub Problem (PSP). RMP is the LP relaxation of the original problem and consist of only the initial columns to start with.

The RMP is an integer program whose solution is not possible to be obtained directly, so its LP relaxation is solved. CG process iterates between this LP relaxed RMP and the PSP. By solving the RMP dual variables are sent to the PSP. The PSP uses these dual variables to find new columns to send back to RMP. If there exists no negative reduced cost pairing, then the solution is optimal for both the RMP and the complete problem. The solutions obtained from CG quite often will be fractional; therefore, B&B algorithm has to be applied so as to get the optimal integer solution.

PSP comprise of finding legal pairing or pairings with negative reduced cost and in the dual linear program a negative reduced cost pairing equals to a violated inequality. In general, either of the aforementioned pairing generation networks (duty period and flight network) is utilized in solving the PSP. In this network, the reduced cost of a pairing equals the length of the path which includes a sequence of nodes corresponding to the flight legs in the pairing. Therefore, PSP can be solved as a *resource constrained shortest path problem (RCSP)*, in which resource constraints ensure the legality of the pairings. Since the number of potential feasible pairings is very large, they do not have to be generated a priori, but only a subset of pairings needs to be created on-the-fly, which is called *delayed* or *dynamic column generation*.

Recall that in order to find the optimal integer solution or at least tighten the integrality gap, CG needs to be combined with a B&B algorithm to improve the polyhedral description of the LP relaxation of the problem. There are three general approaches to achieve this integration. The first approach involves implementing traditional CG before the optimization process begins and then solving the resultant integer program to optimality over the set of all feasible pairings. In second approach, the LP Relaxation of the problem is solved with delayed CG and then B & B is applied to find integer solution over the set of columns generated to solve the LP relaxation. The third approach is the B & P algorithm, which will be explained later in chapter.

As far as the GSPP is considered, the most accepted three mathematical optimization-based solution approaches to the problem are Benders' decomposition, Lagrangian relaxation and B & P. Each of these approaches deal with the problem from a different standpoint.

In Benders' decomposition the number of variables is decreased, but the number of constraints is increased. The constraints are divided into a master problem and a sub problem. The sub problem constraints are added to the master problem to convey the information from the sub problem. In each iteration one set of variables (complicating variables) are fixed and the problem with the remaining variables are called. First restricted problem is solved to optimality and the solution from previous iterations is used to fix the complicating variables in each iteration until the optimal solution to the master problem is confirmed to be an optimal solution to the original problem.

In Lagrangian relaxation, the constraints are divided into two sets: simple and complex. The basic idea is to relax the problem by extracting the complex constraints, replacing them to the objective function and penalizing their violation by Lagrange multipliers. The resultant problem obtained can be solved by a subgradient algorithm. Contrary to CG, which provides an upper bound (primal solution) in each iteration, Lagrangian relaxation generates a lower bound on the optimal integer program value at each iteration of the subgradient algorithm. This algorithm is easy to implement and not computationally complex compared to CG process, but it does not guarantee the optimal value.

B & P algorithm is similar to the B & B technique, but CG is applied at each node of the B & B tree. A RMP is formed in each node of the B & P process. Contrary to B & C, which produces rows (cuts) to improve (tighten) the LP relaxation; B & P generates columns to cut off the current optimal point of the RMP. While B & P algorithm proved to be an exact method to solve large combinatorial problems to optimality, it requires that the RMP and the PSP to be compatible to each other when the model is decomposed. First of these compatibility rules is that the branching rule should not alter the structure of the PSP during the entire process. Second rule is that symmetry in the structure of the model should be avoided. In branching phase, solution space should be partitioned into totally separate sets; otherwise same solution may be encountered in several different nodes of the B & P tree, which is quite harmful for the process. Finally, branching constraints should divide the solution space equally (Ben Amor & Valério de Carvalho, 2005). A survey on B & P is given by Barnhart, Johnson, Nemhauser, Savelsbergh, and Vance (1998).

Also there are some heuristic algorithms that applied to solve the problem. Actually, the most successful ones are based on Lagrangian relaxation, in which it is used to compute the relative value of a pairing in accordance with its likelihood to be selected in the optimal solution. Then, the relative value of a pairing is used in a greedy heuristic algorithm for calculating a solution in each iteration. The final solution is the best covering of flights by the pairings that found in all iterations.

Example

This example is used to demonstrate the basic understanding of the model formulation. Suppose there are 8 flight legs assigned to a particular fleet to serve 4 airports (A, B, C and D). In Table 1, the last column shows the day of the week the flight operates and indicates that the sixth flight leg do not operate on weekends and third flight is not flown on Sunday. Let's assume that A, B and C are the crew bases.

The above flight schedule is translated into a flight (time space) network shown as Figure 4 below;

Let's consider the daily problem and suppose the legal duty periods, which are expressed in terms of flight legs, are as follows (the numbers inside the curly brackets show the flight number/numbers) (airlines.org):

$d_1 = \{1\}$ $d_2 = \{2\}$ $d_3 = \{3\}$ $d_4 = \{4\}$ $d_5 = \{5\}$ $d_6 = \{6\}$ $d_7 = \{7\}$ $d_8 = \{8\}$

$d_9 = \{1, 2\}$ $d_{10} = \{1, 3\}$ $d_{11} = \{2, 4\}$ $d_{12} = \{5, 7\}$ $d_{13} = \{3, 6\}$

The 5 pairings extracted from the duty periods are shown below (these pairings are not exhaustive);

$p_1 = \{d_{10}, d_6\}$ $p_2 = \{d_{11}, d_7\}$

$p_3 = \{d_4, d_8, \text{layover}, d_9\}$

$p_4 = \{d_5, d_8\}$ $p_5 = \{d_{12}, \text{layover}, d_{13}\}$

Assuming pairing costs as $c_1=7$, $c_2=6$, $c_3=14$, $c_4=4$, and $c_5=16$, the integer programming formulation can be written as follows:

$$\text{s.t.}\, x_1 + x_3 = 1 \tag{8}$$

$$x_2 + x_3 = 1 \tag{9}$$

Table 1. Small-scale flight schedule

Flight #	Orig.	Dest.	Dep. Time	Arr. Time	Days of Operation
1	A	B	06:00	09:00	1234567
2	B	C	10:00	12:00	1234567
3	B	C	11:00	13:00	123456
4	C	D	13:00	15:00	1234567
5	A	D	13:00	15:00	1234567
6	C	A	14:00	16:00	12345
7	D	B	16:00	18:00	1234567
8	D	A	16:00	18:00	1234567

Figure 4. Flight network representation

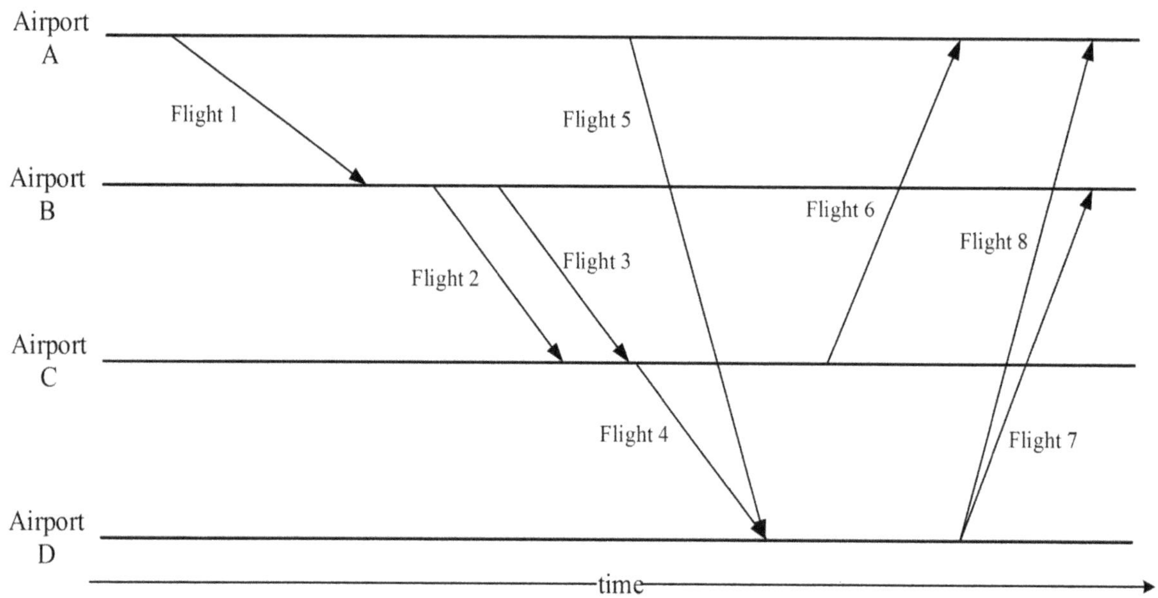

$$x_1 + x_5 = 1 \tag{10}$$

$$x_2 + x_3 = 1 \tag{11}$$

$$x_4 + x_5 = 1 \tag{12}$$

$$x_1 + x_5 = 1 \tag{13}$$

$$x_2 + x_5 = 1 \tag{14}$$

$$x_3 + x_4 = 1 \tag{15}$$

$$x_1, x_2, x_3, x_4, x_5 \in \{0,1\} \tag{16}$$

ROBUST ACP, AN OPTIMIZATION ALGORITHM AND A SIMULATION MODEL

Despite the fact that there are extensive publications dealing with traditional ACP problem, there is small fraction of research related specifically to robust ACP problem. Hence, it is not easy to design a DSF for robust ACP problem from the information given in the existing literature. Also, the literature related to DSSs for crew schedule planning does not include and utilize state-of-the-art methods. Considering the crew scheduling stage, robustness is often considered in the pairing stage instead of rostering stage.

Robust ACP Models

Until recently, there exists only several research studies reported in the literature which address the robustness in ACP problem context and proposed mathematical frameworks for solving the robust problem under different robustness definitions. In a nutshell, there are four major different robustness measures to generate pairings.

Minimizing Expected Pairing Cost

Schaefer, Johnson, Kleywegt, and Nemhauser (2005) replaced and minimized the planned crew cost in the pairings of the traditional ACP problem by an expected operational cost of the pairings which is calculated by the simulation of individual pairings. In the simulation they use only push-back recovery, which assume that departure of each flight leg is delayed until crews and aircraft are available. The main shortfalls of this approach include: the lack of consideration of the interaction between pairings and the difficulties arise while estimating the expected operational pairing cost.

Maximizing Connection Time

Ehrgott and Ryan (2002), and Yen and Birge (2006) measured robustness as the excess connection time above the minimum turn time (MTT). The robustness measure is same in both publications, except the assumption of Yen and Birge (2006) that the flight time is a random variable. A non-robustness penalty value is defined as the difference between MMT and connection time, and the robustness of the pairings are calculated by aggregating all penalties over the connection times in the pairing.

Ehrgott and Ryan (2002) presented a bi-objective optimization approach that attempts to generate pareto-optimal solutions. The bi-objective is to minimize both the planned cost and the non-robustness penalty, in which non-robustness penalty is used to penalize the aircraft changes when the crew ground time is short. They developed a model based on a ε-constraint method and a set partitioning algorithm to solve the model.

Yen and Birge (2006) introduced a two-stage stochastic programming model with recourse, in which recourse problem is employed to calculate the cost of delays. In this two-stage approach non-robust flight connections are discarded on at a time until the operational crew cost is minimized. The calculation of the operational crew cost is done by adding expected recovery cost to planned crew cost. Main shortcoming of the proposed approach is the computationally intractability of the solution method.

Maximizing Crew Swap Opportunities

Once a crew is delayed, it would be a cost-effective solution to not call a reserve or standby crew but swap with another crew who is available at that time. Shebalov and Klabjan (2006) proposed a bi-objective optimization approach to generate crew schedules that has increased flexibility in case of crew recovery. One of the two objective functions is to maximize the number of move-up crews, which are the crews that can be possibly swapped during operations. The other one is to minimize the crew costs which are incorporated as a constraint into the model. Also, a solution method based on the combination of CG and Lagrangian relaxation for the computationally difficult problem was introduced.

Ionescu and Kliewer (2011) also presented a robustness approach based on swapping opportunities of crews and formulated the problem as a two-stage stochastic programming model with recourse. As a solution method, a CG based framework with a constraint generation extension is proposed. So as to evaluate the robustness of the schedules, a scenario based simulation model with recovery actions is used. The model is constructed such that once the crew schedule gets disrupted; the negative impact of the disruptions is expected to be minimal.

Maximizing Utilization of Unit Crewing

Unit crewing is an approach to keep the crew together to work on the same pairing as long as possible. The resultant crew schedule is robust in the sense that it is less likely to delay other flights due to waiting for a member of the crew from a disrupted upstream flight, which is referred to as unit crewed schedule. It is aimed to minimize crew related propagated delay if disruption occurs.

Tam, Ryan, and Ehrgott (2014) suggested a bi-objective optimization approach with an additional objective to maximize unit crewing connection with minimizing the planned crew cost so as to facilitate crew members operate the same sequence of flight legs within the duty periods as much as possible.

Minimizing Propagated Delays Approach

In most of the researches mentioned above, the interaction and interdependence between pairings, and delay propagation have not been captured literally. Soykan and Erol (2014) presented a Bi-Objective GSCP approach to minimize the estimated propagated delay while at the same time maintaining a cost effective solution. In this approach GSCP formulation is preferred to GSPP, because of several reasons. First of all, the airline, which the historical data is collected for experimentation, is a small-scale one and deadheading is utilized as a common practice by the airline. Second, the incurred deadheading cost is lower than lodging and per diem costs, especially when the layover time is too long, so the airlines prefer deadheading from a cost standpoint. Besides these reasons, LP relaxation of GSCP is mathematically far steadier, and thus, easier to solve and construct a feasible integer solution from a LP relaxation solution. In this chapter, GSPP version of this approach is presented, due to the fact that, in practice, medium and large-scaled airlines usually do not include deadheading because of the higher incurred costs and tightly constrained structure (most of the flight legs have narrow time windows to be executed) of the probable pairings.

In this GSPP approach, the objective is to minimize the propagated delays while a certain level of cost is preserved with the help of a given set of historical data. In the developed model there is a trade-off between the crew cost and delay propagation minimization. Thus, a ε-constraint method is employed to transform one of the objectives into a constraint. As an initial stage for the methodology, deterministic ACP problem must be solved in order to find the minimum crew cost level, and propagated delays must be calculated from the given historical data.

Calculating the Propagated Delays

Before describing the mathematical model and solution methodology, calculation of the propagated delays from the historical data needs to be well understood. To this end, flight delay decomposition is demonstrated in Figure 5, which illustrates the relationship between departures, arrivals, and delays of two flights in the same pairing. The solid lines with arrows show the planned flight and the dotted ones display the actual flight, for flight legs *i* and *j*. *PDT, ADT, PAT* and *AAT* are the Planned Departure Time, Actual Departure Time, Planned Arrival Time and Actual Arrival Time, respectively. The Planned Turn Time between flights *i* and *j* (PTT_{ij}) is the time between *PDT* of flight *j* and *PAT* of flight *i*, see Equation 17. *ATT* (Actual Turn Time) is the *ADT* of flight leg *j* minus the *AAT* of flight leg *i*. If *ATT* is smaller than Minimum Turn Time (*MMT*) the crew is considered disrupted, considering the assumption that the same crew is assigned to flight leg *i* and *j*.

As shown in Figure 5, when flight leg *i* is delayed, the slack time ($Slack_{ij}$) between two flights and some part of the *MMT* is consumed, and this causes Propagated Delay (*PD*). Hence, the Total Departure Delay (*TDD*) of flight *j* is consisting of the PD_{ij} and the Independent Departure Delay (*IDD*) of flight *j* itself. Likewise, the Total Arrival Delay (*TAD*) of flight *j* comprises PD_{ij} and the Independent Arrival Delay (*IAD*). Although *IDD* includes only the independent delay before a flight departs, *IAD* includes both *IDD* and the additional independent delay in the air or at the origin and destination airports.

$$PTT_{ij} = PDT_j - PAT_i \tag{17}$$

Figure 5. Flight Delay Decomposition
Source: (Chiraphadhanakul & Eggenberg, 2009)

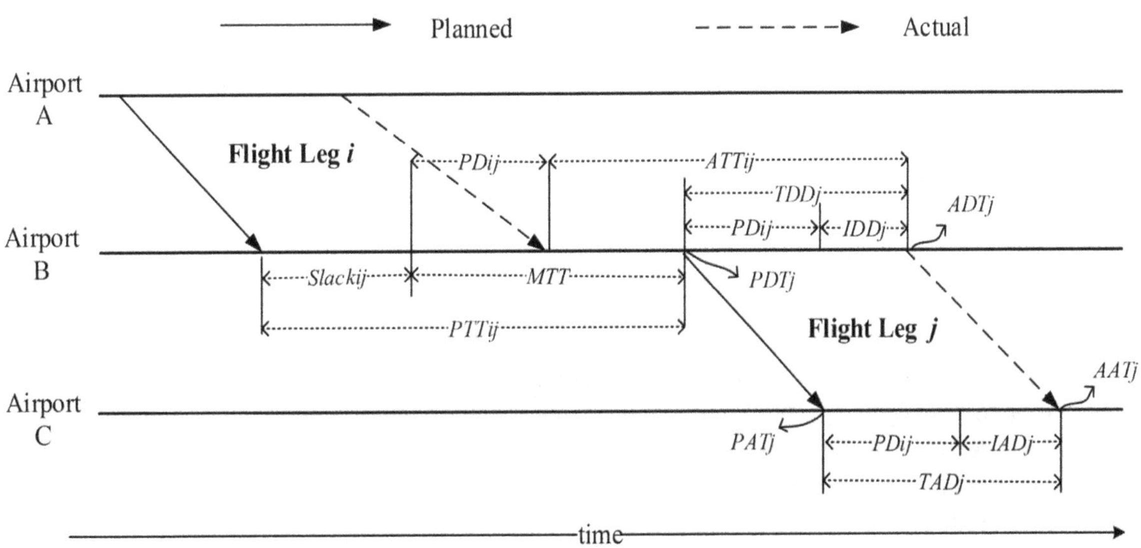

$$Slack_{ij} = PTT_{ij} - MTT_{ij} \tag{18}$$

The following relationships can be calculated to find the propagated delay:

$$TDD_j = \max\left(ADT_j - PDT_j, 0\right) \tag{19}$$

$$TDD_j = max\left(PD_{ij} + IDD_j, 0\right) \tag{20}$$

$$TAD_j = \max\left(AAT_j - PAT_j, 0\right) \tag{21}$$

$$TAD_j = max\left(PD_{ij} + IAT_j, 0\right) \tag{22}$$

$$PD_{ij} = \max\left(TAD_i - Slack_{ij}, 0\right) \tag{23}$$

A Sample Average Approximation method, which is a non-parametric approach, is utilized to exploit historical data to generate the solution, and then the future data is used to evaluate the solution. Each day of operation in the historical data is considered as one instance of a delay scenario (ω), assuming that each delay scenario is equally likely. Therefore, the set of delay scenarios (Ω) cardinality is equal to total day of operations in the set. Each pair of consecutive flight legs *i* and *j* in delay scenario (ω) is the *TAD* of flight leg *i,* which is obtained from historical dataset, and $Slack_{ij}$ is the crew ground slack time between flight legs *i* and *j,* which is obtained from the original flight schedule calculated simply by subtracting the *MMT* from the *PTT*. Propagated delay for each pairing must be calculated by taking into account each consecutive flight legs in any given pairing. For each day of operation, *PD*s can be obtained using *IAD*s, which assumes that the first flight of each day has zero propagated delay.

Mathematical Model

The formulation of the mathematical model is given below.

Define:

- *i,j*: indices for flight legs that need to be covered,
- *k:* an index for a pairing,
- *m:* an index for a crew base,
- *ω:* an index for a delay scenario,
- *F:* the set of flight legs,

- *P:* the set of pairings,
- *M:* the set of crew bases,
- Ω: the set of delay scenarios,
  p^{ω}: the set of delay scenario probabilities ($\omega \in \Omega$),
- c_k: cost of pairing *k*,
- pd^{w}_{ijk}: propagated delay from flight *i* to *j* in pairing *k* in delay scenario ω,
- a_{ik}: =1, if flight leg *i* is covered by pairing *k* and =0 otherwise,
- h_{km}: =1, if crew base *m* is the crew base for pairing *k* and =0 otherwise,
- l_m: minimum number of crew serving in crew base *m,*
- u_m: maximum number of crew serving in crew base *m,*
- Z_{opt}: optimal solution value of the deterministic problem,
- *ε:* robustness factor, which measures how much extra crew cost to be compromised,
- x_k: =1, if pairing *k* is involved in the solution and =0 otherwise,

Objective function:

$$min.\mathbb{E}\left[\sum_{k\in P}\left\{\sum_{i,j\in K}(pd_{ijk})\right\}x_k\right] \tag{24}$$

With the assumption that each delay scenario (ω) is equally likely and probable set of delay scenarios (Ω) has finite cardinality, the expected value of the total propagated delay in a pairing can be calculated as follows:

$$\mathbb{E}\left(pd_{ijk}\right) = \sum_{w\in\Omega} p^{w} pd^{w}_{ijk} \tag{25}$$

In order to capture the trade-off between two conflicting objectives, the problem is formulated as a bi-objective optimization model. Specifically, the objective functions are;

$$min.\sum_{k\in P}\left(x_k \sum_{i,j\in K}\sum_{w\in\Omega} p^{w} pd^{w}_{ijk}\right) \tag{26}$$

$$min.\sum_{k\in P} c_k x_k \tag{27}$$

As shown in Equation 26 and 27, the propagated delay is tried to be minimized while simultaneously conserving a cost-effective solution. Thus, a ε-constraint method is employed and crew minimization objective function is transformed into a constraint, limited by an upper bound. The compact formulation after applying ε-constraint method is shown below;

$$min. \sum_{k \in P} \left(x_k \sum_{i,j \in K} \sum_{w \in \Omega} p^w pd^w_{ijk} \right) \tag{28}$$

$$s.t. \sum_{k \in P} a_{ik} x_k = 1 \forall i \in F \tag{29}$$

$$\sum_{k \in P} h_{km} x_k \geq l_m \forall m \in M \tag{30}$$

$$\sum_{k \in P} h_{km} x_k \leq u_m \forall m \in M \tag{31}$$

$$\sum_{k \in P} c_k x_k \leq \left(1 + \varepsilon\right) Z_{Opt} \tag{32}$$

$$x_k \in \left\{0,1\right\} \forall k \in P \tag{33}$$

The original compact formulation is a GSPP-based formulation consisting of a binary decision variable. The objective function of the model is to minimize expected propagated delays (28) with the flight leg partitioning or set partitioning constraints (29), crew base balancing partitioning (30, 31) and robustness-cost objective limit constraint (32). There is one flight leg partitioning constraint for each flight leg and there is one crew base balancing constraint for each crew base.

The flight leg partitioning constraints (29) ensure that every flight in the given flight schedule is covered by exactly one pairing. Thus, a pairing can only be flown by a crew member and deadheading is not allowed.

The crew base balancing constraints (30, 31) ensure that the number of crews included from each crew base is in the specified limits. Each such constraint represents a crew base limitation for that crew base. The robustness-cost objective limit constraint (32) consists of variables of all the pairings and control the crew cost that it is not exceeding the minimum crew cost too much. But first of all, the deterministic ACP problem needs to be solved to obtain the minimum crew cost of the deterministic problem. With the help of the robustness factor (ε) it is possible to generate multiple solutions with various degrees of robustness. Lastly, constraint (33) ensures that decision variables have binary value.

Solution Approach

The compact formulation which is introduced above is reformulated via the CG technique. In this structure, the flight legs are covered by pairings and balanced crew bases are maintained. A pairing is a unique path in the flight network as governed by the constraints that ensure that at least one pairing covers the flight legs and pairings conform to crew base limitations.

The master problem is formulated as:

$$min. \sum_{k \in P} \left(x_k \sum_{i,j \in K} \sum_{w \in \Omega} p^w pd^w_{ijk} \right) \tag{34}$$

$$s.t. \sum_{k \in P} a_{ik} x_k = 1 \forall i \in F \tag{35}$$

$$\sum_{k \in P} h_{km} x_k \geq l_m \forall m \in M \tag{36}$$

$$\sum_{k \in P} h_{km} x_k \leq u_m \forall m \in M \tag{37}$$

$$x_k \in [0,1] \forall k \in P \tag{38}$$

The PSP is formulated as:

$$\min. \sum_{k \in P} \left\{ x_k \left(\sum_{i,j \in K} \sum_{w \in \Omega} p^w pd^w_{ijk} - \pi A \right) \right\} \tag{39}$$

$$s.t. \sum_{k \in P} c_k x_k \leq (1 + \varepsilon) Z_{Opt} \tag{40}$$

$$x_k \in \{0,1\} \forall k \in P \tag{41}$$

The master problem for robust ACP problem is defined with (34-38) linear program where the columns of constraints are some initial subset of pairings. When the master problem formulation is compared with the compact formulation, it can be seen that the integer restriction on the solution of decision variables has been relaxed, robustness-cost objective limit constraint removed, and the number of variables has been restricted to a reasonable sized subset of all pairings. By applying strong duality theorem, dual values (π) for the optimal solution of the master problem can be obtained. These dual values (π) are used in the objective function of the PSP to generate columns which will be added to the RMP to improve the solution value (39). The constraint of the PSP ensures that the columns obey the robustness-cost objective limit constraint (40). It is vital to attain high performance in solving the PSP, which generates new pairings, because quite often it consumes most of the total computation time.

Pairing Generation

The feasible pairings are generated by using a flight network, in which each flight and legal connections between flights are represented by arcs. Two flights will have a connection arc, if the arrival airport of the first flight is the same as the departure of the second one, and the time between them is legal.

Prior to generation of pairings, a feasibility check needs to be performed to eliminate any missing data or broken connection. It needs to be done by making sure that all nodes have at least one predecessor and one successor node except source and sink node. Pairing enumeration is done with a depth first search (DFS) algorithm in the flight network. In every step of the DFS algorithm, the generated pairings so far need to be checked for legality to confirm the performance of the pairing generation process. In order to improve the quality of the pairings and to accelerate the process, promising connections can be searched first.

Branching Rule

In general, B&P algorithm necessitates custom-built branching rules, which fits the formulation of the problem, because it is required to ensure that PSP does not generate a non-feasible column imposed by the branching rule. The commonly used branching strategy, which has proven successful in B&P algorithm, is follow-on branching. In this kind of branching it is enforced or forbidden two flights to be operated consecutively in a pairing. A follow-on flight pair consists of two flights that can be operated as a subsequence. The main advantage of the follow-on branching is that it reduces the size of the flight network and as a result, solution time improves during the B & P process.

Figure 6 illustrates the flowchart of the proposed B & P algorithm and the main steps of the algorithm are given in Table 2 (Soykan & Erol, 2014):

Simulation Model

Owing to the complex nature of airline operations it is not practical to verify optimization results, and assess the robustness of the obtained solutions and the performance of the proposed approach in an analytical way. Thus, a discrete-event simulation model needs to be employed to simulate airline operations and various unexpected disruptions that are faced during daily airline operations. The simulation model is used as a trace-driven simulation, where the historical data is used as an input.

Figure 6. Flowchart of the solution algorithm
Source: Soykan & Erol, 2014

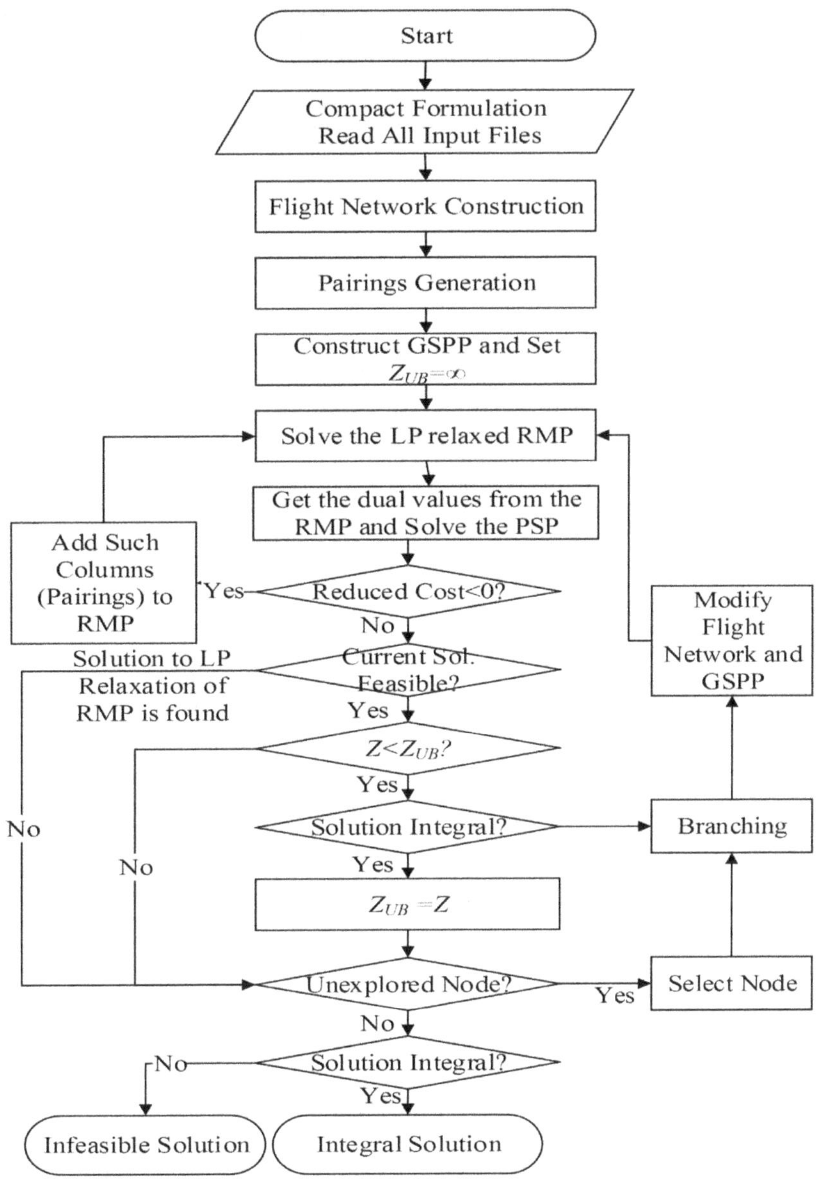

Simulation model assess the robustness of the crew schedule which allows analyzing the OTP of a potential crew schedule based on historical data for delay times. The historical data required to be comprised of at least the following information: actual departure and arrival times, cancellations and delay codes. In order to eliminate seasonal effects of delay times, entire historical data needs to be based on the same season as the potential schedule.

The simulation model represents the airline operations process and functions in accord with the optimization model to generate high quality and robust crew schedules. The optimization model searches the

Table 2. Main steps of the algorithm

Steps	Explanation
Step 1 Compact Formulation and read all input files	Construct the compact formulation and read all the data including the flight leg partitioning, crew base balancing constraints, propagated delays, robustness factor, optimal solution of the deterministic ACP problem, crew rules etc
Step 2 Flight network construction	Build flight networks in line with the flight schedule and crew rules
Step 3 Pairing generation with DFS procedure	Construct pairings considering the feasibility rules and calculate the related cost for each feasible pairing.
Step 4 Construct GSPP Model	Construct the GSPP model with the objective of minimizing propagated delay, flight leg coverage constraints, crew balancing constraints and robustness-cost objective constraint. Set Z_{UP} (the upper bound of the tree) to infinity.
Step 5 Solve the LP relaxed RMP	Solve and find the optimal solution of the LP relaxed RMP.
Step 6 Get the dual values from the RMP and solve the PSP	Get the dual values from the RMP and solve the PSP with the exact or greedy method in order to generate columns with negative reduced cost. From each PSP solution generate certain amount of columns that satisfies the minimum reduced cost threshold. If no such column exists GO TO Step 8, the optimal solution for the LP relaxed RMP is found.
Step 7 Add such columns to the RMP	Add new columns generated by the PSP to the RMP and GO TO Step 5.
Step 8 Check the feasibility of the current solution:	If the current solution is not feasible, prune the node and GO TO Step 13.
Step 9 Compare current objective with the upper bound	If the current objective is not lower than Z_{UP}, prune the node and GO TO Step 13.
Step 10 Check if the current solution is integral	If the current solution is integral, switch Z_{UP} with the current solution objective value, prune any node that has objective value higher than this upper bound and GO TO Step 13. Otherwise, record the current node as the parent node.
Step 11 Branch	Implement follow-on branching.
Step 12 Modify the flight network and GSPP	Modify the flight network and GSPP by removing the related arcs and columns in line with the branch selected and GO TO Step 5.
Step 13 Check if there is unexplored node	If there is an unexplored node, select a recently created unexplored node and GO TO Step 12.
Step 14 Check if the current solution is integral	If the current solution is integral, all tree is explored and the latest solution is the optimal solution. Otherwise, the problem is infeasible.

solution space and provides trial crew schedules for the simulation model to evaluate the robustness so as to measure crew schedule performance. The simulation model is used as a tool to determine potential operational cost benefits under different levels of robustness.

In simulation model, the exponential distribution is used to generate flight delays and the lognormal distribution is used to generate early departure times, and it is assumed that the only source of delay is the aircraft, in other words, the main assumption is that crew always arrive at the airport at their scheduled duty start time.

DECISION SUPPORT FRAMEWORK

Since its introduction in 1970s, DSSs have been successfully applied in many practical transportation systems and continue to attract considerable attention in various planning decisions. By the same token, airlines have been utilizing DSSs heavily with various complexity and functionality. Especially in airline schedule planning process, DSSs play a crucial role in managing scarce resources, in particular the crews and the aircrafts, efficiently and effectively, and they allow decision makers and planners to focus on critical complexities rather than on routine subjects improving the decision quality in a timely manner. However, DSSs are not meant to replace decision makers or schedule planners, but to extend their capabilities, and accordingly support and improve the competence and effectiveness of the planning process. Without utilizing a DSS, it is a great challenge for users to find an optimal or even feasible schedule out of hundreds of flight legs. Moreover, especially in major airlines, multiple individuals with distinct roles participate in schedule planning process and a DSS needs to assist all these roles and deliver capabilities to them.

The architecture of the presented DSF is illustrated in Figure 7, which shows how the components of the DSF are interrelated to each other.

The optimization and simulation models are integrated with a data management and graphical user interface (GUI), and designed into a DSF, which is a generic architecture describing necessary elements of a DSS and their interrelationship, in order to effectively manage crew schedule planning activities for improved robustness. These essential modules of the DSF present the individual functionalities. The schedule planner initiates the generation of crew pairings and controls the model parameters via a GUI module, which visualize the results in a clear and understandable way.

Data management module stores and fetches the data, which is needed to make a decision and perform analysis, from several databases. These database subcomponents include data extracted from internal sources as well as from external sources. Internal data constitutes of two groups of data; static data, and regulations and rules. The two databases are connected through structured data query and input/output functions.

Model management module, basically, allows users to transform the enormous amount of data into information. It supports model construction and analysis, and also provides an evaluation scheme which determines the robustness performance metrics for alternative solutions generated by the optimization model. This module allows the users to generate robust crew schedules and interactively evaluate alternative solutions by presenting potential values of the performance measures. The simulation model and the optimization model continuously exchange information. The optimization model searches the solution space and sends candidate robust solutions to the simulation model. After alternative solutions are generated, the simulation model, which captures the complex and stochastic nature of the airline operations, makes it possible to evaluate alternative robust solutions to the problem and what-if analysis. Since the airline operations are stochastic in nature, the application of a simulation model is quite helpful. With the use of the simulation model, users can conduct "what if?" analysis to determine how robustness performance is affected by changes in the model parameters.

As previously mentioned the main user for the DSF is the airline crew schedule planners, who mainly work in a planning department of an airline where they are responsible for creating crew schedules. The GUI module is the component that provides the connection between users and the DSS, which is the only one the user actually deals with and interacts with the system. The visualization displays the generated crew schedules such a way that only necessary information is included and allows users to

Figure 7. The architecture of the DSF

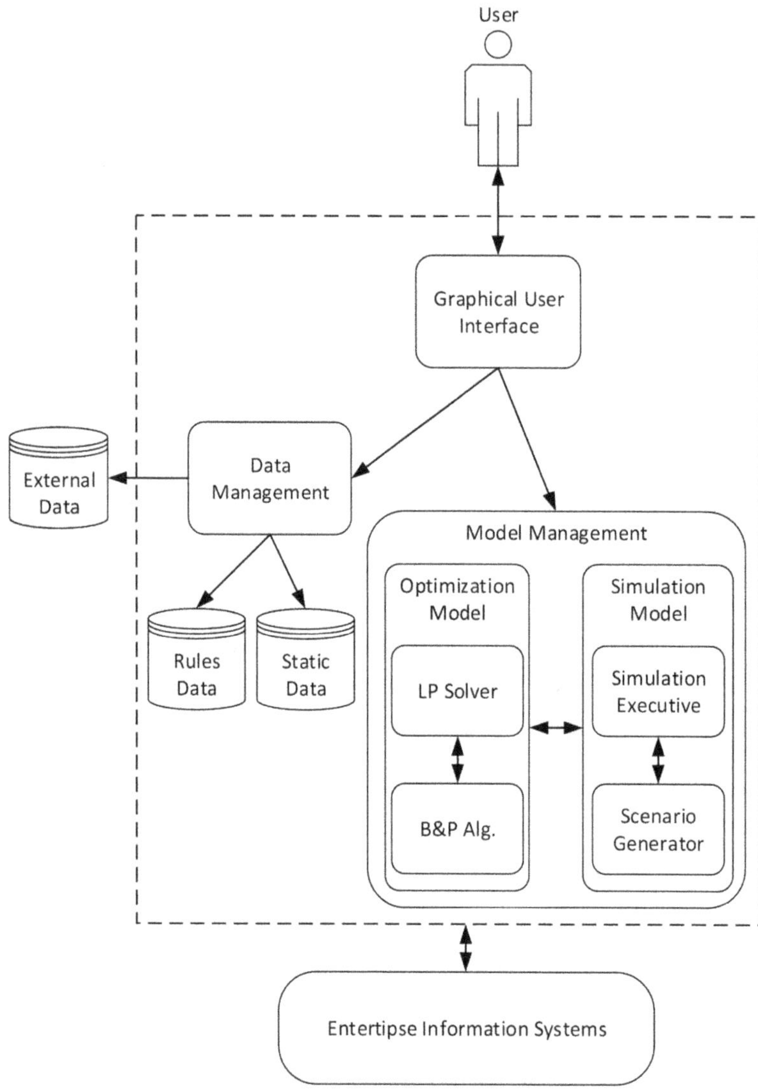

comprehend the underlying information more easily. It enhances the display of the graphics and the graphical schedule editing on a user's local terminal. Also company related constraints and requirements are included and easily changed in the pairing generation process through the GUI. For example; if the problem remains infeasible due to uncovered flight legs, the DSS sends an infeasibility message to the GUI specifying which flight legs are uncovered. Model management module has a set of diagnostic functions that identify uncovered flight legs that cause infeasibility. In this case, if the module is unable to find an analytical solution to move toward feasibility, it sends a message along with several solutions to the built-in GUI for interactive decision making. These solutions consist of changing the parameters of the optimization model, assigning additional duty periods to specific pairings or using additional pairings to cover flight legs left uncovered.

CONCLUSION

ACP problem forms a critical phase in the airline schedule planning process, since crew costs are the second largest component of direct operating costs for airlines, amounting approximately 20% of the total costs. In this chapter, a prototype DSS for airline crew schedule planners is presented, which supports generating robust crew schedules in advance of their actual operations. The framework makes use of a state-of-the-art B&P-based optimization algorithm to generate high quality crew schedules and a simulation model to measure the performance of the schedules for the purpose of robustness. The proposed DSF can be used either as a standalone tool or as part of an integrated or enterprise-level DSS.

The combinatorial nature and complexity of the ACP problem and magnitude of data makes this planning process tedious. Moreover, stochastic nature of the airline operations necessitates crew pairings to be robust enough to respond to changing conditions or irregular events that may happen during operations. As a result, a versatile and powerful DSS needs to be deployed to support robust schedule planning. Although there is extensive number of computer-aided DSSs in all areas of the airline schedule planning process, there is still room for further developments and improvements, especially regarding the robust ACP process. Utilizing B & P algorithm in the underlying optimization model along with a discrete event simulation model as part of a DSF to solve robust ACP problem is a promising approach, which implies that this area is far from exhausted and is suitable for future research.

REFERENCES

AhmadBeygi, S., Cohn, A., Guan, Y., & Belobaba, P. (2008). Analysis of the potential for delay propagation in passenger airline networks. *Journal of Air Transport Management, 14*(5), 221–236. doi:10.1016/j.jairtraman.2008.04.010

Andersson, E., Housos, E., Kohl, N., & Wedelin, D. (1998). *Crew pairing optimization Operations Research in the Airline Industry* (pp. 228–258). New York: Springer US. doi:10.1007/978-1-4615-5501-8_8

Barnhart, C., Cohn, A. M., Johnson, E. L., Klabjan, D., Nemhauser, G. L., & Vance, P. H. (2003). *Airline crew scheduling Handbook of Transportation Science* (pp. 517–560). New York: Springer US.

Barnhart, C., Farahat, A., & Lohatepanont, M. (2009). Airline fleet assignment with enhanced revenue modeling. *Operations Research, 57*(1), 231–244. doi:10.1287/opre.1070.0503

Barnhart, C., Johnson, E. L., Nemhauser, G. L., Savelsbergh, M. W., & Vance, P. H. (1998). Branch-and-price: Column generation for solving huge integer programs. *Operations Research, 46*(3), 316–329. doi:10.1287/opre.46.3.316

Barnhart, C., & Shenoi, R. G. (1998). An approximate model and solution approach for the long-haul crew pairing problem. *Transportation Science, 32*(3), 221–231. doi:10.1287/trsc.32.3.221

Ben Amor, H., & Valério de Carvalho, J. (2005). Cutting Stock Problems. In G. Desaulniers, J. Desrosiers, & M. Solomon (Eds.), *Column Generation* (pp. 131–161). New York: Springer US. doi:10.1007/0-387-25486-2_5

Bertsimas, D., & Sim, M. (2004). The price of robustness. *Operations Research*, *52*(1), 35–53. doi:10.1287/opre.1030.0065

Chiraphadhanakul, V., & Eggenberg, N. (2009). *How to evaluate the robustness of airlines schedules* [Technical Report TRANSP–OR 091006]. École Polytechnique Fédérale de Lausanne, Switzerland.

Desaulniers, G., Desrosiers, J., Dumas, Y., Marc, S., Rioux, B., Solomon, M. M., & Soumis, F. (1997). Crew pairing at air France. *European Journal of Operational Research*, *97*(2), 245–259. doi:10.1016/S0377-2217(96)00195-6

Ehrgott, M., & Ryan, D. M. (2002). Constructing robust crew schedules with bicriteria optimization. *Journal of Multi-Criteria Decision Analysis*, *11*(3), 139–150. doi:10.1002/mcda.321

El Moudani, W., Cosenza, C. A. N., de Coligny, M., & Mora-Camino, F. (2001). A bi-criterion approach for the airlines crew rostering problem. *Paper presented at the Evolutionary Multi-Criterion Optimization*. doi:10.1007/3-540-44719-9_34

Gao, C., Johnson, E., & Smith, B. (2009). Integrated airline fleet and crew robust planning. *Transportation Science*, *43*(1), 2–16. doi:10.1287/trsc.1080.0257

Garey, M. R., & Johnson, D. S. (1979). *Computers and Intractability: A Guide to the Theory of NP-Completeness*. San Francisco, USA: W. H. Freeman and Co.

Gopalakrishnan, B., & Johnson, E. L. (2005). Airline crew scheduling: State-of-the-art. *Annals of Operations Research*, *140*(1), 305–337. doi:10.1007/s10479-005-3975-3

Guest, T. (2007). *A Matter of Time: Air Traffic Delay in Europe* (Vol. 2). Brussels, Belgium: Eurocontrol.

Hoffman, K. L., & Padberg, M. (1993). Solving airline crew scheduling problems by branch-and-cut. *Management Science*, *39*(6), 657–682. doi:10.1287/mnsc.39.6.657

Ionescu, L., & Kliewer, N. (2011). Increasing flexibility of airline crew schedules. *Procedia: Social and Behavioral Sciences*, *20*, 1019–1028. doi:10.1016/j.sbspro.2011.08.111

Kohl, N., & Karisch, S. E. (2004). Airline crew rostering: Problem types, modeling, and optimization. *Annals of Operations Research*, *127*(1-4), 223–257. doi:10.1023/B:ANOR.0000019091.54417.ca

Li, Y., & Tan, N. (2013). Study on Fleet Assignment Problem Model and Algorithm. *Mathematical Problems in Engineering*, 2013.

Liang, Z., & Chaovalitwongse, W. A. (2009). *The Aircraft Maintenance Routing Problem Optimization and Logistics Challenges in the Enterprise* (pp. 327–348). New York: Springer. doi:10.1007/978-0-387-88617-6_12

Liang, Z., Chaovalitwongse, W. A., Huang, H. C., & Johnson, E. L. (2011). On a new rotation tour network model for aircraft maintenance routing problem. *Transportation Science*, *45*(1), 109–120. doi:10.1287/trsc.1100.0338

Lohatepanont, M., & Barnhart, C. (2004). Airline schedule planning: Integrated models and algorithms for schedule design and fleet assignment. *Transportation Science*, *38*(1), 19–32. doi:10.1287/trsc.1030.0026

Mercier, A., & Soumis, F. (2007). An integrated aircraft routing, crew scheduling and flight retiming model. *Computers & Operations Research*, *34*(8), 2251–2265. doi:10.1016/j.cor.2005.09.001

Rasmussen, M. S., Lusby, R. M., Ryan, D., & Larsen, J. (2011). *Subsequence Generation for the Airline Crew Pairing Problem [Technical Report]*. DTU Management.

Saddoune, M., Desaulniers, G., & Soumis, F. (2013). Aircrew pairings with possible repetitions of the same flight number. *Computers & Operations Research*, *40*(3), 805–814. doi:10.1016/j.cor.2010.11.003

Schaefer, A. J., Johnson, E. L., Kleywegt, A. J., & Nemhauser, G. L. (2005). Airline crew scheduling under uncertainty. *Transportation Science*, *39*(3), 340–348. doi:10.1287/trsc.1040.0091

Shebalov, S., & Klabjan, D. (2006). Robust airline crew pairing: Move-up crews. *Transportation Science*, *40*(3), 300–312. doi:10.1287/trsc.1050.0131

Smith, B. C., & Johnson, E. L. (2006). Robust airline fleet assignment: Imposing station purity using station decomposition. *Transportation Science*, *40*(4), 497–516. doi:10.1287/trsc.1060.0153

Soykan, B., & Erol, S. (2014). A Branch-and-Price Algorithm for the Robust Airline Crew Pairing Problem. *The Journal of Defense Sciences*, *13*(1), 37–74.

Tam, B., Ryan, D., & Ehrgott, M. (2014). Multi-objective Approaches to the Unit Crewing Problem in Airline Crew Scheduling. *Journal of Multi-Criteria Decision Analysis*, *21*(5-6), 257–277. doi:10.1002/mcda.1517

Weide, O., Ryan, D., & Ehrgott, M. (2010). An iterative approach to robust and integrated aircraft routing and crew scheduling. *Computers & Operations Research*, *37*(5), 833–844. doi:10.1016/j.cor.2009.03.024

Yen, J. W., & Birge, J. R. (2006). A stochastic programming approach to the airline crew scheduling problem. *Transportation Science*, *40*(1), 3–14. doi:10.1287/trsc.1050.0138

KEY TERMS AND DEFINITIONS

Block Time: The time required from departure to arrival for a particular flight leg.

Deadheading: The activity of carrying crew as passengers to reposition them for better utilization.

Disruption: An actual or impending situation that may cause to interrupt planned normal operations of an airline because one or a series of unexpected events happen.

Flight Duty or Duty Period: A set of consecutive flight legs which normally refers to one day's work of a crew member.

Flight Leg: A non-stop trip between an origin-destination airport pair.

Pairing: A sequence of trips that can be performed by a single crew, starting and ending at the same crew base.

Chapter 2
An Artificial Neural Network Model as the Decision Support System of Ports

Can Elmar Balas
Gazi University, Turkey

ABSTRACT

The new feasibility analysis model proposed in this study for coastal projects consists of three interrelated decision support models: 1) Artificial Neural Network (ANN) to determine the rates and capacity of cargo by considering the economical development of hinterland 2) Queuing model to determine the waiting to service time and the berth occupancy ratios by waiting time modeling of ships using discrete queuing simulation 3) Importance Sampling Monte Carlo (ISMC) to simulate ship arrivals/departures from the quays and to estimate income/expenditure parameters of the coastal project. As a case study, the proposed model was applied to the Iskenderun Pier in Turkey and the future loading/unloading cargo rates of pier were predicted by ANN's. The superiorities of this proposed simulation-ANN model to other classical investment planning methods were the inclusion of uncertainties in the investment parameters like the change of cargo and costs variables in time, and the determination of project benefit/cost with an improved accuracy when compared to classical decision support models.

INTRODUCTION

As 85% of the world and Turkish trade is carried by maritime transportation, new commercial piers and harbors are constructed in Turkey. Iskenderun Pier that is under construction in the organized industry zone of Hatay, will serve for internal and foreign trade of Turkey. The main problem that has been faced in coastal engineering projects of Turkey is to perform a realistic investment planning in order to include uncertainties and risk parameters, which decrease the reliability of cost and income predictions (Balas & Ergin, 2002). In this study for the solution of the problem, ISMC (Melchers, 1999; Balas et al., 2001) was applied and interrelated with ANN's for the investment planning of coastal projects.

DOI: 10.4018/978-1-4666-8648-9.ch002

Time to reach the simulation capacity of the pier was estimated by considering ANN predictions of cargo increase in the bay. The net present values of the investment were determined by modeling income/expenditures of pier activities in ISMC simulations by probability distributions that consider the cargo predictions of the artificial intelligence technique for various economic scenarios. As a result, the annual capacity changes of cargo were determined by neuronets and the annual incomes of quays were computed by simulation considering the occurrence frequency distribution of predicted values.

DECISION SUPPORT MODEL

The ANN sub-model predicted the annual shipment rates in the Iskenderun Bay between the years of 2015 and 2030. This model was interrelated with ISMC simulations, since the maritime trade in the bay was mainly affected in the last decades by the successive wars of Syria and Iraq; therefore, a large extent of uncertainty was inherent in its future predictions. The neural network is a multivariate, nonlinear, nonparametric artificial intelligence system, which consists of a large number of interrelated processing units (neurons) to simulate the human brain learning. ANN's are non-linear prediction models of natural systems, which can be used as alternatives to regressive models in coastal projects (Balas & Koc, 2002; Balas & Balas, 2002; Balas et al., 2004). In recent years, neuronets have been successfully used for analysis of coastal environments, such as for time series processing (Deo & Kumar, 2000), tidal level forecasting (Tsai & Lee, 1999), pattern classification (Deo & Naidu, 1999), wave data assessment (Tsai et al., 1999; Balas et al., 2004) and structural failure predictions (Mase et al., 1995).

The artificial neural network (ANN) is a non-linear robust prediction method that can satisfactorily handle the randomness and uncertainty inherent in data sets and complex natural systems. Neuronets exhibit the characteristics of "biological" networks to simulate the human brain learning and they have been successfully used for the prediction of environmental processes. In summary, artificial neural networks can demonstrate the learning and adaptation capabilities of biological neural systems by predicting stability numbers from experimental data sets considering the change in environmental conditions. On the other hand, fuzzy systems can effectively utilize the uncertainties inherent in human knowledge, but they do not have the capability of learning. Therefore, neural networks have more "generalization ability" (functional approximation capability) than the fuzzy systems in using a database. However if the database is limited and contains qualitative information, fuzzy systems can be alternatively used as design models. The advantage of fuzzy systems is that, they can handle human based information such as experience and judgment, and can consider qualitative data described by language.

Neural networks are simplified models of the human brain. Connections between neurons are axons and dendrites, connection weights represent synapses, and the threshold approximates the activity in the soma. The response of network:

$$y_j = f(\sum_{i=1}^{n} w_i x_i - t_j),$$

where y_j is the output of the jth neuron, x_i is the ith input signal, w_i is the connection weight from the ith input neuron to j^{th} neuron, t_j is the threshold or bias value of the jth neuron and f(x), is a non-linear function.

A multi-layer feed-forward neural network (MFF) has a layered structure as given in Figure 1. Hidden layers exist between the input (first) and the output (last) layer. Each layer has neurons that receive input from a layer below and send their output to units in a layer above. Activation of a hidden unit is a function of the weighted input and the threshold. A supervised learning algorithm with a set of chosen examples called training data set trains the multi-layer feed-forward neural network. Neural networks are classified according to their activation phase as feed forward or recurrent, and according to learning phase as supervised or unsupervised. In the supervised learning algorithm, weight and bias factors were determined by minimizing the convergence criteria. The performance index (∇P) is defined as the mean squared error in the following equation:

$$\nabla P(w) = \frac{1}{zN}\left[\sum_{n=1}^{N} \nabla K(w,n)\right] \tag{1}$$

The weight and bias updates are proportional to the performance index by total instantaneous squared error as follows:

$$K(w,n) = \frac{1}{2}\varepsilon^{T}(n)\varepsilon(n) \tag{2}$$

where, N is the number of input and output vectors, n is the epoch number, ε is the output error at n*th* epoch, z is the number of neurons at the output layer, $\nabla \boldsymbol{K}(w,n)$ is the gradient vector of total instantaneous errors that have components associated with the weights of the hidden and output layers w^h and w^y, respectively for j=1 to s, i=1 to k and l=1 to z, m=1 to s.

Figure 1. Multi layer feed forward ANN model

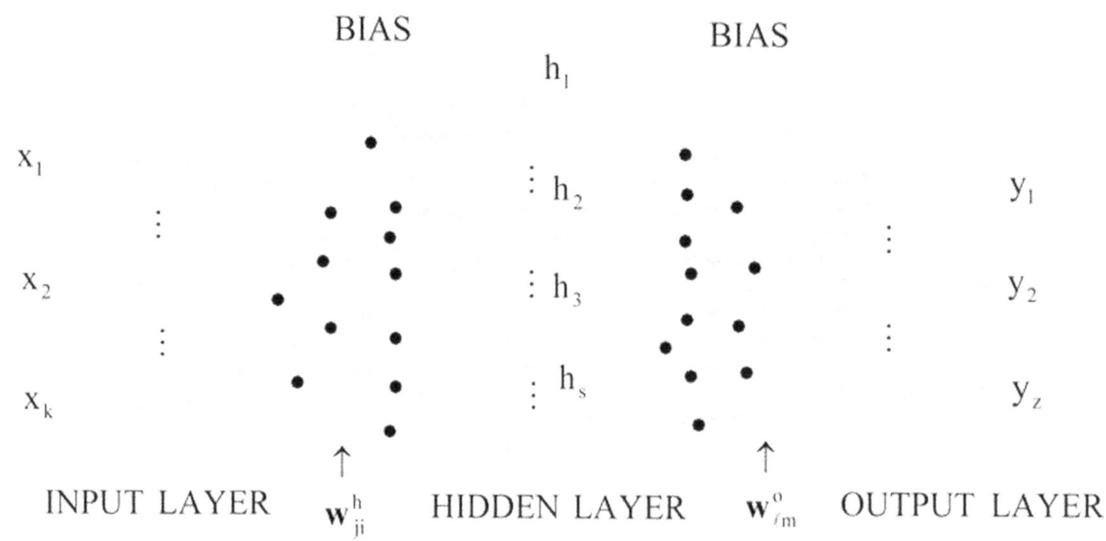

$$\nabla K(\mathbf{w}, n) = \left[\frac{\partial K}{\partial w_{11}^{h}} \cdots \frac{\partial K}{\partial w_{ji}^{h}} \frac{\partial K}{\partial w_{11}^{y}} \cdots \frac{\partial K}{\partial w_{\ell m}^{y}} \right] \tag{3}$$

In the second sub-model, ISMC (Anderson et al., 2000; Cheng & Druzdzel, 2000; Geissler & Chander, 2000; Booth, 2000) was applied to predict the economical parameters of the project, which were defined as random variables. In every single epoch of simulations, ships approached to pier, unloaded their loads at variable times and are charged according to cost tariffs by probability distributions that described these operations and included the future variations in project activities. The capacities and effective working durations of cranes and handling equipment during the loading/unloading processes were taken as random variables considering job requirements. The parameters of probability distributions were determined by investigating the operational conditions in nearby piers of the bay. Modifications were executed on standard deviation and range parameters depending on the operational conditions of the pier by considering the predictions of the management team.

Arrivals of ships considering the operational conditions of pier were simulated; the importance-sampling probability density functions were truncated to include the effect of the indicator function that controlled the arrival of ships outside the specified size limitations, defined as the low occurrence probability events. Sampling was limited to the regions bounded by physical conditions, such as the approach of ships in adjacent piers, which was controlled by correlating the dimensions of the ships. Results of ISMC were also compared with the correlated Monte Carlo and Latin Hypercube Sampling methods to check their accuracy. These three simulation methods of Monte Carlo gave similar results with a maximum deviation of 3% in 30,000 random trials.

CASE STUDY IN TURKEY

The feasibility analysis is carried out for the Iskenderun Pier as a case study, which is located in the organized industry region of Iskenderun that has geographical coordinates of 36.37° North Latitude and 36.07° East Longitude, i.e. located on the Mediterranean coast of Turkey. The pier has 300 m length and 45 m width constructed in front of a fill area of 600 m x 90 m. The project can be revised to increase the length of pier up to 600 m, which is considered in the feasibility study. The rest 68,661 m^2 of the total area is in the organized industry region, and almost 600 m away from the storage area. The pier is at the construction stage and the fill in front of it is in continuation. The 300 m pier alternative is planned to finish at the end of the year 2015.

Maritime Transportation in Iskenderun Bay

The loading and unloading facilities in Iskenderun Bay were analyzed by utilizing the data provided by the Shipping Chamber of Iskenderun, starting from 1988. The database involved the incoming and outgoing dates, agency, name, type and flag of ships; the incoming and outgoing dates of cargo to the harbor, the weight and type of load, the name of harbor where the ship was loaded or unloaded.

The container transport in Iskenderun Bay has been carried out from the Turkish State Railways Harbor. Although the stock area of the harbor is adequate for containerization, only small container ships (1,000-2,000 TEU-twenty feet equivalent unit) can be served, because of the low draft of quays in the harbor. On the other hand, the capacity of container transport in Eastern Mediterranean is substantial and the new pier can have the opportunity to take a significant share of container transportation in Mediterranean by serving to ships with a capacity of 6,000-7,000 TEU. The management of the pier aims to serve vast capacity container ships as a hub-port that fulfils an important demand in the Eastern Mediterranean with appropriate drafts. Iskenderun Pier has the opportunity to take a significant share of transit cargo due to the reconstruction activities of Iraq after the wars, and the domestic maritime transport from SAP region, if the pier starts to its operations as it is planned. The share of container transportation in Eastern Mediterranean depend upon the management and service quality of the pier and is directly a function of its competitive price policy.

Factors Effecting Cargo Rate Predictions

The data between the years of 1988 and 2013 was analyzed and the monthly transport values of harbors in the bay were determined. The specific need to apply ANN can be observed from these figures that, the maritime trade in the Iskenderun Bay was mainly affected in the last decades by the successive wars of Syria and Iraq; and showed a nonlinear variation that depended on the economical and political situation of Middle East. Hence standard statistical methods, like multivariate linear and non-linear regression analyses could not give acceptable future predictions, as they were parametric methods based on the maritime transport rates directly affected by the Syria and Iraq wars. Therefore, ANN's were employed in this study as multivariate, non-linear, nonparametric and robust models, and they were interrelated with ISMC simulation to consider the economical uncertainties associated with the region.

Before Syrian and Iraq wars, Iskenderun Bay was the center of trade throughout the history, as it was the gate of Middle East to the world considering maritime transportation. After these wars, nowadays it gains its significant place in the transportation to Iraq and Iran. However, the available historical data can not reflect the current circumstances that directly effected the maritime transportation and new methodologies are required to estimate the future loading-unloading capacities of the bay by considering the economical situation of the Middle East. The Southeast Anatolian Project (SAP) is another important factor that will affect the future trade of the Iskenderun Pier. The main economical aspects that can affect the project parameters can be listed as follows:

Economic Development of Hinterland

The economic development of Iskenderun and its hinterland can be predicted by considering the GNP of the region. ANN's predicted the future maritime transport rates of the bay by learning the relation between GNP growth of the hinterland and the increase in maritime transportation. ANN's study these relations at their training stages, even if the complex relation is a non-linear one (Balas et al., 2004).

GNP data of Hatay (Iskenderun) and its hinterland were attained from State Planning Organization of Turkey (SPO, 2001). National income accounting reflects the relationship between production, income and expenditures of the economy. National income is one of the most important parameters for development plans and programs in terms of income from economic activities, showing improvements and goals in economy. The income approach measures GNP as the sum of the factor incomes generated in the process of producing final outputs plus indirect taxes net of subsidies plus depreciation. The long-term growth of national income is generally reflected by the trend increase in GNP. Gross Domestic Product

(GDP) is obtained by subtracting the total inputs of production from the total value of goods and services produced by domestic activities of a national economy in a definite period. GNP is calculated by adding the net income value of foreign trade factors to Gross Domestic Product.

Southeast Anatolian Project

SAP region had a population of 5,187,000 in 1990 (SPO, 2001) with an area of 75,358 km^2. The population in 2015 will be estimated as 12,705,000. The main harbors in the area are Mersin and Iskenderun located in the Southeast Mediterranean Region of Turkey. These harbors are connected to the region by highway and railway systems. The container handling capacities of harbors with existing piers are 4,430,000 and 200,000 tones/year for Mersin and Iskenderun, respectively.

SAP is a comprehensive project that depends on agricultural activities of irrigation. The region will develop as an export center of agricultural products and industries with the realization of this project. The regional planning groups determined agricultural and industrial export products by considering the soil-water resources and potentials of the region. Highways and railways will transport the export commodities to Middle East countries at a rate of 771,000 tones/year in 2010. Maritime transport will constitute 25% and 50% of transportations to western regions of Turkey and to European countries, respectively (SPO, 2001).

At present, only 15% of anticipated maritime transportation capacity can be achieved in Iskenderun Bay. However it is estimated that, approximately 6.5 million tones of cargo will be exported from SAP region for domestic and international consumption starting from the year of 2010, and 3.5 million tones of this value is going to be handled by nearby harbors. The share of this quantity between Iskenderun and Mersin harbors is estimated as 1/3 and 2/3, respectively. Since Iskenderun is nearer to SAP region, a gain of 10 $ per container is probable, if harbors and piers in Iskenderun Bay are preferred for the loading/unloading of goods. The total amount of container exported and imported by SAP region is estimated as 2,536,062 tones/year starting from 2010, which constitutes 5.34% of the total amount of cargo (Table 1).

Prediction of Cargo Rates by ANN Decision Support Model

In this study, the loading and unloading rates of cargo in Iskenderun Bay were forecasted by constructing multivariate ANN models designated by ANN1 and ANN2, respectively. To consider the above listed factors in predictions, ANN's that are multivariate, non-linear, nonparametric and robust models were interrelated with ISMC simulation as an alternative technique to conventional statistical methods, such as regressive models. The artificial intelligence methodology can take into account the economi-

Table 1. Total expected maritime transportation between SAP and other regions

Region	Total Transport (Tones/year)	Estimated Maritime Transportation Share %	Estimated Maritime Transportation (Tones/year)
West of Turkey	3,805,086	25	951,271
European countries	561,971	50	280,985
Other countries	2,240,067	100	2,240,067
TOTAL	6,607,124	-	3,472,323

cal developments of the region by defining the GNP as one of the input parameters of ANN structures. Then these capacity increase rates are modeled in the ISMC simulation as random variables defined by probability distributions. The range parameters of these distributions are determined by the estimates of SAP and Iraq reconstruction project.

Two categories of multivariate ANN's were developed in the feasibility model for loading and unloading rates as designated by ANN1 and ANN2, respectively. The network structures of ANN1 and ANN2 were in the form of x2h30y1 and x2h20y1, respectively, with uni-bipolar sigmoid activation functions allocated for hidden and output layers. These multivariate models had two input parameters, which were the monthly changes in cargo and GNP of the hinterland region to estimate next month's cargo rates that were then converted to annual rates. For the training stage of neural networks 60% of 56 data groups, which consisted monthly loading/unloading and GNP rates recorded during the years of 1989-2003, were utilized. The remaining data groups were used for the testing (verification) stage of predictions. The range of training (learning) data groups was selected to include the maximum and minimum values of loading/unloading rates encountered in the data base of Shipping Chamber of Iskenderun, as tabulated in Table 2.

The iterative algorithm of the computations for the ANN sub-model is given in Figure 2. The iterative algorithm of the multi layer feed forward ANN is executed in two phases: 1) Forward activation to produce a solution 2) Backward propagation of the computed error to modify the weights. This process starts at the input layer where each input node or neuron transmits the value received forward to nodes in the hidden layer. Performing the dot product of all values in input nodes and their corresponding intercon-

Table 2. Range of loading/unloading data for training of neural networks

Range	Increase in loading (%)	Increase in unloading (%)	Increase in GNP (%)
Maximum	628.0	620.0	2033.3
Minimum	-95.0	-84.0	-256.4

Figure 2. The flowchart of the computations performed by the ANN sub-model.

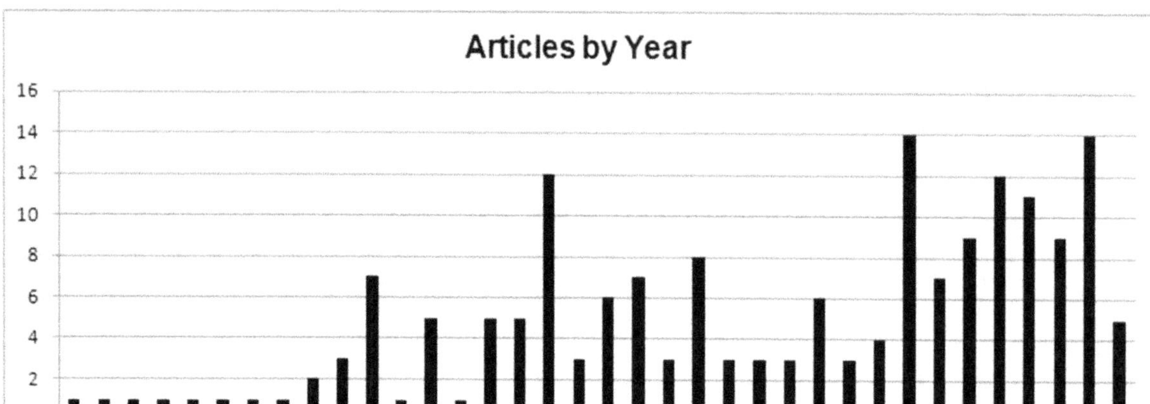

nection weights sums up the collective effect on each of the hidden nodes. The net effect or activation at one hidden node is calculated by using a non-linear transfer function, such as the sigmoidal and/or bipolar sigmoidal functions. The amount of activation obtained from each hidden node is transferred forward to the output layer. The procedure of calculating the net effect is repeated for all hidden nodes and layers. The net effect calculated at the output nodes represents the ANN's solution. In the backward propagation, the error between the ANN and target outputs is used to adjust the interconnection weights. This process starts from the output layers, continues through the hidden layers and ends at the input layer.

Minimum mean square errors of computations at the end of iterations and correlation coefficients were used as the performance criteria for neural network predictions. Neural networks were trained by using the gradient descent with momentum learning algorithm (Coulibaly et al., 2000). Learning rates were used as variables in the learning algorithm, therefore at the end of iterations this parameter was modified according to error function values applying an adaptive learning rate at the training phase of ANN's. At this stage, a white noise consisting of normally distributed random values with mean and standard deviation of 0.0 and 0.1 were added to data groups in order to increase the number of input vectors utilized in training, respectively. The momentum coefficient and the optimum iteration number were determined as 0.5 and 5,000 for networks. The correlation coefficients and minimum mean square errors of computations for trained networks were calculated at the testing stage as R1=0.910, R2=0.964 and MSE1=0.049, MSE2=0.015, respectively (Figure 3).

Results of ANN are compared with the predictions of multivariate linear (LR) and nonlinear (NLR) regression models developed. Traditionally, linear statistical forecasting methods have been widely used in many problems, since they are easy to develop and implemented; however they have the limitation that, problems in nature are generally nonlinear. Using linear regression models to approximate complex nonlinear problems is often not satisfactory, because there is a varying degree of non-linearity in the data, which may not be handled adequately by linear statistical methods. On the other hand, formulation of an

Figure 3. Predicted versus recorded data for loading and unloading rates of cargo at the testing stage of ANN's

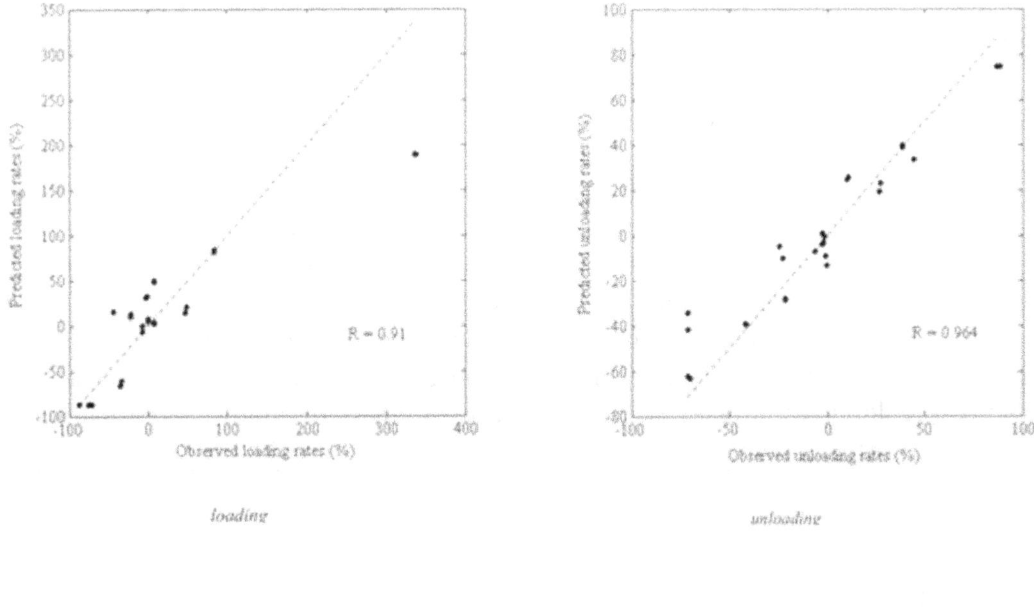

appropriate nonlinear model to a particular data set is a complicated task, since there are many possible nonlinear patterns in the data and nonlinear parametric models developed cannot capture specifically all the non-linear relationships available in data sets. Hence, the nonlinear regression requires deliberate (slow) iterative searching methods to learn the model parameters, and therefore it takes considerably longer time to define nonlinear patterns when compared to ANN models. Recently, ANN's have been proposed as an alternative approach for modeling complex (nonlinear) problems in nature as multivariate, nonlinear, nonparametric and robust stochastic approximations with dynamic feature extraction and can be utilized as very useful forecasting tools, as they are data driven methods (Basheer & Hajmeer, 2000; Zhang, 2001). A large number of successful applications have shown that, ANN's can be applied as robust methods for the modeling of nonlinear problems and are classified as alternative models to traditional regression techniques (Baker & Richards, 1999; Walczak & Sincich, 1999; Janes et al., 2004; Curry et al., 2002; Schepers et al., 2000; Chaloulakou et al., 2003, Razi & Athappilly, 2005).

Cargo rates in Iskenderun Bay were forecasted by constructing two categories of regression models designated by LR1 and NLR1 for loading; LR2 and NLR2 for unloading rates. Predictions of ANN's were compared with the predictions of these multivariate LR and NLR models, in which there were two input parameters of monthly changes in loading/unloading cargo and GNP in order to estimate next month's cargo rates by the regression equations. Monthly values were then converted to annual rates to compare with ANN models. Parameters of LR and NLR models were fitted to the cargo database to obtain the general model equation of:

$$z = \alpha + \beta_1 x + \beta_2 y + \beta_3 x^2 + \beta_4 y^2 + \beta_5 x^3 + \beta_6 y^3 + \beta_7 x^4 + \beta_8 y^4 \quad (4)$$

Regression coefficients of linear and nonlinear models were obtained by using the least squares and the Levenberg-Marquardt (LM) (Jang et. al., 1997) algorithms, respectively, as given in Table 3. Nonlinear regression determines the values of parameters that generate a curve in order to minimize the loss function, i.e. sum of squares of vertical distances of data points on the curve. Given m functions $f_1, ..., f_m$ of n parameters $p_1, ..., p_n$ with $m \geq n$, *in* vector notation: $\mathbf{f}^T=(f_1, ..., f_m)$, and $\mathbf{p}^T=(p_1, ..., p_n)$, the problem consists of finding the parameter vector $\mathbf{p}$ for which the cost function of the fourth order polynomial becomes minimal: $S(\mathbf{p}) = \mathbf{f}^T\mathbf{f}$. As the nonlinear regression model, given a set of empirical data pairs (t_i, y_i), the LM algorithm optimizes the parameters $\mathbf{p}$ of the model curve $c(t|\mathbf{p})$ so that the sum of the squares of the deviations $f_i(\mathbf{p})=y_i - c(t_i|\mathbf{p})$ becomes minimal. At the training stages of regression models, the mean correlation coefficients were determined as R_1=0.164 and R_1=0.611 for linear and nonlinear models, respectively.

Testing results of these models are given in Table 4 for the performance criteria parameters of the mean error (ME), correlation (R) and determination coefficients (r). From Table 4, it is seen that ANN could predict the cargo rates in the bay with a better accuracy than regressive models used in the paper when the testing results of these models were compared. In Figure 4, predictions of ANN1 and NLR1 models compared with the recorded data of loading cargo rates, the extrapolation capability of the fitted NLR1 model and, the predictions of ANN2 and NLR2 models compared with recorded data of unloading cargo rates are illustrated. It can be seen from that, the extrapolation ability of NLR models outside the recorded data range was limited when compared to ANN's.

Table 3. Parameters of NLR and LR models fitted to cargo rate database

Parameter	NLR1	NLR2	LR1	LR2
α	6.47998	2.23477	4.9360	-0.51100
β_1	-4.46345	$-4.03117\text{x}10^{-1}$	0.1420	-0.36900
β_2	$-5.98516\text{x}10^{-1}$	$7.38697\text{x}10^{-1}$	-0.0160	0.01800
β_3	$-2.35476\text{x}10^{-3}$	$-3.86778\text{x}10^{-2}$	-	-
β_4	$-2.79028\text{x}10^{-3}$	$1.64618\text{x}10^{-3}$	-	-
β_5	$5.13248\text{x}10^{-4}$	$5.64121\text{x}10^{-5}$	-	-
β_6	$7.01263\text{x}10^{-6}$	$-7.67135\text{x}10^{-6}$	-	-
β_7	$-1.37989\text{x}10^{-6}$	$3.93969\text{x}10^{-6}$	-	-
β_8	$-2.70664\text{x}10^{-9}$	$3.28777\text{x}10^{-9}$	-	-

Figure 4. Model results

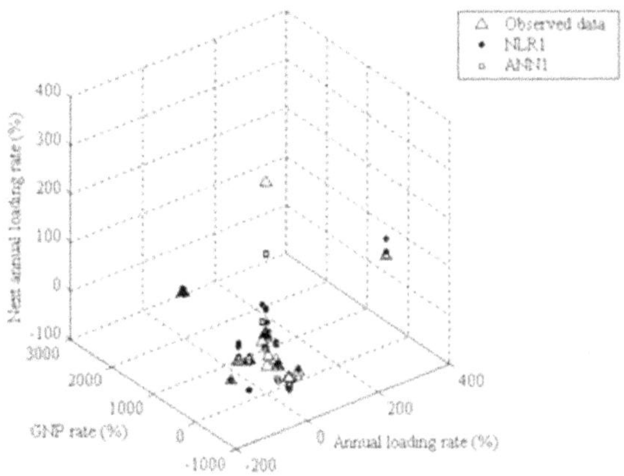

Comparison of the predictions of ANN1 and NLR1 models with the recorded data of loading cargo rates.

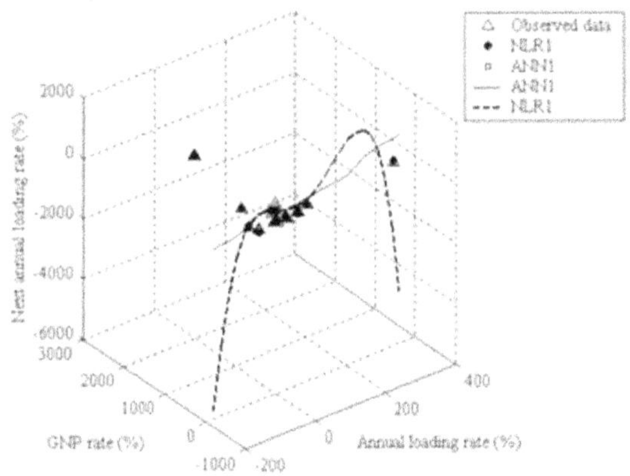

Extrapolation capability of the fitted NLR1 model for the recorded data of loading cargo rates

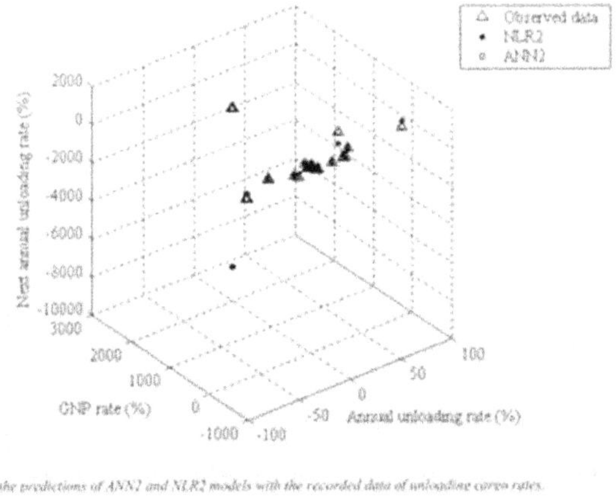

Comparison of the predictions of ANN2 and NLR2 models with the recorded data of unloading cargo rates.

Table 4. Testing results of ANN, LR and NLR models for the performance criteria parameters of mean error (ME), correlation (R) and determination coefficients (r)

Model	R	ME	R
ANN1	0.910	-7.410	0.781
LR1	0.110	21.250	0.010
NLR1	0.590	6.410	0.352
ANN2	0.964	9.290	0.910
LR2	-0.109	18.458	-0.290
NLR2	-0.282	-15.400	-2.310

It can be concluded that, the general performances of ANN's were better in predicting cargo rates when compared to linear and nonlinear multivariate regressive models. Therefore, ANN's forecasts of loading and unloading cargo rates by ANN1 and ANN2 models were employed in the feasibility model, respectively, for the years between 2015 and 2030 to predict the amount of cargo handled at the harbors and piers in Iskenderun Bay (Table 5). At the final stage, incomes and present net values were predicted by the ISMC sub-model.

Simulation of Pier Operations by ISMC Model

This feasibility analysis includes 16 scenarios and simulations can be grouped into four categories (Figure 5):

- In the first simulation case, the pier length is 300 m; there exist 2 quays used by relatively large ships.
- In the second case, the pier has 4 quays used by middle and relatively small-sized ships and has a length of 300 m.
- In the third case, the pier is 600 m long with 4 quays used by two middle-sized and two large ships.
- In the fourth case, the pier is 600 m long with 6 quays used by six middle-sized ships.

Table 5. Increase in loading/unloading rates predicted by ANN and regression models (%)

Year	Increase in loading rate			Increase in unloading rate		
	ANN1	LR1	NLR1	ANN2	LR2	NLR2
2010	47.00	-7.31	66.27	65.00	-9.89	29.93
2014	67.00	2.61	-92.24	66.00	4.85	31.40
2016	87.00	3.99	-19.92	69.00	-0.95	32.30
2020	88.00	4.21	174.84	82.00	1.34	32.97
2030	86.0	4.22	692.50	87.00	0.43	33.36

Figure 5. Simulations

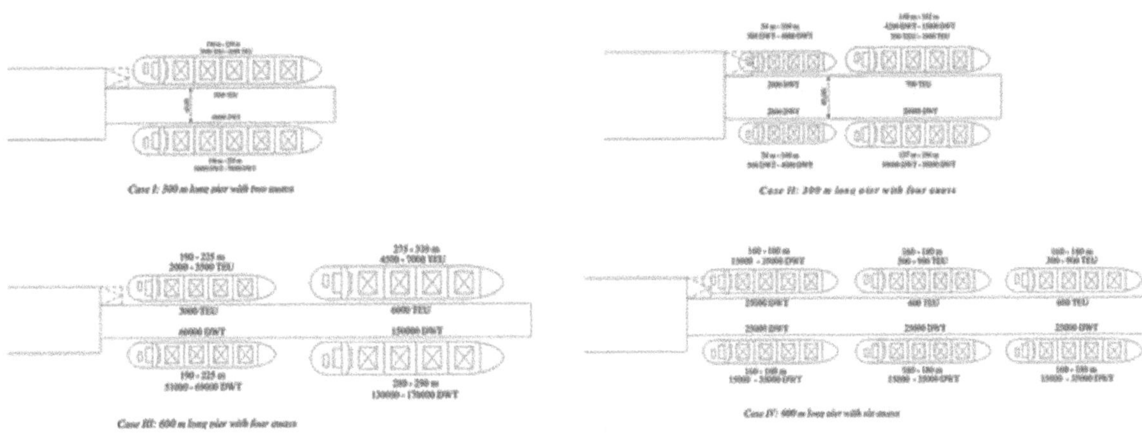

The model applies an iterative algorithm to interrelate the ship queuing and quay operation simulations together with ANN's. The flowchart of iterative ISMC simulation model procedures illustrated in Figure 6 can be summarized as follows:

1. Determine the total investment cost including reclamations, purchase of machinery/equipment and all other investments carried out for project Scenario 1.

The investment cost for the construction of 300 m pier, including reclamations, purchase of machinery/equipment was estimated as 71.28 million $, and 16 million $ was already spent for the construction. By extending the length of the pier to 600 m, large capacity container ships (7,000 TEU) will be handled and the total investment cost will increase to 106.56 million $. The construction of 600 m pier will be completed at the end of 2016 and the pier is going to start to handle cargo at the beginning of 2017.

At the construction stage own resources compensate credit payments. After 2015, when the 300 m pier will start to operate, these payments will be paid from profits and 55.28 million $ credit will be required with an interest payment of 22.25 million $. For the 600 m pier, 90.56 million $ credit will be needed with an interest payment of 36.45 million $. The interest rate of credit is 7% as Libor (London Inter-Bank Offer Rate) and the duration of payment is 8 years starting after the first three years. Libor is the interest rate that the banks charge each other for loans in Eurodollars. This rate is applicable to the short-term international inter-bank market, and applies to very large loans borrowed from one day to five years. This market allows banks with liquidity requirements to borrow quickly from other banks with surpluses, enabling banks to avoid holding excessively large amounts of their asset base as liquid assets.

2. Determine the berth occupancy and waiting to service time ratios by the queuing simulation model. Interrelate the queuing model with ISMC simulation model.

In the feasibility study, the duration that the pier may not operate due to weather conditions in a year was estimated as 20 days by using the long-term deep-water significant wave height statistics obtained from the wind data of Iskenderun. The ratio of the waiting to service time of ships and the berth occupancy ratio are estimated by the waiting time modeling of ships using the discrete queuing simulations, as a

Figure 6. The flowchart for the computations of the feasibility model

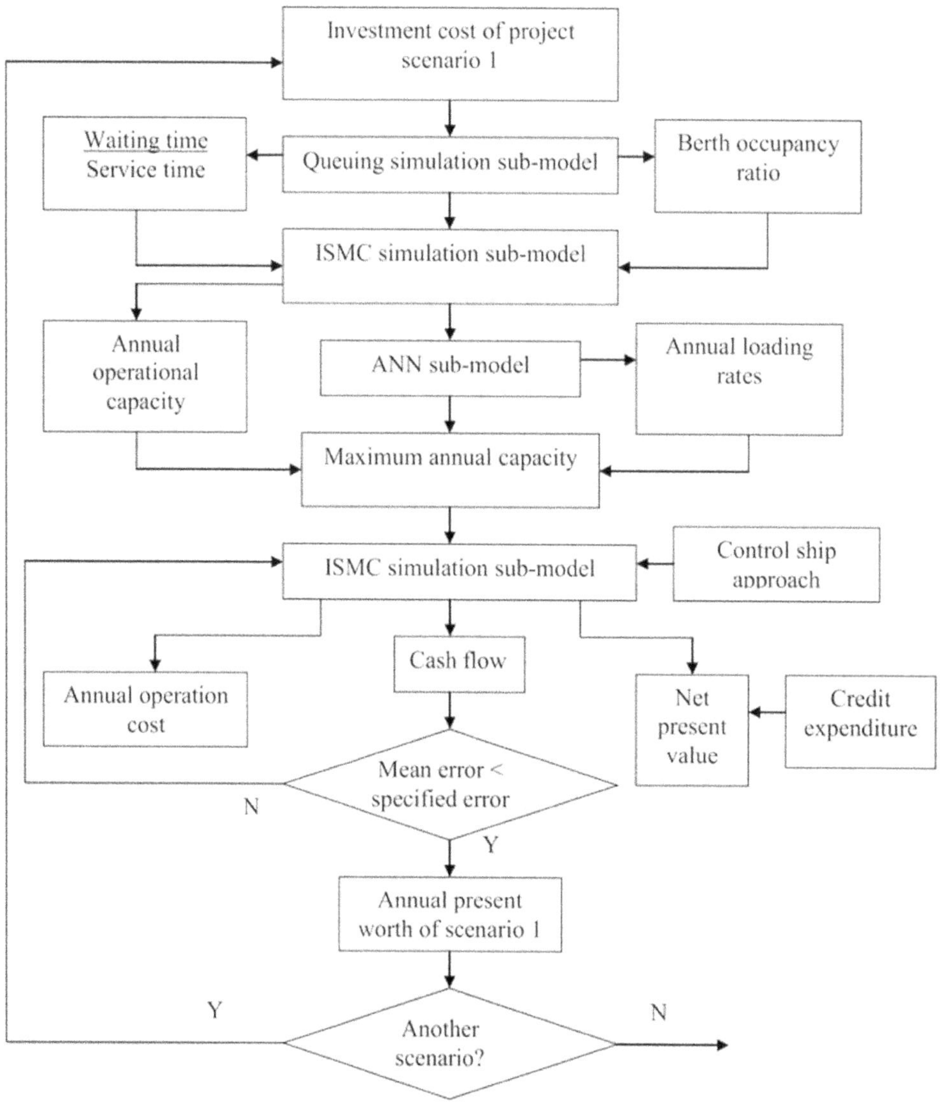

separate simulation model. Waiting time of ships in the queue is represented by the negative exponential Erlang distribution. The queuing model controlled the approach of ships in adjacent quays by correlating the dimensions of vessels. Hence, when the randomly generated ship approached the quay, the ship that had appropriate dimensions could approach to the adjacent quay. The occupancy ratio of quays and the waiting to service time of ships are then modeled as random variables in the initial run of Monte Carlo simulations to interrelate queuing model with ISMC.

3. Estimate the annual operational capacity by the first iterative run of ISMC simulation model.

The capacity of the pier is designated in the paper as the maximum annual loading-unloading volume of the pier determined from the initial discrete simulation of terminals by using queuing theory and is therefore equal to the physically possible yearly utilization amount of the pier for the maximum ratio of waiting to service time predicted by the queuing simulation model. Then the ISMC model simulated initially the loading/unloading activities considering the berth occupancy ratio at quays of Iskenderun Pier determined from the queuing model run, i.e. the waiting time simulation of ships. ISMC model attains probability distributions to the realistic physical capacity of cargo handling in quays determined in step 2 to interrelate simulation with the queuing model.

4. Determine the annual loading/unloading rates by ANN model considering the cargo traffic in the bay and interrelate with the second run of ISMC simulation model.

ANN sub-model predicted the load volume in the bay by using data obtained from State Planning Organization and Turkish Chamber of Shipping. The data were used in the training; testing and prediction stages of ANN and the annual cargo rates in the Iskenderun Bay were forecasted until 2030. At the testing stage, results of Feed Forward Back Propagation Neural Networks were compared with the linear and nonlinear regression models. The parameters of these regression models were determined from the maritime transport data mainly affected by the Gulf and Iraq wars. As the future situation of the region is considerably different now, an expected deviation was observed between the predictions of regression and ANN-ISMC models.

The net present values of the investment were determined by interrelating probability distributions of income and expenditures of pier activities in the second iterative run of ISMC simulations described in step 6 with loading and unloading capacity predictions of artificial intelligence for various economic scenarios. In other words, the annual loading/unloading increase rates (in percentage) determined by the ANN model were modeled in ISMC simulations by probability distributions as random variables to interrelate these model predictions. For example, the cargo rate increase of the 300 m pier in 2016, is represented by the Gauss probability density distribution that has a mean value of μ=29.25% and standard deviation of σ=2%. As a result, ANN predictions were further modeled in ISMC model as random variables represented by probability density functions in order to include uncertainties inherent in these predictions.

5. Compare the annual operational capacity determined in the first run of ISMC with annual loading/unloading rates of ANN. Take the smaller value as the maximum annual capacity for harbor operations in the second group of ISMC iterations.

The annual operational capacity found by the queuing sub-model and first iterative run of ISMC simulation limits the annual loading/unloading rates of the pier predicted by ANN's. The average percentage of quays occupied by ships was predicted by the waiting time modeling of ships using the queuing model in step 2, in the range of 60%-65% according to their load types. This amount is defined in the paper as the physical capacity of the pier and the initial capacities at the starting operation year are taken as 25% and 50% of this value to run the operational scenarios. The operation costs of quays and amortizations are calculated for two different scenarios associated with the pier alternatives of 300 m and 600 m in length, starting to operate in 2015 and 2017, respectively.

First and second operational scenarios indicate that, the pier will start to operate with 25% and 50% of its physical cargo capacity, respectively, having mean occupation (fullness) ratios as determined by queuing simulation in step 2 in the range of 60%-65% according to load types. After it is predicted from the waiting time modeling of ships, the occupation ratio of quays is then modeled as a random variable in the second iterative run of ISMC simulations by the truncated Gauss distribution probability density function, in order to include the uncertainties inherent in the predictions. According to various types of loads different distributions are used, for example the distribution with the mean of 60% had a standard deviation of 6% and a variation range of 45%-70%; and the distribution with the mean of 65% had a standard deviation of 7% and a variation range of 50%-75%. While calculating the incomes and expenditures of the management in the second group of ISMC simulations, the annual cargo rates predicted by ANN sub-model and capacity predictions of queuing sub-model were considered as the upper limit (whichever is smaller) for the yearly cargo handled by the quays.

6. Apply the second iterative run of ISMC simulation model to determine the annual operation costs of quays, amortizations, cash flow and net present value depending on the credit expenditure.

In the second iterative ISMC simulations, 8 scenarios were considered according to initial cargo rates, namely;

- For the pier of 300 m in length, 2 quays exist and initial capacity usage is 25%
- For the pier of 300 m in length, 2 quays exist and initial capacity usage is 50%
- For the pier of 300 m in length, 4 quays exist and initial capacity usage is 25%
- For the pier of 300 m in length, 4 quays exist and initial capacity usage is 50%
- For the pier of 600 m in length, 4 quays exist and initial capacity usage is 25%
- For the pier of 600 m in length, 4 quays exist and initial capacity usage is 50%
- For the pier of 600 m in length, 6 quays exist and initial capacity usage is 25%
- For the pier of 600 m in length, 6 quays exist and initial capacity usage is 50%

The cargo rates predicted for these scenarios at the starting years of 2015 and 2017, and the increase in the rates in forthcoming years can show variations according to the management, service and price politics. Therefore, Gauss (Normal) distribution was used to model the occurrence probabilities of initial capacity estimates, which are represented as random variables in ISMC simulations. For example in Scenario I, the mean value of initial capacity distribution was μ=25% with a standard deviation of σ=2%; and the maximum and minimum values of distribution were predicted as 20% and 33%, respectively.

In these second groups of iterative simulations, ships approach according to probability distributions by considering their weights and load types, unload their cargo at variable durations and are priced according to tariffs. The capacities and effective working durations of cranes and handling equipment during loading/unloading processes are taken as random variables according to job requirements. Probability distributions are also assigned to income variables, such as services for assistance, tugboat, fresh water supply, collection of wastewaters from ships, loading/unloading, shifting, limbo and terminal/warehouse services in order to model the possible variations during the lifetime of the pier.

Berthing and departure durations of ships were modeled by correlated Gauss probability distributions considering the vessel dimensions. For example for the ship that has an average total berthing and departure time of 4 hours, the 10 percentile is indicated as 3 hours in optimistic conditions, and the 90 percentile is defined as 5 hours in inappropriate (pessimistic) handling conditions. The operation time of the pier is modeled by the Gauss probability density distribution with a mean value of 18 hours and standard deviation of 1 hour.

At least two cranes per ship operate at the pier during the cargo handling and cranes of the ship are also utilized. Efficiencies of handling equipment were under their theoretical capacities due to operational conditions. Therefore, efficiency coefficients were modeled by Gauss probability distributions as well, for example the distributions for coal, container, grain and fertilizer had the mean value of 65% with a standard deviation of 1%; and the minimum and maximum values of distribution ranges were predicted as 60% and 70%, respectively.

Variations that can occur in the lifetime capacities of handling equipment are modeled by probability distributions considering main cargo items. For example, the loading/unloading capacity rate of coal was modeled by a Gauss probability distribution with mean and standard deviation values of 900 and 25 tones/hour, respectively. The handling equipment of coal would operate at the quay with at most 1,000 tones/hour and at least 600 tones/hour capacity with respective probabilities in the life time of the pier. The handling capacity of container was modeled by a Gauss probability distribution with a mean value of 30 TEU per hour and a standard deviation of 3 TEU per hour. The handling capacity of the general cargo was modeled by a Gauss probability distribution with a mean value of 100 tones per hour and a standard deviation of 5 tones per hour.

7. Control the approach of ships in adjacent piers by the correlated probability distributions of ship dimensions.

If the ship berthing at the specific quay has a larger size in dimension than the average size of vessels in the simulation, the ship berthing at the adjacent pier is generated smaller than average dimensions. For this purpose, correlation coefficients between the quays are determined to correlate ship lengths and load types between the quays for the modeling of physical conditions in quays. The percent usage of quays throughout the year was determined in step 2 by considering the sharing of cargo and type of the handling at the quay. To consider the changes in the future years, probability distributions were assigned to these variables in ISMC simulations.

8. Go to step 6 and repeat the simulations until the specified standard mean error is satisfied.

In the feasibility study, computer simulations repetitively reproduced ship arrivals, departures and loading/unloading activities of various quays 30,000 times for each of the random variable. The number of iterations was greater than the minimum stability (accuracy) criterion of Monte Carlo simulations (Melchers, 1999) and the specified standard mean error of simulation (ε) was satisfied for each simulation.

9. Determine the Net Present Worth of project scenario when the specified standard mean error is satisfied.

The number of scenarios examined in step 6 is increased to 16 with the assessment of the possibility that the pier can become a free trade zone. For this case the pier operations will be exempt from state taxes, since commodities that are handled at the pier are mainly processed in the free trade zone, as tax-free products. Annual cash flow and net present values were determined in simulations also considering the amount of yearly credit utilized and its interest payment. The results of Scenario 1 are examined as follows:

Scenario 1: The pier length is 300 m and the number of quays is 2. At the end of 30,000 simulations, it is found that for the yearly gross incomes calculated until 2030, the mean value is 22,570,000 $ and the standard deviation is 2,206,000 $. The distribution of annual gross income that will be obtained from this scenario is presented in Figure 7 in million dollars versus its probability of realization calculated from the occurrence frequency in total number of simulations. The distribution includes the main uncertainties related to services, management, cargo rates, and economical developments of hinterland.

10. Go to Step 1 to determine the Net Present Worth of project for another scenario.

Predictions of other project scenarios are presented in Figure 8 in which the variation of net annual present worth value of investment is illustrated.

The loading/unloading rates and incomes of quays were computed by the occurrence frequency distribution of 30,000 predicted values; and were described by statistical parameters such as mean value, standard deviation, and coefficient of variation. The simulation of pier activities by ISMC provided the inclusions of uncertainties inherent in quay operations, such as wave, queuing, labor and machinery conditions.

DISCUSSION OF RESULTS

1. The future loading and unloading capacities of Iskenderun Pier were calculated by utilizing ANN's. The predictions of this technique took into account the increase of GNP by learning its relation with handling capacities of cargo in harbors, which is an important macro economic parameter for

Figure 7. The probabilities of obtaining annual gross income for Case I

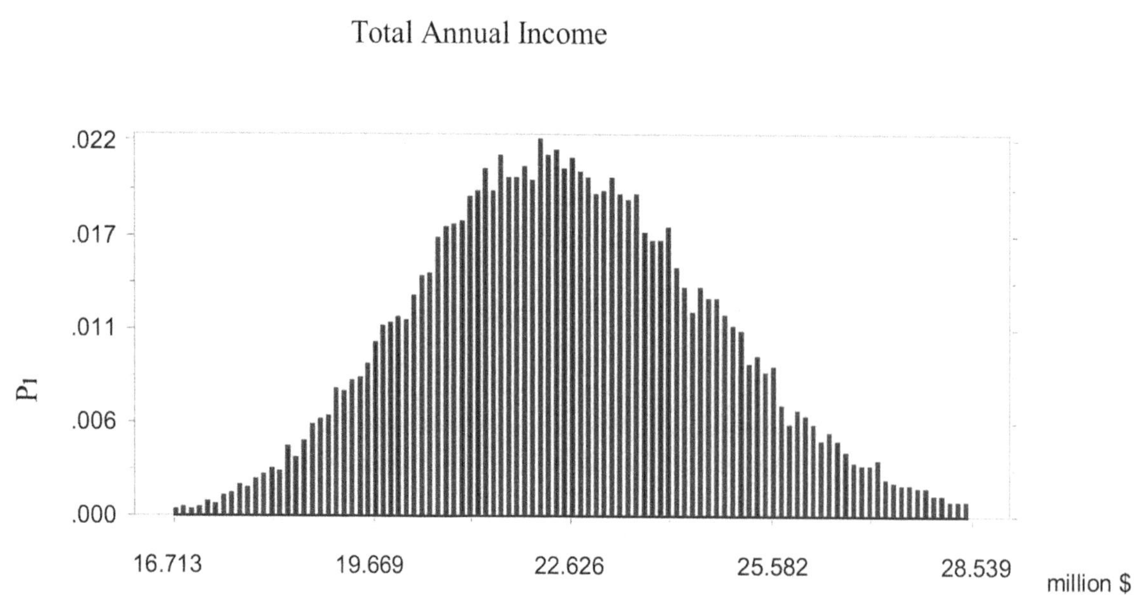

Figure 8. The variation of net present worth values of investment determined by ISMC Simulation (in Million $).

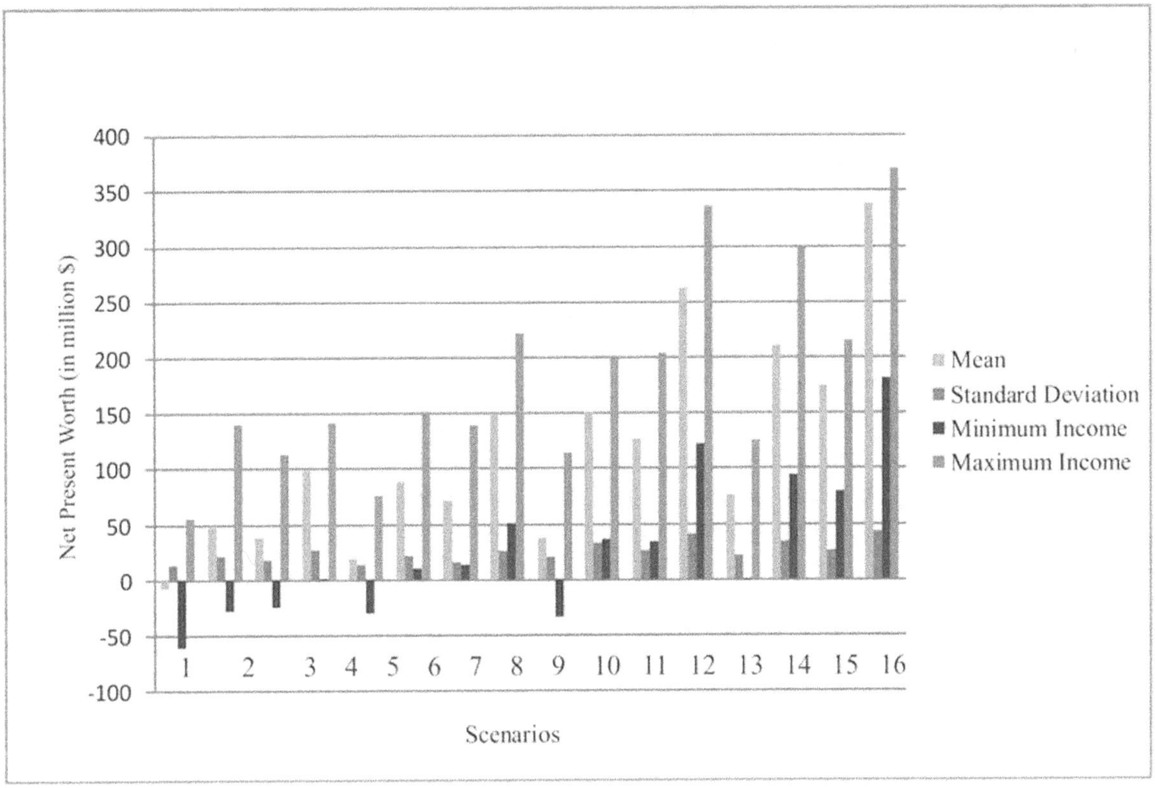

Iskenderun (Hatay) together with its hinterland. The time to reach the simulation capacity of the pier was then estimated by considering ANN predictions of cargo increase in the bay. ANN's were trained and tested by the loading/unloading database of the Chamber of Shipping of Iskenderun.

2. The loading/unloading operations of ships were simulated by probability distributions in ISMC Method that was interrelated with ANN's for the capacity predictions of Iskenderun Pier in time. Loading and unloading rates of cargo in the bay were forecasted by constructing two different ANN structures. The capacity of the Iskenderun Pier was calculated by simulating the berthing of ships to the quays considering the waiting time in the queue. The required time to reach the simulation capacity of the pier was then determined by economical scenarios and considering the loading/ unloading forecasts of ANN's carried out for Iskenderun Bay.
3. There are several nonlinear regression models in the literature and ANN can be used as an alternative to the traditional statistical regression techniques (Yu et al., 2005; Cameron and Windmeijer, 1997). ANN's are universal data forced models; hence transition from a random state to the final model can be achieved through iterative training (Svozil et al., 1997). The nonparametric description of ANN indicates that, neither the operational form of the mean function nor its error distribution can be predetermined (Basheer & Hajmeer, 2000). The number of free parameters is not an indication for prediction capacity or error criteria of the model (Schittenkopf et al., 1997). The number of weighted connections is a measure of complexity of network architecture, which is related to the

nature of the problem and the data available (Mehrotra et al., 1996). The dimensional mean error (ME) and the dimensionless correlation (R) and determination (r) coefficients are used as the error measures of ANN models at the testing stage. Performances of the neural network and regression models are generally determined by multiple correlation (determination) coefficients (Salle et al., 2003; Johnson-Restrepo, et al., 2003; Metting, et al., 2000). The determination coefficient is a statistical parameter that is widely used to determine the fraction of variability in y that can be explained by the variability in x. In other words, it gives the proportion of the variance (fluctuation) of one variable that is predictable from the other variable (Montgomery & Runger, 1999).

4. The annual cargo handling capacities of Iskenderun Pier were simulated in the ISMC Method as random variables, which were modeled by probability distributions of handling operations in quays. Thus, the annual capacity variations were realistically obtained from simulations of quay operations that can be encountered during the economical lifetime of the pier.
5. The loading/unloading operations of quays in the Iskenderun Pier were simulated by attaining probability distributions to the berthing operations of ships in appropriate quays, waiting time in the queue, capacity and size of ships on adjacent quays, efficiency of handling equipment, rate of loading/unloading cargo, and unit cost of these operations, expenditures related, maintenance and taxes. Simulations of all these activities of the pier during economic life were repeated for 16 economic scenarios and the net profit of investment was determined in terms of Net Present Worth (Figure 9). Maintenance cost of equipment was considered during this period as expenditure in simulations. Simulations were interrelated with ANN predictions for the realization of each eco-

Figure 9. Net annual present worth of project scenarios (million $)

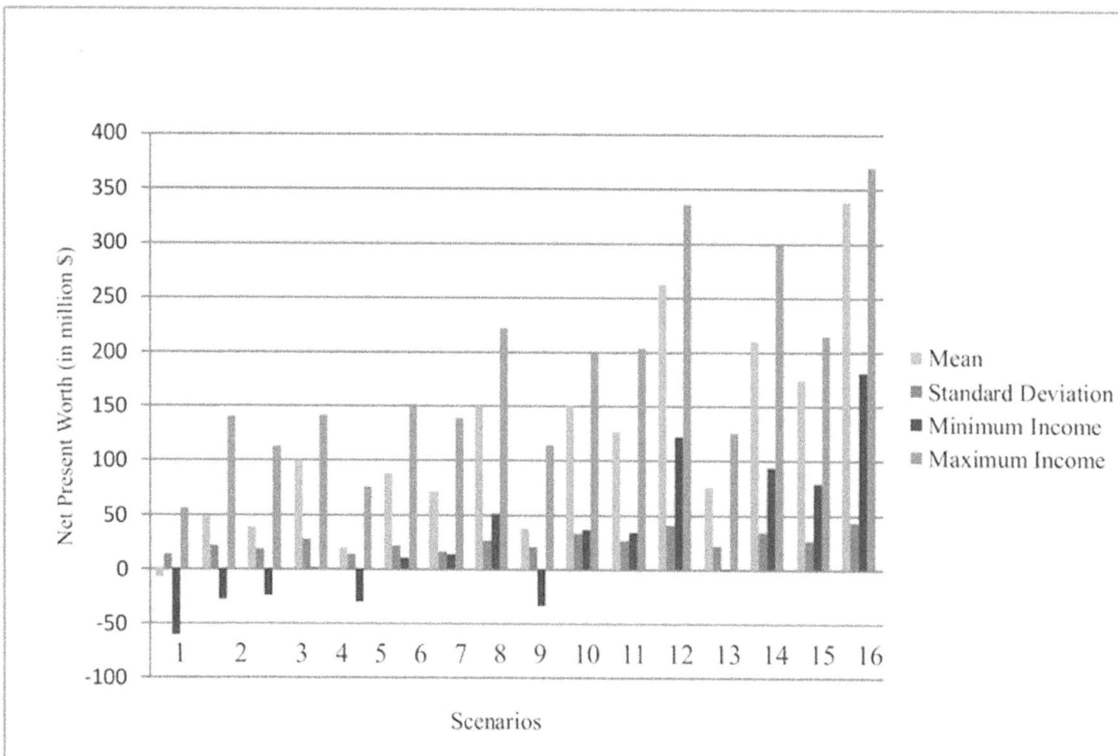

nomic scenario. The average (expected) annual present worth values presented in Figure 9 have the highest occurrence probability on the Normal (Gauss) probability density functions illustrated in Figure 7.

6. The artificial intelligence methodology for feasibility analysis developed in this paper takes into account the economical developments of the region by using GNP as one of the input parameters of ANN structures. Then these capacity increase rates are modeled in the ISMC simulation as random variables defined by probability distributions. The range parameters of these distributions are determined by the estimates of SAP and Iraq reconstruction project.
7. Only in the first scenario the investment had a chance to suffer a deficit, where the pier length was 300 meters, number of quays was 2, the pier was not a free trade zone and its initial capacity was 25% of its operational capacity. In other scenarios, the investment yields a profit, as the new pier can serve for high capacity container ships in the Eastern Mediterranean due to its suitable drafts along quays.

CONCLUSION

The new feasibility analysis model proposed in this study for coastal projects consists of three interrelated sub-models: 1) Artificial Neural Network (ANN) to determine the rates and capacity of cargo by considering the economical development of hinterland 2) Queuing model to determine the waiting to service time and the berth occupancy ratios by waiting time modeling of ships using discrete queuing simulation 3) Importance Sampling Monte Carlo (ISMC) to simulate ship arrivals/departures from the quays and to estimate income/expenditure parameters of the coastal project from a large number of site realizations (simulation epochs) in computer.

In the queuing model, the berthing of ships to the quays was simulated by considering the waiting time and the non-working days in the bay due to weather conditions. The required time to reach the operational capacity in various economical scenarios was determined by the cargo rate predictions of ANN's. To consider the uncertainties inherent in predictions, ANN method which is a multivariate; non-linear, nonparametric model was interrelated with ISMC simulation as an alternative model to conventional statistical methods, such as regressive models.

ANN could predict the cargo rates in the bay with a better accuracy than the regressive models used in the paper when the testing results of these models were compared by using the performance criteria parameters of the mean error (ME), correlation (R) and determination coefficients (r). Therefore, ANN's forecasts of loading/unloading rates were employed in the model for the years between 2015 and 2030 to predict the amount of cargo handled at the harbors and piers in Iskenderun Bay. The artificial intelligence methodology could take into account the economical developments of the region by defining the GNP as one of the input parameters of ANN structures. Then these capacity increase rates are modeled in the ISMC simulation as random variables defined by probability distributions. The range parameters of these distributions are determined by considering the economical factors of the region, such as SAP and Iraq reconstruction project.

For each economical scenario, the ISMC method in which probability distributions were used to describe the ship operations during the economical lifetime of the pier simulated the loading/unloading conditions to realistically estimate the profit of the investment in possible economical scenarios. Uncertainties in cost finance and incomes were included in the simulations and the cost/income parameters of the coastal project were determined.

REFERENCES

Anderson, E. C., Williamson, E. G., & Thompson, E. A. (2000). Monte Carlo evaluation of the likelihood for N-ε from temporally spaced samples. *Genetics*, *156*(4), 2109–2118. PMID:11102399

Baker, B. D., & Richards, C. E. (1999). A comparison of conventional linear regression methods and neural networks for forecasting educational spending. *Economics of Education Review*, *18*(4), 405–415. doi:10.1016/S0272-7757(99)00003-5

Balas, C. E., & Balas, L. (2002). Risk assessment of some revetments in Southwest Wales, United Kingdom. *Journal of Waterway, Port, Coastal, and Ocean Engineering*, *128*(5), 216–223. doi:10.1061/(ASCE)0733-950X(2002)128:5(216)

Balas, C. E., & Ergin, A. (2002). Reliability-Based risk assessment in coastal projects: Case study in Turkey. *Journal of Waterway, Port, Coastal, and Ocean Engineering*, *128*(2), 52–61. doi:10.1061/(ASCE)0733-950X(2002)128:2(52)

Balas, C. E., & Koc, L. (2002). Risk assessment of vertical breakwaters- A case study in Turkey. *China Ocean Engineering*, *16*(1), 123–134.

Balas, C. E., Koc, L., & Balas, L. (2004). Predictions of missing wave data by recurrent neuronets. *Journal of Waterway, Port, Coastal, and Ocean Engineering*, *130*(5), 256–265. doi:10.1061/(ASCE)0733-950X(2004)130:5(256)

Balas, C. E., Williams, A. T., Simmons, S. L., & Ergin, A. (2001). A statistical riverine litter propagation model. *Marine Pollution Bulletin*, *42*(11), 1169–1176. doi:10.1016/S0025-326X(01)00133-3 PMID:11763231

Basheer, I. A., & Hajmeer, M. (2000). Artificial neural networks: Fundamentals, computing, design, and application. *Journal of Microbiological Methods*, *43*(1), 3–31. doi:10.1016/S0167-7012(00)00201-3 PMID:11084225

Booth, T. E. (2000). Adaptive importance sampling with rapidly varying importance function. *Nuclear Science and Engineering*, *136*(3), 399–408.

Cameron, A. C., & Windmeijer, F. A. G. (1997). An R-squared measure of goodness of fit for some common nonlinear regression models. *Journal of Econometrics*, *77*(2), 329–342. doi:10.1016/S0304-4076(96)01818-0

Chaloulakou, A., Saisana, M., & Spyrellis, N. (2003). Comparative assessment of neural networks and regression models for forecasting summertime ozone in Athens. *The Science of the Total Environment*, *313*(1-3), 1–13. doi:10.1016/S0048-9697(03)00335-8 PMID:12922056

Cheng, J., & Druzdzel, M. J. (2000). An adaptive importance-sampling algorithm for evidential reasoning in large Bayesian networks. *Journal of Artificial Intelligence Research, 13*, 155–188.

Coulibaly, P., Anctil, F., & Bobee, B. (2000). Daily reservoir inflow forecasting using artificial neural networks with stopped training approach. *Journal of Hydrology (Amsterdam), 230*(3-4), 244–257. doi:10.1016/S0022-1694(00)00214-6

Curry, B., Morgan, P., & Silver, M. (2002). Neural networks and non-linear statistical methods: An application to the modelling of price quality relationships. *Computers & Operations Research, 29*(8), 951–969. doi:10.1016/S0305-0548(00)00096-4

Deo, M. C., & Kumar, N. K. (2000). Interpolation of wave heights. *Ocean Engineering, 27*(9), 907–919. doi:10.1016/S0029-8018(99)00023-2

Deo, M. C., & Naidu, C. S. (1999). Real time wave forecasting using neural networks. *Ocean Engineering, 26*, 192–203.

Geissler, P. L., & Chandler, D. (2000). Importance sampling and theory of nonequilibrium solvation dynamics in water. *The Journal of Chemical Physics, 113*(21), 9759–9765. doi:10.1063/1.1290136

Janes, K. R., Yang, S. X., & Hacker, R. R. (2004). Single-component modelling of pig farm odour with statistical methods and neural networks. *Biosystems Engineering, 88*(3), 271–279. doi:10.1016/j.biosystemseng.2004.04.004

Jang, J. S. R., Sun, T. S., & Mizutani, E. (1997). *Neuro-fuzzy and soft computing - A computational approach to learn machine intelligence*. USA: Prentice-Hall.

Johnson-Restrepo, B., Pacheco-Londono, L., & Olivero-Verbel, J. (2003). Molecular parameters responsible for the melting point of 1,2,3-diazaborine compounds. *Journal of Chemical Information and Computer Sciences, 43*(5), 1513–1519. doi:10.1021/ci020387k PMID:14502485

Mase, H., Sakamato, M., & Sakai, T. (1995). Neural network for stability analysis of rubble mound breakwaters. *Journal of Waterway, Port, Coastal, and Ocean Engineering, 121*(6), 294–299. doi:10.1061/(ASCE)0733-950X(1995)121:6(294)

Mehrotra, K., Mohan, C. K., & Ranka, S. (1996). *Elements of artificial neural networks*. Cambridge: MIT Press.

Melchers, R. E. (1999). *Structural reliability analysis and prediction* (2nd ed., pp. 90–110). New York, USA: John Wiley& Sons.

Metting, H. J., van Zomeren, P. V., van der Ley, C. P., Coenegracht, P. M. J., & de Jong, G. J. (2000). Comparison of migration modeling in micellar electro kinetic chromatography by linear regression and by use of an artificial neural network. *Chromatographia, 52*(9-10), 607–613. doi:10.1007/BF02789760

Montgomery, D. C., & Runger, G. C. (1999). *Applied statistics and probability for engineers*. New York, USA: John Wiley & Sons.

Razi, M. A., & Athappilly, K. (2005). A comparative predictive analysis of neural networks (NNs), nonlinear regression and classification and regression tree (CART) models. *Expert Systems with Applications*, *29*(1), 65–74. doi:10.1016/j.eswa.2005.01.006

Salle, C. T. P., Guahyba, A. S., Wald, V. B., Silva, A. B., Salle, F. O., & Nascimento, V. P. (2003). Use of artificial neural networks to estimate production variables of broilers breeders in the production phase. *British Poultry Science*, *44*(2), 211–217. doi:10.1080/0007166031000088361 PMID:12828206

Schepers, A. W., Thibault, J., & Lacroix, C. (2000). Comparison of simple neural networks and nonlinear regression models for descriptive modeling of Lactobacillus helveticus growth in pH-controlled batch cultures. *Enzyme and Microbial Technology*, *26*(5-6), 431–445. doi:10.1016/S0141-0229(99)00183-0 PMID:10713218

Schittenkopf, C., Deco, G., & Brauer, W. (1997). Two strategies to avoid overfitting in feedforward networks. *Neural Networks*, *10*(3), 505–516. doi:10.1016/S0893-6080(96)00086-X PMID:18255655

The Eight-Development Plan of Turkey, Transportation, Maritime Transport (in Turkish). (2001). State Planning Organization. Ankara: Republic of Turkey Prime Ministry State Planning Organization.

Svozil, D., Kvasnicka, V., & Pospichal, J. (1997). Introduction to multilayer feed forward neural networks. *Chemometrics and Intelligent Laboratory Systems*, *39*(1), 43–62. doi:10.1016/S0169-7439(97)00061-0

Tsai, C. P., & Lee, T. L. (1999). Back-propagation neural network in tidal-level forecasting. *Journal of Waterway, Port, Coastal, and Ocean Engineering*, *125*(4), 195–202. doi:10.1061/(ASCE)0733-950X(1999)125:4(195)

Tsai, C. P., Shen, J. N., & Kerh, T. (1999). Wave forecasting using neural network model. In B. Kumar & B. H. V. Topping (Eds.), *Artificial Intelligence Applications in Civil and Structural Engineering* (pp. 125–130). Edinburgh, UK: Civil-Comp. Press. doi:10.4203/ccp.62.5.3

UK Trade and Investment (2003). Retrieved from https://www.uktradeinvest.gov.uk/ukti/appmanager/ukti/splash

USAID (U.S. Agency for International Development). (2003). Retrieved from http://www.usaid.gov/iraq/

Walczak, S., & Sincich, T. (1999). A comparative analysis of regression and neural networks for university admissions. *Information Sciences*, *119*(1-2), 1–20. doi:10.1016/S0020-0255(99)00057-2

Yu, R., Leung, P., & Bienfang, P. (2006). Predicting shrimp growth: Artificial neural network versus nonlinear regression models. *Aquacultural Engineering*, *34*(1), 26–32. doi:10.1016/j.aquaeng.2005.03.003

Zhang, G. P., Patuwo, B. E., & Hu, M. Y. (2001). A simulation study of artificial neural networks for nonlinear time-series forecasting. *Computers & Operations Research*, *28*(4), 381–396. doi:10.1016/S0305-0548(99)00123-9

KEY TERMS AND DEFINITIONS

Activation Function: It is a linear or nonlinear function that converts a neuron's weighted input to its output activation.

Back-Propagation Architecture: It is a network which has several neurons that exist in one or more hidden layers.

Layers: They are made up of a number of interconnected neurons.

Learning: It is an iterative process which involves adjustments to the synaptic connections that exist between the neurons.

Machine Learning: It is a type of algorithm that uses computational methods to learn information directly from known data set to make forecasts.

Network: It refers to the interconnections between the neurons in the layers.

Neurons: They are interconnected processing elements or units.

APPENDIX

Notation

$c(\mathbf{t}|\mathbf{p})$ Model curve in regression
$\mathbf{p}$ Parameter vector of function f
R Correlation coefficient
r Determination coefficient
$S(\mathbf{p})$ Cost function of regression
$ Dollar of Unite States of America
$\mathbf{w}$ Weight factor vector in ANN
$x_k h_s y_z$ Structure of ANN
z Number of cells in the output layer of ANN's
μ_{xi} Expected value of random variable
σ_{xi} Unbiased estimate of standard deviation
$\nabla\mathbf{K}$ Gradient of square error function in ANN
$\nabla\mathbf{P}(w)$ Gradient of performance index in ANN

Chapter 3
Efficiency Analysis of Turkey's Transportation System Using Decision Support Model:
Data Envelopment Method

Adnan Sozen
Gazi University, Turkey

Fatih Cipil
Turkish National Police, Turkey

ABSTRACT

Since the European Union has certain targets and criteria that must be fulfilled by its members, Turkey's situation in transportation compared to the EU member countries (23 countries whose data were accessed) was examined through the approaches of decision support models (Data Envelopment Analysis and Malmquist Index). This study investigates whether Turkey utilizes its road, airway and railway transportation indicators efficiently or not within the framework of the European Union (EU) accession process. In addition, it aims to demonstrate Turkey's current position compared to other EU countries by performing relative efficiency analysis on road, airway and railway transportation indicators. The conclusions of this study will help policymakers to determine Turkey's policy objectives for its integration to the EU in terms of transportation indicators.

INTRODUCTION

Traffic is among the prioritized issues for not only Turkey but also other countries. According to the European Transport Safety Council (ETSC), traffic accidents are the main cause of death in all EU countries for those aged 45 and below. It is indicated in various reports published by World Bank that Turkey loses around 2% of its Gross National Product (GNP) every year in traffic accidents, which suggests that the annual socioeconomic cost of traffic accidents in Turkey is nearly 10 billion dollars (Transportation Strategy Report, 2004; Cubuk & Cansiz, 2005, pp. 47-49). Number of vehicles per capita, which is

DOI: 10.4018/978-1-4666-8648-9.ch003

among the indicators of economic welfare and development, has reached its peak density especially in cities in recent years. This situation has unsurprisingly increased traffic problems in and out of towns. Given the fact that the volume of road transportation in Turkey is high, Turkey is among the countries with the highest number of traffic accidents. Statistics indicate that over 4000 people are killed and over 200,000 people are injured in traffic accidents every year in Turkey on average. In order to attain a stable performance in preventing accidents and establishing road safety; the characteristics of vehicles, drivers and roads, which are the three main components of the transportation system, need to be addressed as a whole. More than one factor have an impact on the frequency and violence of traffic accidents such as driver behaviors, characteristics of vehicles and roads, environmental factors and traffic characteristics. The causes of traffic accidents in Turkey are distributed as follows: 96.82% drivers, 2.38% pedestrians, 0.16% passengers, 0.32% vehicles, and 0.32% roads. These statistics suggest that the human factor is responsible for traffic accidents by 99.36% (Beasley, 2004).

Traffic safety is among the issues prioritized by Turkey and other countries in the world. In the EU summit held in Helsinki on 10-11 December 1999, Turkey was admitted to candidacy having equal rights with other candidate countries. Thus began Turkey's accession process to the EU. With this development, Turkey would be required to make extensive adjustments in numerous areas including the transportation industry for a long period. The transportation sector is among the areas on which the creation and implementation of a common policy was put as an obligation in the Rome Treaty. Since the Rome Treaty, which is the foundation of the EU, the primary policy that has been pursued by members is the free circulation of individuals and products. These countries make significant efforts to alleviate the obstacles in their borders. The main components of the common transportation policy addressed within this framework were defined as reducing employment, protecting the environment, being open to competition and providing the best service to the customer with the largest perspective. The main purpose of transportation is to provide the service necessitated by the economic and social targets of national development; with the most convenient qualities demanded by the user; in a way to satisfy national security requirements; in a safe and environmentally-friendly manner; for the cheapest price; using contemporary technologies; and in harmony with international regulations and EU policies; continuously and without any interruption (Aydogdu, 2006).

The safe and comfortable transportation of people and goods within a short period can be achieved by airline transportation. It is well-known that airline transportation accelerates the economic and technological developments in local, regional, national and international scales. Besides, it is thought that airline transportation allows different people to come together and lets them know each. Thus, it also makes social and cultural contributions to the communities.

In recent years, in all around the world, the share of the airline transportation, for people and freight, has increased nearly at the same levels with Gross Domestic Product (GDP). The same relation also holds for Turkey. However, it was observed that, in Turkey, increases and decreases in GDP affected the airline transportation sector more strongly than the other countries. Turkey has been one of the countries with fastest growing airline transportation sector in the recent years. On the other hand, for people and freight, the share of the airline transportation was lower than share of the road transportation. Annually, almost 200 million passengers are transported in Turkey. It is known that nearly 10% of them are transported by means of airline, while this value reaches to 30% in European countries having well-developed airline transportation sectors.

Railway works, which were started in 1856 in Turkey, have been progressing very slowly compared to the rapidly developing technology and rapidly increasing population. Intensive works are being carried out in various fields as part of Turkey's recent integration efforts to the EU. Since the European Union has certain targets and criteria that must be fulfilled by its members, it is necessary to examine Turkey's comparative position in the field of transportation -especially in railways- by considering several variables.

It is aimed in this study to determine Turkey's position among EU member countries and to define its pros and cons in terms of road and railway transportation indicators. Within the scope of this study, relative performance evaluation was performed by applying DEA approach so as to determine if 12 regions of Turkey were using their airline transportation effectively or not. Meanwhile, to monitor the variation of the performance with time, total factor change of the airline transportation was examined by Malmquist Index approach. During EU harmonization process of Turkey, these analysis will help Turkey to follow new developments and tendencies appearing under the "Transportation" headline and present some numerical data to the legislations developed for transportation policies. This way, it is believed that this study will contribute to the efforts towards harmonization. Therefore, a relative efficiency analysis was conducted using the method of Data Envelopment Analysis (DEA) for 27 EU member countries and Turkey by considering their transportation and traffic structures. Thus, Turkey's position among EU member countries with respect to traffic indicators was determined. Based on the concrete findings, the policies that Turkey should develop in order to increase harmonization with the transportation and traffic structures of EU member countries were discussed.

BASIC TRANSPORTATION INDICATORS

As in all other countries, traffic accidents in Turkey are among the top problems. According to a research by the World Bank, the socioeconomic cost of traffic accidents in Turkey constitutes 2.2% of the entire GNP (Bedrettin, 2004). Traffic indicators by years based on the data provided by Turkish Statistics Institute (TUIK) (www.tuik.gov.tr) are given in Figure 1. In recent years, numbers of drivers and vehicles as well as the number of accidents are on the rise. However, deaths caused by accidents slightly declined; which might have stemmed from the fact that most new vehicles are automobiles whose equipments have been designed with an increased focus on safety.

Figure 1. demonstrates the distribution of passengers and cargo with respect to the system of transportation in Turkey. As Figure 2 suggests, both passenger and cargo transportation in Turkey are performed on roads, and thus, researches should primarily focus on the road transportation industry.

The transportation sector constitutes 7% of the EU's Gross Product, 7% of employment, 40% of member countries' investments and 30% of energy consumption. Across the EU, in the last twenty years, an annual demand of 2.3% for goods and of 3.1% for passengers emerged on average. Steps towards the liberalization of the EU's economy such as the elimination of boundaries between countries and the liberalization of maritime transportation have rendered inevitable the creation of a Common Transportation Policy. These steps are of importance in terms of maintaining growth and overcoming problems such as deadlock and market saturation.

In recent years, in all around the world, the share of the airline transportation, for people and freight, has increased nearly at the same levels with Gross Domestic Product (GDP). The same relation also

Figure 1. Traffic indicators in Turkey

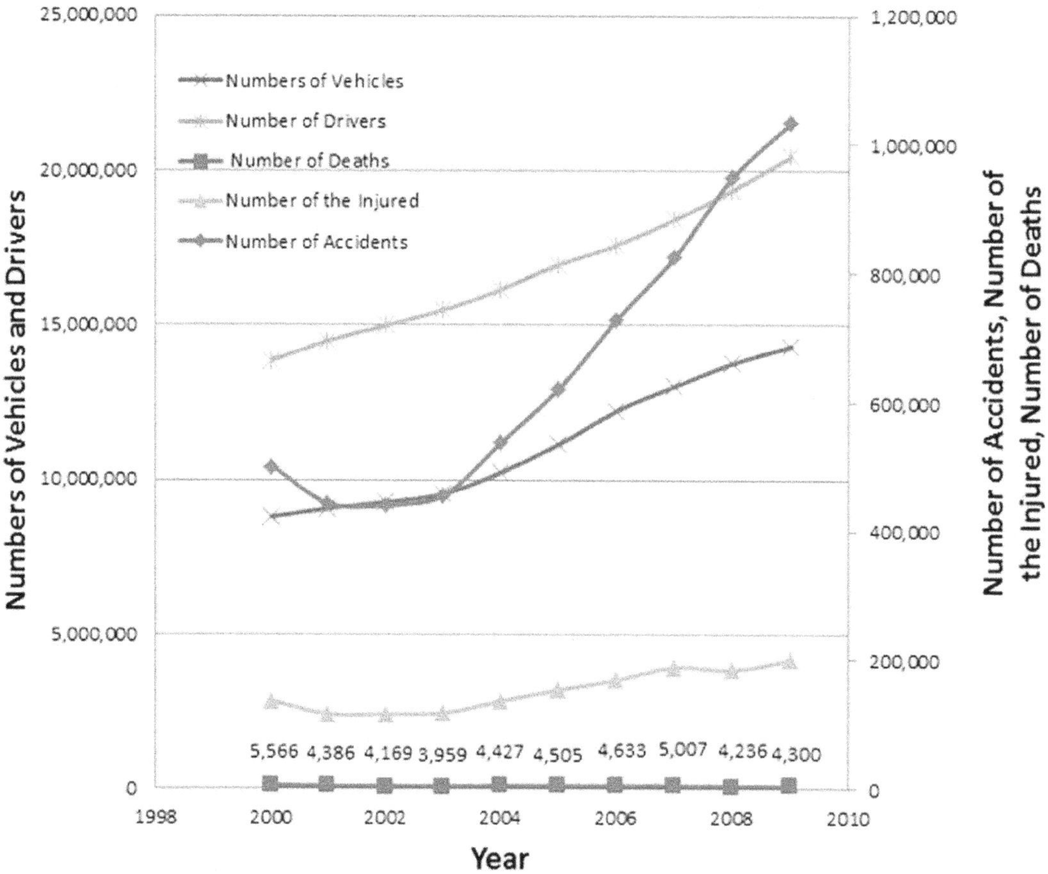

holds for Turkey. However, it was observed that, in Turkey, increases and decreases in GDP affected the airline transportation sector more strongly than the other countries. Turkey has been one of the countries with fastest growing airline transportation sector in the recent years. On the other hand, for people and freight, the share of the airline transportation was lower than share of the road transportation. Annually, almost 200 million passengers are transported in Turkey. It is known that nearly 10% of them are transported by means of airline, while this value reaches to 30% in European countries having well-developed airline transportation sectors.

Similar to the changes observed in the world, there was an increase in airline transportation of people and freight in Turkey between 2004 and 2010, see Figures 2-4. As seen from the graphs, each region has an increasing trend. Especially, the annual variations of Marmara Region are notable, see Figures 3 & 4. It is highly probable that rapid industrialization of this region was the primary reason for this. For the other 11 regions, while the airline transportation demonstrated a decrease in some years, increasing trend was maintained for the period of 2004-2010. In order to have sustainable developments in airline transportation, regional plans should be made and applied correctly. In this study, it was aimed to contribute the development of domestic and international airline transportation in Turkey and to help the development of airline transportation-related strategies about alignment to the EU acquis.

Figure 2. The annual change in the air traffic, for both arrivals and departures, between 2004-2010

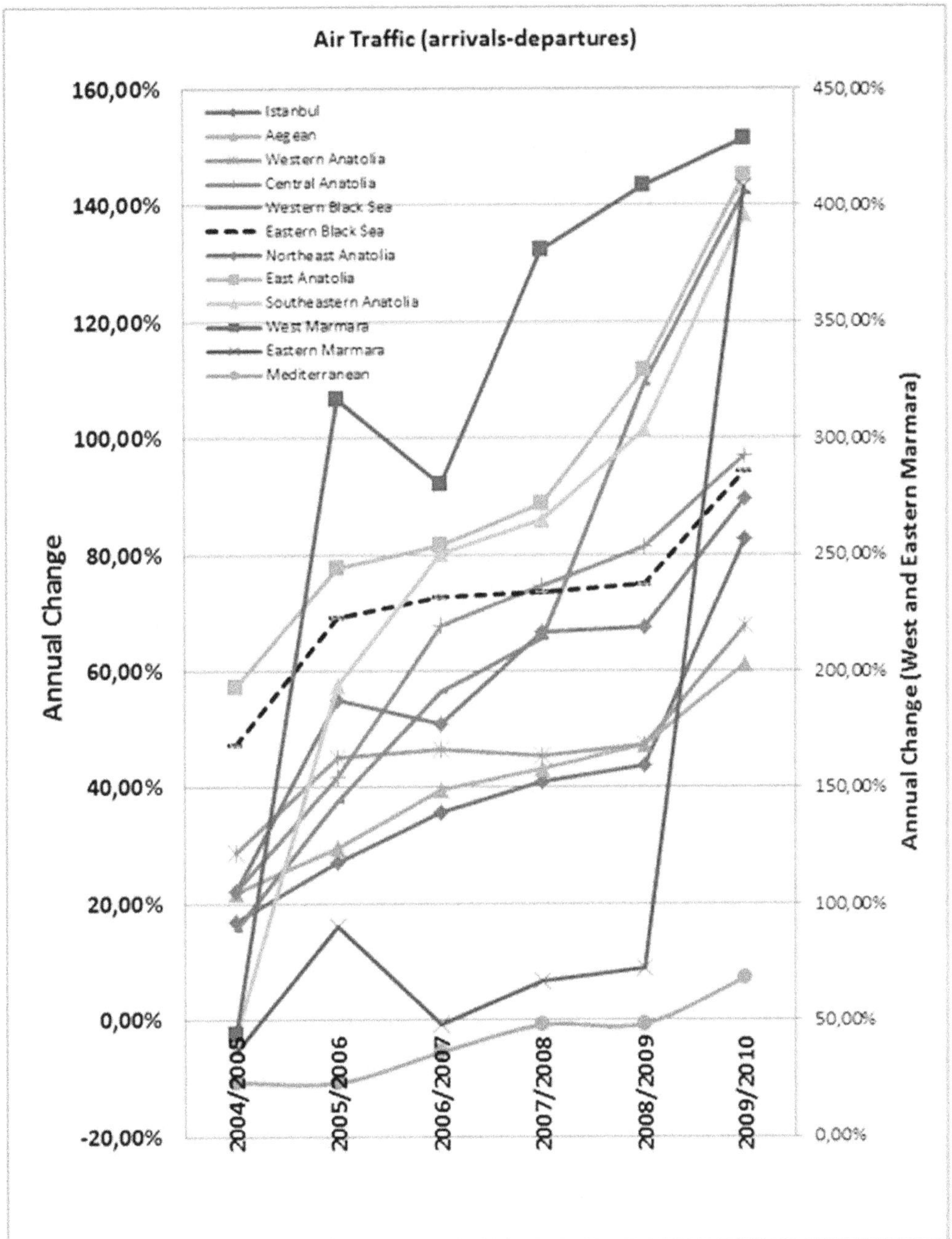

Railway works, which were started in 1856 in Turkey, have been progressing very slowly compared to the rapidly developing technology and rapidly increasing population. Intensive works are being carried out in various fields as part of Turkey's recent integration efforts to the EU. Since the European Union has certain targets and criteria that must be fulfilled by its members, it is necessary to examine Turkey's comparative position in the field of transportation -especially in railways- by considering several variables.

Figure 3. The annual change in the number of passengers between 2004-2010

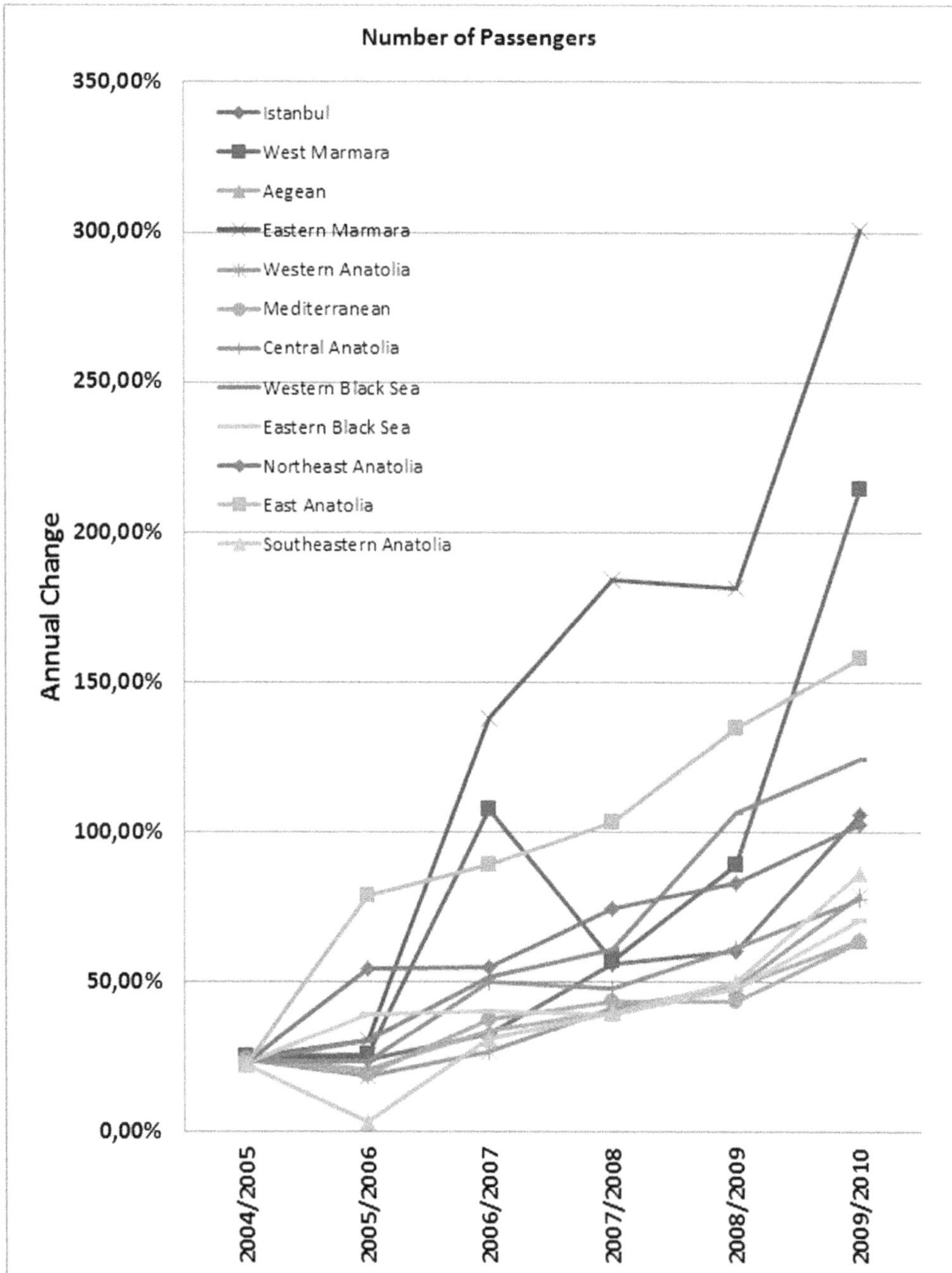

DECISION SUPPORT MODEL (DATA ENVELOPMENT ANALYSIS AND MALMQUIST INDEX)

Data envelopment analysis (DEA) is a nonparametric technique to calculate productive efficiency of decision making units, comparable homogeneous organizations (Emrouznejad et al, 2008, pp. 151-157; Ray et al., 2004; Cook et al., 2009; Charnes et al., 1978; Sozen et al., 2012, pp. 192-202; Sozen et al.,

Figure 4. The annual change in the freight carried (tone) by airline in different regions of Turkey

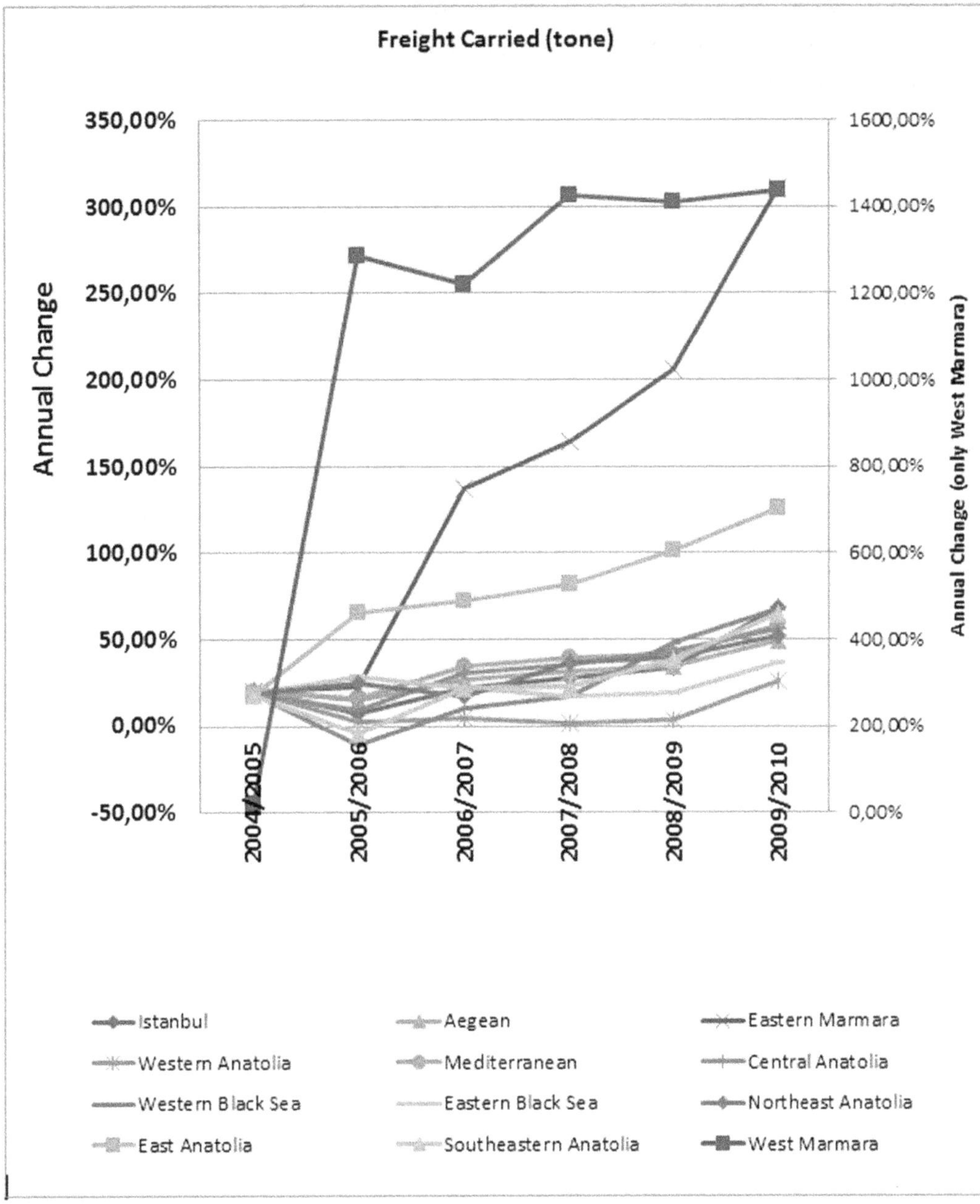

2010, pp. 6194-6203). DEA is commonly applied to estimate the relevant technology or production function over the production possibility set defined as the feasible combinations of inputs and outputs. Here, the production function the maximum amount of output for a specified set of inputs, given the existing technology. Since it affects decision-making, the efficiency score for the given decision-making unit can be calculated.

The Malmquist index (MI) is an index based on production function used to measure productivity change over time (Malmquist, 1953, pp. 209-242; Caves et al., 1982, pp.1393-1414; Fare et al., 1994, pp. 66-83). In the output-oriented MI, the production technology (P) shows the transformation of inputs x into output y in time t. The output possibility set is then;

$$P_t(x) = \{y_t, x_t \text{ can produce } y_t\} \quad (1)$$

Figure 5 illustrates production technology for period t and t+1. Technical efficiency of a DMU can be measured by an output distance function, which measures the distance of an economy from the production function. The output distance function can be related to production technology through the following equation:

$$d_t(x_t, y_t) = \min\{\theta: (x_t, y_t/\theta) \in P_t(x)\} \quad (2)$$

where, θ defines a set of real numbers. The minimum of θ needs to be found such that input/output combination is a part of the production technology. This is illustrated in Figure 5 as one input and one output assumption exhibiting constant return Turkey's efficiency changes scale (CRS). In the figure, points A and A' represent the input/output combinations of DMU_o in periods t and t+1, respectively.

Figure 5. DEA and MI decomposition with the Efficient Frontier (EF)

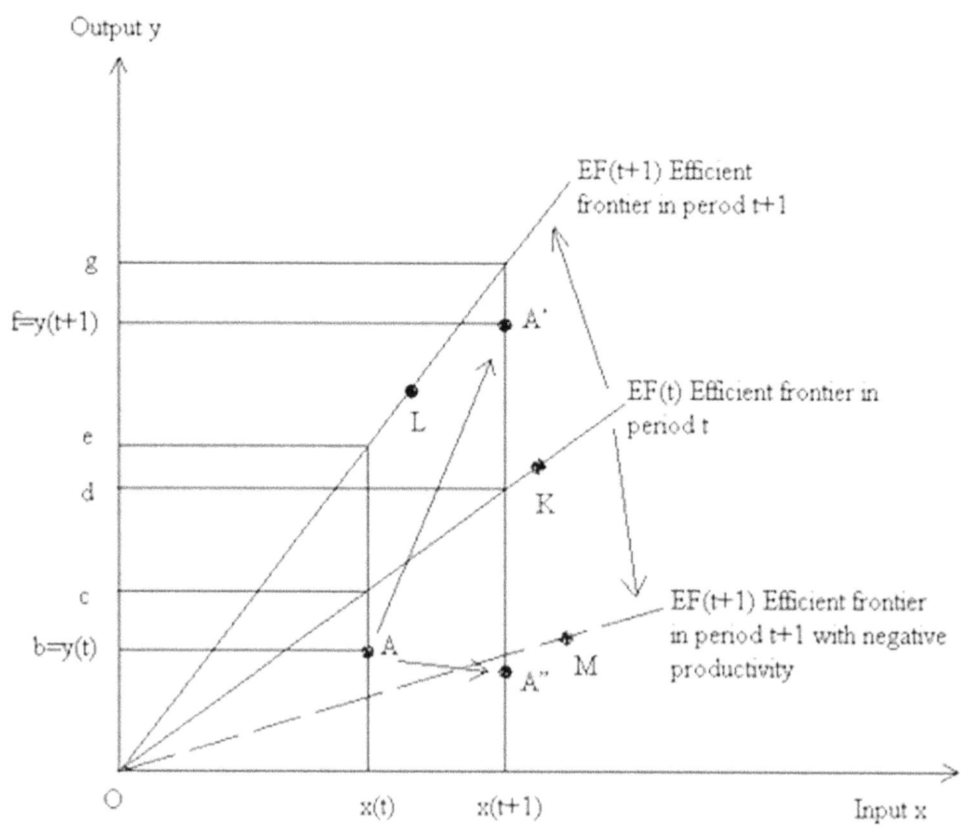

The efficient frontier (EF) is determined by the best performing unit. When the economy is operating at point A, it produces output b. This shows that x_t, y_t is technically inefficient, since the efficiency is measured by Oc/Ob (CCR (CRS) DEA model, inefficient). If production was on the frontier, then the efficiency would be 1 (CRS DEA model, efficient). Assume that at time t+1, point A shifted to point A' and the efficient frontier EF(t) shifted to EF(t+1). Then the efficiency change in DMU_0 is the ratio of the efficiency at t+1 to the efficiency at t. DEA CCR output-oriented Malmquist TFP index can be constructed based on t and t+1 technology as;

$$MI_0(x_t, y_t, x_{t+1}, y_{t+1}) = \sqrt{\frac{d_t^o(x_{t+1}, y_{t+1})}{d_t^o(x_t, y_t)} x \frac{d_{t+1}^o(x_{t+1} y_{t+1})}{d_{t+1}^o(x_t, y_t)}} \quad (3)$$

or graphically,

$$MI_0(t, t+1) = \sqrt{\frac{Of / Od}{Of / Og} x \frac{Ob / Oc}{Ob / Oe}} \quad (4)$$

Fare et al. (1983) defined that MI >1 indicates productivity gain; MI <1 indicates productivity loss and MI=1 indicates the status quo from period t to t+1, i.e., no change in productivity. They showed that this index is equivalent to:

$$MI_0(x_{t+1}, y_{t+1}, x_t, y_t) = \frac{d_{t+1}^o(x_{t+1}, y_{t+1})}{d_t^o(x_t, y_t)} \sqrt{\frac{d_t^o(x_{t+1}, y_{t+1})}{d_{t+1}^o(x_{t+1}, y_{t+1})} x \frac{d_{t+1}^o(x_t y_t)}{d_{t+1}^o(x_t, y_t)}} \quad (5)$$

As seen from the equation, the index is composed of two parts. The first term on the right measures the change in technical efficiency, also called Catch-Up (CU) index, which is the distance function from time period t to t+1. The second term measures the technical change, also called Frontier Shift (FS) index, between time period t and t+1. Two components of the MI as in the last equation are represented by:

$$\text{Efficiency change (EFFCH)} = \frac{Of / Od}{Ob / Oc} \quad (6)$$

$$\text{Technical change (TECCH)} = \sqrt{\frac{Of / Od}{Of / Og} x \frac{Ob / Oc}{Ob / Oe}} \quad (7)$$

The condition CU>1 suggests that DMU_0 has moved closer to the period EF (t+1) than to the period EF (t). CU=1 and CU<1 indicate the same distance or more has been covered respectively. The condition FS>1 shows that there is more output than input, indicating a positive shift in the frontier. A value of FS equal to one and less than 1 demonstrates no shift and negative shift, respectively (Hashimoto & Haneda, 2008, pp. 1829-1836).

In CRS model, any radial increase in input vector, the increase of all input components at the same rate, creates a radial increase in the output vector. CCR DEA model cannot be applied globally in many economies, since it assumes that the output and input change by the same amount. Therefore, it is rather restrictive. As an extension of this approach, production technology may exhibit increasing, constant and diminishing returns to scale on the production frontiers, which are reflected in variable returns to scale (VRS) assumption, see Figure 5. In decreasing return to scale (DRS) model, any radial increase in the input vector creates a radial increase in the output vector at a lower rate. In increasing return to scale (IRS) model, any radial increase in the input vector creates a radial increase in the output vector at a higher rate.

Suppose that, in an economic situation where production frontier moved to OM from OL at time t+1, see Figure 5. In this case, DMU has a positive change in technical efficiency but a negative progress. In VRS technology, efficiency change is described as the product of pure efficiency change (PECH) and scale efficiency change (SECH):

$$\text{EFFCH} = \text{PECH x SECH} \tag{8}$$

$$\text{PECH} = \frac{d_{t+1}^{v}(x_{t+1}, y_{t+1})}{d_{t}^{v}(x_{t}, y_{t})} \tag{9}$$

$$\text{SECH} = \frac{d_{t+1}^{c}(x_{t+1}, y_{t+1})}{d_{t+1}^{v}(x_{t+1}, y_{t+1})} / \frac{d_{t}^{c}(x_{t}, y_{t})}{d_{t}^{v}(x_{t}, y_{t})} \tag{10}$$

where, the subscripts ***c*** and ***v*** represent CRS and VRS technologies, respectively (Chen et al., 2008, pp. 580-583).

Various DEA models have been proposed measuring efficiency in different ways. These models are basically categorized into two: input-oriented and output-oriented. In order to find out the input-oriented efficiency of a DMU, examination of possibilities to reduce the input(s) without reducing the output(s) should be evaluated. A model is configured to determine how much the input use of a DMU decrease in a 100% efficiency condition at a constant output level. The condition for pare to efficiency is realized when it is not possible to reduce the amount of any input without increasing some other input or reducing some output.

In contrast to input-oriented efficiency, to determine the output-oriented efficiency of a DMU, a model is configured to determine how much output increase in a 100% efficiency condition at a given input level. In this approach, pare to efficiency condition occurs when it is not possible to increase the amount of any output without decreasing some other output or increasing some input. See Figure 5 for demonstration of the input- and output-oriented efficiencies of DMUs.

APPLICATIONS

In this study, road, air and rail transport activities were undertaken separately integrate the transport activity was allowed evaluation. In general, these concepts are difficult to be distinguished from each other in transportation.

Road Efficiency

In this study, DEA, which is a method used in many studies to evaluate performance, was employed in order to determine Turkey's performance in traffic indicators in comparison with EU member countries. Using the DEA models of CRS (CCR) and VRS (BCC), a section (cross-section) analysis was performed and efficiency scores were ranked. In the study; the main road traffic indicators were used as variables whereas Turkey and 27 EU member countries were regarded as decision making units (DMU) (Statistical Pocket Book, 2011). To determine Turkey's efficiency, three different efficiency models were formulated taking into consideration nine input and various output variables. These models were analyzed according to their input and output values, as presented in Table 1. The aim here is;

Model 1: Determining the efficiency that takes the number of accidents as the output and the main road indicators as the input

Table 1. Input-output variables used in the analysis

Main Road Traffic Indicators	MODEL 1 (Input/ Output)	MODEL 2 (Input/ Output)	MODEL 3 (Input/ Output)
Length of Motorways	I	I	I
Length of Road Network-Main or National Roads	I	I	I
Length of Road Network-Secondary or Regional Roads	I	I	I
Motorization	I	I	I
Passenger Cars-Stock of Registered Vehicles	I	I	I
Buses and Coaches-Stock of Registered Vehicles	I	I	I
Goods Vehicles-Stock of Registered Vehicles	I	I	I
Road share of inland freight transport	I	I	I
Passenger Cars- New Vehicle Registrations	I	I	I
Road Fatalities	-	O	-
Road Fatalities Country Rankings-Fatalities per Million Inhabitants	-	-	-
Road Fatalities Country Rankings-Fatalities per 10 Billion pkm	-	-	O
Road Fatalities Country Rankings-Fatalities per Million Passenger Cars	-	-	O
Number of Accidents Involving Personal Injury	O	-	-
Injuries-Numbers	-	O	-

Model 2: Determining the efficiency that takes the number of casualties on road as the output and the main road indicators as the input

Model 3: Determining the efficiency that takes the number of deaths per vehicle and per kilometre as the output and the main road indicators as the input

It measures the number of accidents in an outcome-oriented and the efficiency in an input-oriented way; that is, it is the method that focuses on the degree that those inputs that minimize output should be decreased. Moreover, both CRS and VRS efficiency scores were calculated for each model. That is, it is determined in the CRS model whether increasing the amount of input influences the output at the same rate or not. In the VRS model, inputs and outputs are considered as variables.

DMUs' (countries') both CRS and VRS efficiencies according to the Model 1 analysis results conducted for determining accident performance are presented in Table 2.

By performing input-oriented DEA for Model 1; the DMU that ensures minimum number of accidents with its existing inputs was determined. If the efficiency score found through input-oriented model is 1 or 100% and if the inputs or outputs of this DMU do not have mixed inefficiency, this country is deemed efficient. Countries with efficiency scores of higher or lower than 100 are not efficient. Yet another meaning of the efficiency score of an inefficient country is the following: this unit can become efficient if it decreases its inputs as much as its efficiency score. According to the Model 1 CRS analysis results, 13 countries including Belgium, Germany, Italy and Turkey have an efficiency score of 100%. According to the Model 1 VRS analysis results, on the other hand, 19 countries are efficient such as Romania, Denmark and France. These countries were ranked using super efficiency analysis.

This ranking is presented in Table 2. The table ranks the countries that produce the highest amount of outputs using the lowest amount of inputs. In the ranking of countries' efficiencies based on CRS analysis, Belgium ranked first whereas Turkey ranked fourth. Romania, on the other hand, ranked first based on VRS analysis. Countries that are not found efficient in neither of the analyses need to decrease their inputs as much as the weights of the input values of their reference countries presented in Table 3 (for CRS and VRS).

Table 3 also illustrate the times efficient countries became references for inefficient ones (shaded in grey). For example, based on CRS analysis results (for Model 1), Turkey became a reference as an efficient country for 11 inefficient countries (Table 3). These countries, including Bulgaria, Denmark and Spain, need to decrease their inputs as much as the multiplication of Turkey's input values by their weights given in brackets. Based on VRS analysis results, similarly, Turkey became a reference for Cyprus and Greece. In other words, inefficient countries should take the capacity of efficient countries to convert their inputs into outputs as a reference. They should decrease their inputs as much as the values given in blankets nearby the references in both tables.

According to the performance analysis results of Model 2, which takes the number of casualties in accidents as output and road transportation indicators as input, both CRS and VRS efficiencies of DMUs are presented in Table 3.

By performing input-oriented DEA for Model 2; the DMU that ensures minimum number of casualties with its existing inputs was determined. If the efficiency score found through input-oriented model is 1 or 100 and if the inputs or outputs of this DMU do not have mixed inefficiency, this country is deemed efficient. Countries with efficiency scores of higher or lower than 100 are not efficient. Yet

Table 2. Analysis results for all Model

	MODEL 1		MODEL 2		MODEL 3		*Super-efficiency ranking for efficiency countries in Model 1*		*Super-efficiency ranking for efficiency countries in Model 2*		*Super-efficiency ranking for efficiency countries in Model 3*	
DMU	CRS	VRS	CRS	VRS	CRS	VRS	CRS	VRS	CRS	VRS	CRS	VRS
BE	100.00	100.00	100.00	100.00	100.00	100.00	374.88	474.81	471.04	602.05	180.90	307.39
BG	82.66	100.00	100.00	100.00	100.00	100.00	-	142.83	185.26	185.77	152.36	182.30
CZ	87.43	92.00	84.41	92.82	50.28	76.53	-	109.15	-	-	-	-
DK	23.83	91.69	46.12	92.19	44.98	91.69	-	-	-	-	-	-
DE	100.00	100.00	100.00	100.00	15.85	69.86	246.04	298.31	226.23	800.00		
EE	46.19	100.00	57.02	100.00	100.00	100.00	-	-	-	182.39	162.11	202.22
IE	54.36	88.94	43.29	86.55	50.72	86.55	-	134.50	-	-	-	-
EL	54.38	68.91	85.53	86.37	57.05	67.37	-	202.77	-	-	-	-
ES	77.37	85.14	95.08	96.23	19.68	62.40	-	146.88	-	-	-	-
FR	66.32	83.70	100.00	100.00	21.64	65.25	-	125.17	167.97	173.54	-	-
IT	100.00	100.00	100.00	100.00	21.22	55.46	143.49	-	165.10	167.89	-	-
CY	54.60	93.92	47.15	93.90	100.00	100.00	-	178.23	-	-	120.54	184.25
LV	100.00	100.00	77.58	100.00	100.00	100.00	249.45	112.71	-	174.47	196.36	198.48
LT	63.07	100.00	100.00	100.00	87.36	100.00	-	117.54	126.10	152.96	-	109.15
LU	29.43	100.00	47.65	100.00	100.00	100.00	-	139.29	-	154.33	204.30	226.79
HU	100.00	100.00	100.00	100.00	100.00	100.00	141.35	-	173.55	173.97	105.09	105.18
MT	73.37	100.00	100.00	100.00	100.00	100.00	-	757.79	101.68	757.79	720.66	757.79
NL	100.00	100.00	100.00	100.00	40.23	100.00	103.92	252.15	123.12	132.31		111.77
AT	100.00	100.00	100.00	100.00	52.96	77.52	113.94	144.19	112.22	120.24	-	-
PL	100.00	100.00	100.00	100.00	62.47	75.18	112.87		251.88	800.00	-	-
PT	100.00	100.00	100.00	100.00	100.00	100.00	174.81	181.56	142.96	144.38	100.93	110.16
RO	100.00	100.00	100.00	100.00	100.00	100.00	196.04	800	233.55	241.35	306.80	800.00
SI	100.00	100.00	100.00	100.00	100.00	100.00	124.50	-	136.09	157.19	118.72	119.35
SK	59.77	100.00	75.75	100.00	100.00	100.00	-	118.89	-	126.74	116.83	125.03
FI	39.00	80.06	38.67	80.60	35.27	80.06	-	-	-	-	-	-
SE	49.43	82.22	54.83	83.87	23.41	78.54	-	124.98	-	-	-	-
UK	100.00	100.00	100.00	100.00	9.39	62.56	131.50	-	122.21	126.15	-	-
TR	100.00	100.00	100.00	100.00	100.00	100.00	206.49	-	399.71	451.72	186.19	205.69

another meaning of the efficiency score of an inefficient country is the following: this unit can become efficient if it decreases its inputs as much as its efficiency scores. According to the Model 2 CRS analysis results, 16 countries including Belgium, Germany, France, Italy and Turkey have an efficiency score of 100. According to the Model 2 VRS analysis results, on the other hand, 20 countries are efficient such as Romania, Poland, Hungary and France.

Table 3. Weights of inefficient countries based on CRS and VRS analysis results compared to efficient countries and the number of efficient countries' being references

	DMU	CRS Number for efficient countries to be references (dark) Name and weight of reference country for inefficient countries (light)	VRS Number for efficient countries to be references (dark)Name and weight of reference country for inefficient countries (light)
1	BE	0	3
2	BG	LV (0.53) HU (0.13) TR (0.01)	0
3	CZ	DE (0.01) IT (0.05) LV (0.34) RO (0.22)	EE (0.28) IT (0.06) LV (0.52) RO(0.13) UK (0.01)
4	DK	DE(0.00) PT (0.05) SI (0.13) TR (0.00)	LV (0.53) SK (0.47)
5	DE	3	4
6	EE	LV (0.11) AT (0.01) RO (0.02)	5
7	IE	IT (0.02) PL (0.00) PT (0.04) RO (0.05) TR (0.00)	BE (0.06) EE (0.20) MT (0.21) PL (0.01) SK (0.53)
8	EL	20 (0.02) PT (0.20) RO (0.23) TR (0.00)	DE (0.01) LV (0.19) HU (0.16) RO (0.04) SK (0.58) TR (0.01)
9	ES	IT (0.34) PT (0.39) TR (0.01)	DE (0.03) IT (0.34) SK (0.62)
10	FR	IT(0.34)	DE (0.02) IT (0.29) LV (0.29) SK (0.40)
11	IT	6	3
12	CY	LV (0.04) HU (0.03) PT (0.03) TR (0.00)	BE (0.01) EE (0.11) MT(0.83) PL (0.00) PT (0.01) SK(0.04) TR(0.00)
13	LV	7	5
14	LT	LV (0.43) HU (0.05) TR (0.00)	0
15	LU	AT (0.02) TR (0.00)	0
16	HU	4	1
17	MT	LV (0.02) HU (0.00) PT (0.01) TR (0.00)	2
18	NL	0	0
19	AT	5	0
20	PL	2	2
21	PT	9	2
22	RO	7	2
23	SI	1	0
24	SK	IT (0.01) LV(0.07) AT (0.02) PT (0.04) RO (0.08)	8
25	FI	IT(0.01) AT (0.02) PT (0.07) RO (0.05) TR(0.00)	BE (0.12) EE (0.83) PT (0.04) SK (0.01)
26	SE	DE (0.00) AT (0.37) PT (0.06) RO (0.05) TR(0.00)	DE (0.04) EE(0.24) LV (0.45) SK(0.27)
27	UK	0	1
28	TR	11	2

These countries were ranked using super efficiency analysis. This ranking is presented in Table 3. The table ranks the countries that produce the highest amount of outputs using the lowest amount of inputs. In the ranking of countries' efficiencies based on CRS analysis, Belgium ranked first whereas Turkey ranked second.

Based on VRS analysis, on the other hand, Poland ranked first whereas Turkey ranked fifth. Countries that are not found efficient in neither of the analyses need to decrease their inputs as much as the weights of the input values of their reference countries presented in Table 4 (for CRS and VRS).

Table 4 also illustrate the times efficient countries became references for inefficient ones (shaded in grey). For example, based on CRS analysis results (for Model 2), Turkey became a reference as an efficient country for 5 inefficient countries (Table 4). These countries, including Czech Republic and Cyprus, need to decrease their inputs as much as the multiplication of Turkey's input values by their weights given in brackets.

Based on VRS analysis results, similarly, Turkey did not become a reference for other countries although it was an efficient country. In other words, inefficient countries should take the capacity of efficient countries to convert their inputs into outputs as a reference. They should decrease their inputs as much as the values given in blankets nearby the references in both tables.

By performing input-oriented CRS and VRS analyses with Model 2; it was aimed to minimize the number of deaths and the injured. If the efficiency score found through input-oriented model is 1 or 100 and if the inputs or outputs of this DMU do not have mixed inefficiency, it is concluded that the main road traffic indicators of this country cause less casualties in accidents than those of other countries, and this country is deemed efficient. Countries with efficiency scores of higher or lower than 100 are not efficient. Yet another meaning of the efficiency score of an inefficient country is the following: this unit can become efficient if it decreases its inputs, that is its main road traffic indicators, as much as its efficiency score. For this reason, the input-oriented inefficiency score is called contraction coefficient. For example, Czech Republic is not efficient for Model 2 according to CRS analysis with the efficiency score of 84.41 . In order to become efficient, this country needs to reduce its inputs (e.g. number of passengers/vehicles or rate of motorization) by 15.59 .

According to the performance analysis results of Model 3, which takes the number of deaths on roads per one million vehicles for passenger transportation and the number of deaths on roads per one billion passengers-km as outputs and road transportation indicators as input, both CRS and VRS efficiencies of DMUs are presented in Table 5.

By performing input-oriented DEA for Model 3; the DMU that ensures minimum number of deaths on roads with its existing inputs was determined. If the efficiency score found through input-oriented model is 1 or 100 and if the inputs or outputs of this DMU do not have mixed inefficiency, this country is deemed efficient. Countries with efficiency scores of higher or lower than 100 are not efficient. Yet another meaning of the efficiency score of an inefficient country is the following: this unit can become efficient if it decreases its inputs as much as its efficiency score. According to the Model 3 CRS analysis results, 14 countries including Belgium, Hungary, Lithuania, Bulgaria and Turkey have an efficiency score of 100 . According to the Model 3 VRS analysis results, on the other hand, 16 countries are efficient such as Romania, Belgium and Turkey. These countries were ranked using super efficiency analysis. This ranking is presented in Table 9. The table ranks the countries that produce the highest amount of outputs using the lowest amount of inputs. In the ranking of countries' efficiencies based on CRS and

Table 4. Weights of inefficient countries based on CRS and VRS analysis results compared to efficient countries and the number of efficient countries' being references

	DMU	CRS Number for efficient countries to be references (dark)Name and weight of reference country for inefficient countries (light)	VRS Number for efficient countries to be references (dark)Name and weight of reference country for inefficient countries (light)
1	BE	2	4
2	BG	2	4
3	CZ	IT (0.06) HU (0.11) RO (0.20) TR (0.00)	BG (0.06) EE(0.25) IT (0.06) LV (0.08) MT (0.19) PL (0.01) RO(0.14) SK (0.20)
4	DK	BG (0.14) PL (0.02) RO (0.02) SI (0.20)	LV (0.47) SI (0.04) SK (0.49)
5	DE	1	1
6	EE	HU (0.04) RO (0.04) TR (0.00)	4
7	IE	BE (0.03) IT (0.00) AT (0.03) PT (0.02) RO (0.06)	BE(0.02) MT (0.28) SK (0.70)
8	EL	PL (0.03) PT (0.31) RO (0.37)	BE (0.00) MT (0.19) PL (0.04) PT (0.26) RO (0.36) RO (0.14)
9	ES	FR (0.08) IT (0.35) PL (0.16) RO (0.05) SI (0.24)	BG (0.25) 10 (0.09) IT (0.35) PL (0.06) RO(0.11) SI (0.14)
10	FR	1	1
11	IT	5	2
12	CY	HU (0.02) PL (0.00) PT (0.02) RO (0.01) TR (0.00)	BE (0.01) BG (0.01) EE (0.10) MT (0.83) PT (0.01) SK (0.04)
13	LV	BG (0.28) RO (0.00)	3
14	LT	0	1
15	LU	BE(0.01) RO (0.01) SI (0.01)	0
16	HU	6	0
17	MT	0	5
18	NL	0	0
19	AT	3	0
20	PL	5	3
21	PT	5	2
22	RO	11	3
23	SI	5	3
24	SK	IT(0.01) HU(0.06) AT(0.02) RO(0.10) SI (0.11) TR(0.00)	5
25	FI	IT(0.01) HU(0.01) PL (0.02) PT(0.03) RO(0.04) SI (0.16)	BE (0.14) BG(0.07) EE (0.74) LT (0.02) MT (0.03)
26	SE	DE(0.00) HU(0.06) AT (0.39) PT(0.06) TR (0.00)	DE (0.06) EE (0.38) LV (0.34) SK (0.22)
27	UK	0	0
28	TR	5	0

Table 5. Weights of inefficient countries based on CRS analysis results compared to efficient countries and the number of efficient countries' being references

	DMU	CRS Number for efficient countries to be references (dark)Name and weight of reference country for inefficient countries (light)	VRS Number for efficient countries to be references (dark)Name and weight of reference country for inefficient countries (light)
1	BE	2	2
2	BG	10	0
3	CZ	BG (0.03) LV (0.26) RO (0.10) SK (0.26)	LV(0.41) RO(0.16) SK(0.43)
4	DK	BG (0.09) LV (0.34) RO (0.02)	LV (0.53) SK (0.47)
5	DE	RO (0.15)	LV (0.78) RO (0.22)
6	EE	0	2
7	IE	BG (0.02) CY (0.15) SI (0.05) SK (0.35)	BE (0.02) MT (0.28) SK (0.70)
8	EL	BG (0.30) LV (0.11) RO (0.09) SK (0.34)	LV (0.23) RO (0.11) SK (0.65) TR(0.01)
9	ES	BG (0.11) LV (0.08) RO (0.11)	LV (0.36) RO (0.36) SK (0.28)
10	FR	BG (0.05) LV (0.18) RO(0.10)	LV (0.57) RO (0.31) SK (0.12)
11	IT	LV (0.04) RO(0.25)	LV (0.67) RO (0.33)
12	CY	1	0
13	LV	12	11
14	LT	BG (0.20) LV (0.51) LU(0.00) MT(0.03)	0
15	LU	2	0
16	HU	0	0
17	MT	2	1
18	NL	BE(0.03) BG (0.15) LV (0.02) LU (0.10) SI (0.01)	0
19	AT	LV (0.51) MT(0.11)	EE (0.40) LV (0.60) SK (0.00)
20	PL	BG (0.48) LV (0.23) RO (0.09)	LV (0.53) RO (0.06) SK (0.30) TR (0.10)
21	PT	0	1
22	RO	10	9
23	SI	3	0
24	SK	4	10
25	FI	BE (0.05) BG (0.08) LV(0.14) SI (0.14)	BE (0.12) EE (0.83) PT (0.04) SK(0.01)
26	SE	LV(0.19) RO(0.01) SK (0.10)	LV (0.65) RO (0.03) SK (0.32)
27	UK	RO (0.12)	LV (0.47) RO (0.53)
28	TR	0	2

VRS analyses, Turkey ranked sixth. Countries that are not found efficient in neither of the analyses need to decrease their inputs as much as the weights of the input values of their reference countries presented in Table 10 (for CRS and VRS). Table 9 also illustrate the times efficient countries became references for inefficient ones (shaded in grey). For example, based on CRS analysis results (for Model 3), Turkey could not become a reference for other countries although it was efficient (Table 10). According to VRS analysis results, on the other hand, Turkey became a reference as an efficient country for Poland and Greece (Table 11).

Through three different input-oriented models (Model 1-3) and two different DEA models (CRS and VRS), it was concluded that Turkey is more efficient than EU countries in terms of converting its road transportation indicators into useful outputs (Table 5). Turkey had an efficiency score of 100 in all the three models we worked on. Along with Turkey, Romania and Slovenia are other efficient countries. That is, their road transportation indicators caused less number of accidents, less number of casualties and less number of accidents with casualties within residential areas. In addition, they managed to keep the share of transportation within GNP in its maximum with respect to main indicators; which suggests that they utilized traffic indicators efficiently.

Railway Efficiency

The data set used is composed of information from collected statistics regarding a sample of 23 EU countries and Turkey for the period of 2004-2007. Study material has been constituted by means of a thorough study of the Turkish Railway Annual Report. In this study, the input and output variables selected for the Turkish Railway performance evaluation are presented in Table 6. Linear transformation was applied to the first three variables to assess accurately; all the three variables were transformed (1/variable) by subtracting +ε from the maximum value of the variable. Thus, the disadvantage of large input values was prevented for these three variables. ε is a positive small value.

It is necessary to identify inputs (resources) and outputs (transformation of resources) in frontier models. Various criteria might be taken into consideration to select them. Availability, which is an empirical criterion, is the first of them. Secondly, the literature survey, which is a way of ensuring the validity of the research, is another criterion to be considered. The expert opinion to be taken from relevant individuals is the last criterion.

In order to estimate the production frontier for the years 2004 to 2007, panel data obtained from TSR on 24 countries was used. We respected the DEA convention that the minimum number of DMUs is three times greater than the number of inputs plus output (24 observations > 3 (1+4)).

In this section, efficiencies of 23 EU countries and Turkey in terms of railway indicators have been calculated through output-oriented VRS and output-oriented CRS models, and the results were evaluated. As indicated before, those countries that have efficiency scores of more than 100 are not efficient. In order for such a DMU to become efficient, it should increase its output as much as its efficiency

Table 6. Railway indicators used in the analysis.

Variables	Input/Output
Railway per 1000 km^2 (km) (RPKM)	I
Travelling frequency of people by railway (TFPR)	I
Staff per length of lines (SPLL)	I
Total Operating Expenditure (TOE)	I
Total Operating Revenue (TOR)	O

score, and compensate excess inputs and shortages in outputs (mix inefficiency). Another way for an inefficient DMU to become efficient is to equate its inputs and outputs to those of the reference DMUs with the percentages given in the table.

Table 7 show the results regarding the efficiencies of railways for CRS and VRS models. The mean TE's (i.e. the mean of the technical inefficiency for the 24 selected countries) are equal to 0.731 and 0.796 in 2007 for CRS and VRS model, respectively. The mean SE in 2007 is 0.925. According to Table 7, results of the CRS efficiency analysis; the UK, Germany, Switzerland and Czech Republic are efficient in all of the selected years (2004-2007). In other words, they became reference countries in the selected years for inefficient countries. According to the VRS efficiency results (Table 7), the UK, Germany, Luxembourg, Belgium, Switzerland, Czech Republic and Bulgaria are efficient in all of the analyzed years. The average TE had slightly decreased from 2004 to 2007. Concerning the returns to scale, in 2007, 19 of 24 countries have increasing returns to scale, i.e. they would benefit from a proportional expansion of all the considered inputs (Table 7).

Table 7. CRS, VRS technical and scale efficiency of 24 countries with returns scale.

Countries	TE_{CRS}	TE_{VRS}	SE	
Turkey	0.461	0.474	0.973	irs
United Kingdom	1.000	1.000	1.000	-
Luxembourg	0.681	1.000	0.681	irs
Greece	0.201	0.217	0.923	irs
Ireland	0.700	1.000	0.700	irs
Portugal	0.470	0.546	0.860	irs
Germany	1.000	1.000	1.000	-
Denmark	0.764	0.845	0.904	irs
Italy	0.784	0.785	0.999	irs
Netherlands	0.736	0.743	0.990	irs
Spain	0.648	0.648	0.999	irs
Belgium	1.000	1.000	1.000	-
France	0.933	0.933	1.000	-
Switzerland	1.000	1.000	1.000	-
Norway	0.678	0.723	0.938	irs
Austria	0.883	0.901	0.981	irs
Sweden	0.463	0.468	0.989	irs
Finland	0.487	0.507	0.961	irs
Czech Rep.	1.000	1.000	1.000	-
Slovakia	0.983	1.000	0.983	irs
Hungary	0.671	0.703	0.954	irs
Bulgaria	0.658	1.000	0.658	irs
Croatia	0.710	0.978	0.727	irs
Romania	0.626	0.634	0.987	irs
Mean	0.731	0.796	0.925	-

Table 8. Individual Malmquist indexes

Countries	2004-2005					2005-2006					2006-2007				
	effch	techch	pech	sech	tfpch	effch	techch	Pech	sech	tfpch	effch	techch	pech	sech	tfpch
Turkey	0.894	1.204	0.847	1.003	1.022	1.099	0.897	1.237	0.889	0.986	0.912	1.004	0.831	1.097	0.915
United Kingdom	1.000	1.030	1.000	1.000	1.030	1.000	1.172	1.000	1.000	1.172	1.000	1.070	1.000	1.000	1.070
Luxembourg	1.000	0.821	1.000	1.000	0.821	1.000	1.154	1.000	1.000	1.154	0.681	1.030	1.000	0.681	0.701
Greece	0.642	1.469	0.536	1.197	0.943	1.422	0.725	1.844	0.771	1.031	1.564	1.004	1.286	1.216	1.570
Ireland	0.650	1.547	1.126	0.577	1.006	1.544	0.652	1.000	1.544	1.006	1.041	1.004	1.000	1.041	1.045
Portugal	0.825	1.205	0.882	0.936	0.995	1.282	0.801	1.431	0.896	1.026	1.019	1.004	0.943	1.081	1.023
Germany	1.000	0.877	1.000	1.000	0.877	1.000	1.164	1.000	1.000	1.164	1.000	0.884	1.000	1.000	0.884
Denmark	1.112	1.006	1.109	1.003	1.119	0.798	1.212	1.000	0.798	0.967	1.020	1.004	0.845	1.207	1.024
Italy	0.962	1.015	0.961	1.001	0.976	0.804	1.067	0.804	1.000	0.858	1.015	1.041	1.016	0.999	1.057
Netherlands	0.982	1.022	1.000	0.982	1.003	0.782	1.200	0.774	1.010	0.938	0.958	1.064	0.961	0.998	1.020
Spain	1.103	1.027	1.104	0.999	1.132	0.833	1.223	0.828	1.005	1.018	0.942	1.072	0.943	0.999	1.010
Belgium	0.888	0.230	1.000	0.888	0.204	0.992	1.035	1.000	0.992	1.027	1.135	3.980	1.000	1.135	4.519
France	1.024	1.010	1.023	1.001	1.034	0.962	1.042	0.962	1.000	1.002	0.982	1.009	0.982	1.000	0.990
Switzerland	1.000	3.252	1.000	1.000	3.252	1.000	1.031	1.000	1.000	1.031	1.000	1.101	1.000	1.000	1.101
Norway	0.879	1.322	0.686	1.280	1.161	1.200	0.854	1.374	0.873	1.025	0.946	1.004	0.869	1.089	0.949
Austria	1.030	1.018	1.043	0.987	1.048	0.938	1.078	0.951	0.986	1.011	0.969	1.058	0.955	1.014	1.025
Sweden	1.282	1.334	1.275	1.005	1.709	0.436	0.827	0.458	0.952	0.360	1.062	1.004	1.022	1.039	1.066
Finland	0.916	1.104	0.931	0.985	1.011	1.027	0.980	1.121	0.916	1.007	1.000	1.004	0.938	1.066	1.004
Czech Rep,	1.000	2.882	1.000	1.000	2.882	1.000	1.331	1.000	1.000	1.331	1.000	1.005	1.000	1.000	1.005
Slovakia	0.977	1.022	0.979	0.998	0.999	0.946	1.171	0.945	1.001	1.108	1.191	1.123	1.196	0.995	1.337
Hungary	0.834	1.064	0.857	0.973	0.887	0.936	1.207	0.963	0.972	1.131	1.199	1.004	1.163	1.031	1.204
Bulgaria	0.791	1.228	1.000	0.791	0.971	1.080	0.965	1.000	1.08	1.043	1.094	1.004	1.000	1.094	1.098
Croatia	0.796	1.321	0.792	1.004	1.051	1.253	0.787	1.454	0.862	0.987	1.036	1.004	0.978	1.059	1.040
Romania	0.609	1.113	0.608	1.002	0.678	1.215	1.174	1.286	0.944	1.426	0.985	1.004	0.939	1.049	0.989
MEAN	0.909	1.138	0.932	0.975	1.035	0.996	1.014	1.026	0.971	1.010	1.021	1.080	0.99	1.032	1.103

Table 9. Excesses in inputs and shortfalls in outputs in the analysis, according to the CRS and VRS results for 2007

DMU	CRS Number of being the reference (for efficient countries); and peer and weight (for inefficient countries)	VRS Number of being the reference (for efficient countries); and peer and weight (for inefficient countries)
Turkey	UK (0.19)	Ireland(0.05) Slovakia(0.95)
United Kingdom	20	10
Luxembourg	UK (0.08)	8
Greece	U.K. (0.15)	Ireland (0.41) Slovakia(0.59)
Ireland	U.K.(0.07)	5
Portugal	U.K. (0.11)	Luxembourg(0.13) Ireland (0.53) Slovakia (0.34)
Germany	5	3
Denmark	U.K. (0.21)	UK (0.13) Luxembourg (0.77) Slovakia (0.11)
Italy	U.K. (0.47) Germany (0.22)	U.K.(0.40) Germany (0.22) Slovakia (0.38)
Netherlands	U.K.(0.62)	U.K.(0.55) Luxembourg (0.21) Slovakia (0.24)
Spain	U.K.(0.84)	U.K.(0.80) Slovakia (0.20)
Belgium	U.K.(0.04) Germany (0.11)	0
France	U.K.(0.19) Germany (0.64)	U.K.(0.15) Germany (0.64) Slovakia (0.21)
Switzerland	0	0
Norway	U.K.(0.16)	Luxembourg (0.07) Ireland (0.24) Slovakia (0.69)
Austria	U.K.(0.34) Germany (0.09)	U.K.(0.25) Luxembourg (0.49) Germany (0.09) Slovakia (0.17)
Sweden	U.K. (0.33)	U.K.(0.16) Slovakia (0.84)
Finland	U.K.(0.20)	U.K. (0.02) Luxembourg (0.16) Slovakia (0.82)
Czech Rep.	0	0
Slovakia	U.K. (0.19)	15
Hungary	U.K.(0.36)	U.K. (0.28) Luxembourg (0.61) Slovakia (0.11)
Bulgaria	U.K.(0.08)	0
Croatia	U.K.(0.08)	Luxembourg (0.28) Ireland (0.67) Slovakia(0.05)
Romania	U.K.(0.30)	U.K. (0.13) Slovakia (0.87)

Malmquist TFV indexes were decomposed through the DEA technique by using the DEAP Version 2.1 software developed by Tim Coelli (Coelli, 1996).

DEA renders possible the estimation of total factor productivity change (tfpch) in the form of the Malmquist index. The results are demonstrated in Table 14, with the Malmquist index, denoted tfpch, decomposed into technically efficiency change (tech) (the diffusion or catch-up component) and technologically efficiency change (techch) (the innovation or frontier-shift component). In addition, technically efficiency change is split into pure efficiency change (pech) and scale-efficiency change (sech). The railways are ranked (Figure 6) according to the results of the column 5 in Table 8.

The multiplication of the technical efficiency change and the technological change yields the total factor productivity (TFP). The multiplication of the pure efficiency change and the scale efficiency change yields the technical efficiency change. For each index, scores more than 1 indicate increases

Table 10. Results of the super efficiency analysis

	2007		2006		2005		2004	
DMU	CRS	VRS	CRS	VRS	CRS	VRS	CRS	VRS
Turkey	216.78	210.89	197.62	175.19	217.21	216.54	184.42	182.18
U.K.	75.79	75.38	73.07	72.99	94.58	77.78	73.30	71.17
Luxembourg	146.94	big	131.92	big	117.19	big	111.51	big
Greece	498.57	460.30	779.65	591.77	1108.23	1082.64	711.84	585.23
Ireland	142.95	big	148.85	big	229.80	big	149.40	112.60
Portugal	212.76	183.03	216.87	172.65	278.02	247.11	229.46	218.02
Germany	27.27	23.25	26.40	22.94	29.60	25.76	26.28	22.08
Denmark	130.97	118.33	133.56	big	106.56	big	118.54	110.44
Italy	127.48	127.36	129.43	128.64	104.10	104.03	100.16	99.83
Netherlands	135.96	134.54	130.31	128.32	101.85	big	98.86	97.81
Spain	154.42	154.26	145.45	145.26	121.09	120.42	133.52	132.95
Belgium	114.19	big	115.14	big	113.38	big	105.11	big
France	107.16	107.14	105.19	105.11	101.18	101.17	103.62	103.53
Switzerland	28.98	big	29.30	big	24.69	big	30.60	big
Norway	147.51	138.38	139.52	120.23	167.41	165.22	147.11	113.41
Austria	113.23	110.36	109.72	104.48	102.90	100.79	105.97	104.82
Sweden	216.02	213.63	229.43	218.42	62.44	52.11	128.15	127.48
Finland	205.30	197.30	205.41	185.12	210.93	207.50	193.29	193.12
Czech Rep.	56.08	big	57.98	big	78.60	big	87.15	big
Slovakia	101.74	85.32	121.18	109.67	114.63	big	111.99	big
Hungary	149.13	142.33	178.87	165.60	167.49	157.92	139.61	135.32
Bulgaria	152.08	Big	166.41	big	179.72	big	142.12	big
Croatia	140.76	102.29	145.81	99.73	182.72	144.73	145.41	115.20

Table 11. Input and output variables used in the analysis

Indicators	MODEL 1 (Input/ Output)	MODEL 2 (Input/ Output)	MODEL 3 (Input/ Output)
Total Air traffic (arrivals-departures)	I		
Number of passengers (total)	O		
Freight carried (total)	O		
Domestic Air traffic (arrivals-departures)		I	
Number of passengers (domestic)		O	
Freight carried (domestic)		O	
International Air traffic (arrivals-departures)			I
Number of passengers (international)			O
Freight carried (international)			O

Figure 6. Turkey's efficiency changes between 2004-2007.

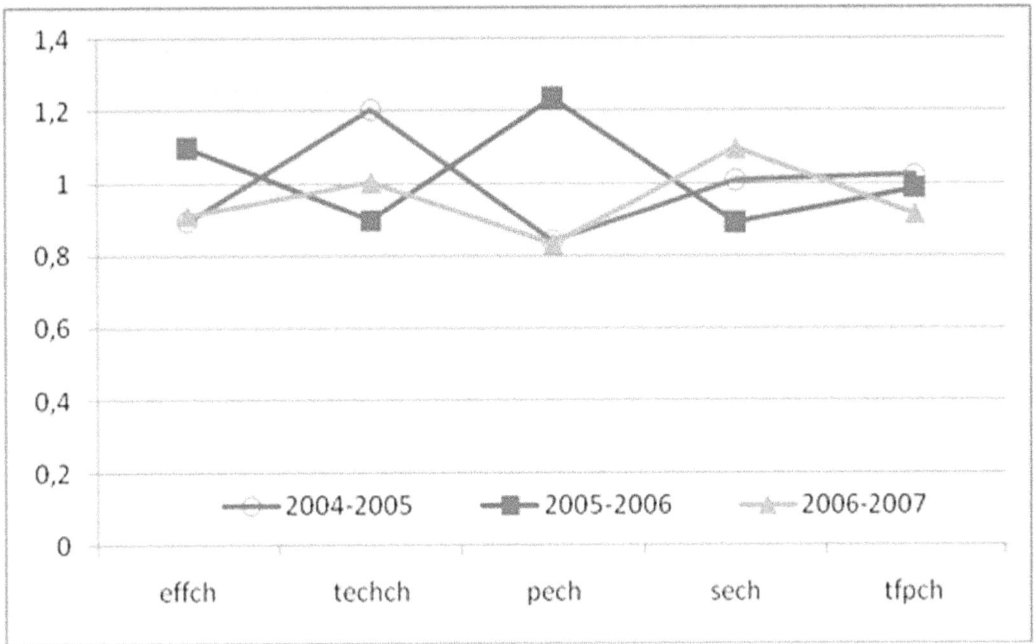

and scores less than 1 indicate decreases over the previous period. According to the Malmquist Index analysis, those countries whose total factor productivity change (tfpch) scores are equal to 1 have made no progress or regress in terms of their inputs and outputs. On the other hand, a tfpch score of more than 1 implies that the country has increased its total factor productivity while a tfpch score of less than 1 indicates that the country has decreased its total factor productivity. The individual Malmquist indexes are presented in Table 8. The change in efficiency (effch) between 2004 and 2005 is on average equal to 0.909, while the mean of techch is 1.138. The average change in tfpch is equal to 1.035 for the same years. Similarly, changes between the years of 2005-2006 and 2006-2007 are 0.996, 1.021, 1.014, 1.08 and 1.01, 1.103 for effch, techch and tfpch, respectively. For 14 countries, we have obtained an increase in the tfpch between the four years considered, while productivity has decreased in 10 countries. In Table 8, it can be seen that the total productivity change score (the Malmquist index presented in the last column) is higher than 1 for all countries, except Turkey, Luxembourg, Germany, France, Norway and Romania; showing that most of the railway operators realized gains in total productivity factor in 2007. The mean Malmquist scores are 1.035, 1.01 and 1.103 for the selected period, which, since it is higher than 1, signifies that for the majority of the countries, tfpch increased in all periods. However, there are six countries with a Malmquist index below 1, indicating that these countries need to improve their productivity. Turkey's tfpch scores in the above-mentioned years were 1.022, 0.986 and 0.915; lagging behind the averages of the examined 24 countries (1.035, 1.010, 1.103). Figure 9 illustrates Turkey's performance changes by year. As the figure demonstrates, although Turkey's tfpch score reached to 1.022 > 1 between 2004-2005 which suggests that Turkey's ability to convert its inputs into outputs had increased, the total factor productivity declined in the periods of 2005-2006 and 2006-2007 with tfpch scores of 0.985 and 0.915<1, respectively. So, TSR could not use its inputs positively to produce outputs from 2005 to 2007. The change in the technical efficiency score (effch, column1 each period) is defined as the diffusion of the best-practice technology in the management of the activity and is attributed to investment planning, technical experience and management and organization in the railways. For 2007, it can be seen that it is more than 1 for sixteen countries, suggesting that technical efficiency increased in 2007 over 2006. However, for some other countries, the change in technical efficiency is below 1, which implies that they experienced regression in this respect in the 2006-2007 periods.

In terms of the efficiency change (effch), Turkey achieved a score more than 1 between the years of 2005-2006; however, she could not improve it between the years of 2006-2007 (Table 8). Belgium achieved the highest increase. Greece, on the other hand, increases the effch score at each turn of year. Slovakia, Hungary and Belgium increased their effch scores between 2006 and 2007. Greece, Belgium and Bulgaria increased their effch scores constantly. Technological change (column 2 each period in Table 8) is the outcome of innovation, i.e. the adoption of new technologies, by best-practice railways. For the 2006-2007 period, its mean value is 1.08, and this index is over 1 for all countries except Germany. This implies that innovation has improved in the period for almost all countries, meaning that these countries invested in new technologies (methodologies, procedures and techniques) and in the relevant commensurate skill upgrades.

In terms of technical efficiency (techch) change, Turkey was able to use the techch score towards advancement since she achieved scores more than 1 in the periods other than the period of 2005-2006 (Table 8). Turkey achieved a techch score below the average score of the other countries. This implies that Turkey, compared to the other countries, could not improve its innovation adequately in terms of the variables considered. The decomposition of the score for the change in technical efficiency into productivity efficiency change (pech) (column 3 each period in Table 8) and scale efficiency change (sech) (column 4 each period in Table 8) shows mixed results, with some countries obtaining gains in both areas while some others obtaining gains in only one of them but losses in the other. The improvement in the productivity efficiency change, which implies an improvement in managerial skills, shows that investments were made in organizational factors associated with the management of railways, such as a better balance between inputs and outputs, best-practice initiatives, more accurate reporting, an improvement in quality, and so forth.Turkey, in terms of the productivity efficiency change (pech), could not use the pech score towards advancement since she achieved scores less than 1 in the periods other than the period of 2005-2006 (Table 8). Turkey's pech score lagged behind the average of the other countries. The scale efficiency, which is the consequence of size, increases in the 2006-2007 period for many countries, thanks to the improvement in capacity utilization. It is important to note that the mean value of technical efficiency improvement is 1.021 (mean), the mean value of productivity change is 0.99 (mean) and the mean value of scale-efficiency change is 1.032. Table 8 illustrates the changes in the scale efficiency (sech) scores of Turkey and several EU countries by year. As the tables demonstrate, Turkey has improved in terms of scale efficiency between the years of 2006-2007 by achieving a score more than 1. However, the change in the other years was not stable.

Table 8 illustrates the total factor productivity changes of Turkey and several EU countries. Slovakia and Greece were the countries that achieved the highest change in total factor productivity in the last year. Turkey's scores are below the average of the other countries. Also, we identify the reference or corner points of the facet (with positive weights) onto which Turkey were projected (Table 9). The UK is the peer of Turkey (with the weight 0.19) as CCR. In addition, the peers of Turkey are Ireland (with the weight 0.05) and Slovakia (with the weight 0.95) as VRS. These are the points with efficient performances that are used to evaluate Turkey's performance since their mix of inputs and outputs are similar to that of Turkey. The peers can serve for a policymaker as benchmarks to be used for a possible improvement of Turkey's performance especially in terms of its poor record in transportation indicators. According to the results of the CCR analysis in Table 9, the UK, Germany, Belgium, Switzerland and Czech Republic are the efficient countries in 2007. The UK became the reference for 19 inefficient countries and Germany became the reference for four inefficient countries. In other words, inefficient countries should take the UK and Germany as their models to use the inputs and outputs in their railway indicators, since the UK and Germany have been efficient in converting their inputs into outputs.

Table 9. , on the other hand, demonstrates the results of VRS efficiency analysis. According to these results, Luxembourg, Ireland, Germany, Switzerland, Czech Republic and Bulgaria are efficient countries. Among them, the UK became the reference for ten countries while Luxembourg for eight, Ireland for five, Germany for three and Slovakia became the reference for 15 countries. The results of the super efficiency analysis are presented in Table 10. Countries that are marked as bold in the table are the countries whose efficiencies are equal to 1. These countries (efficient countries) were ranked in terms of efficiency through the A-P super efficiency method. As the table shows, VRS efficiency has not been significant in the rankings (many countries had big scores).

CRS has become more determining in super efficiency. For example; the UK, Germany, Switzerland and Czech Republic, which were the efficient countries in 2007 in terms of CRS efficiency, were ranked with the A-P super efficiency method as follows: the UK, Czech Republic, Switzerland and Germany. Inefficient countries might take this ranking into consideration when selecting reference countries

Airway Efficiency

Data Envelopment Analysis (DEA), was employed to determine the air transportation performance of 12 regions of Turkey. Using the DEA models of CRS (CCR) and VRS (BCC), a section (cross-section) analysis was performed and subsequently efficiency scores were ranked. In calculations, the main air transport indicators were used as variables and regional areas were regarded as decision making units (DMU). In order to determine the efficiency, three different efficiency models were formulated taking into consideration nine input and various output variables. These models were analyzed according to their input and output values, as presented in Table 11. The aim of each model is;

Model 1: Determining the efficiency that uses the total air traffic as input and yields the number of total passengers and total freight carried as output,

Model 2: Determining the efficiency that uses the domestic air traffic as input and yields the number of domestic passengers and domestic freight carried as output,

Model 3: Determining the efficiency that uses the international air traffic as input and yields the number of international passengers and international freight carried as output

CRS and VRS output-oriented models were used to determine technical efficiency by maximizing outputs at a given input. The CRS efficiency scores of the regions for each model are given in Figure 7.

Figure 7. Efficiency scores for CRS output-oriented Model 1

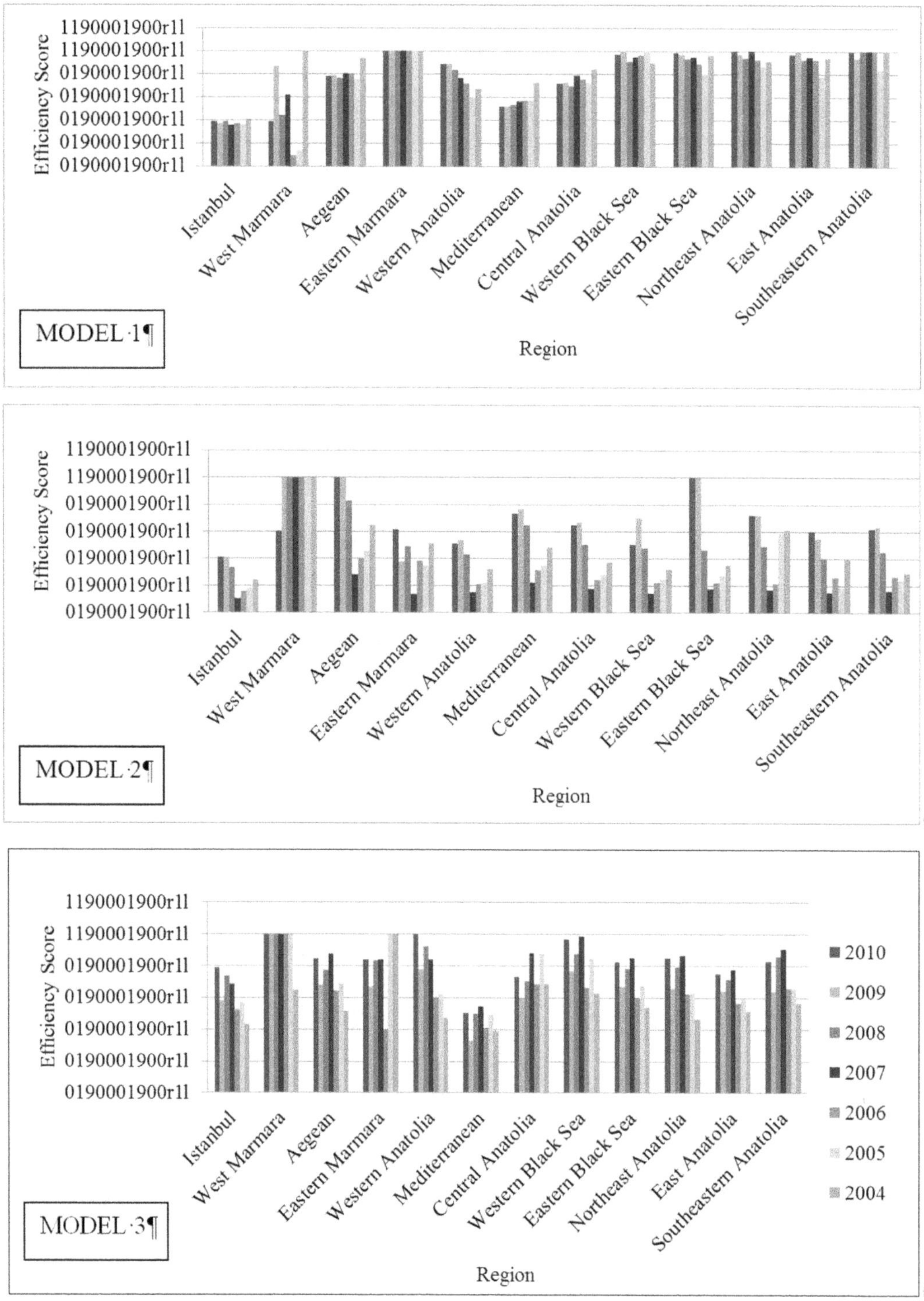

It should be noted that, in determination of the efficiencies, VRS efficiency analysis did not yield reliable results. For this reason, it was more reasonable to compare CRS efficiency scores. According to CRS efficiency scores of Model 1, the efficient region was the Eastern Marmara for the selected period. On the other hand, according to CRS efficiency scores of Model 2 and Model 3, the efficient region was the Western Marmara. Note that, year 2010 for Model 2 and year 2004 for Model 3 did not fit these results.

Malmquist Index results of the Model 1 are given in Table 12 for each efficiency change. Note that, the year 2005 was considered as the base year in these diagrams. According to effch given in Table 12, the region with most apparent efficiency change was the Western Marmara. While this highly industrialized region demonstrated a decrease in 2006-2007, a nearly constant efficiency change was calculated for it for the rest of the period. This can be considered as the effect of the economic crises occurred in 2007. The region, whose efficiency change altered in the positive direction with respect to the other years, was the Western Black Sea. The variation of techch for Model 1 is given in Table 12. As seen from the figure, the most significant changes in the techch were observed in the Western and Eastern Marmara Regions. In contrast to decreases in 2007 and 2009, these regions showed an increase in 2010. The other regions were in a decreasing trend between 2008 and 2010. While in Model 1 changes in pech and techch were similar to each other, the efficiency change of West Marmara Region was in the positive direction for 2009- 2010 period, see Table 12. The efficiency change values of Mediterranean and Eastern Marmara Regions were 0 for all years. This shows that these regions maintained their efficiencies in the whole selected period. According to scale efficiency change values of Model 1, the efficient region was the Eastern Marmara Region and the efficiency changes were 0, see Table 12. As the scale efficiency of the Eastern Marmara region was 1 for all years, it was taken as the reference region for others. All other regions increased their efficiency changes in 2010. When the total productivity change values of the

Table 12. Malmquist Index summary of regions for Model 1

Region	effch	techch	pech	sech	tfpch
Istanbul	0.993	1.064	1	0.993	1.056
West Marmara	0.853	0.909	1	0.853	0.775
Aegean	0.969	1.068	1	0.969	1.035
Eastern Marmara	1	0.989	1	1	0.989
Western Anatolia	1.048	1.06	1	1.048	1.111
Mediterranean	0.946	1.068	1	0.946	1.01
Central Anatolia	0.974	1.065	0.974	1	1.037
Western Black Sea	1.013	1.069	1.013	1	1.084
Eastern Black Sea	1.004	1.065	1.004	0.999	1.069
Northeast Anatolia	1.015	1.065	1.006	1.009	1.081
East Anatolia	1.005	1.066	0.997	1.008	1.071
Southeastern Anatolia	0.999	1.068	1	0.999	1.067
MEAN	0.984	1.045	0.999	0.984	1.028

Model 1 are examined, see Table 12, it can be seen that tfpch values of the Mediterranean Region was positive for all years examined. Total productivity changes values of both Marmara Regions were more variable than others. In Model 1, the average tfpch values of Western and Eastern Marmara Regions for the selected period were found to be smaller than 1, i.e., there was a decreasing trend.

Meanwhile, total factor productivity change values for both model are given in Table 13, respectively. It can be revealed form Table 13 that regions with efficiency change values were smaller than 1 demonstrated decreasing trend in their efficiencies. Detailed analysis of Table 13, obtained by taking 2005 as the base year, shows that efficiency value were fluctuating for Western Marmara Region with a decreasing trend. While the efficiency results were fluctuating for Eastern Region, an increasing trend was maintained. A similar result was also obtained for Model 3 (domestic efficiency), see Table 13.

Table 13. The efficiency change scores for Model 2

REGION		Model 2						Model 3				
		2004-2005	**2005-2006**	**2006-2007**	**2007-2008**	**2008-2009**	**2009-2010**	**2004-2005**	**2005-2006**	**2006-2007**	**2007-2008**	**2008-2009**
Istanbul	effch	0.749	0.839	0.66	3.31	1.228	0.999	1.315	0.921	1.301	1.074	0.79
West Marmara		1	1	1	1	1	0.601	1.547	1	1	1	1
Aegean		0.702	0.886	0.695	2.94	1.208	1	1.343	0.93	1.363	0.882	0.879
Eastern Marmara		0.678	1.104	0.358	3.572	0.77	1.623	1	0.394	2.121	0.997	0.801
Western Anatolia		0.696	0.945	0.711	2.879	1.236	0.953	1.319	0.971	1.387	1.099	0.845
Mediterranean		0.713	0.905	0.697	2.969	1.181	0.952	1.259	0.829	1.324	0.916	0.66
Central Anatolia		0.743	0.858	0.727	2.92	1.321	0.97	1.271	0.782	1.286	0.806	0.849
Western Black Sea		0.784	0.898	0.645	3.288	1.475	0.719	1.344	0.79	1.486	0.887	0.876
Eastern Black Sea		0.784	0.809	0.779	2.682	2.154	1	1.239	0.896	1.41	0.918	0.855
Northeast Anatolia		0.957	0.362	0.792	2.897	1.471	1.004	1.35	0.98	1.393	0.915	0.831
East Anatolia		0.468	1.393	0.557	2.74	1.372	1.097	1.147	0.962	1.37	0.92	0.901
Southeastern Anatolia		0.812	1.133	0.605	2.753	1.408	0.983	1.159	1.002	1.374	0.949	0.741
Istanbul	techch	1.431	1.116	1.406	0.345	0.842	1.016	0.899	1.249	0.774	1.134	1.239
West Marmara		2.155	1.087	1.075	0.844	0.187	0.739	2.564	0.309	0.3	2.105	1.213
Aegean		1.431	1.116	1.406	0.345	0.842	1.017	0.899	1.249	0.774	1.134	1.239
Eastern Marmara		1.431	1.116	1.406	0.345	0.724	1.016	0.892	1.249	0.524	1.134	1.239
Western Anatolia		1.431	1.116	1.406	0.345	0.809	1.018	0.899	1.249	0.774	1.134	1.239
Mediterranean		1.431	1.116	1.406	0.345	0.842	1.016	0.899	1.249	0.774	1.134	1.239
Central Anatolia		1.431	1.116	1.406	0.345	0.832	1.017	0.899	1.249	0.774	1.134	1.239
Western Black Sea		1.431	1.116	1.406	0.345	0.798	1.017	0.899	1.249	0.774	1.134	1.239
Eastern Black Sea		1.733	1.116	1.406	0.391	0.513	1.007	0.899	1.249	0.774	1.134	1.239
Northeast Anatolia		1.431	1.116	1.406	0.345	0.753	1.015	0.899	1.249	0.774	1.134	1.239
East Anatolia		1.431	1.116	1.406	0.345	0.785	1.017	0.899	1.249	0.774	1.134	1.239
Southeastern Anatolia		1.431	1.116	1.406	0.345	0.835	1.017	0.899	1.249	0.774	1.134	1.239

continued on following page

Table 13. Continued

REGION		Model 2						Model 3				
		2004-2005	2005-2006	2006-2007	2007-2008	2008-2009	2009-2010	2004-2005	2005-2006	2006-2007	2007-2008	2008-2009
Istanbul	pech	1	1	1	1	1	1	1	1	1	1	1
West Marmara		1	1	1	1	1	0.622	1	1	1	1	1
Aegean		1	1	1	1	1	1	1	1	1	0.917	1.054
Eastern Marmara		0.825	0.958	0.502	2.113	0.849	1.397	1	0.482	2.074	0.871	0.852
Western Anatolia		0.827	1.125	1.223	0.933	0.815	0.886	0.961	1.047	1	1.047	1
Mediterranean		1	1	1	1	1	1	1	0.758	1.012	0.972	0.737
Central Anatolia		1.045	0.94	1.043	1.046	1.116	0.973	1	1	0.926	0.827	0.998
Western Black Sea		1.078	0.92	0.946	1.279	1.302	0.701	1.058	1.009	1.028	0.948	1.029
Eastern Black Sea		1	0.66	1.515	1	1	1	0.978	0.994	1.018	0.91	1.009
Northeast Anatolia		1.135	0.391	1.076	1.487	1.739	1	1.064	1.165	0.989	0.935	0.978
East Anatolia		0.189	5.292	0.317	1.495	1.339	1.046	0.939	1.093	0.987	0.916	1.063
Southeastern Anatolia		1.038	1.016	1.131	1.036	1.224	0.987	0.897	1.115	1	0.933	0.875
Istanbul	sech	0.749	0.839	0.66	3.31	1.228	0.999	1.315	0.921	1.301	1.074	0.79
West Marmara		1	1	1	1	1	0.966	1.547	1	1	1	1
Aegean		0.702	0.886	0.695	2.94	1.208	1	1.343	0.93	1.363	0.962	0.834
Eastern Marmara		0.822	1.153	0.713	1.691	0.907	1.162	1	0.818	1.023	1.145	0.94
Western Anatolia		0.841	0.84	0.581	3.086	1.515	1.076	1.373	0.928	1.387	1.05	0.845
Mediterranean		0.713	0.905	0.697	2.969	1.181	0.952	1.259	1.094	1.308	0.942	0.895
Central Anatolia		0.711	0.912	0.697	2.79	1.183	0.996	1.271	0.782	1.388	0.975	0.851
Western Black Sea		0.727	0.975	0.682	2.571	1.132	1.026	1.271	0.783	1.445	0.936	0.851
Eastern Black Sea		0.784	1.225	0.514	2.682	2.154	1	1.268	0.901	1.385	1.009	0.848
Northeast Anatolia		0.843	0.926	0.736	1.949	0.846	1.004	1.27	0.842	1.409	0.979	0.85
East Anatolia		2.477	0.263	1.754	1.833	1.025	1.049	1.221	0.88	1.387	1.005	0.847
Southeastern Anatolia		0.783	1.115	0.535	2.656	1.151	0.996	1.292	0.899	1.374	1.018	0.846
Istanbul	tfpch	1.072	0.936	0.928	1.143	1.035	1.014	1.182	1.15	1.007	1.218	0.979
West Marmara		2.155	1.087	1.075	0.844	0.187	0.444	3.967	0.309	0.3	2.105	1.213
Aegean		1.004	0.988	0.978	1.015	1.018	1.017	1.208	1.161	1.055	1	1.09
Eastern Marmara		0.971	1.232	0.503	1.234	0.557	1.65	0.892	0.493	1.111	1.13	0.993
Western Anatolia		0.995	1.054	1	0.994	1	0.97	1.187	1.213	1.074	1.247	1.048
Mediterranean		1.021	1.01	0.98	1.025	0.994	0.967	1.133	1.035	1.025	1.038	0.818
Central Anatolia		1.064	0.957	1.021	1.008	1.1	0.986	1.143	0.976	0.995	0.914	1.052
Western Black Sea		1.122	1.002	0.907	1.135	1.177	0.731	1.209	0.986	1.15	1.006	1.085
Eastern Black Sea		1.358	0.902	1.095	1.048	1.105	1.007	1.115	1.119	1.092	1.041	1.06
Northeast Anatolia		1.37	0.404	1.113	1.001	1.109	1.019	1.215	1.224	1.079	1.037	1.03
East Anatolia		0.67	1.554	0.783	0.946	1.077	1.115	1.031	1.201	1.06	1.043	1.116
Southeastern Anatolia		1.163	1.265	0.851	0.951	1.176	1	1.043	1.251	1.063	1.077	0.918

CONCLUSION

Attaining a high level of performance is a key factor of success for every organization. To this end; it is today an important managerial instrument to define the criteria necessary for development, improve the existing performance and understand why some units in the organization are operating inefficiently. In the study, using Data Envelopment Analysis that is a technique for measuring efficiency and total factor productivity, performance was demonstrated subjectively.

The analysis is based on a DEA model that renders possible the incorporation of multiple inputs and outputs in the determination of relative efficiencies. Benchmarks are provided in order for poorly performing railway operators to improve their operations. We conclude that the examined 24 countries' railway operations exhibit on average improvements in technical efficiency and in technological change. The improvement in technological change is higher than the improvement in technically efficiency. Turkey needs to balance between the basic railway inputs and outputs in order to achieve her objectives to become the corridor between the EU and Asian countries especially through railways, to join the EU and to become a leading country in the world economy. This study provides concrete data to policymakers in Turkey about how the transportation -and especially the railway- inputs and outputs should be used efficiently.

It was aimed to identify performance of airline transportation and use the obtained results in preparation of different strategies and policies for harmonization of Turkey to EU. Meanwhile, it is expected that the subjective results presented in this study will be helpful to increase the level of performance of airline transportation sector in Turkey.

DEA is used to determine the management performance of many companies as well as to specify policies developed to improve them. In this study, it was aimed to ascertain the relative performance of the basic airline indicators and by this way the utilization efficiency of the airline transportation in Turkey was identified. Using subjective calculation results, it was aimed to inform reader about the strategies and targets to be followed so that the performances of the transportation indicators were maintained in their highest level.

REFERENCES

Aydogdu, A. (2006). *Transportation and energy policies that Turkey should* European Commission. *develop in order to reduce the number of traffic accidents in the EU accession period* [Unpublished master's thesis]. Gazi University, Institute of Science, Ankara.

Beasley, J. A. (2004). OR Notes - Data Envelopment Analysis. Retrieved from http://people.brunel.ac.uk/~mastjjb/jeb/or/dea.html

Bedrettin, M. (2004). *Traffic Law and Fundamental Traffic Knowledge* (in Turkish). Ankara: General Directorate of Security.

Caves, D. L., Christensen, L., & Diwert, W. E. (1982). The economic theory of index numbers and the measurement of input, output and productivity. *Econometrica, 50*(6), 1393–1414. doi:10.2307/1913388

Charnes, A., Cooper, W. W., & Rhodes, E. (1978). Measuring the efficiency of decision making units. *European Journal of Operational Research*, *2*(6), 429–444. doi:10.1016/0377-2217(78)90138-8

Chen, P. C., Yu, M. M., Chang, C. C., & Hsu, S. H. (2008). Total factor productivity growth in China's agricultural sector. *China Economic Review*, *19*(4), 580–583. doi:10.1016/j.chieco.2008.07.001

Coelli, T. J. (1996). *A guide to DEAP version 2.1: 1996 data envelopment analysis program* (CEPA working papers No:8/96). Retrieved from November 25, 2014, from http://www.une.edu.au/econometrics/cepa.htm

Cook, W. D., & Seiford, L. M. (2009). Data envelopment analysis (DEA) – Thirty years on. *European Journal of Operational Research*, *192*(1), 1–17. doi:10.1016/j.ejor.2008.01.032

Cubuk, K., & Cansiz, O. F. (2005). Energy status between Transportation Systems in Turkey. *Paper presented at the Energy Efficiency Conference*. Ministry of Energy and Natural Resources, Ankara.

Emrouznejad, A., Parker, B. R., & Tavares, G. (2008). Evaluation of research in efficiency and productivity: A survey and analysis of the first 30 years of scholarly literature in DEA. *Socio-Economic Planning Sciences, 42*(3), 151-157.

EU transport in figures Statistical Pocket Book 2011. (2011). Belgium.

Fare, R., Grosskoppf, S., Norris, M., & Zhang, Z. (1994). Productivity growth, technical progress and efficiency change in industrialized countries. *The American Economic Review*, *84*(1), 66–83.

Hashimoto, A., & Haneda, S. (2008). Measuring the change in R&D efficiency of Japanese pharmaceutical industry. *Research Policy*, *37*(10), 1829–1836. doi:10.1016/j.respol.2008.08.004

Malmquist, S. (1953). Index numbers and indifference surfaces. *Trabajos de Estadistica*, *4*(2), 209–242. doi:10.1007/BF03006863

Ray, S. C. (2004). *Data Envelopment Analysis: Theory and Techniques for Economics and Operations Research*. UK: Cambridge University Press. doi:10.1017/CBO9780511606731

Sözen, A., Alp, I., & Ozdemir, A. (2010). Assessment of operational and environmental performance of thermal power plants in Turkey by using data envelopment analysis. *Energy Policy*, *38*(10), 6194–6203. doi:10.1016/j.enpol.2010.06.005

Sözen, A., Alp, I., & Kilinc, C. (2012). The efficiency assesstment of the hydro-power plants in Turkey by using data envelopment analysis. *Renewable Energy*, *46*, 192–202. doi:10.1016/j.renene.2012.03.021

Transportation Strategy Report (in Turkish) (pp. 23–30). (2004Ministry of Transportation. Ankara.

KEY TERMS AND DEFINITIONS

Data Envelopment Analysis (DEA): It is currently receiving attention as a non-parametric tool which is used to calculate the efficiency of a homogeneous set of organizations, i.e., decision-making units (DMUs).

The DEA Model: It is used to estimate the production-possibility frontier, which is the boundary of the production possibilities set defining the set of input-output combinations.

The Production Possibility Frontier: It is the most important part of the production possibilities set since it reflects the highest amount of output than can be produced from a given amount of input. It is a flexible and important tool offering many advantages over other efficiency measurement methods. The method, which does not specify a relation between input and output, gives no measurement errors. One of the most important benefits of the method is that it is capable of handling multiple inputs and outputs simultaneously.

Chapter 4
Determination of Work Zone Capacity Using ELM, MPMR and GPR

Sangeeta Roy
Vit University, India

J. Jagan
Vit University, India

Pijush Samui
Vit University, India

ABSTRACT

This article examines the capability of Extreme Learning Machine (ELM), Minimax Probability Machine Regression (MPMR) and Gaussian Process Regression (GPR) for determination of Work Zone Capacity. Number of lanes, number of open lanes, work zone layout, length, lane width, percentage trucks, grade, speed, work intensity, darkness factor, and proximity of ramps have been adopted as inputs of ELM, MPMR and GPR. ELM has excellent generalization performance, rapid training speed and little human intervention. MPMR is developed based on the concept of minimax probability machine classification. It does not assume any data distribution. GPR is a probabilistic, and non-parametric model. In GPR, different kinds of prior knowledge can be applied. This article describes a comparative study between the ELM, MPMR and GPR models.

INTRODUCTION

In the past 40 years there has been a rise in traffic which has by far outdone the economic growth and shows no signs of retarding its growth rate. This disproportionate growth in traffic is leading to a voracious consumption of our resources and rapid degradation of environment. Sustainable mobility is the need of the hour. Some of the major reasons for this unprecedented growth of traffic are underlined in the following points.

DOI: 10.4018/978-1-4666-8648-9.ch004

1. **Urban Sprawl:** It can be defined as the unplanned expansion of human population from central urban areas into previously remote and city outskirts. This is mainly done to facilitate the influx from rural areas. This gives rise to certain land use patterns such as single use zoning (i.e. situation such that commercial, residential, institutional and industrial are separated causing extensive use of automobiles), low density zoning, job sprawl, spatial mismatch. All these together have a major impact on increasing rates of traffic growth
2. **Economic Growth:** With the rise of world economy the middle class is seeing unprecedented rise in income as well as easier access to personal automobiles due to decline in interest rates of banks. This has led to lesser use of public transport increasing road traffic.
3. **Trade and Commerce:** In the recent past there has been tremendous growth in trade integration and global trade has increased at roughly three times the rate of global aggregate output. Increased trade gives rise to increased freight traffic resulting in more numbers of heavy trucks, lorries etc.

Increase in population is a key factor in the rise of traffic levels. Population increase is mainly due to two reasons natural increase and migration. Migration is the long term settlement in certain areas after dislocation from some other. Migration is basically from rural to urban areas. This is mostly due to better socio economic condition of cities, better job opportunities, better education facilities, better health conditions and sanitation. Cities also provide a wider array of trade opportunities. This huge influx causes urban sprawl and various other problems which give rise to higher traffic rates. Some of them are discussed as under.

1. Cheaper availability of labour increases the number of infrastructure project which in turn increase freight traffic.
2. Increased workforce causes rapid industrialization causing greater traffic rates of traffic growth.

Due to single use zoning the distances between various zones has increased generating longer and more number of trips. The commuter and shopping traffic has also increased which accounts for a major chunk of private traffic movements. Relocation of workplaces from urban to regional sites leads to larger distances to commute.

Transportation distances have increased considerably in recent past signalling an increase in freight volumes as well as traffic. Economic slowdowns did have a negative impact on freight traffic but the overall trend has been of growth.

Some other reasons for increased traffic growth are enlisted as under:

1. Speculation about future government policies sometimes causes premature growth of certain regions which in turn increase traffic
2. Due to the increase in nuclear families in cities there is a larger cities wastes vertical space and result in horizontal growth causing more urban sprawl affecting traffic patterns.
3. Government policies do not allow vertical construction in places with lesser accessibility causing further urban sprawl especially in the cases of old cities.
4. Increase in transportation infrastructure such as expressways, highways etc too cause both traffic congestion as well as rapid outgrowth.

5. Increased urban economy as well as job avenues commands the need of higher transportation facilities increasing the number of vehicles plying the road.
6. Lack of planning policies with foresight.

With the rise in today's traffic demands as well as number of vehicles and transportation facilities there has a sharp rise in work zones. A work zone may be defined as an area of a highway where construction, maintenance, or utility work activities that can be identified by warning signs/signals/indicators, including those on transport devices (e.g., signs, flashing lights, diversions for traffic, barriers, pavement markings, flagmen, warning signs and arrow boards mounted on the vehicles in a mobile maintenance activity). Work zones also include roadway sections where there is on-going, moving (mobile) work activity such as lane line painting or roadside mowing only if the beginning of the on-going, moving (mobile) work activity is designated by warning signs or signals. Work zones are a necessity of modern day world. The indispensability of work zones can be outlined in the following points:

1. The recent unprecedented growth in traffic has resulted in a large number of vehicles plying the road. This in turn causes rapid deterioration of the roadways as they were not designed to sustain such large amounts of traffic for prolonged periods of time. So there is requirement of more frequent repair work.
2. Due to heavy freight traffic the roads tend to be damaged severely requiring repair work. This is more pronounced in highways and expressways rather than the narrow arterial network of the city centre.
3. Weather too plays a significant role in the formation of work zones. With the changing weather patterns there is non-cyclic behaviour of precipitation which leads to water logging causing deterioration of road network. Other environmental factors such as snow, direct sunlight etc also cause deterioration of roads.
4. Due to land zoning the land use pattern has changed a lot. The decentralization of workplaces has caused a rise in office commuter traffic from residential to industrial areas. This leads to the frequenting of previously unused road networks increasing the need for their maintenance.
5. To facilitate the exponentially increasing number of cars and other traffic there is a need for infrastructure development. The increased construction of flyovers, bridges etc increase the requirements of work zones due to the following reasons:
    i. Lack of foresight in planning leaves such structures redundant in places causing increased congestion in other places. This creates a twofold problem. Firstly it causes congestion in already overworked roadways and secondly there is the big question of maintenance of these redundant structures.
    ii. The entire duration of building these require the rerouting of traffic and in some places creations of diversions. The rerouted traffic creates higher density of traffic increasing the requirement of maintenance whereas diversions both permanent as well as temporary too require a lot of maintenance and repair work.
6. The existing network of roadways grows increasingly insufficient in accommodating the number of vehicles. So further requirement of roads is there causing increase in the demand for work zones.

A work zone also known as temporary traffic control zone is an area of a highway where construction, maintenance, or utility work activities that can be identified by warning signs/signals/indicators, including those on transport devices (e.g., signs, flashing lights, channelizing devices, barriers, pavement markings, flagmen, warning signs and arrow boards mounted on the vehicles in a mobile maintenance activity). Work zones also include roadway sections where there is on-going, moving (mobile) work activity such as lane line painting or roadside mowing only if the beginning of the on-going, moving (mobile) work activity is designated by warning signs or signals.

The various problems which presaged the increased number of work zones in today's date also created a myriad of complicated challenges for the various traffic agencies involved in work zone management and estimation. The creation of a work zone changes the road user conditions which then lead to increased number of accidents. The biggest concern in the creation and management of work zones is to ensure the safety of motorists plying the roads. Work zones require reduction in speed which when not complied with causes road accidents. Sometimes enforcement agencies might also be required for traffic calming effect. Safe work zones are a result of careful and meticulous traffic management.

The placement, duration and maintenance of work zones are a multi-faceted problem. The highway is not closed thus work zones can be defines as spatial and temporal restrictions on the roadway which negatively impacts traffic flow conditions. The major concern of most traffic management and planning agencies is the placement of long term. Many factors are at play here and hence it requires detailed analysis of various factors some of which are enlisted as under:

1. The traffic volumes of the area must be analysed carefully to discern whether a traffic diversion is required or not. It also helps in estimating the length of queue, possible delay in traffic and maximum allowable speed limit.
2. The proper analysis of peak traffic hours is also required to correctly schedule the operational hours of the work zone to provide safety to the workers working there. It also maintains stable and efficient traffic conditions.
3. The time of formation of a work zone should not clash with any major festival of the area or public holiday to prevent the occurrence of accidents on the work zone. It is for the protection of both the workers and the motorist public in general.
4. The work zone layout should be such that it causes least possible hassle to motorists.
5. The behaviour of the drivers plying the road is also a key factor in estimation of work zones. For instance if the drivers are well versed in traffic rules it would do to have a lower allowable speed limit whereas if the drivers do not care about traffic rules it would cause long queues, general traffic mismanagement.

Safety is another area of concern in management and planning of work zones. Deaths and injuries at work zones are caused by a variety factors

1. Speeding traffic
2. Inadequate visibility of signs
3. Poor road surface condition
4. Inadequate traffic control
5. Improper management of material
6. Equipment and personnel in work zones
7. Not paying attention to work zone signs or flaggers indicating slow down
8. Distraction by cellular phone calls, conversations and activities at roadside

The normally framed criterion for determination of effectiveness of work zone establishment and management is the span of the concatenation on the upstream side and the delay experienced by motorists plying the road. Conventionally, work zone engineers rely on their judgement and past experiences in designing and planning work zones. But this is not an exhaustive solution to the problem at hand. In this chapter we are trying to analyse the data gathered from work zones to create a reliable model to map traffic flows and work zone characteristics so as to substantially reduce queue lengths and delay time. For such a model to exist and work efficiently and usefully in practise it should possess certain characteristics. Some of which are enlisted below:

1. The traffic flow must follow the simple and original principle. Complex traffic flows are inflexible for practical purposes. More so a complex model would have a smaller data set causing more error in operation. A simpler model is easier to calibrate.
2. Major factors which affect the traffic flow have to be considered. Say, work zone capacity should be determined from an input of work zone characteristics.
3. The model should be tractable so that it can be easily accessed and drawn-out for the different work zone traffic control plots.

In this, we look for an intelligent decision support system to the benefits of work zone engineers makes traffic management plans efficiently. Earlier empirical studies were done to make estimations regarding the work zones but it was hardly ideal as the scope tends to be very limited and it also at times loses objectivity while analysing multiple, simultaneously occurring complex situations. There has been a long history of different models being used at different times to predict work zone capacities. The most popular ones which were used earlier to calculate delay times and queue lengths were based on two traffic approaches viz. shockwave analysis and queuing analysis. Shockwave analysis in layman terms is the study of shockwaves originating on the boundaries separating regions with different flow conditions to identify congested and uncongested zones. It is of deterministic nature. On the other hand, queuing analysis can be both deterministic as well as probabilistic; can be both macroscopic and microscopic. The stochastic, microscopic queuing model is considered to be the most accurate but it has a shortcoming. The models accuracy totally dependents on the behaviour of human-vehicle-environment.

Chien & Schonfeld (2001) used a four-lane divided highway zone for developing an enhanced model for estimating the optimal length for that work zone. The objective functions are

1. Total cost, also contains the user cost;
2. Mishap cost; and
3. Office maintenance cost.

The model assumes that work zone capacity is consistent and autonomous of the work zone components. The potency and fitness of four traffic planning software tools for the evaluation of traffic management plans had been assessed by Islam & Seneviratne (1993). Sadegh et al.(1988) had presented a simulation for work zones on arterial road networks which is usually present within city limits. The computer program QUEWZ (Queue and User Cost Evaluation of Work Zones) evaluates the sequence lengths and additional user costs for work zones on freeways (Memmott & Dudek 1984; Krammes et al., 1987; Kermode et al., 1970). The conservation of flow principle had been used by this model to reckon the queue distance and user costs for various lane closure compositions and work agenda. The capacity of the work zone is computed from observational speed-flow-density relationships and autonomous of the work zone components, such as work zone layout and work intensity. The accuracy of all the above stated models is totally dependent on the accuracy of traffic capacity and demand values. The accuracy of all the above stated models is totally dependent on the accuracy of traffic capacity and demand values. The annual average daily traffic (AADT), is inadequate for this purpose. We can obtain demand estimates for shorter durations from AADT considering we know the daily, weekly, seasonal factors. These give us a very real estimate of the behaviour and usage patterns of public in unrestricted conditions and help us in calculating the demand after imposition of work zones. Some other factors which cause a reduction in demand after placement of work zones are enlisted as under:

1. Number of motorists choosing alternate routes
2. Number of motorists changing schedule
3. Number of motorists cancelling their trip due to presence of work zone
4. Number of motorists changing mode of transportation

Recently, artificial neural networks (ANN) have won popularity in designing intricate pattern recognition and estimation problems which are not suitable for typical mathematics. This popularity was mainly due to its not requiring any equation formation and acting as universal function estimators. In absence of some data or presence of faulty data also it can continue running due to its parallel nature. Reprogramming can also be done easily without changing the original program structure. But the use of ANN has its disadvantages too. Some of those disadvantages are as under:

1. Requirement of large data set for training which requires supervision increasing runtime and requires greater processing power.
2. Overtly deviating data can cause the model to over train causing it to make wrong predictions.

To overcome these setbacks, Radial Basis Function Neural Networks (RBFNN) is used. It is basically a three layer network in which the second or hidden layer performs a non-linear mapping from the input space into a (usually) higher dimensional space in which the patterns become linearly separable. The final layer performs a simple weighted mean to get a linear output. In this model the activation functions are radial basis functions. Its advantage lies in faster training of the model. It has been used to model work zones (Karim & Adeli, 2003; Asim & Hojjat, 2003; Chien & Tang, 2014). But it has certain disadvantages too. Even though it is faster to train it is comparatively slower to process outputs than multi-layer perceptrons.

In this article, we are using three different approaches to make a suitable model for determination of work zone capacity. We are comparing outputs using three algorithms namely Extreme Learning Machine (ELM), Minimax Probability Machine Regression (MPMR) and Gaussian Process Regression (GPR) for determination of Work Zone Capacity. The database has been taken from the work of Karim & Adeli (2003). ELM is developed by Huang et al. (2006). It is the modified version of single hidden layer feed foreword network. MPMR is developed base on minimax probability machine classification (Lanckriet et al., 2002). GPR is a probabilistic model (Williams, 1998; Quiñonero & Rasmussen, 2005; Seeger, 2004; Gramacy & Lee, 2008; Ko & Fox, 2009; Nguyen et al., 2009; Girolami & Rogers, 2006; Zhou et al., 2008). A comparative study has been presented between the developed ELM, MPMR and GPR models.

GAUSSIAN PROCESS REGRESSION

A Gaussian process is a stochastic process X_t, $t \in T$, for which any finite linear combination of samples has a Gaussian distribution. A linear functionalapplied to the sample function X_t will give a normally distributed result. It can be written as X ~ GP(m,K), which means the random function X is scattered as a GP with mean function m and covariance function K. It can sometimes be assumed that the random variables X_t have mean zero; this greatly simplifies calculations without loss of generality and allows the mean square properties of the process to be entirely determined by the covariance function K.

A GP can be formally defined as a set of random variables and finite number of which can form a Gaussian distribution. It can also be considered as a distribution over functions. Each random variable can be concluded to be a stochastic function f(x) at a certain point x in Gaussian space. A GP can be specified in terms of its mean and covariance factor. We define mean function m(x) and co variance function k (x, x') as given

$$m(x) = E\left[f(x)\right] \tag{1}$$

$$k(x,x') = E\left\{\left[f(x) - m(x)\right]\left[f(x') - m(x')\right]\right\} \tag{2}$$

Therefore GP can be written as

$$f(x) \sim GP\left[m(x), k(x,x')\right] \tag{3}$$

One advantage of GP regression model is that the computation of larger data sets does not interfere with the distribution of smaller sets due to its marginalization property (Stegle et al., 2008; Vasudevan et al., 2009; Zhou et al., 2010; Marquand et al., 2010; Pillonetto et al., 2011) . Let $f(x) = \Phi(x)^T w$ be a linear regression model with prior $w \sim N(0,\Sigma_p)$. Therefore mean and co variance are given by

$$E\left[f(x)\right] = \Phi(x)^T E\left[w\right] = 0, \tag{4}$$

$$E\left[f(x)f(x')\right] = \Phi(x)^T E\left[ww^T\right]\Phi(x') = \Phi(x)^T E_p\Phi(x') \tag{5}$$

Thus f(x) and f(x') are jointly Gaussian. Thus, $f(x_1)$, f(x2),...., f(xn) are jointly Gaussian for n input parameters. Thus for every positive definite co variance function can be expressed in the form of a linear combination of Gaussian shaped basis functions. From the test data we consider any random no. of points X* in a matrix form and generate a random GP f*.

$$f^{*} \sim N\left[0, K\left(X^{*}, X^{*}\right)\right] \tag{6}$$

According to this prior in case of training outputs f, and test outputs f* we have

$$\begin{bmatrix} f \\ f^* \end{bmatrix} \sim N\left[0, \begin{pmatrix} K(X,X) K(X,X^*) \\ K(X^*,X) K(X^*,X^*) \end{pmatrix}\right] \tag{7}$$

If there are n training data and n* testing data then K(X, X*) gives the n x n* matrix of all covariances figured out for all pairs of testing and training data, and in the same way for the other entries K(X, X), K(X∗, X∗) and K(X∗, X). To obtain the posterior dispersion over functions we need to curb this joint prior distribution to contain only those functions which agree with the observed data points.

It is conventional for more sensible modelling situations that we do not have access to function values themselves, but only noisy versions thereof y = f(x) + ε.[8] Assuming additive independent identically distributed Gaussian noise ε with variance σ_n^2, the prior on the noisy observations becomes

$$Cov(y_p, y_q) = k(x_p, x_q) + \sigma_n^2 \delta_{pq}$$

or

$$Cov(y) = k(\mathrm{X}, \mathrm{X}) + \sigma_n^2 1$$

where δ_{pq} is a Kronecker delta which is one if p = q and zero otherwise. It follows from the independence assumption about the noise, that a diagonal matrix is added, in comparison to the noise free case.

If there are n training data and n* testing data then K(X, X*) provides the n x n* matrix of all covariances figured out for all duo of testing and training data, and relatively for the other entries.

In prediction with noisy data we have y = f(x) + ε. Assuming the noise to be identically distributed Gaussian noise, $\mathrm{Cov(y)} = \mathrm{k(X,X)} + \tilde{A}_n^2 \mathrm{I}$

Therefore,

$$\begin{bmatrix} y \\ f^* \end{bmatrix} \sim N\left[0, \begin{pmatrix} K(X,X) + \sigma_n^2 & K(X,X^*) \\ K(X^*,X) & K(X^*,X^*) \end{pmatrix}\right] \tag{8}$$

We have the conditioning statement as

$$f^* \mid \mathrm{X}^*, X, f \sim N\left[K(X^*,X) K(X,X)^{-1} f, K(X^*,X^*) - K(X^*,X) K(X,X)^{-1} K(X,X^*)\right] \tag{9}$$

Therefore the predictive distribution for Gaussian process regression with noisy data is given as,

$$f^* \mid \mathrm{X}, \mathrm{y}, \mathrm{X}^* \sim N\left(\overline{f^*}, \mathrm{cov}(f^*)\right) \tag{10}$$

$$\overline{f^*} \cong E\left[f^* \mid \mathrm{X}, \mathrm{y}, \mathrm{X}^*\right] = K\left(X^*, X\right)\left[K\left(X, X\right) + \sigma_n^2 \,\mathrm{I}\right]^{-1} y, \tag{11}$$

$$Cov\left(f^*\right) = K\left(X^*, X^*\right) - K\left(X^*, X\right)\left[K\left(X, X\right) + \sigma_n^2 \,\mathrm{I}\right]^{-1} K\left(X, X^*\right) \tag{12}$$

Graphically, a GPR can be represented as series of nodes closely related by GPs which processes the given input variables in a Gaussian function space.

Foretelling with Gaussian processes is certainly not a new case, notably for time series investigations; the fundamental concept goes back at least as far as the time series work of Wiener (1949) and Kolmogorov (1941) in the 1940's. Certainly, Lauritzen (1981) talk over the relevant work done by the Danish astronomer T. N. Thiele dating from 1880.

The advantages of Gaussian Processes for Machine Learning are:

1. The prediction interpolates the observation (generally for regular correlation models).
2. The prediction is probabilistic (Gaussian) so that one can compute empirical confidence intervals and exceedance probabilities that might be used to refit (online fitting, adaptive fitting) the prediction in some region of interest.
3. Versatile: different linear regression models and correlation models can be specified. Common models are provided, but it's also possible to specify custom models provided they are stationary.

The disadvantages of Gaussian Processes for Machine Learning include:

1. It is not sparse. It uses the whole samples/features information to perform the prediction.
2. It loses efficiency in high dimensional spaces – namely when the number of features exceeds a few dozen. It might give poor performance and loses computational efficiency.
3. Classification is only a post-processing method, meaning that one first need to solve a regression problem by providing the complete scalar float precision output y of the experiment one attempt to model.

EXTREME LEARNING MACHINE

Extreme Learning Machine (Huang et al., 2006) (ELM) is a currently popular neural network architecture based on random projections. It is a learning algorithm for the training of Single Layer Feed forward Neural Networks (SLFN) and has one hidden layer with random weights, and output layers whose weights are determined analytically. Both training and prediction are fast compared with many other nonlinear methods. It was developed as a method for faster training of the model and to remove overfitting and concentration of local minima problems. In this work we point out that ELM, although introduced as a fast method for training a neural network, is in some sense closer to a kernel method in its operation. A fully trained neural network has learned a mapping such that the weights contain information about the training data. ELM uses a fixed mapping from data to feature space. This is similar to a kernel method, except that instead of some theoretically derived kernel, the mapping ELM uses is random. Therefore, the individual weights of the ELM hidden layer have little meaning, essential information about the weights is captured by their variance.

Extreme Learning Machine proposed by Huang et al. (2006) is a learning algorithm for the training of Single Layer Feedforward Neural Networks (SLFN). It randomly chooses the input weights and analytically determines the output weights of SLFN. It was developed as a method for faster training of the model and to remove overfittingand concentration of local minima problems. The ELM used for single hidden layerfeedforward neural network training can adaptively set the hidden layer node number andrandomly assign for the input weights and hidden layer biases, the output layer weights can then be obtained by the least square method, the whole learning process can be done in one mathematical change without iteration (Ding et al., 2013).

ELM network is fast for training datasets, provided the number of hidden units is small. Speed of training as the whole, slows down with increasing the number of runs. Model selection may require considerable number of repetitions. First factor to consider is the inherent randomness of ELM results. If average performance of ELM is to be considered, any runs must be repeated several times, which adds to the computational burden. Complexity of the model selection is determined by the number of tuning parameters, since all the sensible combinations need to be considered.

First parameter is the number of hidden units. The only theoretical upper limit for the number of hidden units to try is N (which is enough for 0 training error). At this limit, computing pseudo inverse corresponds to ordinary inversion of an $N \times N$ matrix, with a complexity of $O(N^3)$. In practice, smaller upper limits are usually used. Generally, somewhat arbitrary fixed values have been used for weight variance, and model selection has only considered the number of hidden units. Test runs show that also the weight variance has a noticeable effect on results, and should thus be considered a tuning parameter. Generally, small training error and good generalization may be contradictory goals.

ELM is proved (Huang et al., 2006; Wang et al., 2011) to be able to reproduce the training data exactly if the number of hidden units equals or exceeds the number of data points. This behaviour, although important in proving computational power of ELM, is generally not desirable in modeling. A model should generalize, rather than exactly memorizing the training data. This view is indirectly acknowledged in practical ELM work, where the number of hidden units is much smaller than N. This may prevent ELM network from overfitting to the training data, a factor usually not discussed in ELM literature.

Generalization ability of ELM is attributed to the fact that computing output layer weights by pseudo inverse achieves a minimum norm solution. Generalization ability of a neural network is in (Bartlett, 1998) shown to relate to small norm of weights. However, Bartlett's work considers the neural network as whole, not only the output layer. Although ELM minimizes the norm of output layer weights, norm of the hidden layer weights depends on the variance parameter, and does not change in ELM training. In the hidden layer, generalization ability is related to the operating point of hidden unit activations. A model with small hidden layer weights is nearly linear, and generalizes well. A highly non-linear model, produced by large weights, is more prone to overfitting. Therefore, conclusions about generalization ability of ELM should not be based on the output weights only.

Extreme Learning Machine proposed by Huang et al. (2006) is a learning algorithm for the training of Single Layer Feedforward Neural Networks (SLFN). It randomly chooses the input weights and analytically determines the output weights of SLFN. It was developed as a method for faster training of the model and to remove overfitting and concentration of local minima problems. The ELM used for single hidden layer feed forward neural network training can adaptively set the hidden layer node number and randomly assign for the input weights and hidden layer biases, the output layer weights can then be obtained by the least square method, the whole learning process can be done in one mathematical change without iteration (Ding et al., 2013; Huang & Chen, 2006; Huang et al., 2006; Zhu et al., 2005; Feng et al., 2009; Sun et al., 2007).

Let us consider the following datasets (D).

$$D=\{x_i, y_i\}_{i=1}^{N} \tag{13}$$

where x is the input, y is the output and N is the number of datasets.

In a single hidden layer feed forward networks (SLFN), the relation between input and output is as under:

$$\sum_{i=1}^{\bar{N}} \beta_i g\left(w_i . x_j + b_i\right) = y_{ji}, j = 1,......,\mathrm{N} \tag{14}$$

where w_i is the weight vector between the ith neuron in the hidden layer and the input layer, b_i gives the bias of the ith neuron in the hidden layer, x_i is the jth input data vector, g(.) is an activation function of the hidden neuron, β_i is the weight vector between the ith hidden neuron and the output layer, $\bar{N}$ is number of hidden nodes, N is number of training samples and y_j means the target vector of the jth input data.

Hence the above equation can be written in the following way.

$$\mathrm{H}^2 = \mathrm{T} \tag{15}$$

where

$$H = \begin{bmatrix} g\left(w_1.x_1 + b\right).......g\left(w_{\bar{N}}.x_1 + b_{\bar{N}}\right) \\ g\left(w_1.x_1 + b\right).......g\left(w_{\bar{N}}.x_N + b_{\bar{N}}\right) \end{bmatrix}_{N\times\bar{N}},$$

$$\beta = \begin{bmatrix} \beta_1^T \\ . \\ . \\ \beta_{\bar{N}}^T \end{bmatrix}_{\bar{N}\times m}$$

And

$$T = \begin{bmatrix} y_1^T \\ . \\ . \\ y_{\bar{N}}^T \end{bmatrix}_{N\times m}$$

where H is the hidden layer output matrix of the network (Huang & Babri, 1998; Huang, 2003; Huang et al., 2012; Miche et al., 2010; Wang et al., 2005; Huang et al., 2008).

In this learning algorithm (i.e. ELM) the values of w_i and b_i are not tuned during training. Random values are assigned for w_i and b_i according to the theorem given Huang et al. (2003) that as long as the number of hidden nodes is large enough, when the activation function f (x) is infinitely differentiable at any interval the parameters of the network do not all need to adjust. The value of β is determined from the following equation.

$$\beta = H^{-1}T \tag{16}$$

where H^{-1} is the Moore-Penrose generalized inverse (Serre, 2002) of the hidden layer output matrix H.

Since the input weights and the hidden layer biases are in randomly assigned according the given input samples, training of the single layer feed forward neural network is converted into a linear problem which can easily be calculated by least square regression analysis.

Some of the advantages of using ELM are mentioned as under:

1. From the obtained value of β, it is evident this algorithm has minimal training error
2. It has much better generalization properties than previous algorithms
3. Since the value of β is unique it can avoid producing the local optimal solution.
4. It has much faster training speeds
5. It is suitable for many non linear functions as well as kernel functions.

MINIMAX PROBABILITY MACHINE REGRESSION

The basic framework of this regression model a regression model can be expressed as maximizing the minimum probability of future predictions being within some bound of the true regression function (Strohmann & Grudic, 2002, Sun et al., 2009; Liu et al., 2006). For this model to be functional in real life situations it should make minimal assumptions about the functions and processes underlying the true regression. In this model developed by us MPMR uses kernel function for prediction of output (y). In MPMR, the relation between input(x) and y is given below.

$$y = \sum_{i=1}^{N} \beta_i K\left(x_{i,} x\right) + b \tag{17}$$

where N is the number of datasets, $K(x_i,x)$ is kernel function, β_i and b are outputs of MPMR.

One data set is obtained by shifting all of the regression data +ε along the output variable axis. The other dataset is obtained by shifting all of the regression data -ε along the output variable axis. Regression model is the classification boundary between these two classes. More details of MPMR are given by Strohmann & Grudic (2002).The datasets are scaled between 0 and 1. Radial basis function

($\exp\left\{-\frac{(x_i - x)(x_i - x)^T}{2\sigma^2}\right\}$, where σ is the width of radial basis function) has been adopted as kernel function for the MPMR model.

METHODOLOGY

In this article the estimation of capacity of work zones is done by the three computational models explained above. The primary factors which impact work zones estimation has been considered as the input parameters for the models. Those factors are as under:

1. Number of lanes (x_1)
2. Number of open lanes (can vary from one to the maximum number of lanes existing in each direction)
3. Work zone layout (it is used to identify between the three most commonly practised work zone layouts which are lane shifting, lane merging and crossover. For computational modelling, these 3 layouts are determined by numbers 0.1, 0.5, and 0.9, respectively)
4. Work zone length
5. Lane width
6. Percentage of trucks (includes trucks, buses and all other kinds of heavy vehicles)
7. Grade
8. Work zone velocity
9. Work intensity (it is not a parameter which can be quantified easily. Hence for computational ease, in this article it has been widely branched as low, medium or high rely upon labour, equipments used etc. For computational modelling, these 3 work intensity groups are determined by 0.1, 0.5, and 0.9, respectively.)
10. Darkness factor
11. Proximity of ramps or interchanges (The proximity of ramps bring about disturbances in traffic flow, minimizing the number of vehicles moving through the work zone. This criterion is modelled as a binary two-valued parameter with values of 0 (no ramps) or 1 (ramps exist in proximity to the work zone)

To develop ELM, MPMR and ELM, dataset has been divided into two groups namely training data set and testing data set.

Training Dataset: This is required to develop the models. This article uses 70% of dataset as training dataset.

Testing Dataset: This is required to verify the developed model. The remaining 30% datasets have been used as testing dataset.

The dataset is normalized between 0 and 1. Radial basis function has been used as covariance function for the GPR. The performance of ELM, MPMR and GPR has been assessed in terms of Coefficient of Correlation(R). For a good model, the value of R should be close to one. MATLAB has been used to develop the ELM, MPMR and GPR models.

RESULTS AND DISCUSSIONS

For developing the MPMR model, the design value of ε and width (σ) of radial basis function have been determined by trial and error approach. The design values of ε and σ are 0.001 and 0.4 respectively. The performance of training and testing datasets has been depicted in Figure 1.

As shown in figure 1, the value of R is close to one for training as well as testing datasets. So, the developed MPMR predicts work zone capacity reasonably well.

To frame GPR, the design values of Gaussian noise and σ have been determined by trial and error approach. The design values hence obtained of Gaussian noise and σ are 0.003 and 0.5 respectively. The performance of training and testing datasets has been conveyed in Figure 2.

Figure 2 publicized the value of R which is near to one for training and testing datasets. So, the developed GPR shows its capability for the prediction of work zone capacity.

In order to develop ELM, radial basis function has been adopted as an activation function. The developed ELM gives best performance for 12 hidden nodes. Therefore, the number of hidden nodes is set to 12. Figure3 shows the performance of training and testing dataset respectively.

As revealed in Figure 3, the value of R is close to one for training as well as testing datasets. Therefore, the developed ELM proves its ability for prediction of work zone capacity.

A comparative study has been carried out between the developed MPMR, ELM and GPR models. The developed ELM, MPMR and GPR gave almost same performance. ELM can avoid problem of overfitting, local minima, and improper learning rate. It takes less training time compare to the other model. MPMR has control over future prediction. However, GPR and ELM have no control over future prediction. GPR assumes that the dataset should follow Gaussian distribution. ELM and MPMR do not assume any data distribution.

Figure 1. Performance of MPMR

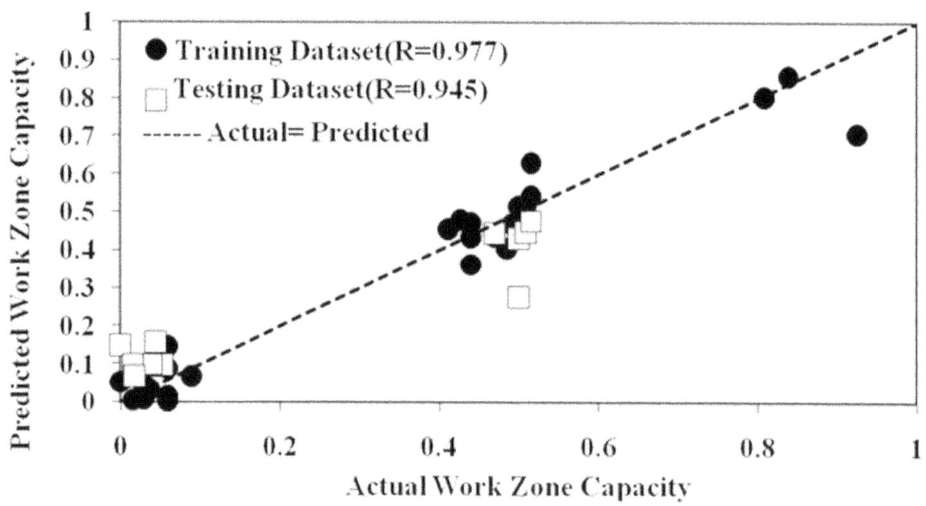

Figure 2. Performance of GPR

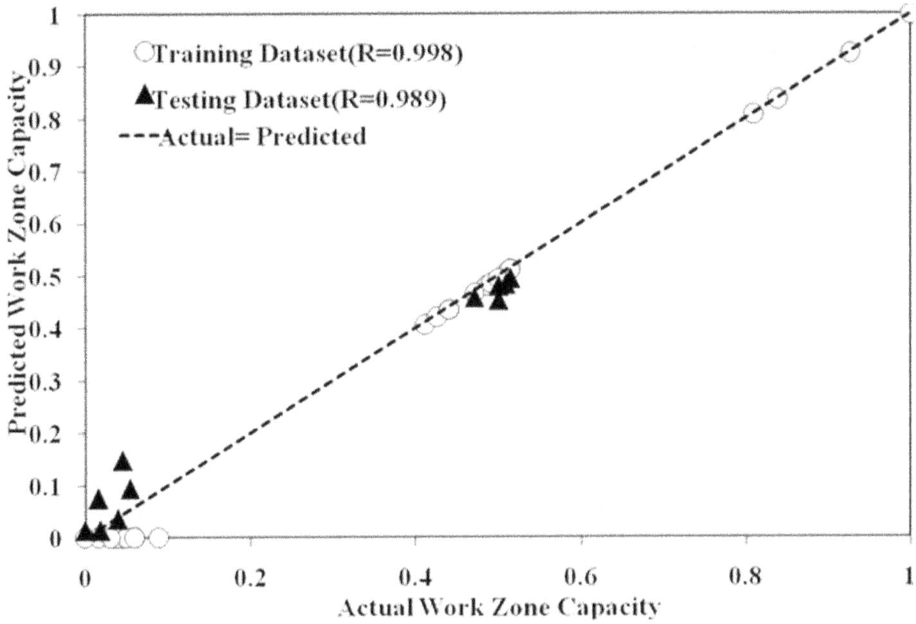

Figure 3. Performance of ELM

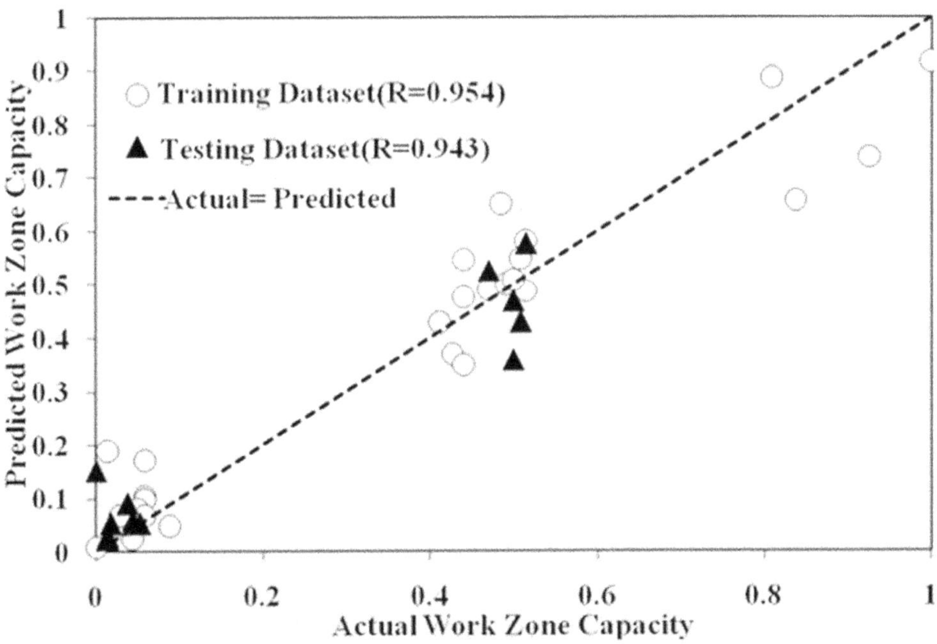

CONCLUSION

This article has described ELM, MPMR and GPR for prediction of work zone capacity. The described methodology was executed and the performance of ELM, MPMR and GPR obtained were excellent and encouraging. The developed models can be used for efficient transportation planning. Users can use the developed models as quick tools for prediction of work zone capacity. Moreover, ELM, MPMR and GPR methods can be used as a solvent for distinguished transportation problems.

REFERENCES

Bartlett, P. L. (1998). The sample complexity of pattern classification with neural networks: The size of the weights is more important than the size of the network. *IEEE Transactions on Information Theory*, *44*(2), 525–536. doi:10.1109/18.661502

Chien, S., & Schonfeld, P. (2001). Optimal work zone lengths for four-lane highways. *Journal of Transportation Engineering*, *127*(2), 124–131. doi:10.1061/(ASCE)0733-947X(2001)127:2(124)

Chien, S. I., & Tang, Y. (2014). Scheduling highway work zones with genetic algorithm considering the impact of traffic diversion. *Journal of Advanced Transportation*, *48*(4), 287–303. doi:10.1002/atr.213

Ding, S., Zhao, H., Zhang, Y., Xu, X., & Nie, R. (2013). *Extreme learning machine: algorithm, theory and applications*. Springer Science Business Media Dordrecht.

Feng, G., Huang, G. B., Lin, Q., & Gay, R. (2009). Error minimized extreme learning machine with growth of hidden nodes and incremental learning. [PubMed]. *IEEE Transactions on Neural Networks*, *20*(8), 1352–1357. doi:10.1109/TNN.2009.2024147

Girolami, M., & Rogers, S. (2006). Variational Bayesian multinomial probit regression with gaussian process priors. *Neural Computation*, *18*(8), 1790–1817. doi:10.1162/neco.2006.18.8.1790

Gramacy, R. B., & Lee, H. K. H. (2008). Bayesian treed Gaussian process models with an application to computer modelling. *Journal of the American Statistical Association*, *103*(483), 1119–1130. doi:10.1198/016214508000000689

Huang, G. B. (2003). Learning capability and storage capacity of two hidden- layer Feedforward networks. [PubMed]. *IEEE Transactions on Neural Networks*, *14*(2), 274–281. doi:10.1109/TNN.2003.809401

Huang, G. B., & Babri, H. A. (1998). Upper bounds on the number of hidden neurons in feedforward networks with arbitrary bounded nonlinear activation functions. [PubMed]. *IEEE Transactions on Neural Networks*, *9*(1), 224–229. doi:10.1109/72.655045

Huang, G. B., & Chen, L. (2006). Convex incremental extreme learning machine. *Neurocomputing*, *70*(16-18), 3056–3062. doi:10.1016/j.neucom.2007.02.009

Huang, G. B., Li, M. B., Chen, L., & Siew, C. K. (2008). Incremental extreme learning machine with fully complex hidden nodes. *Neurocomputing*, *71*(4-6), 576–583. doi:10.1016/j.neucom.2007.07.025

Huang, G. B., Zhou, H., Ding, X., & Zhang, R. (2012). Extreme learning machine for regression and multiclass classification. [PubMed]. *IEEE Transactions on Systems, Man, and Cybernetics. Part B, Cybernetics*, *42*(2), 513–529. doi:10.1109/TSMCB.2011.2168604

Huang, G. B., Zhu, Q. Y., & Siew, C. K. (2006). Extreme learning machine: a new learning scheme of feed forward neural networks. *Proceedings of IEEE International Joint Conference on Neural Networks*, *2*, 985–990.

Huang, G. B., Zhu, Q. Y., & Siew, C. K. (2006). Extreme learning machine: Theory and applications. *Neurocomputing*, *70*(1-3), 489–501. doi:10.1016/j.neucom.2005.12.126

Islam, M. N., & Seneviratne, P. N. (1993). Work-zone traffic management with transportation planning software. *Canadian Journal of Civil Engineering*, *20*(3), 471–479. doi:10.1139/l93-061

Karim, A., & Adeli, H. (2003). Radial Basis Function Neural Network for Work Zone Capacity and Queue Estimation. *Journal of Transportation Engineering*, *129*(5), 494–503. doi:10.1061/(ASCE)0733-947X(2003)129:5(494)

Karim, A., & Adeli, H. (2003). Radial basis function neural network for work zone capacity and queue estimation. *Journal of Transportation Engineering*, *129*(5), 494–503. doi:10.1061/(ASCE)0733-947X(2003)129:5(494)

Kermode, R. H., & Myyra, W. A. (1970). Freeway Lane Closure. *Traffic Engineering, 40*(5).

Ko, J., & Fox, D. (2009). GP-BayesFilters: Bayesian filtering using Gaussian process prediction and observation models. *Autonomous Robots*, *27*(1), 75–90. doi:10.1007/s10514-009-9119-x

Krammes, R. A., Dudek, C. L., & Memmott, J. L. (1987). Computer model for evaluating and scheduling freeway work-zone lane closures. *Transportation Research Record, 1148*, 18–24.

Lanckriet, G. R. G., Ghaoui, L. E., Bhattacharyya, C., & Jordan, M. I. (2002). A robust minimax approach to classification. *Journal of Machine Learning Research*, *3*, 555–582.

Lauritzen, S. L. (1981). Time series analysis in 1880: A discussion of contributions. *T. N. Thiele. International Statistical Review*, *49*(3), 319–331. doi:10.2307/1402616

Marquand, A., Howard, M., Brammer, M., Chu, C., Coen, S., & Mourão Miranda, J. (2010). Quantitative prediction of subjective pain intensity from whole-brain fMRI data using Gaussian processes. [PubMed]. *NeuroImage*, *49*(3), 2178–2189. doi:10.1016/j.neuroimage.2009.10.072

Memmott, J. L., & Dudek, C. L. (1984). Queue and user cost evaluation of work zones (QUEWZ). *Transportation Research Record*, *979*, 12–19.

Miche, Y., Sorjamaa, A., Bas, P., Simula, O., Jutten, C., & Lendasse, A. (2010). OP-ELM: Optimally pruned extreme learning machine. [PubMed]. *IEEE Transactions on Neural Networks*, *21*(1), 158–162. doi:10.1109/TNN.2009.2036259

Nguyen-Tuong, D., Seeger, M., & Peters, J. (2009). Model learning with local Gaussian process regression. *Advanced Robotics*, *23*(15), 2015–2034. doi:10.1163/016918609X12529286896877

Pillonetto, G., Chiuso, A., & De Nicolao, G. (2011). Prediction error identification of linear systems: A nonparametric Gaussian regression approach. *Automatica*, *47*(2), 291–305. doi:10.1016/j.automatica.2010.11.004

Quiñonero Candela, J., & Rasmussen, C. E. (2005). A unifying view of sparse approximate Gaussian process regression. *Journal of Machine Learning Research*, *6*, 1939–1959.

Sadegh, A., Radwan, A. E., & Rouphail, N. M. (1988). ARTWORK: A simulation model of urban arterial work zones. *Transportation Research Record*, *1163*, 1–3.

Seeger, M. (2004). Gaussian processes for machine learning. [PubMed]. *International Journal of Neural Systems*, *14*(2), 69–106. doi:10.1142/S0129065704001899

Serre, D. (2002). *Matrices: Theory and Applications*. Springer-Verlag.

Stegle, O., Fallert, S. V., MacKay, D. J. C., & Brage, S. (2008). Gaussian process robust regression for noisy heart rate data. [PubMed]. *IEEE Transactions on Bio-Medical Engineering*, *55*(9), 2143–2151. doi:10.1109/TBME.2008.923118

Strohmann, T. R., & Grudic, G. Z. (2002). *A Formulation for minimax probability machine regression* (T. G. Dietterich, S. Becker, & Z. Ghahramani, Eds.). Advances in Neural Information Processing Systems Cambridge, MA: MIT Press.

Sun, J., Bai, Y., Luo, J., & Dang, J. (2009). Modelling of a chaotic time series using a modified minimax probability machine regression. *The Chinese Journal of Physiology*, *47*(4), 491–501.

Sun, Z. L., Au, K. F., & Choi, T. M. (2007). A neuro-fuzzy inference system through integration of fuzzy logic and extreme learning machines. [PubMed]. *IEEE Transactions on Systems, Man, and Cybernetics. Part B, Cybernetics*, *37*(5), 1321–1331. doi:10.1109/TSMCB.2007.901375

Vasudevan, S., Ramos, F., Nettleton, E., & Durrant Whyte, H. (2009). Gaussian process modeling of large-scalwe terrain. *Journal of Field Robotics*, *26*(10), 812–840. doi:10.1002/rob.20309

Wang, Y., Cao, F., & Yuan, Y. (2011). A study on effectiveness of extreme learning machine. *Neurocomputing*, *74*(16), 2483–2490. doi:10.1016/j.neucom.2010.11.030

Wang, Y. H., Li, Y., Yang, S. L., & Yang, L. (2005). Classification of substrates and inhibitors of P-glycoprotein using unsupervised machine learning approach. [PubMed]. *Journal of Chemical Information and Modeling*, *45*(3), 750–757. doi:10.1021/ci050041k

Williams, C. K. I. (1998). Prediction with Gaussian processes: From linear regression and beyond. In M. I. Jordan (Ed.), *Learning and inference in graphical models* (pp. 599–621). Kluwer Academic Press; doi:10.1007/978-94-011-5014-9_23

Zhou, P., Chen, X., Wu, Y., & Shang, Z. (2010). Gaussian process: An alternative approach for QSAM modeling of peptides. [PubMed]. *Amino Acids*, *38*(1), 199–212. doi:10.1007/s00726-008-0228-1

Zhou, P., Tian, F., Chen, X., & Shang, Z. (2008). Modeling and prediction of binding affinities between the human amphiphysin SH3 domain and its peptide ligands using genetic algorithm-Gaussian processes. [PubMed]. *Biopolymers*, *90*(6), 792–802. doi:10.1002/bip.21091

Zhu, Q. Y., Qin, A. K., Suganthan, P. N., & Huang, G. B. (2005). Evolutionary extreme learning machine. *Pattern Recognition*, *38*(10), 1759–1763. doi:10.1016/j.patcog.2005.03.028

Zunxiong, L., Xin, X., & Zhang, D. (2006). Predict chaotic time series using minimax probability machine regression. *Information Technology Journal*, *5*(3), 529–533. doi:10.3923/itj.2006.529.533

KEY TERMS AND DEFINITIONS

Artificial Neural Network: Artificial Neural Network is a structure framed with densely interconnected with artificial neurons that performs the parallel computations for data processing and knowledge representations.

Extreme Learning Machine: Extreme Learning Machine is based on empirical risk minimization theory which is used for single hidden layer feedforward neural networks the slow training speed and over-fitting problems.

Gaussian Process Regression: It is defined as the distribution over functions, and inference taking place directly in the space of functions.

Minimax Probability Machine Regression: It is the method used for maximizing the minimum probability of future predictions being within some bound of the true regression function.

Prediction: The action of forecasting something is called as prediction.

Transportation: The process of shipping or moving an item from one place to another place is called transportation.

Work Zone Capacity: The work zone capacity determination is done for scheduling the construction activities in a proper way, so that it can avoid unnecessary congestion of traffic.

Chapter 5
The Evolution of Logistics Hubs and a Conceptual Framework for Logistics Hubs Location Decisions

Serpil Erol
Gazi University, Turkey

Aykut Süreyya Duyguvar
Turkish Military Academy, Turkey

ABSTRACT

Hubs are special centers that used for switching, transferring and handling freight or passengers in transportation systems. The hub location problem deals with locating hub facilities and allocating demand nodes to hubs in order to route the traffic between origin–destination pairs. Aiming at minimizing the total costs, maximizing utilization of transporters, maximizing the service level, etc., the flow between O–D pairs is dispatched through some selected intermediate logistics centers (called hub nodes) and links connecting the hubs. In this work, a conceptual framework is proposed for hub location decision problem in transportation networks since the previous reviews are out of date. Several models, methods and decision tools are evaluated. Also future trends are added to represent a new point of view for global scale. Since the models and solution approaches are complicated, it is not practical to catch optimum results, and some relevant algorithms will be represented to solve hub models in this chapter.

INTRODUCTION

Today's business world needs a challenging logistics strategy. This strategy enables managing shipments across globally spread supply and demand nodes within performance expectations. To fulfill the needs of customers around the world, intermodal logistics networks give some opportunities (Arnold et al., 2004; Gooley, 1997). In this network, a parcel may be sent through one or more modes of transportation from the origin to destination node to smoothly by the intermodal containers (Crainic et al.,

DOI: 10.4018/978-1-4666-8648-9.ch005

2007; Slack, 1990; Slack, 2001). Existing transportation infrastructure would affect the design and the management of intermodal logistics network and location of transfer points can differ according to the cost structure. (Warsing et al., 2001). Selection of a hub location is one of the hardest efforts in design for the intermodal logistics networks.

For the recent years, a hub network has transformed for intermodal activities. The economy of scale is the primary motivation behind the growth the use of hub based intermodal networks (Slack, 1990). Since an intermodal logistics network is an augmentation of its separate single mode, it is normal that the hub network has developed as the most suitable logistics network structure for intermodal transportation (Bookbinder & Fox, 1998).

In a hub-and-spoke logistics network, the consolidation of shipments between the transfer points enables huge cost savings as expected. This consolidation brings economies-of-scale and density. Economies-of-scale are provided by the putting the less-than-truckload shipments into containerized shipments which reduce the unit transportation cost. The economies of density are provided by high load factors for the road/rail/air shipments through predetermined distances. In this manner, cost structure becomes non-linear since the unit transportation cost is a non-increasing function of the volume shipped. Intermodal rail and air freight have relatively different cost structure from the road transportation costs.

The logistics hub network design requires decision of the number and location of logistics hubs and the assignment of nodes (demand or origin) that are served by each hub. In the context of this chapter, a logistics hub is a transshipment center or facility that receives, consolidates or breaks down, and dispatches parcels. A logistics hub might have access to road, rail, or air terminals to handle the freights that may be operated by other service providers or third parties. The decision of mode (road, rail, or air) for moving freight between origin to destination via hubs is determined by the tradeoff between the service level expected by the customer and the transportation cost burdened. The intermodal flow of freights or parcels is managed through the services offered by the carriers that have capability on each mode (Ishfaq & Sox, 2010).

The performance of a logistics hub network can be based on different metrics such as cost, service level, time, reliability, safety, and flexibility (Beuthe & Bouffioux, 2008). In this research, conceptual framework has been developed through a comprehensive literature review and case studies for logistics hubs. More specifically, the models represented in this research focus on minimizing total network costs while satisfying maximum service levels.

BACKGROUND

One of the trending subjects in facility location researches is the hub location problem. There exist several application areas for the hub location problem in logistics network; therefore, this chapter is focused on representing this problem to researchers. This chapter is composed of three parts: provides literature review, mathematical models/algorithms and solution approaches which are optimizing logistics services such as, movement of people, commodities and information between origin and destination. Each origin-destination pair is connected via one or more hubs (see Figure 1). The freights carried from i to j may not be same with the freights carried from j to i.

Figure 1. A logistics hub network

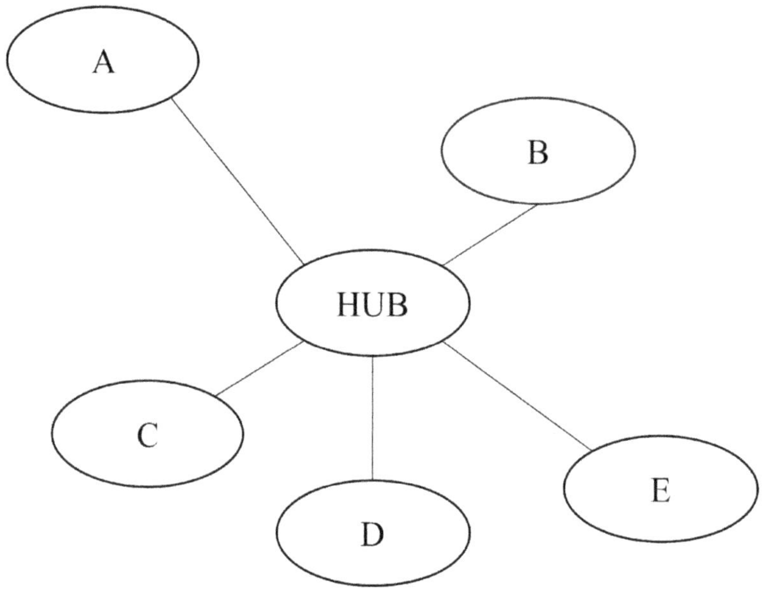

If there are N nodes (origin or destination points), there exist $N(N-1)$ origin-destination pairs of nodes in a logistics network which is called a mesh network. It should be underlined that i-j pair may differ from j-i. A sample mesh network with 5 nodes is represented in Figure 2. Assuming that we have freight handling and forwarding activities in this network, there exist 20 paths between the nodes.

Figure 2. A mesh logistics network

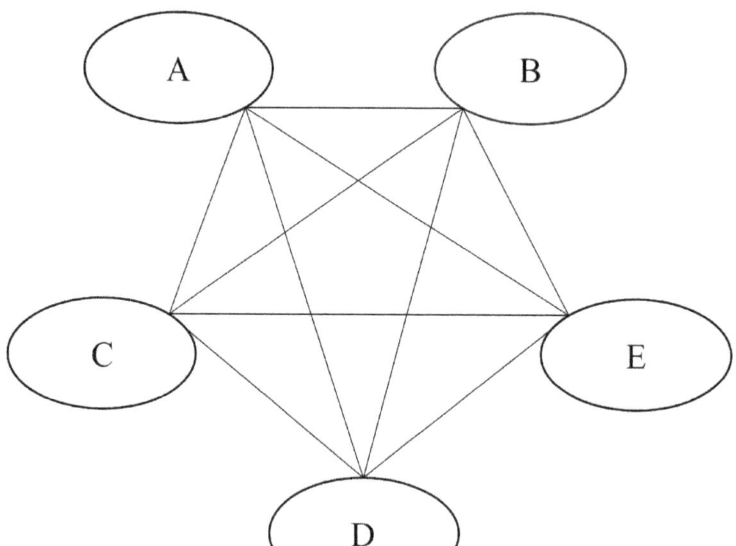

If we determine one of the nodes as a hub and then connect all of the other nodes to the hub, there will be *2(N-1)* paths between all origin-destination node pairs. This network is named *hub-and-spoke network* and also represented in Figure 3.

In this network, there exist 8 paths between the nodes. Thus, the resources can be allocated more effectively and efficiently, and more nodes can be served with less resource via a hub network than mesh network.

The main loss of hub network is that more than one path is mandated to send freight between each nodes via one or more hubs. Non-hub nodes are allocated to a hub to be served.

In multi hub networks, the each non-hub node is connected at least one hub and hub nodes are fully connected to one another such as Figure 4. In Figure 4, there are 10 cities and two hubs. The amount of freight sent from one hub to another is greater than the freight moved from each non-hub node to that hub.

It is not effective way to connect every pair of nodes to each other. So, hubs are set to deal with this problem. Hubs bring economic advantage by establishing more qualitative paths between the hubs (Camargo et al., 2008). In logistics networks, it is economic to establish a highway roads or high speed railways with several lines between hub nodes, and get effective movement of freights, and, thus, savings in time and cost (Hekmatfar & Pishvaee, 2009).

The hub location problem is a fresh subject and has been discussed in 1980s by number of researchers, since Hakimi (1964) first published a paper on hub location. In 80s, the first paper on this subject was published by Toh & Higgins (1985).

Many application areas of hub problem are given below:

- Hakimi (1964) and Berman et al. (2007) studied hub locations for emergency services.
- Berman et al. (2007) represented the hub locations for helicopters used for search and rescue.

Figure 3. Transformation of a mesh logistics network to a hub network

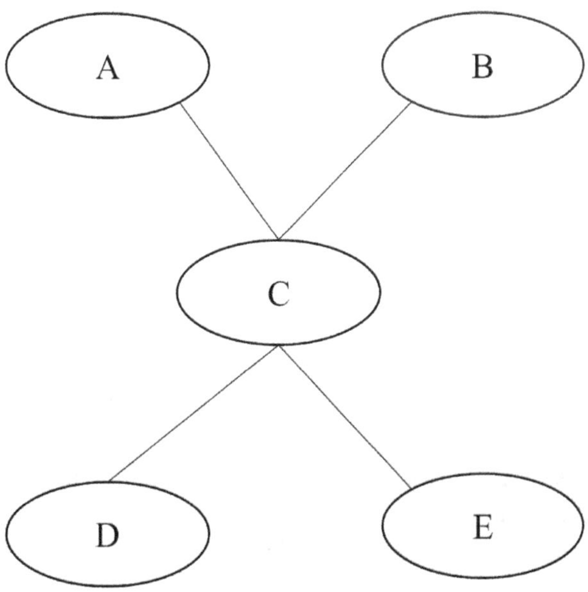

Figure 4. A multi hub logistics network

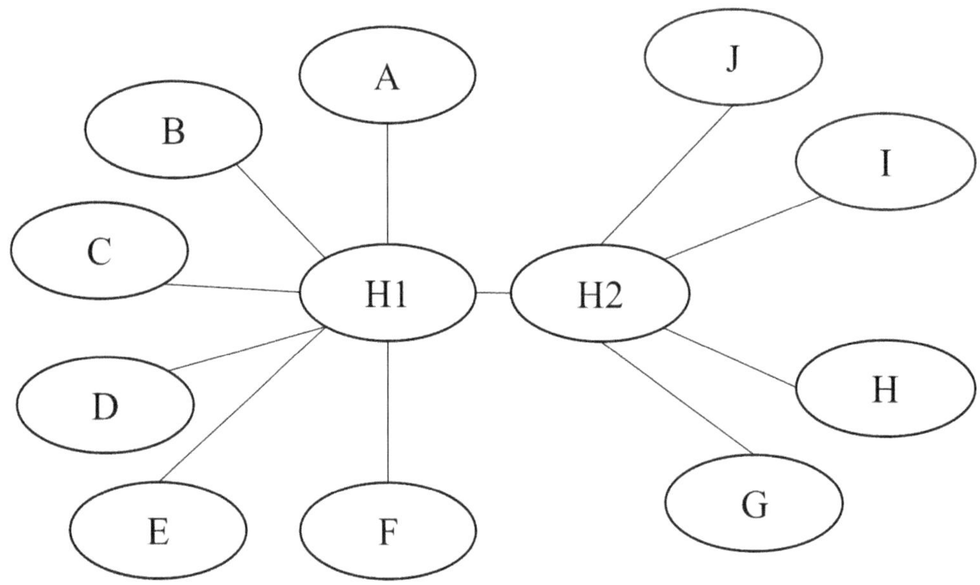

- Toh et al. (1985), Shaw (1993), Aykin (1995), Jaillet et al. (1996), Bania et al. (1998), Sasaki et al. (1999), Martin & Roman (2003) and Adler & Hashai (2005) highlighted airways hub location problems.
- Don et al. (1995), Lumsdenk et al. (1999), Aversa et al. (2005), Baird (2006), Yaman et al. (2007), Cunha & Silva (2007) and Eiselt (2007) developed transportation models in hub location problem.
- Kuby & Gray (1993), Krishnamoorthy et al. (1994), Ernst & Krishnamoorthy (1996) and Ebery et al. (2000) formulated fast parcel services and post-delivery issues in HLP.
- Lee et al. (1996) and Klincewicz (1998) studied communication systems and data networks.
- Campbell et al. (2002) used HLP for fast consumer goods.

To solve small size hub problems, integer programming optimization methods can be useful. However, for larger sized models, heuristic methods or Meta heuristic methods should be developed and used to approach the optimum result. In the recent years, more meta heuristic methods, number of methods to solve such problems has been increased since the computational capabilities improved.

- Ostresh (1975) offered a procedure (each customer node is assigned to the closest origin) to solve the two-hub model.
- O'Kelly (1987) represented the P-hub median problem (P-HLMP) as a quadratic integer programming problem. Since pHLMP includes high complexity, two relaxation heuristics method have been formulation to reach the closest solution.
- Klincewicz (1991) described exchange heuristics and compared the solution with clustering heuristics and enumeration heuristics.
- Klincewicz (1992) studied Tabu search algorithm.
- Aykin (1994) studied on Lagrangian relaxation for HLP with capacity limitation (CHLP).
- Skorin-Kapov & Skorin-Kapov (1994) used Tabu search methods.

- Aykin (1995) solved uncapacitated single allocation hub location problem with installed fixed cost simulated annealing.
- Campbell (1996) proposed greedy interchange using Alflo, Maxflo.
- Abdinnour-Helm & Venkataramanan (1998) discussed genetic algorithm methods.
- Sohn (1998) offered shortest path method.
- Abdinnour-Helm(1998) represented tabu search and genetic algorithm.
- Ernst & Krishnamoorthy (1999) developed solution methods for capacitated single allocation hub location problem (CHLP-S).
- Sasaki et al. (1999) stated greedy algorithm for one stop logistics network. Benchmark between the branch and bound and greedy algorithms have underlined and the greedy algorithm required less computational time.
- Ebery et al. (2000) developed an algorithm using shortest path approach to solve CHLP-M, and used B & B algorithm for solution.
- Pamuk et al. (2001) formulated a Tabu search method for P-hub center problems (P-HLCP).
- Marianov et al. (2003) studied Tabu search algorithm for airlines hub location problem. This method works best for networks of up to 50 nodes with 4 or 5 hubs.
- Bollapragada et al. (2005) developed a new network-planning model and offered an effective greedy solution heuristic for the capacitated hub maximum-covering location problem with multi allocations (CHMCLP-M).
- Martine Labbe et al. (2005) developed a branch and cut algorithm for uncapacitated single allocation hub location problem (UHLP-S). In this paper, the connected hubs are called backbone and the connected non-hub nodes is called access network.
- Topcuoglu et al. (2005) offered a genetic algorithm method.
- Chen (2007) offered a mixed method, Simulated Annealing and Tabu search.
- Canovas et al. (2007) used a heuristic method, based on a dual-ascent technique for hub location problem (UHLP-M) without capacity limitation.
- Cunha & Silva (2007) proposed a genetic algorithm approach provides an efficient local improvement procedure for each generated individual of the population. Less-than-truckload (LTL) flows in Brazil are handled. Uncapacitated HLP with single allocation (UHLP-S) is used for the problem.
- Kratica et al. (2007) developed a modified MILP formulation for the CHLP for UHLMP-S and proposed two genetic algorithms for solution. Computational study demonstrated that the first EA provided optimal solutions for all small scale HLPs.
- Rodriguez et al. (2007) formulated simulated annealing for capacitated hub location problem (CHLP).
- Thomadsen & Larsen (2007) studied the design of two-layered fully interconnected networks. They formulated a set partitioning model and used column generation approach.
- Wagner (2007) developed a clustering method using Tabu search and genetic algorithm for traveling salesman problem (TSP).
- Camargo et al. (2008) used a bender decomposition method for uncapacitated multiple allocation hub location problem (UHLP-M).
- Costa et al. (2008) formulated a bi-criteria integer linear programming model for the capacitated single allocation hub location problems (CHLP-S).
- Rodriguez-Martin et al. (2008) proposed a MILP model and used two B&C algorithms based on decomposition methods for a CHLP-M.

- Yaman (2008) developed PHLMP-S model and offered a heuristics solution which is based on Lagrangian relaxation and local search.
- Wagner (2008) offered linear programming relaxation for P-UHLMP-M.
- Wang (2008), introduces combining hub-and-spoke transportation system among regional port with consolidation and dispersing transportation systems between ports and their own hinterlands by using a nonlinear optimization model for two-stage logistics system.
- Alumur et al. (2009) defined the single allocation incomplete p-hub median, the incomplete hub location with fixed costs, the incomplete hub covering, and the incomplete p-hub center network design problems, and introduce efficient mathematical formulations for these problems with $O(n^3)$ variables.
- Eiselt & Marianov (2009) formulated the competitive hub location problem in which customers have gravity-like utility functions. In the resulting probabilistic model, customers choose an airline depending on a combination of functions of flying time and fare. The (conditional) follower's hub location problem is solved by means of a heuristic concentration method.
- Contreras et al (2010) represented a mathematical programming formulation using branch and cut algorithm to solve the model for the THLP (Tree of Hubs Location Problem).
- Correia et al (2010) study an extension of the classical CHLP-S problem in which the size of the hubs is part of the decision making process.
- Ishfaq & Sox (2010) studied for an intermodal hub logistics network operated by a logistics service provider serving interregional area. A Tabu search meta-heuristic is developed a mathematical optimization approach deals with the p-hub median model for interacting hub location-allocation problems to the domain of intermodal logistics.
- Kratica et al. (2011) addressed the capacitated hub location problem (CHLP), a variant of the capacitated hub problem. They proposed two evolutionary algorithms (EAs) that use binary encoding and standard genetic operators adapted to the problem.
- Gelareh & Nickel (2011) offered an indexed formulation for the UHLP-M customized for city logistics and liner shipping network design. They formulated benders decomposition method and greedy heuristics. They introduced Australian Post (AP) dataset.
- Vasconcelos et al. (2011) proposed a new hub location model defined by the minimization of costs. The main contribution of this work is to permit the analysis of a hub-and-spoke network operated under ''decentralized management''. They present an integer programming formulation, subject the new model to experiments with an intermodal general cargo network in Brazil.
- Alumur et al. (2012) introduced the multimodal single allocation hub location and hub network design problem. They proposed linear mixed integer programming model for this problem and enhance model via a set of effective valid inequalities and an efficient heuristic.
- Hwang & Lee (2012) developed a uncapacitated single allocation p-hub maximal covering model to maximize the logistics covered, where the logistics of demand is said to be covered if the distance between two nodes is less than or equal to the specified range in consideration of the distance savings between hubs. Integer programming formulation has been represented and used CAB dataset. Distance based allocation and volume based allocation heuristics were suggested.
- Camargo & Miranda (2012) addressed a HLP-S under traffic. An effective generalized Benders decomposition algorithm has been deployed, enabling the solution of large scale instances in reasonable time.

- Puerto et al. (2013) introduced a new formulation for the median HLP-S problem and a branch-and-bound-and-cut (B & B & Cut) based algorithm to solve optimally this model.
- Sender & Clausen (2013) represented a capacitated multiple allocation hub location problem, which arose from a network design problem in German wagonload traffic.
- Davari et al. (2013) offered an incomplete hub-covering network design problem, where the exact locations of demands were unknown and estimated as fuzzy variables.
- Yu et al. (2013) studied the hub location problem in urban public transit network design, which includes two phases: to determine the candidate nodes and to optimize the distribution of transit hubs. Mathematical model, aiming at maximizing served populations per construction cost, was proposed to optimize multi-hub location based on the selected candidate nodes.
- Lüer-Villagra & Marianov, (2013) formulated and solve a new hub location and pricing problem, describing a situation in which an existing transportation company operating in a hub network, and a new company wants to enter into the same market, using an incomplete hub and spoke network.
- Oktal & Ozger (2013) developed mixed integer linear programming model introducing additional constraints to the traditional model of UHLP-M.
- Rodríguez Martín et al. (2014) studied the hub location and routing problem (HLRP). They tried to find the location of hubs, the allocation of nodes to hubs, and the routing among the nodes allocated to the same hubs, with the aim of minimizing the total transportation cost. They proposed a mixed integer programming formulation for this problem and strengthen it with valid inequalities. They devised separation routines for these inequalities and develop a branch-and- cut algorithm which is tested on CAB and AP instances from the literature.
- Duyguvar & Erol (2014) used uncapacitated multiple allocation p-Hub Set Covering Location Problem to analyze the effectiveness and sustainability of current logistics centers in Turkey and tried to find the optimum locations for the new logistics hubs.
- Mohammadi et al. (2014) addressed a novel sustainable hub location problem (SHLP) in which two new environmental-based cost functions accounting for air and noise pollution of vehicles are incorporated. To cope with uncertain data incorporated in the model, a mixed possibilistic–stochastic programming approach was proposed to construct the crisp counterpart. A simulated annealing (SA) and an imperialist competitive algorithm (ICA) with a new solution representation were developed to solve real-sized instances whose performances are compared with a proposed lower bound.
- O'Kelly (2014) provided a data based analysis of FedEx air freighter activities and fuel use from selected hub locations.
- Lin et al. (2014) develop a modified mixed integer programming model for the intermodal terminal location problem (ITLP) to increase computation efficiency. Greedy randomized adaptive search procedure (GRASP) and the attribute based hill climber (ABHC) heuristics are presented to solve the problem more efficiently while obtaining near optimal solutions.

Literature review of HLP is represented briefly in *Table 1*. Papers have been categorized by corresponding problem type and solving method.

Table 1. Literature review of HLP

Authors	Model	Solution Approach
O'Kelly (1986)	1-HLP-S	Nearest distance
Ostresh (1975)	2-HLP	Location–allocation (shortest route)
Kratica et al. (2011)	CHLP	Evolutionary algorithms
Ebery et al. (2000)	CHLP-M	Nearest distance – B & B
Marin (2005)	CHLP-M	Integer linear programming
Rodriguez et al. (2007)	CHLP-M	SA
Rodriguez-Martin & Salazar-Gonzalez (2008)	CHLP-M	MILP with B & C based on decomposition
Costa et al. (2008)	CHLP-S	Bi-criteria ILP
Wang, C. (2008)	CHLP-S	Linear programming
Yaman et al. (2007)	CHLP-S	TS (with greedy algorithm) – B & C
Bollapragada et al. (2005)	CHMCLP-M	Greedy algorithm (maximum covering)
Oktal & Ozger (2013)	CMAHLP	MILP
Sender & Clausen (2013)	CMAHLP	Local search heuristics
Correia et al (2010)	CSAHLP	LP relaxation
Correia et al (2010)	CSAHLPM	LP relaxation
Rodriguez Martin et al. (2014)	HLRP	MILP with B & C
Lin et al. (2014)	ITLP	GRASP & ABHC
Klincewicz (1991)	P-HLMP	Clustering
Klincewicz (1992)	P-HLMP	TS and greedy algorithm
O'Kelly (1987)	P-HLMP	Nearest distance
Campbell (1991)	P-HLMP-M	Integer programming
Campbell (1994)	P-HLMP-M	Integer programming
Campbell (1996)	P-HLMP-M	Greedy- interchange
Ishfaq & Sox (2010)	P-HLMP-M	TS
Sasaki et al. (1999)	P-HLMP-M	B&B algorithm and greedy algorithm
Sohn & Park (1998)	P-HLMP-M	Nearest distance
Sung & Jin (2001)	P-HLMP-M	Clustering (dual-based approach)
Wagner (2008)	P-HLMP-M	LP relaxation
Alumur et al. (2009)	P-HLMP-S	LP relaxation
Aykin (1994)	P-HLMP-S	B & B (lagrangian relaxation)
Davari et al. (2013)	P-HLMP-S	Variable neighborhood search
Ebery (2001)	P-HLMP-S	MILP
Kratica et al. (2007)	P-HLMP-S	GA
Lüer-Villagra & Marianov, (2013)	P-HLMP-S	GA
Skorin-Kapov & Skorin-Kapov (1994)	P-HLMP-S	TS
Yaman (2008)	P-HLMP-S	Lagrangian relaxation and local search
Skorin-Kapov et al. (1996)	P-HLMP-S/M	LP
Eiselt & Marianov (2009)	P-HLP-M	Heuristic algorithm
Yu et al. (2013)	P-HLP-M	Integer programming

continued on following page

Table 1. Continued

Authors	Model	Solution Approach
Pirkul & Schilling (1998)	P-HLP-S	Lagrangian relaxation
Hwang & Lee (2012)	P-UMHLP-S	Distance based allocation heuristics Volume based allocation heuristics
Alumur et al. (2012)	SAHLP	Heuristic algorithm
Camargo & Miranda (2012)	SAHLP	Benders decomposition
Puerto et al. (2013)	SA-OM-HLP	B&B&Cut
Mohammadi et al. (2014)	Sustainable HLP	SA and imperialist competitive algorithm
Contreras et al (2010)	THLP	B & C
Camargo et al. (2008)	UHLP-M	Benders decomposition
Canovas et al. (2007)	UHLP-M	B & B
Gelareh & Nickel (2011)	UHLP-M	Benders decomposition
Klincewicz (1996)	UHLP-M	B & B
Marianov & Serra (2003)	UHLP-M	TS
Mayer & Wagner (2002)	UHLP-M	Dual ascent approach
Pamuk & Sepil (2001)	UHLP-M	TS (with greedy local algorithm)
Thomadsen & Larsen (2007)	UHLP-M	B & B
Vasconcelos et al. (2011)	UHLP-M	Integer programming
Wagner (2007)	UHLP-M	Clustering (GA & TS)
Duyguvar & Erol (2014)	UHLP-M	Integer programming
Abdinnour-Helm (1998)	UHLP-S	GA/TS
Abdinnour-Helm & Venkataramanan (1998)	UHLP-S	GA
Aykin (1995)	UHLP-S	SA (greedy interchange) B & B
Chen (2007)	UHLP-S	TS and SA
Cunha & Silva (2007)	UHLP-S	GA
Labbe et al. (2005)	UHLP-S	Branch & cut (B & C)
O' Kelly (1992)	UHLP-S	Heuristic algorithm
Topcouglu et al. (2005)	UHLP-S	GA

MODELS AND ALGORITHMS

Basic Models in Hub Location Problem

HLP (Single Allocation)

In the single hub location problem, each non-hub node is connected to only one hub. That's why the problem is called single allocation. Problem is formulated by O'Kelly (1987). Objective function is Mini-Sum, and the backbone network (the hub nodes) are connected together, and each non-hub node is allocated to one of a hub in that backbone network. It is predetermined that how many hubs will be set. Number of hubs is represented as p. Between origin and destination nodes one or two hubs are used. There is no setup cost and no capacity limitation for the hubs to be set. Decision variables are binary.

1. Assumptions:
   a. The objective of the model is MiniSum.
   b. There is only one hub node.
   c. The number of hub is 1.
   d. No setup cost and no capacity limitation.
   e. Decision variables 0–1.
2. Notations:
   h_{ij}: flow from origin *i* to destination *j*
   C_{ij}: cost of transportation from origin *i* to destination *j*
3. Decision variables:
   X_j = a hub is located at node *j*
   Y_{ij} = node *i* is connected to a hub located at node *j*
4. The UHLP-S model is represented below:

$$Min\sum_i\sum_j\sum_k h_{ik}(C_{ij}+C_{jk})y_{ij}y_{kj} \tag{1}$$

Subject to

$$\sum_j X_j = 1 \tag{2}$$

$$Y_{ij} - X_j \leq 0 \quad \forall\, i,j \tag{3}$$

$$X_j = 0,1 \qquad \forall\, j \tag{4}$$

$$Y_{ij} = 0,1 \quad \forall\, i,j \tag{5}$$

The objective function (1) minimizes the total cost of transportation through hub *j*. Equation (2) constraints choosing only one hub location. Equation (3) forces that demand node *i* can be connected to the hub at *j* if it has been settled a hub at *j*. Equations (4) and (5) are binary constraints.

The objective function is quadratic since it requires the product of decision variables. If non-hub node *i* is allocated to a hub *j*, then all nodes *k(k, i)* should be allocated to the hub *j*. The objective function (1) can be revised as below:

$$\sum_i \sum_j \sum_k h_{ik}(C_{ij} + C_{jk}) y_{ij} y_{kj} = \sum_i \sum_j C_{ij} y_{ij} \left(\sum_k h_{ik} \right) + \sum_j \sum_i C_{ji} y_{ij} \left(\sum_k h_{ik} \right) = \sum_i \sum_j C_{ij} y_{ij} \left(O_i + D_i \right) \tag{6}$$

where O_i: Outflow of i; D_i: Inflow of i.

P-Hub Location Problem

In the pHLP the number of hubs to locate is known and denoted as p, and each non-hub node is allocated to only one hub node (O'Kelly 1987).

1. Assumptions:
   a. The objective of the model is MiniSum.
   b. The number of hubs to locate is p.
   c. There should be one or two hub between origin and destination.
   d. No setup cost and no capacity limitation.
   e. Decision variables 0–1.
2. Notations:
   a. Notations are the same as HLP-S.
   b. The discount factor for movements between hubs represented as α, and its value is $0 < \alpha < 1$. Since cost of movements between two hubs is expected to be less than the cost of a hub node and a non-hub node transportation. C_{ij} is multiplied by discount factor α.
   c. P = the number of hubs to locate.
3. Decision variables are the same as HLP-S.
4. The pHLP-S model is represented below:

$$Min \sum_i \sum_k C_{ik} Y_{ik} \left(\sum_j h_{ij} \right) + \sum_k \sum_i C_{ki} Y_{ik} + \alpha \sum_i \sum_j \sum_k \sum_m h_{ij} C_{km} y_{ik} y_{jm} \tag{7}$$

Subject to

$$\sum_j y_{ij} = 1 \qquad \forall i, \tag{8}$$

$$\sum_j x_j = P \tag{9}$$

$$y_{ij} - x_j \leq 0 \qquad \forall_{i,\, j} \tag{10}$$

$$x_j = 0,1 \qquad \forall_j \tag{11}$$

$$y_{ij} = 0,1 \qquad \forall_{i,j} \tag{12}$$

The objective function (7) minimizes the total transportation cost related to the *P* hubs locations and the non-hubs allocated to them. The transportation cost includes movements from non-hubs to hubs, hubs to non-hubs and between the hubs. Equation (8) forces that each non-hubs node is allocated to one and only hub. Equation (9) declares that the number of hub nodes is *p*. Equation (10) requires that non-hub nodes cannot be linked to each other directly without using a hub. Equations (11) and (12) are 0-1 binary constraints.

pHLP with Multi Allocation

Campbell (1991) formulated pHLP-M like P-median problem and modelled with a linear objective function. It is permitted that each non-hub to be allocated to more than one hub as seen *Figure 5.*

1. Assumptions: pHLP-M assumptions are the same as pHLP with single allocation. Only exception is that each non-hub node can be allocated to more than one hub.

Figure 5. Multiple allocation

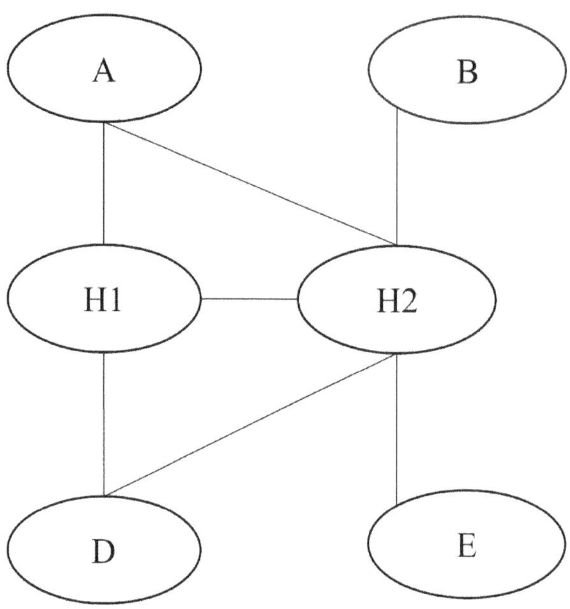

2. Notations: Notations are the same as pHLP with single allocation. Only exception is C_{ij} which is represented below:

$$C_{ij}^{km} = Cost\ of\ transportation\ from\ i\ to\ j\ via\ hubs\ k\ and\ m. \qquad (13)$$
$$C_{ij}^{km} = C_{ik} + \alpha C_{km} + C_{mj}$$

3. Decision variables:

X_j :a hub is located at node j
Z_{ij}^{km} : flow from origin i to destination j uses hubs at candidates sites k and m.

4. The pHLP-M model is represented below:

$$Min \sum_i \sum_j \sum_k \sum_m C_{ij}^{km} h_{ij} Z_{ij}^{km} \qquad (14)$$

Subject to

$$\sum_k x_k = P \qquad (15)$$

$$\sum_k \sum_m Z_{ij}^{km}{}_j = 1 \ \forall\, i, j \qquad (16)$$

$$Z_{ij}^{km} \leq \ x_m \ \ \forall\ i, j, k, m \qquad (17)$$

$$Z_{ij}^{km} \leq \ x_k \ \ \forall\ i, j, k, m \qquad (18)$$

$$Z_{ij}^{km} \geq 0 \ \forall\ i, j, k, m \qquad (19)$$

$$x_k = 0{,}1 \ \ \forall k \qquad (20)$$

The objective function (14) minimizes the weighted total transportation cost. The equation (15) ensures that number of hubs to be located is p. The equation (16) forces that each origin-destination pair (i, j) is required to be allocated to one of hub pair. The equations (17) and (18) state that transportation from i to j is not available unless there exist hubs at k and m. Equations (19) and (20) are non-negativity and binary constraints.

The hardest part of pHLP-M is the number of variables. If every nodes is a candidate for location of a hub, there will be $O(N^4)$ of such variables. A model with 32 nodes brings almost one million variables with itself.

pHLP Including Fixed Costs

O' Kelly (1992) represented pHLP including fixed costs of hubs to locate.

1. Assumptions: The pHLP with fixed cost model assumptions are similar to previous ones except two things:
   - The number of hubs to locate is not determined.
   - Locating a hub requires a *fixed cost*.
2. Notations:Two more notations added to the ones of pHLP:
   B = a weight on the fixed costs to evaluate the tradeoff between fixed costs and transportation costs.
   f_k = the set up cost of a hub to locate at k.
3. Decision variables: The decision variables are the same as pHLP.
4. pHLP with fixed cost model:

The pHLP with fixed cost model has two differences as represented below:

- Equation (9) is removed since the number of hubs to locate is not determined.
- The term (21) is added to the objective function:

$$B\sum_{k} f_x x_k$$

P-HLP with Single Allocation

pHLP-S is like pHLP-M except it is not allowed that each non-hub to be allocated to more than one hub as seen *Figure 6*.

1. Assumptions: The pHLP-S model assumptions are similar to pHLP-M, but the exceptions are represented below:
   - Each spoke hub is allocated to only one hub.
   - All of the variables are binary variables (0–1).
2. Notations: Notations of the model are the same as pHLP-M.
3. Decision variables: Following notation is added:

Figure 6. Single allocation

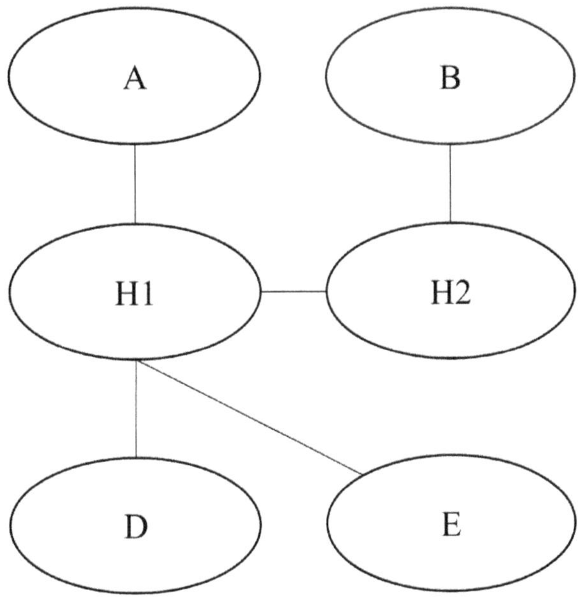

Y_{ik} = non-hub node *i* is assigned to a hub node k.

4. The pHLP-S model is represented below:

$$Min \sum_i \sum_j \sum_k \sum_m C_{ij}^{km} h_{ij} Z_{ij}^{km} \tag{22}$$

Subject to

$$\sum_k X_k = P \tag{23}$$

$$\sum_k \sum_m Z_{ij}^{km} = 1 \quad \forall\ i, j \tag{24}$$

$$Y_i k X_k \ \leq X_k \quad \forall\ i, k \tag{25}$$

$$\sum_k Y_{ik} = 1 \qquad \forall\ i \tag{26}$$

$$Y_{ik} + Y_{jm} - 2Z_{ij}^{km} \geq 0 \quad \forall\ i, j, k, m \tag{27}$$

$$X_k = \quad 0,1 \qquad \forall\ k \tag{28}$$

$$Y_{ik} = \quad 0,1 \qquad \forall\ i, k \tag{29}$$

$$Z_{ij}^{km} = \quad 0,1 \qquad \forall\ i, j, k, m \tag{30}$$

Equations (22)–(24) are similar to (14)–(16). Equations (25) force that node i cannot be allocated to a hub k unless there exists a hub at location k. Equations (26) ensure that each non-hub node i should be allocated to only one hub. Equations(27) refer that the freight from origin node i to destination node j cannot be forwarded via hub k and hub m, unless i is allocated to a hub k and node j is allocated to hub m. Equations (28)–(30) are binary and non-negativity constraints.

Fixed Cost of Linking a Node to a Hub

In this model, fixed cost of linking a non-hube node to any hub is inspected and it is not forced to each non-hub node to be allocate to a single hub. Campbell (1994) studied this model and represented the extensions below:

1. Assumptions: The assumptions of fixed cost of linking a non-hub node to any hub model are like pHLP.
2. Notations: Following notation is added to pHLP:
   g_{ik}: the fixed cost of allocating non-hub node i to a hub at candidate k.
3. Decision variables: Decision variables are the same as pHLP-S.
4. Model: The mathematical model of linking a non-hub with a fixed cost to a hub is like pHLP. Following equation (31) should be added.

$$\sum_i \sum_k g_{ik} Y_{ik} \tag{31}$$

Minimum Flow Between Nodes and Hub Connection Problem

Campbell (1994) formulated flow between any link between nodes and hubs to exceed some minimum value. It is not forbidden to allocate a non-hub to more than one hub. This model is like pHLP-S too.

1. Assumptions: The assumptions of this model are like pHLP-S. A minimum flow on links is added to the model.
2. Notations: Following notation is added to show the minimum flow between i and k.:
   l_{ik}: the minimum flow between node i and hub k.
3. Decision Variables: Decision variables are the same as pHLP-S.
4. Objective Function and its Constraints

Following equations (32) and (33) should be included:

$$Y_{ik} + Y_{jm} - 2Z_{ij}^{km} \geq 0 \quad \forall\ i, j, k, m \tag{32}$$

$$\sum_{m}\sum_{j} h_{ij} Z_{ij}^{km} + \sum_{P}\sum_{s} h_{pi} Z_{Pi}^{sk} \geq L_{ik} Y_{ik} \quad \forall i, k \tag{33}$$

Equations (32) refer that the freight from origin i to destination j cannot be forwarded via hub k and hub m, unless i is allocated to a hub k and j is allocated to hub m. The equation (33) represent that the sum of flow *from i to j via k and m* and flow from any hub to i via hub k should be greater than or equal to the minimum flow (L_{ik}) between node i and hub k.

Capacitated HLP

Campbell (1994) represented capacitated HLP that the total flows, in and out, for a hub should be less than or equal to a fixed value.

1. Assumptions: The assumptions are the same as pHLP. Capacity limitations are added to the model for each hub node.
2. Notations: Following notation is added to refer the capacity of each hub node k:
   θ_k: the capacity of a hub at candidate node k.
3. Decision Variables: Decision variables are the same as pHLP.
4. The model of capacitated HLP: The equation (34) is added to pHLP model:

$$\sum_{m}\sum_{i}\sum_{j} h_{ij} Z_{ij}^{km} + \sum_{s}\sum_{i}\sum_{j} h_{ij} Z_{ij}^{sk} \geq \theta_k X_{ik} \quad \forall\ k \tag{34}$$

The first term of (34) refers incoming flow, second term refers outgoing flow for each hub k. Right hand side of (34) represents the capacity of hub k.

P-Hub Center Location Problem

Campbell (1994) represented the pHCLP. This is an essential issue for its applications. An emergency facility location problem or the crisis management center location problem can be handled by this model.

The P-hub center location problem is similar to P-center location problem. The objective of a HCLP is to find a set of hub nodes that minimizes the maximum cost of each origin/destination pair.

1. Assumptions: The pHCLP model assumptions are like median pHLP, but additionally, X_k variables are relaxed and the objective function is MiniMax.
2. Notations: The notations of the model are like median pHLP.
3. Decision variables: Decision variables are also like median pHLP.
4. The model of pHCLP is represented below:

$$Min\,Max\left\{C_{ij}^{km} h_{ij} Z_{ij}^{km}\right\} \tag{35}$$

$Subject\ to$

$$\sum_{k} X_k = P \tag{36}$$

$$\sum_{k}\sum_{m} Z_{ij\ j}^{km} = 1\ \forall\ i, j \tag{37}$$

$$Z_{ij}^{km} \leq X_k\ \forall\ i, j, k, m \tag{38}$$

$$Z_{ij}^{km} \leq X \forall\ i, j, k, m \tag{39}$$

$$0 \leq X_k \leq 1 \qquad \forall\ k \tag{40}$$

$$0 \leq Z_{ij}^{km} \leq 1 \qquad \forall\ i, j, k, m \tag{41}$$

Equation (35) ensures minimizing the maximum cost of flow between each non-hub nodes via hubs. Equations (36)–(41) are like median pHLP. Campbell et al. (2007) offered a new approach for pHCLP and inspected the complexity of the model and then formulated algorithms for solution.

Covering Issue of HLP

Covering issue of HLP is first handled by Campbell (1994). He formulated this problem and there exist two approach: hub set covering location model and hub maximal covering location model.

The approach of a hub covering problem is explained below:

The cost of transportation from node *i* to node *j* via hubs *m, k* where the origin i and destination j are covered by *(m, k)*, should be less than or equal to a certain fixed value, as Equations (42)

$$C_{ij}^{km} \leq \gamma_{ij}$$

1. Hub set covering location model: Hub set covering location is represented below.:
   a. Assumptions: The assumptions of hub set covering problem are like median pHLP. The number of hubs to locate is not determined. There exists a fixed cost of locating a hub.
   b. Notations: The notations of the model are represented below:

F_k : *the fixed cost of locating a hub ink.*
C_{ij}^{km} : *unit cost from i to j via hubs k and m.*
γ_{ij} : *highest cost from i to j.*
V_{ij}^{km} : *hubs m and k covering non − hubs i and j.*

   c. Decision variables: Decision variables of the model are like median pHLP.
   d. The model of hub set covering location problem is represented below:

$$Min \sum_{i} F_k X_k \quad (43)$$

Subject to

$$Z_{ij}^{km} \leq X_k \qquad \forall\ i, j, k, m \quad (44)$$

$$Z_{ij}^{km} \leq X_m \qquad \forall\ i, j, k, m \quad (45)$$

$$\sum_{k} \sum_{m} V_{ij}^{km} Z_{ij}^{km} \geq \qquad 1 \qquad \forall\ i, j \quad (46)$$

$$0 \leq X_k \leq 1 \qquad \forall k \tag{47}$$

$$0 \leq Z_{ij}^{km} \leq 1 \qquad \forall i, j, k, m \tag{48}$$

The objective function of the model Equations (43) tries to minimize the total hub location costs. Equations (46) force that all of nodes are included via set of hubs. The rest of the equations were explained before in previous models.

2. Hub maximal covering location model:
   a. Assumptions: The assumptions of hub maximal covering location model are like median pHLP. The number of hubs to locate is determined. The fixed cost of locating a hub is not taken into account.
   b. Notations: Following notation is added:
      h_{ij} = the amount of transportation from i to j.
   c. Decision variables: Decision variables are like median pHLP where number of hubs is determined.
   d. The model of hub maximal covering location problem is represented below:

$$Min \sum_{i}\sum_{j}\sum_{m} h_{ij} Z_{ij}^{km} V_{ij}^{km} \tag{49}$$

Subject to

$$\sum_{k} X_k = P \tag{50}$$

$$\sum_{k}\sum_{m} Z_{ij}^{km} = 1 \quad \forall i, j \tag{51}$$

$$Z_{ij}^{km} \leq X_k \qquad \forall i, j, k, m \tag{52}$$

$$Z_{ij}^{km} \leq X_m \qquad \forall\, i, j, k, m \tag{53}$$

$$0 \leq X_k \leq 1 \qquad \forall\ k \tag{54}$$

$$0 \leq Z_{ij}^{km} \leq 1 \qquad \forall\ i, j, k, m \tag{55}$$

The objective function of Equation (49) tries to maximize the satisfied demand. The rest of the equations were explained before in previous models.

Solutions and Recommendations for Hub Location Problem

There are many algorithms developed or used by researcher to deal with the complexity of hub location problems. In the following, some applicable algorithms are explained.

Some Applicable Algorithms

O'Kelly (1987) formulated two heuristic methods to solve HLP-S with no capacity limitation.

First method requires each non-hub node is linked to the closest hub. It is regarded that when $\alpha = 0$, the third part of the nonlinear equation (7) will be zero and the problem is converted to a P-Median Problem. This method is useful while $\alpha < 0.5$.

In the second heuristic method all ways to linking spokes nodes to the closest or second closest hub nodes are inspected. If we have N nodes to select P hubs, there are *N!/P!.N!* alternatives. For each one of the *N!/P!.N!* alternatives, all of the 2^{N-P} ways to linking a node to the closest and second closest hubs are examined. Increasing number of nodes brought complexity together in that model.

The second heuristic method provides better solution than the first heuristic. Also, the second heuristic brings a closer upper bound than the first method.

Greedy Heuristic Algorithm

Sasaki et al. (1999) offered greedy algorithm to solve the 1-stop uncapacitated HLP-M. One stop forces that the flow between each origin-destination, only one hub node is used. This approach is suitable for relatively minor networks such as Japan domestic airline network.

Genetic Algorithm

Topcuoglu et al. (2005) formulated this algorithm to deal with uncapacitated HLP-S with fixed cost. A genetic algorithm (GA) is a search algorithm for calculating the near-optimal solutions in large spaces, which is inherited from genetics of population. Holland (1975) presented this algorithm for the first time. Genetic algorithms have been applied to a large set of problems in various fields in the literature. There exist two books presented by Goldberg (1989) and Mitchell (1998) for comprehensive GA examples.

Benders Decomposition Method

Camargo et al. (2008) offered this method based on a partitioning method for solving mixed linear and nonlinear integer programming problems. Camargo et al. (2008) studied the problem through partitioning it into two simpler parts: an integer model, known as MP, and a linear model, known as SP.

The MP is a approximated version of the primary problem with a set of integer variables and its associated constraints, while SP is the original problem with the values of the integer variables temporarily fixed by the MP.

The algorithm provides solution by each one of the two simpler problems iteratively, one at a time. At each step, a new constraint, called as Benders cut, is added to the MP. This new constraint is originated by the dual problem of the SP. The algorithm continues until the objective function value of the optimal solution to the MP is equal to that of the SP.

FUTURE RESEARCH DIRECTIONS

There are untouched fields for HLP research for the future. One of the subjects is to consider high quality connections HLP backbone design and the flow from hubs to spokes. This subject may cover strategic and operational decisions together, and enables decision maker coordinate wisely. Similarly, another research area is about the relation between flow and communication hub location. In all of these areas, there is a large literature that may be reviewed to formulate more comprehensive hub location models. One of the interesting subjects is dealing with the sustainability of in hub location models (Campbell & O'Kelly, 2012).

Also there are many research fields related to the environment, such as specific air, ground, or water flow systems and hub operations, including the unique characteristics and constraints of each flow modes (Campbell & O'Kelly, 2012).

Future trends of HLP:

- Communications networks due to the growing of transmission of huge data between nodes (Campbell & O'Kelly, 2012).
- Humanitarian logistics applications is new trend and the models of this research area can be applied to HLP (Farahani et al., 2013).
- The sustainability issues can be added economic modeling approach of HLP. Social and environmental research fields can be used to formulate a sustainable HLP model.
- Dynamic HLP is another real life application for HLP modelling field. A number of researches state that hubs' locations are no longer optimal due to the changing business environment, politics, social factors and other parameters in time (Alder & Hashai, 2005).
- Global HLP network design is another topic which is needed to be studied in near future (Farahani et al., 2013).
- More stochastic and uncertain cases should be studied to cover real life applications (Farahani et al., 2013).
- Obnoxious facility location problem is another area which is applicable in HLP (Farahani et al., 2013).

- Multimodal HLPs is emerging area which includes strategic network design of hub network according to the modes of transportation. (Farahani et al., 2013).
- There is relatively less researches dealing large problems. Hybrid methods, especially hybrid meta-heuristics can be developed to handle large HLPs. Moreover, exact methods like Benders' decomposition, column generation, branch and cut, etc. are still important.
- Another important aspect of location modelling of urban logistics facilities is to incorporate the change in traffic flow on transport networks, such as using traffic assignment procedures. However, there have been few studies considering traffic assignment within location modelling. With the progress of urbanization in the world, about half of the population is concentrated within urban areas in 2010 according to the statistics of United Nations and it is predicted that this will increase to 70% by 2050. A number of mega-cities with populations of over 10 million have appeared and there have been various issues due to dense populations in mega-cities, including traffic congestion, air pollution, crashes and large energy consumption. These issues are very complicated and innovative methodologies are required to overcome (Taniguchi et al., 2014).
- Emissions by freight vehicles often generate harmful effects on the health of residents. A number of models should be developed relating to emissions by freight operations, including network modelling. A traffic network model was developed to evaluate the impacts of policies for reducing greenhouse gas emissions from road freight vehicles in Sydney (Taylor et al., 2005). The best policy options identified for reducing GHG emissions from freight transport were, 'higher load factors' and 'best fuel technology'. Emission models were created for various vehicle and fuel types. Assignment procedures were developed for combining trip matrices for truck types obtained from a previous commercial vehicle survey with passenger vehicles. A link based function was applied that incorporated the effects of the number of vehicles by type as well as their speed, loads and fuel used. A load correction factor was used to adjust the base emissions to estimate the effects of vehicles carrying higher loads. This allowed the trade-off between the number of vehicles and the loads carried to be investigated. This topic is related to sustainability of HLP network either.
- Traffic congestion has been found to lead to substantial increases in CO_2 emissions for freight vehicles (Barth & Boriboonsomsin, 2008; Frey et al. 2008; Figliozzi, 2011). TRL (1999) have developed relationships between emissions, travel distance and travel speeds for large trucks. Hub location issues should be developed through emissions. Continuing rapid population and economic growth, typically seen in megacities, has led to significant increase in freight movements in transport networks, causing traffic congestion and environmental damage. This enhances the need to accurately estimate the freight movements as well as to forecast the expected future truck flows and underlying commodity flows (Taniguchi et al., 2014). In addition, optimizing the location of logistics facilities is important in megacities, not only in the case of emergencies but at any time, since it always strongly influences land use and traffic flows and incidentally produces external diseconomies such as traffic congestion and air pollution.

CONCLUSION

HLP have been studied almost last three decades. There exist various researches starting with O'Kelly's papers. HLPs deal with the location of hubs and routing of transportation from origins to destinations via the hubs. Hub networks are very useful in large-scale networks such as global freight or passenger movements and the flow of communication and data systems. The capability of hub networks are consolidating/concentrating flows while decreasing costs, increasing service quality, and handling the uncertainty. Transportation hub location problems involve locating hub facilities to handle flows of freight or passengers, whereas telecom hub networks provide for the movement of electronic messages.

In this chapter detailed information about HLPs is represented. The similarity of HLPs and network design problems is underlined. Strong inspirations for application areas of HLPs are also discussed. Key methodological approaches are explained for modelling HLPs and some relating solution algorithms are also provided. To motivate researchers and to draw their attention to HLP, recent emerging areas are listed and discussed.

Since the beginning, HLP has become a substantial area of location research—and is still hot topic between the researchers. Even though the motivations for future hub location research may be different from the past, it is a fact that HLP researches are going to grow in near future and beyond due to the increasing flow of people, commodities and data in global scale.

REFERENCES

Abdinnour-Helm, S. (1998). A hybrid heuristic for the uncapacitated hub location problem. *European Journal of Operational Research*, *106*(2-3), 489–499. doi:10.1016/S0377-2217(97)00286-5

Abdinnour-Helm, S., & Venkataramanan, M. A. (1998). Solution approaches to hub location problems. *Annals of Operations Research*, *78*, 31–50. doi:10.1023/A:1018954217758

Adler, N., & Hashai, N. (2005). Effect of open skies in the Middle East region. *Transportation Research*, *39*, 878–894.

Alder, N., & Hashai, N. (2005). Effect of open skies in the Middle East region. *Transportation Research*, *39*, 878–894.

Alumur, S., Kara, B., & Karasan, O. (2009). The design of single allocation incomplete hub networks. *Transportation Research Part B: Methodological*, *43*(10), 936–951. doi:10.1016/j.trb.2009.04.004

Alumur, S., Kara, B., & Karasan, O. (2012). Multimodal hub location and hub network design. *Omega*, *40*(6), 927–939. doi:10.1016/j.omega.2012.02.005

Alumur, S., & Kara, B. Y. (2008). Network hub location problems: The state of the art. *European Journal of Operational Research*, *190*(1), 1–21. doi:10.1016/j.ejor.2007.06.008

Arnold, P., Peeters, D., & Thomas, I. (2004). Modelling a rail/road intermodal transportation system. *Transportation Research Part E, Logistics and Transportation Review*, *40*(3), 255–270. doi:10.1016/j.tre.2003.08.005

Aversa, R., Botter, R. C., Haralambides, H. E., & Yoshizaki, H. T. Y. (2005). A mixed integer programming model on the location of a hub port in the East Coast of South America. *Maritime Economics & Logistics*, *7*(1), 1–18. doi:10.1057/palgrave.mel.9100121

Aykin, T. (1994). Lagrangian relaxation based approaches to capacitated hub-and-spoke network design problem. *European Journal of Operational Research*, *79*(3), 501–523. doi:10.1016/0377-2217(94)90062-0

Aykin, T. (1995). Networking policies for hub-and-spoke systems with application to the air transportation system. *Transportation Science*, *29*(3), 201–221. doi:10.1287/trsc.29.3.201

Baird, A. J. (2006). Optimizing the Container transshipment hub location in northern Europe. *Journal of Transport Geography*, *14*(3), 195–214. doi:10.1016/j.jtrangeo.2004.12.004

Bania, N., Bauer, P., & Zlatoper, T. U. (1998). Air passenger service: A taxonomy of route network, hub location, and competition. *Logistics and Transportation Review*, *34*(1), 53–74. doi:10.1016/S1366-5545(97)00037-9

Barth, M., & Boriboonsomsin, K. (2008). Real-world carbon dioxide impacts of traffic congestion. *Transportation Research Record*, *2058*, 163–171. doi:10.3141/2058-20

Beuthe, M., & Bouffioux, C. H. (2008). Analysing qualitative attributes of freight transport from stated orders of preference experiment. *Journal of Transport Economics and Policy*, *42*, 105–128.

Bollapragada, R., Camm, J., Rao, U. S., & Wu, J. (2005). A two-phase greedy algorithm to locate and allocate hubs for fixed-wireless broad bound access. *Operations Research Letters*, *33*(2), 134–142. doi:10.1016/j.orl.2004.05.007

Bookbinder, J. H., & Fox, N. S. (1998). Intermodal routing of Canada–Mexico shipments under NAFTA. *Transportation Research Part E, Logistics and Transportation Review*, *34*(4), 289–303. doi:10.1016/S1366-5545(98)00017-9

deCamargo, R. S., & Miranda, G. (2012). Single allocation hub location problem under congestion: Network owner and user perspectives. *Expert Systems with Applications*, *39*(3), 3385–3391. doi:10.1016/j.eswa.2011.09.026

deCamargo, R. S., Miranda, J. G., & Luna, H. P. (2008). Benders decomposition for the uncapacitated multiple allocation hub location problem. *Computers & Operations Research*, *35*(4), 1047–1064. doi:10.1016/j.cor.2006.07.002

Campbell, A. M., Lowe, T. J., & Zhang, L. (2007). The P-hub center allocation problem. *European Journal of Operational Research*, *176*(2), 819–835. doi:10.1016/j.ejor.2005.09.024

Campbell, J. F. (1991). *Hub location problems and the p-hub median problem* [Working Paper 91-06-21]. Center for Business and Industrial Studies. University of Missouri, St. Louis, MO.

Campbell, J. F. (1994). Integer programming formulations of discrete hub location problems. *European Journal of Operational Research*, *72*(2), 387–405. doi:10.1016/0377-2217(94)90318-2

Campbell, J. F. (1996). Hub location and P-hub median problem. *Operations Research*, *44*(6), 923–935. doi:10.1287/opre.44.6.923

Campbell, J. F., Erns, A., & Krishnamoorthy, M. (2002). Hub location problems. In Z. Drezner & H. Hammacher (Eds.), *Facility location: applications and theory*. Berlin: Springer. doi:10.1007/978-3-642-56082-8_12

Campbell, J. F., & O'Kelly, M. E. (2012). Twenty-five years of hub location research. *Transportation Science*, *46*(2), 153–169. doi:10.1287/trsc.1120.0410

Canovas, L., Garcia, S., & Marin, A. (2007). Solving the uncapacitated multiple allocation hub location problem by means of a dual-ascent technique. *European Journal of Operational Research*, *179*(3), 990–1007. doi:10.1016/j.ejor.2005.08.028

Chen, J. F. (2007). A hybrid heuristic for the uncapacitated single allocation hub location problem. *Omega*, *35*(2), 211–220. doi:10.1016/j.omega.2005.05.004

Contreras, I., Fernández, E., & Marín, A. (2010). The tree of hubs location problem. *European Journal of Operational Research*, *202*(2), 390–400. doi:10.1016/j.ejor.2009.05.044

Correia, I., Nickel, S., & Saldanha-da-Gama, F. (2010). Single-assignment hub location problems with multiple capacity levels. *Transportation Research Part B: Methodological*, *44*(8-9), 1047–1066. doi:10.1016/j.trb.2009.12.016

Correia, I., Nickel, S., & Saldanha-da-Gama, F. (2010). The capacitated single-allocation hub location problem revisited: A note on a classical formulation. *European Journal of Operational Research*, *207*(1), 92–96. doi:10.1016/j.ejor.2010.04.015

Costa, M. G., Captivo, M. E., & Climaco, J. (2008). Capacitated single allocation hub location problem-A bi-criteria approach. *Computers & Operations Research*, *35*(11), 3671–3695. doi:10.1016/j.cor.2007.04.005

Crainic, T. G., Kim, K. H., Barnhart, C., & Laporte, G. (2007). Intermodal Transportation, Chapter 8 in Transportation. In C. Barnhart and G. Laporte (Eds.), Handbooks in Operations Research and Management Science (pp. 467-537). North-Holland, Amsterdam.

Crainic, T. G., & Laporte, G. (1997). Planning models for freight transportation. *European Journal of Operational Research*, *97*(3), 409–438. doi:10.1016/S0377-2217(96)00298-6

Cunha, C. B., & Silva, M. R. (2007). A genetic algorithm for the problem of configuring a hub-and-spoke network for a LTL trucking company in Brazil. *European Journal of Operational Research*, *179*(3), 747–758. doi:10.1016/j.ejor.2005.03.057

Daskin, M. S. (1995). *Network and discrete location: Models, algorithms, and applications*. New York: Wiley. doi:10.1002/9781118032343

Davari, S., Zarandi, M. H., & Turksen, I. B. (2013). The incomplete hub-covering location problem considering imprecise location of demands. *Scientia Iranica E*, *20*(3), 983–991.

Don, T., Harit, S., English, J. R., & Whicker, G. (1995). Hub and spoke networks in truckload trucking: Configuration testing and operational concerns. *Logistics and Transportation*, *31*, 209–237.

Duyguvar, A.S. & Erol, S. (2014). Evaluation of logistics hubs in Turkey and location model proposal for new sustainable logistics hubs. Proceedings of *3rd National Logistics & Supply Chain Congress* (pp. 261-271).

Ebery, J. (2001). Solving large single allocation p-hub problems with two or three hubs. *European Journal of Operational Research*, *128*(2), 447–458. doi:10.1016/S0377-2217(99)00370-7

Ebery, J., Krishnamoorty, M., Ernst, A., & Boland, N. (2000). The capacitated multiple allocation hub location problem: Formulation and algorithms. *European Journal of Operational Research, 120*(3), 614–631. doi:10.1016/S0377-2217(98)00395-6

Eiselt, H. A. (2007). Locating landfills – optimization vs. reality. *European Journal of Operational Research, 179*(3), 1040–1049. doi:10.1016/j.ejor.2005.11.039

Eiselt, H. A., & Marianov, V. (2009). A conditional p-hub location problem with attraction functions. *Computers & Operations Research, 36*(12), 3128–3135. doi:10.1016/j.cor.2008.11.014

Ernst, A., & Krishnamoorthy, M. (1996). Efficient algorithms for the uncapacitated single allocation P-hub median problem. *Location Science, 4*(3), 139–154. doi:10.1016/S0966-8349(96)00011-3

Farahani, R., Hekmatfar, M., Arabani, A., & Nikbakhsh, E. (2013). Hub location problems: A review of models, classification, solution techniques, and applications. *Computers & Industrial Engineering, 64*(4), 1096–1109. doi:10.1016/j.cie.2013.01.012

Figliozzi, M. A. (2011). The impacts of congestion on time-definitive urban freight distribution networks CO2 emission levels: Results from a case study in Portland, Oregon. *Transportation Research Part C, Emerging Technologies, 19*(5), 766–778. doi:10.1016/j.trc.2010.11.002

Frey, H., Rouphail, N., & Zhai, H. (2008). Link-based emission factors for heavy-duty diesel trucks based on real-world data. *Transportation Research Record, 2058*, 23–332. doi:10.3141/2058-04

Gelareh, S., & Nickel, S. (2011). Hub location problems in transportation networks. *Transportation Research Part E, Logistics and Transportation Review, 47*(6), 1092–1111. doi:10.1016/j.tre.2011.04.009

Geoffrion, A. (1976). The purpose of mathematical programming is insight, not numbers. *Interfaces, 7*(1), 81–92. doi:10.1287/inte.7.1.81

Goldberg, D. E. (1989). *Genetic algorithms in search, optimization and machine learning. MA*. Reading: Addison-Wesley.

Gooley, T. B. (1997). One location, many options. *Logistics Management and Distribution Report, 37*(11), 81–89.

Hakimi, S. L. (1964). Optimum location of switching centers and the absolute centers and medians of a graph. *Operations Research, 12*(3), 450–459. doi:10.1287/opre.12.3.450

Hekmatfar, M., & Pishvaee, M. (2009). *Hub location problem in facility location: Concepts, models, algorithms and case studies* (pp. 243–270). Berlin: Physica-Verlag. doi:10.1007/978-3-7908-2151-2_11

Holland, J. H. (1975). *Adaptation in natural and artificial systems*. Michigan: University of Michigan Press.

Hwang, Y. H., & Lee, H. Y. (2012). Uncapacitated single allocation p-hub maximal covering problem. *Computers & Industrial Engineering, 63*(2), 382–389. doi:10.1016/j.cie.2012.03.014

Ishfaq, R., & Sox, C. (2010). Intermodal logistics: The interplay of financial, operational and service issues. *Transportation Research Part E, Logistics and Transportation Review, 46*(6), 926–949. doi:10.1016/j.tre.2010.02.003

Jaillet, P., Song, G., & Yu, G. (1996). Airline network design and hub location problems. *Location Science, 4*(3), 195–212. doi:10.1016/S0966-8349(96)00016-2

Klincewicz, J. G. (1991). Heuristics for the P-hub location problem. *European Journal of Operational Research, 53*(1), 25–37. doi:10.1016/0377-2217(91)90090-I

Klincewicz, J. G. (1992). Avoiding local optima in the P-hub location problem using tabu search and grasp. *Annals of Operations Research, 40*(1), 283–302. doi:10.1007/BF02060483

Klincewicz, J. G. (1996). A dual algorithm for the uncapacitated hub location problem. *Location Science, 4*(3), 173–184. doi:10.1016/S0966-8349(96)00010-1

Klincewicz, J. G. (1998). Hub location in backbone tributary network design: A review. *Location Science, 6*(1-4), 307–335. doi:10.1016/S0966-8349(98)00042-4

Kratica, J., Milanovi'c, M., Stanimirovi'c, Z., & Tosic, D. (2011). An evolutionary-based approach for solving a capacitated hub location problem. *Applied Soft Computing, 11*(2), 1858–1866. doi:10.1016/j.asoc.2010.05.035

Kratica, J., Stanimirovic, Z., Tosic, D., & Filipovic, V. (2007). Two genetic algorithms for solving the uncapacitated single allocation p-hub median problem. *European Journal of Operational Research, 182*(1), 15–28. doi:10.1016/j.ejor.2006.06.056

Krishnamoorthy, M., Mills, G., & Sier, D. (1994). *Strategic configuration of the mail processing network: location–allocation modeling-stage 1* [Technical Report DMS-C94/9]. Division of Mathematics and statistics CSIRO, Australia.

Kuby, M. J., & Gray, R. G. (1993). Hub network design problem with stoppers and feeders: Case of federal express. *Transportation Research, 27*, 1–12.

Labbe, M., Yaman, H., & Gourdin, E. (2005). A branch and cut algorithm for hub location problems with single assignment. *European Journal of Operational Research, 102*, 371–405.

Lee, Y., Lim, B., & Park, J. (1996). A hub location problem in designing digital data service networks: Lagrangian relaxation approach. *Location Science, 4*(3), 185–194. doi:10.1016/S0966-8349(96)00009-5

Lin, C., Chiang, Y., & Lin, S. (2014). Efficient model and heuristic for the intermodal terminal location problem. *Computers & Operations Research, 51*, 41–51. doi:10.1016/j.cor.2014.05.004

Lüer-Villagra, A., & Marianov, V. (2013). A competitive hub location and pricing problem. *European Journal of Operational Research, 231*(3), 734–744. doi:10.1016/j.ejor.2013.06.006

Lumsden, K, ., Dallari, F., & Ruggeri, R. (1999). Improving the efficiency of the hub and spoke system for the SKF European distribution network. *International Journal of Physical Distribution & Logistics Management, 29*(1), 50–66. doi:10.1108/09600039910253878

Marín, A. (2005). Formulating and solving splittable capacitated multiple allocation hub location problems. *Computers & Operations Research, 32*(12), 3093–3109. doi:10.1016/j.cor.2004.04.008

Marianov, V., & Serra, D. (2003). Location models for airline hubs behaving as M/D/c queues. *Computers & Operations Research, 30*(7), 983–1003. doi:10.1016/S0305-0548(02)00052-7

Martín, J. C., & Roman, C. (2003). Hub location in the South-Atlantic airline market. *Transportation Research Part A, Policy and Practice*, *37*(10), 865–888. doi:10.1016/S0965-8564(03)00060-0

Mayer, G., & Wagner, B. (2002). HubLocator: An exact solution method for the multiple allocation hub location problem. *Computers & Operations Research*, *29*(6), 715–739. doi:10.1016/S0305-0548(01)00080-6

Mitchell, M. (1998). *An introduction to genetic algorithms*. Cambridge, MA: MIT Press.

Mohammadi, M., Torabi, S. A., & Tavakkoli-Moghaddam, R. (2014). Sustainable hub location under mixed uncertainty. *Transportation Research Part E, Logistics and Transportation Review*, *62*, 89–115. doi:10.1016/j.tre.2013.12.005

O'Kelly, M. E. (1986). The location of interacting hub facilities. *Transportation Science*, *20*(2), 92–106. doi:10.1287/trsc.20.2.92

O'Kelly, M. E. (1987). A quadratic integer program for the location of interacting hub facilities. *European Journal of Operational Research*, *32*(3), 393–404. doi:10.1016/S0377-2217(87)80007-3

O'Kelly, M. E. (1992). Hub facility location with fixed costs. *Papers in Regional Science*, *71*, 292–306.

O'Kelly, M. E. (2014). Air freight hubs in the FedEx system: Analysis of fuel use. *Journal of Air Transport Management*, *36*, 1–12. doi:10.1016/j.jairtraman.2013.12.002

Oktal, H., & Ozger, A. (2013). Hub location in air cargo transportation: A case study. *Journal of Air Transport Management*, *27*, 1–4. doi:10.1016/j.jairtraman.2012.10.009

Ostresh, L. M. J. (1975). An efficient algorithm for solving the two center location – allocation problem. *Journal of Regional Science*, *15*(2), 209–216. doi:10.1111/j.1467-9787.1975.tb00921.x

Pamuk, F. S., & Sepil, C. (2001). A solution to the hub center problem via a single-relocation algorithm with Tabu search. *IIE Transactions*, *33*(5), 399–411. doi:10.1080/07408170108936838

Pirkul, H., & Schilling, D. A. (1998). An efficient procedure for designing single allocation hub and spoke systems. *Management Science*, *44*(12 Part 2), s235–s242. doi:10.1287/mnsc.44.12.S235

Puerto, J., Ramos, A. B., & Rodríguez-Chía, A. M. (2013). A specialized branch & bound & cut for single-allocation ordered median hub location problems. *Discrete Applied Mathematics*, *161*(16-17), 2624–2646. doi:10.1016/j.dam.2013.05.035

Rodríguez, V., Alvarez, M. J., & Barcos, L. (2007). Hub location under capacity constraints. *Transportation Research Part E, Logistics and Transportation Review*, *43*(5), 495–505. doi:10.1016/j.tre.2006.01.005

Rodríguez-Martín, I., Salazar-González, J., & Yaman, H. (2014). A branch-and-cut algorithm for the hub location and routing problem. *Computers & Operations Research*, *50*, 161–174. doi:10.1016/j.cor.2014.04.014

Sasaki, M., Suzuki, A., & Drezner, Z. (1999). On the selection of hub airports for an airline hub-and-spoke system. *Computers & Operations Research*, *26*(14), 1411–1422. doi:10.1016/S0305-0548(99)00043-X

Sender, J., & Clausen, U. (2013). Heuristics for solving a capacitated multiple allocation hub location problem with application in German wagonload traffic. *Electronic Notes in Discrete Mathematics*, *41*, 13–20. doi:10.1016/j.endm.2013.05.070

Shaw, S. L. (1993). Hub structures of major US passenger airline. *Journal of Transport Geography*, *1*(1), 47–58. doi:10.1016/0966-6923(93)90037-Z

Skorin-Kapov, D., & Skorin-Kapov, J. (1994). On Tabu search for the location of interacting hub facilities. *European Journal of Operational Research*, *73*(3), 502–509. doi:10.1016/0377-2217(94)90245-3

Skorin-Kapov, D., Skorin-Kapov, J., & O'Kelly, M. (1996). Tight linear programming relaxations of uncapacitated p-hub median problems. *Proceedings of the National Decision Science Conference (Vol. 94*, pp. 582–593). doi:10.1016/0377-2217(95)00100-X

Slack, B. (1990). Intermodal transportation in North America and the development of inland load centers. *The Professional Geographer*, *42*(1), 72–83. doi:10.1111/j.0033-0124.1990.00072.x

Slack, B. (2001). Intermodal transportation. In A. Brewer, K. J. Button, & D. A. Hensher (Eds.), *Handbook of Logistics and Supply Chain Management* (1st ed., p. 700). Pergamon.

Sohn, J., & Park, S. (1998). Efficient solution procedure and reduced size formulation for P-hub location problem. *European Journal of Operational Research*, *108*(1), 118–126. doi:10.1016/S0377-2217(97)00201-4

Sung, C. S., & Jin, H. W. (2001). Dual-based approach for a hub network design problem under non-restrictive policy. *European Journal of Operational Research*, *132*(1), 88–105. doi:10.1016/S0377-2217(00)00114-4

Taniguchi, E., Thompson, R. G., & Yamada, T. (2014). Recent trends and innovations in modelling city logistics. *Procedia: Social and Behavioral Sciences*, *125*, 4–14. doi:10.1016/j.sbspro.2014.01.1451

Taylor, M. A. P., Zito, R., Smith, N., & D'Este, G. (2005). Modelling the impacts of transport policies to reduce greenhouse gas emissions from urban freight transport in Sydney. *Journal of the Eastern Asia Society for Transportation Studies*, *6*, 3135–3150.

Thomadsen, T., & Larsen, J. (2007). A hub location problem with fully interconnected backbone and access networks. *Computers & Operations Research*, *34*(8), 2520–2531. doi:10.1016/j.cor.2005.09.018

Toh, R. S., & Higgins, R. C. (1985). The impact of hub and spoke network centralization and route monopoly on domestic airline profitability. *Transportation Journal*, *24*, 16–27.

Topcuoglu, H., Corut, F., Ermis, M., & Yilmaz, G. (2005). Solving the uncapacitated hub location using genetic algorithms. *Computers & Operations Research*, *32*(4), 967–984. doi:10.1016/j.cor.2003.09.008

Vasconcelos, A. D., Nassi, C. D., & Lopes, L. A. S. (2011). The uncapacitated hub location problem in networks under decentralized management. *Computers & Operations Research*, *38*(12), 1656–1666. doi:10.1016/j.cor.2011.03.004

Wagner, B. (2007). An exact solution procedure for a cluster hub location problem. *European Journal of Operational Research*, *178*(2), 391–401. doi:10.1016/j.ejor.2006.02.011

Wagner, B. (2008). A note on location of hubs in a competitive environment. *European Journal of Operational Research, 184*(1), 57–62. doi:10.1016/j.ejor.2006.10.057

Wang, C. (2008). Optimization of hub-and-spoke two-stage logistics network in regional port cluster. *Systems Engineering – Theory &Practice, 28*(9), 152–158.

Warsing, D. P., Souza, G. C., & Greis, N. P. (2001). Determining the value of dedicated multimodal cargo facilities in a multi-region distribution network. *European Journal of Operational Research, 133*(1), 81–93. doi:10.1016/S0377-2217(00)00185-5

Yaman, H. (2008). Star P-hub median problem with modular arc capacities. *Computers & Operations Research, 35*(9), 3009–3019. doi:10.1016/j.cor.2007.01.014

Yaman, H., Kara, B. Y., & Tansel, B. C. (2007). The latest arrival hub location problem for cargo delivery systems with stopovers. *Transportation Research Part B: Methodological, 41*(8), 906–919. doi:10.1016/j.trb.2007.03.003

Yu, B., Zhu, H., Cai, W., Ma, N., Kuang, Q., & Yao, B. (2013). Two-phase optimization approach to transit hub location – the case of Dalian. *Journal of Transport Geography, 33*, 62–71. doi:10.1016/j.jtrangeo.2013.09.008

KEY TERMS AND DEFINITIONS

Hub: A transport hub (also transport interchange) is a place where passengers and cargo are exchanged between vehicles or between transport modes. The facilities that are servicing many origin-destination pairs as transformation and tradeoff nodes, and are used in traffic systems and telecommunications.

Hub-and-Spoke: The model is named after a bicycle wheel, which has a strong central hub with a series of connecting spokes. Traffic moves along spokes connected to the hub at the center. The model is commonly used in industry, in particular in transport, telecommunications and freight, as well as in distributed computing, where it is known as a star network.

Logistics Center: A Logistics Centre is the hub of a specific area where all the activities relating to transport, logistics and goods distribution – both for national and international transit – are carried out, on a commercial basis, by various operators.

Node: Nodes of a distribution network are generated by factories, depots, distribution centers, retailers and customers, as well. Configuration at the level of the distribution system concerns primarily the problem of location of the nodes in a geographic space and distribution of capacity among the nodes.

p-Center Problem: The p-center problem seeks the location of p facilities. Each demand point receives its service from the closest facility. The objective is to minimize the maximal distance for all demand points.

p-Median Problem: p-Median problem interests in finding the location of p facilities to serve demand nodes so that the transportation cost is minimized. There are no capacity constraints at the facilities.

Supply Network: A supply network is a pattern of temporal and spatial processes carried out at facility nodes and over distribution links, which adds value for customers through the manufacturing and delivery of products.

Section 2

Decisions in Urban Transportation

Chapter 6
A Web Based Decision Support System (DSS) for Individuals' Urban Travel Alternatives

Ebru V. Ocalir-Akunal
Gazi University, Turkey

ABSTRACT

The progress in web based DSS technologies in recent decades has given individuals many opportunities to make their business environments and daily lives so easier that could never been imagined before. Use of traffic information systems are getting wider. In this study, a decision support system for individuals' travel alternatives between any origin and destination points in urban areas is described. The original study is derived from a project prepared for the Ankara Municipality (Turkey). The described project gave individuals opportunity to select among all travel alternatives under different criteria, both for public transport riders and car users. It is a web based DSS, which uses GIS. The Floyd-Warshall algorithm has been used to solve all-pairs shortest path problem. Static data are based on base map, addresses, point of interests, lines, stops, schedules, etc. The dynamic data are received from the online vehicle tracking system. The design of the project is given to help other researchers design their projects according to the requirements specific for their own cities.

INTRODUCTION

The progress in web based DSS technologies, in recent decades has given individuals many opportunities to make their business environments and daily lives easier in ways that would have been unimaginable until recently. The contribution of Geographical Information Systems (GIS) to these systems is also significant, in that it provides the possibility of involving the geographical attributes and spatial location to any decision making process.

Nowadays, many public authorities have reformed their web sites to better provide online information and services to their residents. In addition to many other types of online services, use of city information systems and traffic information systems are increasing within this medium. For example, all over the

DOI: 10.4018/978-1-4666-8648-9.ch006

world major city authorities are publishing their city's maps on their web sites and are utilizing query possibilities. In some cities, traffic cameras supply online information on some of the most problematic junctions and arterials. Traffic jams can be followed and observed via constantly updated traffic volume maps. On most of these web pages, information on public transport lines are also listed and marked on these maps for ease of viewing. Some of these traffic volume maps help with managing queries by using shortest path algorithms. However, written literature about these applications is currently very limited, like most of the other intelligent traffic solutions. Commercial secrets can be brought forward as a possible reason for this lack of literature and resources on such applications; however a more important reason is that any progress with these technologies is very challenging and resource intensive. Once these applications are initially implemented, the unlimited imagination and demands of the users of these systems contribute in creation of many special projects. It should be remembered that, although most of these web applications looks similar at first sight, each of them is designed for a specific problem determined by their city dynamics, which changes the chosen algorithms, and that makes them unique. The user interface (UI) hence the appearance is similar on many of these applications however the algorithms and processes that are running in the background are very different with each of them.

Decision support systems are fed by many algorithms from computer science. Shortest path algorithms are among the most important ones in solving transportation problems.

In this study, a case study is given with some details. The original study is derived from a project prepared for the Ankara Metropolitan Municipality (Turkey). The described project is a decision support system (DSS) (http://map.ego.gov.tr:8080/ego/index.aspx) that aims to give individuals the opportunity of choice among all travel alternatives under different criteria, both for public transport riders and car users.

Decision support techniques, thus decision support systems are designed to support groups or individuals. The introduced system here is designed for individual DSSs.

The organization of the rest of the chapter is as follows:

In the next section, background information from existing literature is given. A literature survey on fields of applications of Intelligent Transport Systems (ITS) is given. General information on shortest path problems with an emphasis on the shortest travel time between all pairs of points provides some details about the algorithms used in the project. Web based decision support systems are also explained briefly.

Literature survey is followed by the case study itself. In this section, a project prepared for the Ankara Municipality is described. A list of Municipality's demands is given to better define the problem that is to be solved. The design of the system, its capabilities and examples with screenshots are also included.

The conclusion section includes a general evaluation of the system in addition to the suggestions for further research.

BACKGROUND

Intelligent Transport Systems (ITS)

Intelligent transportation systems (ITS) are advanced applications which provide smart use of the transort systems to users. The innovative services related to different modes of transport provide users the chance to be better informed about the transport system(s) (e.g. transport modes, travel alternatives, road conditions, etc.) and the decision makers to better manage the complicated situations (traffic management, toll collection, traffic jams, emergency responses, monitoring of the overall system). Although ITS may refer to all modes of transport, EU Council Directive 2010/40/EU (7 July 2010) defines ITS as systems in which information and communication technologies are applied in the field of road transport, including infrastructure, vehicles and users, and in traffic management and mobility management, as well as for interfaces with other modes of transport

The EU Council Directive 2010/40/EU (7 July 2010) supports the deployment of the innovative transport technologies across Europe. The first priorities would be traffic and travel information, the eCall emergency system and intelligent truck parking (http://ec.europa.eu/transport/themes/its/road/action_plan/).

The FTA (Federal Transit Administration) of the USA created the Advanced Public Transportation Systems (APTS) program to encourage the use of technology to improve the quality and usefulness of public transportation and ridesharing services as part of the US Department of Transportation's initiative in Intelligent Transportation Systems (ITS) (Casey et al., 1991).

ITS are expected to contribute to a cleaner, safer and more efficient transport system. They are tailored to the specific safety needs of road users. Human factors, knowledge and principles are incorporated into the design, deployment and evaluation of these systems (Regan et al., 2001).

ITS serve as a DSS in many research areas such as (Nwagboso, 1997);

- traffic management and control (and smart transport systems) (Mounce et al., 2013; Debnath et al., 2014; Jarašūnienė & Batarlienė, 2013),
- public transport systems (Satunin & Babkin, 2014; Castellanos & Fruett, 2014; Adamski, 2014; Kostakos et al., 2013: Sun et al., 2013; Schietekat & Booysen 2013; Cavar et al., 2008),
- freight operation systems (Sternberg & Andersson, 2014; Sicilia-Montalvo et al., 2013; Zuazola et al., 2013; Sarwi & Kuwahara, 2008; Gutiérrez et al., 2010),
- mobile telecommunications systems (Boccardo et al., 2014; Giuli et al., 2013; Cheng et al., 2014; Drilo et al., 2009; Robin-Jouan et al., 2005)
- debiting (Catling, 1990) and smart card systems (Abbas, 2014; Liu et al., 2009; Jakubauskas, 2006),
- user information systems (Viktorsson, 2013; Cantarella, 2013) and
- advanced vehicle safety systems (Nallaivarothayan et al., 2013; Conesa et al., 2013).

The proposed case study in this chapter is an example of ITS. As it mainly serves to public transport users, with better traveler information, including pretrip information, it can also be called as APTS.

Shortest-Path Problems

Planning shortest paths using algorithms is one of the most powerful methods for modeling combinatorial optimization problems in transportation planning as well as computer programming, communication networks, circuit design, bioinformatics, internet node traffic, social networking, and graph analysis in general (Planken et al., 2012).

The shortest path problem appears either directly or as a subsidiary problem in many applications such as;

- vehicle routing problems (Contardo et al., 2014; Martinelli et al., 2014; Letchford et al., 2014; Xiong & Wang, 2014; Gauvin et al., 2014; Jiang & Xie, 2014, Muter et al., 2014; Tas et al., 2014, Masson, 2014),
- some investment problems (Sullivan et al., 2014; Ergun et al., 2014; Qui & Yagan, 2014; Kang et al., 2014; Redekopp et al., 2013; Ruan et al., 2013),
- some stock problems (Kristianto et al., 2014; Solyali & Sural, 2012; Pinto, 2012),
- some problems of dynamic programming with discrete states and discrete time (Wikborg & Lee, 2013; Chang & Kang, 2012; Nishi & Maeno, 2010; Schoenemann & Cremers, 2010; Pratikakis et al., 2010; Aksakalli, 2007),
- network optimization problems (Ma et al., 2014; Zhang et al., 2014; Yates et al., 2014),
- telecommunication networks (Kajdanowicz et al., 2014; Santos et al., 2013; Caliskan, 2012; Kajdanowicz et al., 2012),
- some problems of artificial intelligence (Tsatcha et al., 2014; Shah et al., 2013; Achour et al., 2013; Machuca et al., 2012),
- pattern recognition (Liagouras et al., 2014; Goncalves & Bruno, 2013; Amancio et al., 2012),
- problems of labyrinths (Amancio et al., 2011; Haranczyk & Sethian, 2009; Yan & Yuan, 2003),
- mathematical games (Nikolova & Stier-Moses, 2014; Liang et al., 2010) and
- some methods for the digital signal processing (Chen & Hu, 2014; Gupta et al., 2013).

The first examples for the use of shortest path algorithms in transportation field (Steenbrink, 1974), (Van Vliet, 1978) and (Gallo & Pallottino, 1984) are cited in Pallottino & Scutella (1997). There is no best algorithm for every kind of transportation problem, so the research on this field has shifted towards the design and implementation of shortest path procedures which are able to individually capture the peculiarities of the problems that are under consideration (Pallottino & Scutella, 1997).

The components of a path planning problem are (Díaz-Báñez, 2008, p. 497);

- the environment (e.g., a network, a graph, a tree or a geometric domain);
- constraints (obstacle avoidance, short length, number of turns depending on the specific application) and
- a combination of criteria (e.g., optimizing distances or cost functions).

In search for good algorithms, Dreyfus (1967, 1969) presented five discrete shortest-path problems:

1. determining the shortest path between two specified nodes of a network;
2. determining the shortest paths between all pairs of nodes of a network;
3. determining the second, third, etc., shortest path;
4. determining the fastest path through a network with travel times depending on the departure time; and (5) finding the shortest path between specified endpoints that passes through specified intermediate nodes.
    - *The shortest path between two specified nodes of a network* is a min-cost flow problem. Given a set of N nodes, numbered arbitrarily from 1 to N, and the N x N matrix D, whose element d_{ij} represents the length of the directed arc connecting node i to node j, find the path connecting node 1 and node N of shortest length (Dreyfus, 1967).

- *The all pairs shortest path problem* is considered to be very fundamental for many fields such as transportation, networking, and social science. In computer science the problem is to compute the shortest paths between all pairs of vertices *G* assuming that edge costs are nonnegative real values (Han, 2004, p.245).
- Sometimes it is desirable to know *the value of the second (and third, etc.) shortest path* through a network. For example, suppose that some complex quantitative (or even qualitative) feature characterizes paths, and the shortest path possessing this additional attribute is sought. By ignoring the special aspect in question and ordering paths from shortest to longest, the best path with the additional feature can sometimes be determined fairly efficiently.
- *The fastest path through a network with travel times depending on the departure time is* finding the fastest path between two points where the time of travel between origin i and destination j depends on the time of departure from i.
- *The shortest path between specified endpoints that passes through specified intermediate nodes* is a problem which was reported by solution of this problem was reported by Saksena & Kumar (1966).

Around the 1960's, Edsger Dijkstra (Dijkstra, 1959), Robert Floyd (Floyd, 1962) and Stephen Warshall (Warshall, 1962) published papers on algorithms for solving shortest path problems. These algorithms are named after their inventors (Höfner & Möller, 2012).

Dijkstra's *label algorithm* which was proposed in 1959 is accepted as one of the best shortest path algorithms. Among its very wide range of applications, multi-point routing, surveying and mapping science, the shortest path of logistics and transport, the intelligent transportation system and the expressway network toll collection are worth considering (Shu-Xi, 2012, p.1187). It solves a single-source shortest path problem for a (finite) graph with non-negative edge costs by iteration, producing a shortest-path tree. The original algorithm keeps a set of approximations to shortest paths from a source *s* to the other nodes, improving those approximations in each iteration of the algorithm. To do so, it defines two sets of nodes, one for which the shortest path problem has already been solved and one which contains all other nodes (Höfner& Möller, 2012).

Floyd-Warshall algorithm solves all pair's shortest path problems.

There are also other algorithms which have been developed for solving the shortest path problems (Majhi & Chaupattanayak, 2013, p.57): The *single-destination shortest path problem*, tries to find out shortest paths from all vertices in the directed graph to a single destination vertex *v*. This can be reduced to the *single-source shortest path problem* by reversing the arcs in the directed graph. The *all-pairs shortest path problem* finds out shortest paths between every pair of vertices *v*, *v'* in the graph. *Bellman–Ford algorithm* is a shortest path algorithm that solves single source shortest path problems if the edge weights may be negative. A *search algorithm* is a shortest path algorithm that solves the single pair shortest path using heuristics to speed up the search. It has better performance than the Dijkastra's algorithm. *Johnson's algorithm* finds the shortest paths between all pairs of vertices in a sparse directed graph. *Perturbation theory* finds the locally shortest path.

The Shortest Travel Time between All Pairs of Points: Floyd-Warshall Algorithm

We need an algorithm for computing the shortest paths between all pairs of vertices in a directed graph to solve our problem, which is an instance "all-pairs shortest paths" problem. We could iterate through every pair of points and compute the shortest path between each using a single-source shortest-path algorithm. The problem can be solved by Dijkstra's algorithm by running it from each vertex of the graph (Peres, et al., 2013, p.1-2). However, Floyd-Warshall algorithm is an easier way to solve our problem.

The Floyd–Warshall algorithm, which is an example of dynamic programming, is a graph analysis algorithm for finding shortest paths in a weighted, directed graph. A single execution of the algorithm will find the shortest paths between all pairs of vertices. The use of Floyd–Warshall algorithm covers a large spectrum from shortest paths in directed graphs, to transitive closure of directed graphs, optimal routing, fast computation of path finder networks, etc.

The natural representation for a graph in the Floyd-Warshall algorithm is an adjacency matrix. Assume that we are given a directed graph $G= (V,E)$ and that the vertices V are numbered $1,2,\ldots,n$. Further assume that we are given a matrix $C[i,j]$ that tells us the cost of edge (i,j). If there is no edge $C[i,j]$ then we assume $C[i,j]$ is set to infinity. We assume all other costs are nonnegative.

The Floyd-Warshall algorithm computes a cheapest-cost array A, where $A[i,j]$ gives the cheapest cost of any path that there are no negative cost cyles in the graph. It computes the cost matrix of the cheapest paths between all pairs of a directed graph $G= (V,E)$ in $O(|V|^3)$ time and $O(|V|^2)$ space (Aho, 2004, p.62).

Floyd-Warshall algorithm uses a matrix of lengths $D_0\{0\}$ as its input. If there is an edge between nodes i and j, than the matrix $D_0\{0\}$ contains its length at the corresponding coordinates. The diagonal of the matrix contains only zeros. If there is no edge between edges i and j, than the position (i, j) contains positive infinity. In other words, the matrix represents lengths of all paths between nodes that do not contain any intermediate node (Figure 1).

With each iteration of Floyd-Warshall algorithm this matrix is recalculated, so it contains leng ths of paths among all pairs of nodes using gradually enlarging set of intermediate nodes. The matrix $D_1\{1\}$, which is created by the first iteration of the procedure, contains paths among all nodes using exactly one (predefined) intermediate node. $D_2\{2\}$ contains lengths using two predefined intermediate nodes. Finally the matrix $D_n\{n\}$ uses an intermediate nodes.

Figure 1. Graph and matrix of lengths in Floyd-Warshall algorithm

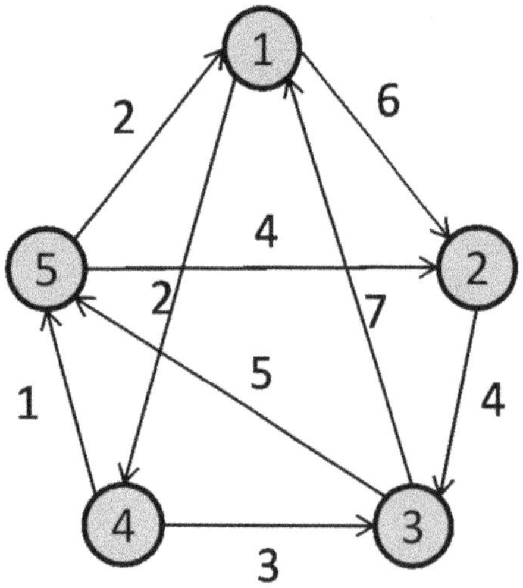

$$\begin{bmatrix} 0 & 6 & \infty & 2 & \infty \\ \infty & 0 & 4 & \infty & \infty \\ 7 & \infty & 0 & \infty & 5 \\ \infty & \infty & 3 & 0 & 1 \\ 2 & 4 & \infty & \infty & 0 \end{bmatrix}$$

Web Based DSSs

Since the early 1970's, decision support systems (DSS) technology and applications have evolved significantly (Shim et al., 2002). Modern decision support systems (DSSs) equipped with computerized systems, support decision tasks such as information gathering, model building, sensitivity analysis, collaboration, alternative evaluation and decision implementation in addition to business processes and ad hoc analyses (Bhargava & Power, 2004).

By the mid-1990s, several academic researchers and software developers had realized that the World Wide Web and Internet technologies created new opportunities for building and deploying new generation decision support systems (Bhargava et al., 2007). Since then, considerable progress has been observed in the development of Web-based analytical applications. Web technologies have provided decision support systems platform-independent, remote and distributed computation and exchange of complex media information (Bhargava & Power, 2004). It is expected that, during the 21st century, with the development and spread of the Internet, the Web, and the overall telecommunications technology to result in organizational environments that will increasingly become more global, complex, and connected (Shim et al., 2002).

Some researchers (Courtney, 2001; Mitroff & Linstone (1993), cited in Shim et al., 2002) suggest that DSS researchers should embrace a more comprehensive view of organizational decision making and develop decision support systems capable of handling much "softer" information and much broader concerns than the mathematical models and knowledge-based systems compared to the past. This challenging change should give a very important role to DSSs in the future.

Bonczek et al. (1981) conceptualized the DSSs as consisting of three components—the user interface, a knowledge processing system, and a knowledge base. This model led to a wide stream of research in the area of DSS. DSS applications today can be utilized from a single machine or can operate from within a client server environment. The use of DSSs on the web brought three major changes: change in the user community, changes in problem domains, and changes in the underlying technology architecture (Bharati & Chaudhury, 2004, p.188).

Bhargava & Power (2004) list the frameworks or typologies for decision support that have become more widespread and sophisticated because of Web technologies:

- Data-driven DSS help managers organize, retrieve, and analyze large volumes of relevant data using database queries and OLAP techniques.
- Model-driven DSS use formal representations of decision models and provide analytical support using the tools of decision analysis, optimization, stochastic modeling, simulation, statistics, and logic modeling.
- Communication-driven DSS rely on electronic communication technologies to link multiple decision makers who might be separated in space or time or to link decision makers with relevant information and tools. The Web has expanded this technology.
- Knowledge-driven DSS can suggest or recommend actions to managers. The Web helps deliver this type of DSS to a much broader audience of decision makers.
- Document-driven DSS integrate a variety of storage and processing technologies to provide managers document retrieval and analysis.

For about 20 years, the integration of the Internet and GIS (Geographic Information Systems) has also created new research areas such as Web-based GIS, Internet GIS, and web distributed GIS services (Dragicevic, 2004) which helped development of web-based and GIS supported DSSs. DSSs can now incorporate both geographic information systems (GIS) functionalities such as spatial data management, cartographic display and analytical modeling capabilities, a flexible user interface, and complex spatial data structures (Sugumaran & Sugumaran, 2007).

Some examples to online web based transport and traffic information systems are given: The City of Calgary (located in Canada) uses a city information system which includes within it categories such as cemeteries, education buildings and libraries, health and public safety information, land use and developments, etc. Traffic information systems, as a subsystem of this system contains information on construction detours, traffic incidents, bus routes, central train stations, construction projects, residential parking zones and snow/icing ban zones. The system allows filter the layers (http://www.calgary.ca/_layouts/cocis/calgarymap). The second example is the city and county of Honolulu, which publishes traffic cameras of city online. The third example is also a DSS, in which many city maps worldwide can be seen. The logic of the design is similar to the case study in Ankara, however much of the details cannot be reached (https://www1.honolulu.gov/cameras/trafficzones1.htm).

Challenges in Web based DSS's Development

Although web technologies have contributed a lot to the practice of decision support systems (e.g., platform independent, remote and distributed computation, exchange of complex multimedia information, reduced software development, maintenance, and distribution costs, etc.) there are still some challenges to be overcome (Bhargava & Power, 2004):

Consistency of Data, Information & Data: Research challenges should include how to keep high consistency between the data, information, and knowledge when they are transformed to different formats and how to maintain efficiency when there are huge quantities of data being processed. It is an important problem especially when graphical representation of data is involved (Liu et al., 2010).

Integration of qualitative methods: Qualitative modelling methods are becoming increasingly important for DSS's and web based technologies should involve qualitative aspects of the decision process: experience, intuition, judgment, and specialist expertise (Liu et al., 2010).

Performance: Performance is one of the most important limitations in the development of Web based DSS's. Most of them have very complex structures and take substantial amount of time to run on the server. Use of mobile tools, mobile e-services, and wireless Internet protocols should be developed (Bhargava & Power, 2004).

Social and Behavioral Challenges: Decision support applications have first been designed for industrial and organizational users, who have a professional interest in using the applications (Bhargava & Power, 2004). However, the challenges in technology have extended the coverage to every kind of consumers, e.g. travelers, students, housewives, etc., that means any person with access to a smart phone. Research should be done to improve the technologies to become more user-friendly.

A WEB BASED DSS DESIGN

Study Area

Ankara (Turkey) is a metropolitan city (capital of Turkey) with a population exceeding 4.5 million. The city currently has also rail transit modes, however the primary means of public transport is provided through the utilization of city bus and minibus. City bus and minibus services are operated by both public and private entrepreneurs. The public transport in Ankara is mainly a hub-and-spoke system, in which most of the lines converge on the city center. In this complex structure, it is difficult for an individual to find the most optimal path from his origin to his destination. In addition, the bus lines are always subject to modifications, which make planning travel a difficult ordeal, even for the local citizens. Individuals need an opportunity to be able select among all travel alternatives under different criteria, both for public transport riders and car users.

Definition of the Problem

The Ankara Municipality's definition of problem rose from a number of issues that they identified during their daily public transport operations.

The main problem at the beginning was the lack of digital data. Although the Municipality has developed a bus information system by itself, that system was not able to meet the requirements defined.

The main reason was the lack of information about bus stop signs. The service area is divided into 5 service regions, which are operated by different sub-units. Each sub-unit operates a certain number of bus lines. In most cases these lines cross over more than one service sub-area. The problem was, instead of having a list of real bus stops on the area, the municipality collected information given by each of the sub-units and then counted the number of stops in each line using this data. This meant that there was an issue of overlap and most transfer points were counted more than once during the data collection.

Another issue surfaced together with the bus stop overlap: The bus stops did not have any specific ID numbers. They were traditionally named after the nearest POI (point of interest), such as schools and hospitals, etc. This method of identification created duplicate and missing names for stops. These POI names brought with them a cataloging problem, with many long and complex names being used to identify bus stops, record keeping had to be very precise as a single letter missing from the name due to clerical error would make matching that bus stop with meaningful data impossible. A systematic numbering for the bus stops was therefore absolutely necessary.

The bus stops in that former system had not been digitalized with any acceptable accuracy. As a matter of fact, each of the bus line routes were digitized by the usage of a GPS device, carried by a traveler-observer, and bus stops on this line were marked along the locations where the bus made stops. This method of data collection meant that the number of lines on any single transfer point was also the number marked points by traveler-observers, and in real world they never overlapped. The traveler can mark any point within a possible marking area as a bus stop. This would not have been a problem if the lines were to be considered independent of one another however, the lines needed to overlap each other at certain locations within the system in order to allow transfers. There is also one bus stop sign in each bus stop area and it should be carefully digitized. The transfer can only be followed if different lines are combined to this single point in space. In Figure 2, the problem and the necessary situation of the bus stops/stations are described. In the former situation, a group of points (each time a traveler/observer

Figure 2. The accuracy problem of the bus stops/transfer points

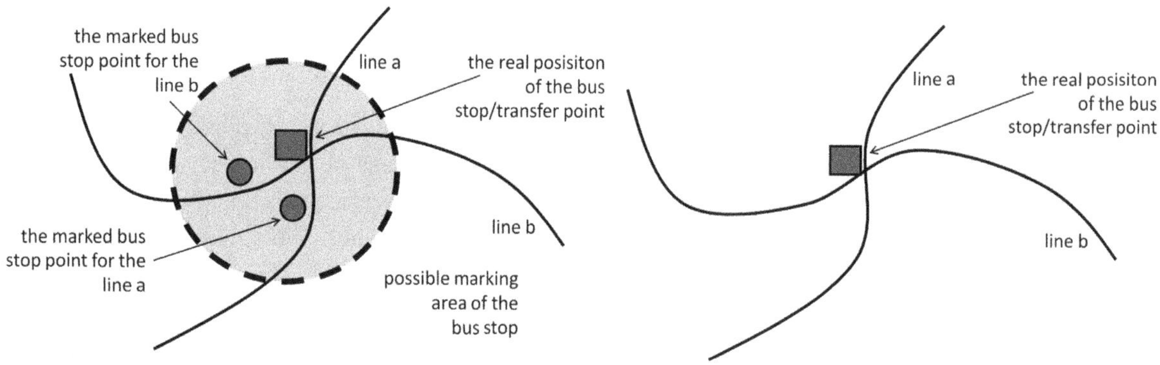

a) the former situation *b) the situation after the project*

marks a digital sign when he/she visits a bus stop) defined a bus-stop. In this case it was impossible to bridge a relationship with the system which needs a single defined bus stop for a possible transfer point. The situation after the implementation of the project shows the definition of just one certain for every bus-stop.

Another problem was that the lines were subject to seasonal changes during operation; these changes were known only by the operators. The lines carried the same line numbers. As a matter of fact, computers need a better explanation to help decision makers. That means, even if very small differences in the route exist, the lines should be defined and digitized independently.

To summarize; the Municipality had a real location based database problem as databases are the main foundations of healthy decision making systems; this issue created an important handicap. The proposed system introduced a good alternative that is supported by an intensive field study.

As a result, problem to be solved is;

- The travel times (and the travel costs) need to be scheduled between two points (origin and destination) on urban space,
- the shortest travel time (and minimum travel cost) between all pairs of points should be computed,
- All possible alternatives should be listed and ranked as a decision support system.

With these constraints;

- Travel modes are all kinds of public transit alternatives.
- Transfers are allowed, together with walking possibility.

Phases of the Study

An extensive field study had to be run in order to build up the system. The phases of the study:

1. Defining coordinates of the physical objects and lines.

The system has originally been designed for a certain number of stops and lines but it also allows online updates, which means the original numbers in the system has already been changed, however Table 1 still gives an idea of the attributes of the original system. The geographical positions of the stops and the stations, together with the taxicab stands have been defined by DGPS (Differential Global Positioning System) with a 50 cm accuracy level. Aerial photos were also used in that phase.

The lines have also been digitalized with the GPS data collected by traveler-observers. However, the stop information has been received from DGPS measurements.

2. Coordination of the collected information with other relevant data

Other data, e.g., time tables, vehicle fleet, sub-unit borders and information, etc., were connected to the previously collected information in the database. Giving a systematic ID number to each of the bus-stop was also done in this phase. The database is also enriched by the GIS information. An important contribution came from the vehicle tracking system of the Municipality; the information of online positions of the public transport vehicles is connected to the database.

3. Design of the system

The experts tried to design the best alternative for the defined problem. An all pairs shortest path problem is needed to be solved.

The all pairs shortest path problem is solved for origins and destinations, which can be;

- A certain address,
- A certain public transport stop,
- A certain point on map.

Normally, in such a graph, bus stops are considered as the nodes of the graph. In our model, line numbers are the nodes and the routes are the edges (Figure 3). There is also a weighting for the relationships among the lines, which is defined according to the frequencies.

Table 1. Some attributes of the proposed system

		Number
Stops and Stations	Bus stops	8072
	First stops/central points	46
	Rail system stations	51
	Minibus main stops	70
	Taxicab stands	619
Lines	Public Travel modes Rail modes (suburban train, HRS and LRS) Private and public operated buses	3 494
	Paratransit Minibus	201

Figure 3. Design of the graph

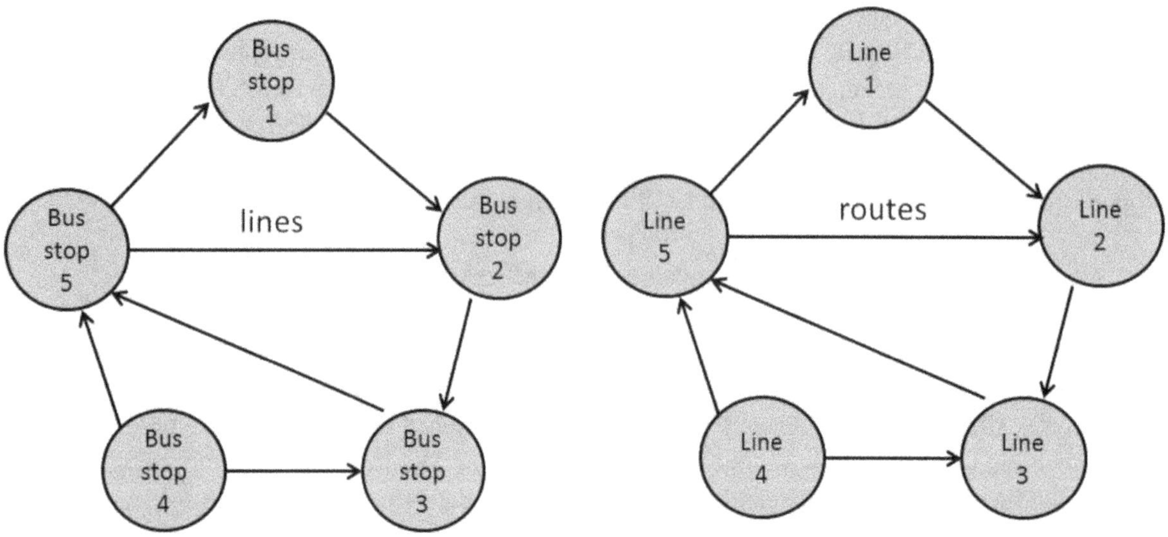

a) A typical graph for public transport system $G = (V_{bus\ stops}, E_{lines})$

b) The graph designed for the project $G = (V_{lines}, E_{routes})$

To determine the transfer points in the system for the selected routes, the criteria are:

- The traveler should spend minimum travel time in total
- The traveler should walk shortest distance during transfers,
- The traveler should spend maximum time when he/she once gets into the vehicle.

Total travel time and travel cost are calculated, and a list of possible travel alternatives, which is ranked according to these criteria, is submitted to transport user.

The application is as follows: The user selects an origin point, either on the map or by entering the address into a search bar. The stops within a certain area from that point are identified and linked with the related lines. A similar procedure is done for the destination point. The relationship(s) between line pairs are searched on the graph. Direct connections and transfers are listed separately. These relationships are determined via Floyd-Warshall algorithm. Finally, the alternative(s) with minimum cost (max frequency) is (are) considered.

The physical system is designed for municipality's needs and opportunities. The critical point in the system is the transfer points which enhance connectivity of the overall system. This connectivity allows the DSS make queries for shortest path with transfer possibilities.

An important problem of the users of this system was the very rapid changes in bus line supply. Municipality makes radical changes in bus lines, their schedules and even their names when needed. Although these changes are given in lists in Municipality's webpage, for a single transport user it is nearly impossible to combine this change to other related links in mind. In addition, the vehicle tracking system does not only follow the vehicle's movements on urban space. The system uses the pre-digitalized bus lines as a guide and controls if the bus follows its pre-defined route. For that reason, it is very important to keep the data update. An update center in municipality continuously updates any change in the system. Some staff in municipality is charged to update the lines as an update unit. These changes are immediately published to web again and the system is refreshed. The system and the information are stored in municipality's server.

In Figure 4, an example to the web page application of the DSS is given. After calling the application's web page (http://map.ego.gov.tr:8080/ego/index.aspx), two points on map are marked by right-click of mouse as origin and destination and we ask the system list all possible route alternatives between these two points. It is possible to filter these routes with 500 meters walking possibility for transfers. The list is given in a separate window. If we move our mouse on the list, we can see all details of every route with total fares. As a transport user, we can choose any of these alternatives. In the example, we chose the first route on the list. It is a combination of two lines, one a rail system line and a bus line, with a transfer point. We can follow the instructions given in the list to reach to our destination point.

The system also gives some other query possibilities. For example, if we click on any bus stop on our route, we can get some additional information. By that way we can see all of the corresponding bus lines in that bus stop. Let say, after we chose our route from our origin to destination, by clicking on the nearest bus stop, we can see our line together with other lines, and again by clicking our line, we can see the nearest bus on the route via vehicle tracking system. As a matter of fact, we see all of the buses on the way in the selected line and by that way we can make a travel plan in short term. The buses are defined by their plate numbers.

A similar application is also made for the car trips. The user can choose two points on map again as an origin and a destination point, and a list of all possible routes are listed only for car trips. The user can chose any of them and the chosen route is plotted on map. The distance and possible travel duration are also given.

Mobile application has also been written for android and ios devices. The application can be downloaded free to any smart telephone. This time the transport user has also the possibility to query "the nearest bus stop to his/her positon" as a smart application.

According to municipality's demands, some additional applications, like address search or poi (point of interest) search are also included in the system. These applications can also be used to find out any origin or destination point, or simply serve as a search tool for users.

CONCLUSION

The described project gives individuals opportunity to select among all travel alternatives under different criteria, both for public transport riders and car users. Satellite connection to vehicle tracking system allows users get online information about the lines and the vehicles. The possible user interfaces of the system are; internet access to web page, mobile application by android and iOS devices and SMS to cell phones, which first two are in use.

The submitted DSS gives travelers the opportunity to evaluate;

- for public transport,
    - all public travel modes (rail system alternatives, bus systems, minibuses),
    - list of all possible alternative routes,
    - direct links, with transfers and with pedestrian links,
    - expected duration of the listed trips,

Figure 4. A web page application for selected origin and destination points

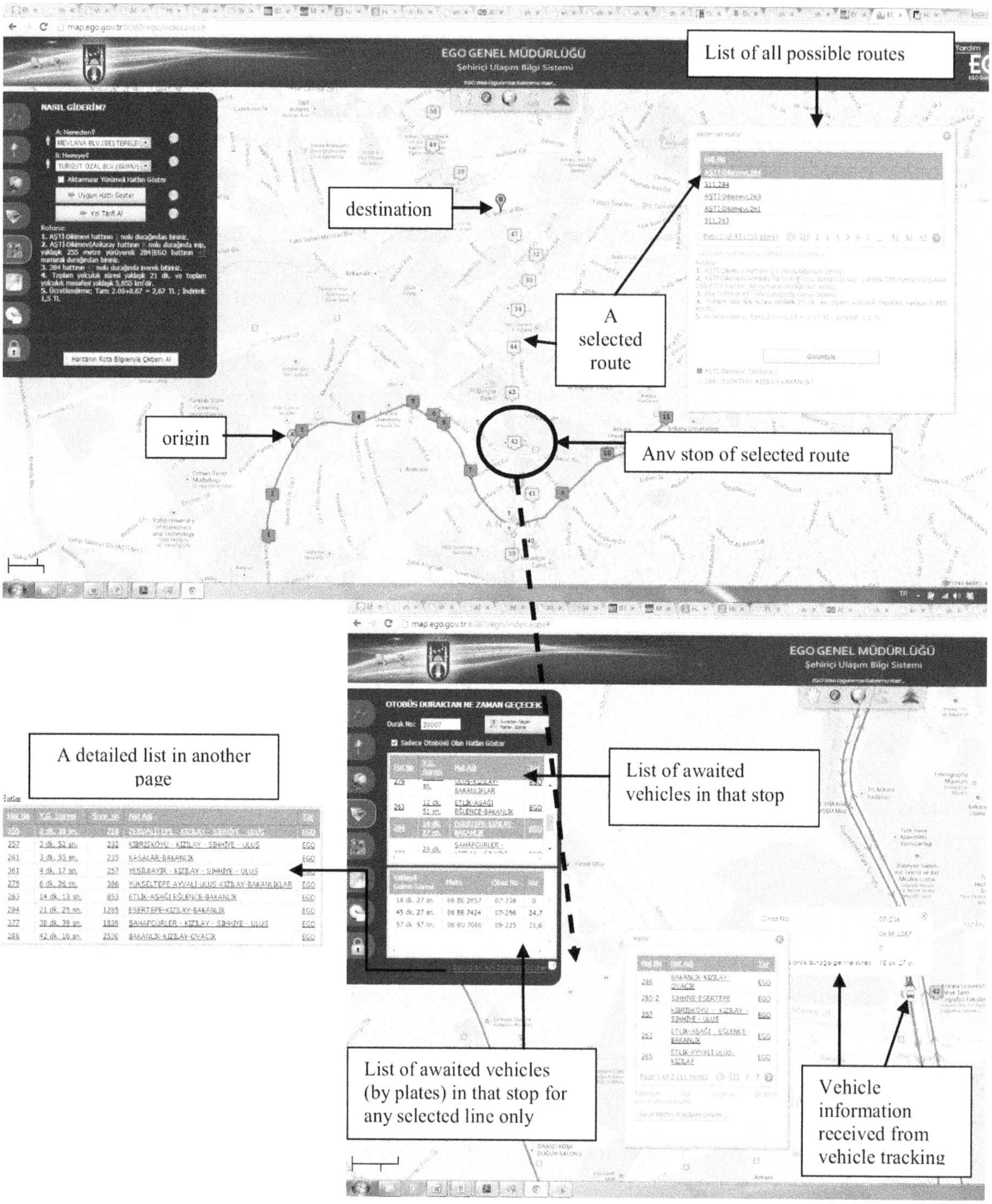

    - monetary cost of each trip,
    - online tracking of all of the vehicles on the line (satellite connection to vehicle tracking system),
- for each public transport stop (all have a user ID),
    - list of all lines with time schedules,
    - transfer possibilities,
    - list of nearest vehicles approaching (satellite connection to vehicle follow-up system)
- and for car trips,
    - list of all possible alternatives,
    - expected duration of the listed trips,
    - print the selected route with directives.

The graph and the decision support system are designed according to the client's needs. However, some extensions in future are possible with the designed system.

- The system uses a static distance (500 m walking distance) matrix for transfers. This can be transformed to a dynamic structure.
- The system is supported by multi-language possibilities, however the service language is only limited to Turkish, which is an important obstacle for tourists. The service should be given in also other languages.
- Although the design of the system allows it, smart stops issue is not welcome by the client. Although in the first phase web and mobile applications are built, the system should be extended to smart stops.
- The system gives online information about the current situation. Time schedules of the lines are also given in lists but at the moment traveler should make his/her own plan for a possible future travel. However, the system should allow make a travel plan for any future travel time. This is also possible with the system design.

The proposed ideas can be applied to the transport net of another city. The design of the project is given to help other researchers design their projects according to the requirements specific for their own cities. The success of these projects would be determined by the quality of the collected data.

REFERENCES

Abbas, M. (2014). Challenges in implementation of TVM (Ticket Vending Machine) in developing countries for mass transport system: A study of human behavior while interacting with ticket vending machine-TVM. In *Design, User Experience, and Usability. User Experience Design for Everyday Life Applications and Services* (pp. 245–254). Switzerland: Springer International Publishing. doi:10.1007/978-3-319-07635-5_24

Achour, N., Lakhdari, A., & Ferguene, F. (2013). Motion planning for car-like robots using hierarchical genetic algorithms. In S. I. Ao, L. Gelman, D. W. Hukins, A. Hunter, & A. M. Korsunsky (Eds.), *World Congress on Engineering - WCE 2013, Vol II, Lecture Notes in Engineering and Computer Science* (pp. 787-791). Hong Kong China. IAENG Press.

Adamski, A. (2014). DISCON: public transport dispatching robust control. *Procedia: Social and Behavioral Sciences*, *111*, 1206–1216. doi:10.1016/j.sbspro.2014.01.155

Aho, A. V. (2004). Computer representations of graphs. In J. L. Gross, J. Yellen, & P. Zhang (Eds.), *Handbook of Graph Theory* (p. 62). USA: CRC Press.

Aksakalli, V. (2007). The BAO* algorithm for stochastic shortest path problems with dynamic learning. *Proceedings of the 46th IEEE Conference on Decision and Control* (Vol 1-14, pp. 2460-2465). New York. IEEE. doi:10.1109/CDC.2007.4434409

Amancio, D. R., Aluisio, S. M., Oliviera, O. N., & Costa, L. D. (2012). Complex networks analysis of language complexity. *Europhysics Letters, 100*(5).

Amancio, D. R., Oliveira, O. N. Jr, & Costa, L. D. (2011). On the concepts of complex networks to quantify the difficulty in finding the way out of labyrinths. *Physica A-Statistical Mechanics and Its Applications*, *390*(23-24), 4673–4683. doi:10.1016/j.physa.2011.06.079

Bharati, P., & Chaudhury, A. (2004). An empirical investigation of decision-making satisfaction in web-based decision support systems. *Decision Support Systems*, *37*(2), 187–197. doi:10.1016/S0167-9236(03)00006-X

Bhargava, H. K., & Power, D. J. (2004). Decision support systems and Web technologies: A status report. *Americas Conference on Information Systems (AMCIS)*. Boston.

Bhargava, H. K., Power, D. J., & Sun, D. (2007). Progress in web-based decision support technologies. *Decision Support Systems*, *43*(4), 1083–1095. doi:10.1016/j.dss.2005.07.002

Boccardo, P., Arneodo, F., & Botta, D. (2014). Application of Geomatic techniques in Infomobility and Intelligent Transport Systems (ITS). *European Journal of Remote Sensing*, *47*, 95–115. doi:10.5721/EuJRS20144707

Bonczek, R. H., Holsapple, C. W., & Whinston, A. B. (1981). *Foundations of decision support systems*. Sandiego, CA: Academic Press.

Calişkan, C. (2012). A faster polynomial algorithm for the constrained maximum flow problem. *Computers & Operations Research*, *39*(11), 2634–2641. doi:10.1016/j.cor.2012.01.010

Cantarella, G. E. (2013). Day-to-day dynamic models for intelligent transportation systems design and appraisal. *Transportation Research Part C, Emerging Technologies*, *29*, 117–130. doi:10.1016/j.trc.2012.03.005

Casey, R. F., Labell, L. N., Prensky, S. P., & Schweiger, C. L. (1991). *Advanced public transportation systems: the state of the art (No. UMTA-MA-06-0196-91-2)*. Umta.

Castellanos, J. C., & Fruett, F. (2014). Embedded system to evaluate the passenger comfort in public transportation based on dynamical vehicle behavior with user's feedback. *Measurement*, *47*, 442–451. doi:10.1016/j.measurement.2013.08.068

Catling, I. (1990). System convergence—The potential for an integrated road transport environment. *Transportation*, *17*(3), 285–299. doi:10.1007/BF02127040

Cavar, I., Baric, D., & Kavran, Z. (2008). Assessment of proposed business plans of on-trip and pre-trip information in the LRT system (public urban transit) of the city of Zagreb. In Urban Transport XIV: Urban Transport and the Environment in the 21st Century.

Chang, C. L., & Kang, S. J. (2012). Using the reinforcement learning scheme to the priority-based routing and call admission control in WDM networks. *Journal of Internet Technology*, *13*(5), 793–801.

Chen, J. A., & Hu, J. H. (2014). Energy-efficient digital signal processing via voltage-overscaling-based residue number system. *IEEE Transactions on Very Large Scale Integration (VLSI). Systems*, *21*(7), 1322–1332.

Cheng, T. K., Leu, J. S., Chen, I. X., Wu, J. L. C., & Zhu, Z. Y. (2014). Practical context awareness information system for wireless vehicular networks based on a core telematics platform. *International Journal of Distributed Sensor Networks*, 2014.

Conesa, J., Cavas-Martínez, F., & Fernández-Pacheco, D. G. (2013). An agent-based paradigm for detecting and acting on vehicles driving in the opposite direction on highways. *Expert Systems with Applications*, *40*(13), 5113–5124. doi:10.1016/j.eswa.2013.03.026

Contardo, C., Cordeau, J.-F., & Gendron, B. (2014). An exact algorithm based on cut-and-column generation for the capacitated location-routing problem. *INFORMS Journal on Computing*, *26*(1), 88–102. doi:10.1287/ijoc.2013.0549

Council Directive 2001/29/EC on the framework for the deployment of Intelligent Transport Systems in the field of road transport and for interfaces with other modes of transport [2001] OJ L 207, 1-13.

Courtney, J. F. (2001). Decision making and knowledge management in inquiring organizations: Toward a new decision-making paradigm for DSS. *Decision Support Systems*, *31*(1), 17–38. doi:10.1016/S0167-9236(00)00117-2

Debnath, A. K., Chin, H. C., Haque, M. M., & Yuen, B. (2014). A methodological framework for benchmarking smart transport cities. *Cities (London, England)*, *37*, 47–56. doi:10.1016/j.cities.2013.11.004

Díaz-Báñez, J. M. (2008). Optimization for continuous shortest paths in transportation. In F. Erdogdu (Ed.), *Optimization In Food Engineering* (pp. 497–519). USA: Taylor & Francis.

Dijkstra, E. (1959). A note on two problems in connexion with graphs. *Numerische Mathematik*, *1*(1), 269–271. doi:10.1007/BF01386390

Dragicevic, S. (2004). The potential of Web-based GIS. *Journal of Geographical Systems*, *6*(2), 79–81. doi:10.1007/s10109-004-0133-4

Dreyfus, S. E. (1967). *An appraisal of some shortest path algorithms*. USA: Memorandum. Rand Corporation.

Dreyfus, S. E. (1969). An appraisal of some shortest path algorithms. *Operations Research*, *17*(3), 395–412. doi:10.1287/opre.17.3.395

Drilo, B., Saric, D., & Filjar, R. (2009, May). The role of telecommunications in development of new-generation intelligent transport systems. Proceedings of *Wireless Communication, Vehicular Technology, Information Theory and Aerospace & Electronic Systems Technology, 1st International Conference* (pp. 125-127). IEEE. doi:10.1109/WIRELESSVITAE.2009.5172435

Ergun, H., Rawn, B., Belmans, R., & Van Hertem, D. (2014). Technology and topology optimization for multizonal transmission systems. *IEEE Transactions on Power Systems*, *29*(5), 2469–2477. doi:10.1109/TPWRS.2014.2305174

Floyd, R. (1962). Algorithm 97: Shortest path. *Communications of the ACM*, *5*(6), 345. doi:10.1145/367766.368168

Gade, D., & Kuecuekyavuz, S. (2013). Formulations for dynamic lot sizing with service levels. *Naval Research Logistics*, *60*(2), 87–101. doi:10.1002/nav.21519

Gauvin, C., Desaulniers, G., & Gendreau, M. (2014). A branch-cut-and-price algorithm for the vehicle routing problem with stochastic demands. *Computers & Operations Research*, *50*, 141–153. doi:10.1016/j.cor.2014.03.028

Giuli, D., Paganelli, F., Cuomo, S., & Cianchi, P. (2013). Toward a Cooperative Approach for Continuous Innovation of Mobility Information Services. *IEEE Systems Journal*, *7*(4), 669–680. doi:10.1109/JSYST.2013.2255521

Goncalves, W. N., & Bruno, O. M. (2013). Combining fractal and deterministic walkers for texture analysis and classification. *Pattern Recognition*, *46*(11), 2953–2968. doi:10.1016/j.patcog.2013.03.012

Gupta, V., Mohapatra, D., Raghunathan, A., & Roy, K. (2013). Low-power digital signal processing using approximate adders. *IEEE Transactions on Computer-Aided Design of Integrated Circuits and Systems*, *32*(1), 124–137. doi:10.1109/TCAD.2012.2217962

Gutiérrez, V., Izaguirre, M., Pérez, J., Muñoz, L., López, D., & Sánchez, M. (2010, July). Ambient Intelligence in Intermodal Transport Services: A Practical Implementation in Road Logistics. Proceedings of *Sensor Technologies and Applications (SENSORCOMM) Fourth International Conference* (pp. 203-209). IEEE. doi:10.1109/SENSORCOMM.2010.107

Han, Y. (2004). Improved algorithm for all pairs shortest paths. *Information Processing Letters*, *91*(5), 245–250. doi:10.1016/j.ipl.2004.05.006

Haranczyk, M., & Sethian, J. A. (2009). Navigating molecular worms inside chemical labyrinths. *Proceedings of the National Academy of Sciences of the United States of America*, *106*(51), 21472–21477. doi:10.1073/pnas.0910016106 PMID:20018716

Höfner, P., & Möller, B. (2012). Dijkstra, Floyd and Warshall meet Kleene. *Formal Aspects of Computing*, *24*(4-6), 459–476. doi:10.1007/s00165-012-0245-4

Jakubauskas, G. (2006). Improvement of urban passenger transport ticketing systems by deploying intelligent transport systems. *Transport*, *21*(4), 252–259.

Jarašūnienė, A., & Batarlienė, N. (2013). Lithuanian road safety solutions based on intelligent transport systems. *Transport*, *28*(1), 97–107. doi:10.3846/16484142.2013.782895

Jiang, N., & Xie, C. (2014). Computing and analyzing mixed equilibrium network flows with gasoline and electric vehicles. *Computer-Aided Civil and Infrastructure Engineering*, *29*(8), 626–641.

Kajdanowicz, T., Indyk, W., Kazienko, P., & Kukul, J. (2012). Comparison of the Efficiency of MapReduce and Bulk Synchronous Parallel Approaches to Large Network Processing. In *12th IEEE International Conference on Data Mining Workshops* (pp. 218-225). Brussels, Belgium. doi:10.1109/ICDMW.2012.135

Kajdanowicz, T., Kazienko, P., & Indyk, W. (2014). Parallel processing of large graphs. *Future Generation Computer Systems-The International Journal of Grid Computing and Escience*, *32*, 324–337. doi:10.1016/j.future.2013.08.007

Kang, J. E., Chow, J. Y. J., & Recker, W. W. (2014). On activity-based network design problems. *Transportation Research Part B: Methodological*, *57*, 398–418. doi:10.1016/j.trb.2013.05.009

Kostakos, V., Camacho, T., & Mantero, C. (2013). Towards proximity-based passenger sensing on public transport buses. *Personal and Ubiquitous Computing*, *17*(8), 1807–1816. doi:10.1007/s00779-013-0652-4

Kristianto, Y., Gunasekaran, A., & Jiao, J. (2014). Logical reconfiguration of reconfigurable manufacturing systems with stream of variations modelling: A stochastic two-stage programming and shortest path model. *International Journal of Production Research*, *52*(5), 1401–1418. doi:10.1080/00207543.2013.843798

Letchford, A. N., Nasiri, S. D., & Oukil, A. (2014). Pricing routines for vehicle routing with time windows on road networks. *Computers & Operations Research*, *51*, 331–337. doi:10.1016/j.cor.2014.06.022

Liagouras, G. A., Sayegh, A. A., & Koutsakis, P. (2014). A new location-aware calendar-based application for dynamic minimum path trip planning. *Wireless Personal Communications*, *78*(1), 29–44. doi:10.1007/s11277-014-1732-0

Liang, Z., Chaovalitwongse, W. A., Cha, M., & Moon, S. B. (2010). Redundant multicast routing in multilayer networks with shared risk resource groups: Complexity, models and algorithms. *Computers & Operations Research*, *37*(10), 1731–1739. doi:10.1016/j.cor.2009.12.009

Liu, L., Hou, A., Biderman, A., Ratti, C., & Chen, J. (2009, October). Understanding individual and collective mobility patterns from smart card records: A case study in Shenzhen. Proceedings of *Intelligent Transportation Systems 12th International IEEE Conference* (pp. 1-6). IEEE. doi:10.1109/ITSC.2009.5309662

Liu, S., Duffy, A. H. B., Whitfield, R. I., & Boyle, I. M. (2010). Integration of decision support systems to improve decision support performance. *Knowledge and Information Systems*, *22*(3), 261–286. doi:10.1007/s10115-009-0192-4

Ma, Y., Wang, H. W., Xie, Y., & Guo, M. (2014). Path planning for multiple mobile robots under double-warehouse. *Information Sciences*, *278*, 357–379. doi:10.1016/j.ins.2014.03.058

Machuca, E., Mandow, L., de la Cruz, J. L. P., & Ruiz-Sepulveda, A. (2012). A comparison of heuristic best-first algorithms for bicriterion shortest path problems. *European Journal of Operational Research*, *217*(1), 44–53. doi:10.1016/j.ejor.2011.08.030

Majhi, S. K., & Chaupattanayak, S. (2013). Some interesting and novel results from analysis of shortest path searching algorithms. *International Journal of Advances in Computer Science and Cloud Computing*, *1*(1), 57–62.

Martinelli, R., Pecin, D., & Poggi, M. (2014). Efficient elementary and restricted non-elementary route pricing. *European Journal of Operational Research*, *239*(1), 102–111. doi:10.1016/j.ejor.2014.05.005

Masson, R., Ropke, S., Lehuede, F., & Peton, O. (2014). A branch-and-cut-and-price approach for the pickup and delivery problem with shuttle routes. *European Journal of Operational Research*, *236*(3), 849–862. doi:10.1016/j.ejor.2013.08.042

Mitroff, I. I., & Linstone, H. A. (1993). *The unbounded mind: Breaking the chains of traditional business thinking*. New York: Oxford Univ. Press.

Mounce, R., Hollier, G., Smith, M., Hodge, V. J., Jackson, T., & Austin, J. (2013). A metric for pattern-matching applications to traffic management. *Transportation Research Part C, Emerging Technologies*, *29*, 148–155. doi:10.1016/j.trc.2012.04.019

Muter, I., Cordeau, J.-F., & Laporte, G. A. (2014). Branch-and-price algorithm for the multidepot vehicle routing problem with interdepot routes. *Transportation Science*, *48*(3), 425–441. doi:10.1287/trsc.2013.0489

Nallaivarothayan, H., Ryan, D., Denman, S., Sridharan, S., Fookes, C., & Rakotonirainy, A. (2013). Detecting anomalous events at railway level crossings. *Proceedings of the Institution of Mechanical Engineers. Part F, Journal of Rail and Rapid Transit*, *227*(5), 539–553. doi:10.1177/0954409713501296

Nikolova, E., & Stier-Moses, N. E. (2014). A mean-risk model for the traffic assignment problem with stochastic travel times. *Operations Research*, *62*(2), 366–382. doi:10.1287/opre.2013.1246

Nishi, T., & Maeno, R. (2010). Petri net decomposition approach to optimization of route planning problems for AGV systems. *IEEE Transactions on Automation Science and Engineering*, *7*(3), 523–537. doi:10.1109/TASE.2010.2043096

Nwagboso, C. O. (1997). Advanced vehicle and infrastructure systems. Computer application, control and automation. Chapter 1. Introduction to intelligent transport systems. UK: Wiley (John) & Sons.

Pallottino, S., & Scutellà, M. G. (1997). Dual algorithms for the shortest path tree problem. *Networks*, *29*, 125–133. doi:10.1002/(SICI)1097-0037(199703)29:2<125::AID-NET7>3.0.CO;2-L

Peres, Y., Sotnikov, D., Sudakov, B., & Zwick, U. (2013). All-pairs shortest paths in $O(n^2)$ time with high probability. *Journal of the ACM*, *60*(4), 1-25.

Pinto, R. (2012). Stock rationing under service level constraints in a vertically integrated distribution system. *International Journal of Production Economics*, *136*(1), 231–240. doi:10.1016/j.ijpe.2011.11.023

Planken, L., de Weerdt, M., & Van der Krogt, R. (2012). Computing all-pairs shortest paths by leveraging low treewidth. *Journal of Artificial Intelligence Research*, *43*, 353–388.

Pratikakis, N. E., Realff, M. J., & Lee, J. H. (2009). Controlled exploration of state space in off-line ADP and its application to stochastic shortest path problems. *Computers & Chemical Engineering*, *33*(12), 2111–2122. doi:10.1016/j.compchemeng.2009.06.012

Qiu, Y., & Yagan, O. (2014). Robustness in clustering-based weighted inter-connected networks. *The European Physical Journal B*, *87*(4), 89. doi:10.1140/epjb/e2014-40466-1

Redekopp, M., Simmhan, Y., & Prasanna, V. K. (2013, May 20-24). Optimizations and Analysis of BSP Graph Processing Models on Public Clouds. Proceedings of *IEEE 27th International Parallel and Distributed Processing Symposium*. Boston.

Regan, M. A., Oxley, J. A., Godley, S. T., & Tingvall, C. (2001). *Intelligent transport systems: Safety and human factors issues*. Royal Automobile Club of Victoria.

Robin-Jouan, Y., Ehrlich, J., Guillaumin, B., Delarche, M., & Dutech, M. (2005, April). Transport-specific communication services: Safety-based or critical applications for mobiles and cooperation with infrastructure networks. *Annales des Télécommunications*, *60*(3-4), 405–440.

Ruan, Z., Hui, P., Lin, H., & Liu, Z. H. (2013). Risks of an epidemic in a two-layered railway-local area traveling network. *The European Physical Journal B*, *86*(1), 13. doi:10.1140/epjb/e2012-30292-x

Saksena, J. P., & Kumar, S. (1966). The routing problem with 'K' specified nodes. *Operations Research*, *14*(5), 909–913. doi:10.1287/opre.14.5.909

Santos, D., de Sousa, A., Alvelos, F., & Pioro, M. (2013). Optimizing network load balancing: An hybridization approach of metaheuristics with column generation. *Telecommunication Systems*, *52*(2), 959–968. doi:10.1007/s11235-011-9604-3

Sarvi, M., & Kuwahara, M. (2008). Using ITS to improve the capacity of freeway merging sections by transferring freight vehicles. *Intelligent Transportation Systems. IEEE Transactions on*, *9*(4), 580–588.

Satunin, S., & Babkin, E. (2014). A multi-agent approach to Intelligent Transportation Systems modeling with combinatorial auctions. *Expert Systems with Applications*, *41*(15), 6622–6633. doi:10.1016/j.eswa.2014.05.015

Schietekat, J. M., & Booysen, M. J. (2013, July). Detection of reckless driving in the Sub-Saharan informal public transportation system using acceleration-sensing telematics. Proceedings of EUROCON, (pp. 597-601). IEEE. doi:10.1109/EUROCON.2013.6625042

Schoenemann, T., & Cremers, D. (2010). A combinatorial solution for model-based image segmentation and real-time tracking. *IEEE Transactions on Pattern Analysis and Machine Intelligence*, *32*(7), 1153–1164. doi:10.1109/TPAMI.2009.79 PMID:20489221

JayadevaShah, S., Bhaya, A., Kothari, R., & Chandra, S. (2013). Ants find the shortest path: A mathematical proof. *Swarm Intelligence*, *7*(1), 43–62. doi:10.1007/s11721-013-0076-9

Shim, J. P., Warkentin, M., Courtney, J. F., Power, D. J., Sharda, R., & Carlsson, C. (2002). Past, present, and future of decision support technology. *Decision Support Systems*, *33*(2), 111–126. doi:10.1016/S0167-9236(01)00139-7

Shu-Xi, W. (2012). The Improved Dijkstra's Shortest Path Algorithm and Its Application. *Procedia Engineering*, *29*, 1186–1190. doi:10.1016/j.proeng.2012.01.110

Solyal, O., & Sural, H. (2012). The one-warehouse multi-retailer problem: Reformulation, classification, and computational results. *Annals of Operations Research*, *196*(1), 517–541. doi:10.1007/s10479-011-1022-0

Steenbrink, P. A. (1974). *Optimization of transport networks*. London: J. Wiley.

Sternberg, H., & Andersson, M. (2014). Decentralized intelligence in freight transport—A critical review. *Computers in Industry*, *65*(2), 306–313. doi:10.1016/j.compind.2013.11.011

Sugumaran, V., & Sugumaran, R. (2007). Web-based spatial decision support systems (WEBSDSS): Evolution, architecture, examples and challenges. *Communications of the Association for Information Systems*, *19*, 844–875.

NiSullivan, B., Tang, Y., & Marquis, C. (2014). Persistently learning: How small-world network imprints affect subsequent firm learning. *Strategic Organization*, *12*(3), 180–199. doi:10.1177/1476127014543772

Sun, X., Lu, H., & Wu, J. (2013). Bus detection based on sparse representation for transit signal priority. *Neurocomputing*, *118*, 1–9. doi:10.1016/j.neucom.2013.02.006

Taş, D., Gendreau, M., Dellaert, N., van Woensel, T., & de Kok, A. G. (2014). Vehicle routing with soft time windows and stochastic travel times: A column generation and branch-and-price solution approach. *European Journal of Operational Research*, *236*(3), 789–799. doi:10.1016/j.ejor.2013.05.024

Tsatcha, D., Saux, E., & Claramunt, C. A. (2014). Bidirectional path-finding algorithm and data structure for maritime routing. *International Journal of Geographical Information Science*, *28*(7), 1355–1377. doi:10.1080/13658816.2014.887087

Van Vliet, D. (1978). Improved shortest path algorithms for transport networks. *Transportation Research*, *12*(1), 7–20. doi:10.1016/0041-1647(78)90102-8

CanoViktorsson, C. (2013). Traffic Radio as a Precursor to Smart Travel Planning Systems: The Challenge of Organizing "Collective Intelligence". *Journal of Urban Technology*, *20*(4), 43–55. doi:10.1080/10630732.2013.863439

Warshall, S. (1962). A theorem on Boolean matrices. *Journal of the ACM*, *9*(1), 11–12. doi:10.1145/321105.321107

Wikborg, U., & Lee, T. E. (2013). Noncyclic scheduling for timed discrete-event systems with application to single-armed cluster tools using pareto-optimal optimization. *IEEE Transactions on Automation Science and Engineering*, *10*(3), 699–710. doi:10.1109/TASE.2012.2217128

Xiong, G., & Wang, Y. (2014). Best routes selection in multimodal networks using multi-objective genetic algorithm. *Journal of Combinatorial Optimization*, *28*(3), 655–673. doi:10.1007/s10878-012-9574-8

Yan, Z., & Yuan, C. W. (2003). Ant colony optimization for navigating complex labyrinths. In G. Y. Wang, Q. Liu, Y. Y. Yao, & A. Skowron (Eds.), *Rough Sets, Fuzzy Sets, Data Mining, and Granular Computing, Lecture Notes in Artificial Intelligence, Volume:2639* (pp. 445–448). Berlin: Springer Verlag. doi:10.1007/3-540-39205-X_74

Yates, J., Wang, X. H., & Chen, N. N. (2014). Assessing the effectiveness of k-shortest path sets in problems of network interdiction. *Optimization and Engineering*, *15*(3), 721–749. doi:10.1007/s11081-013-9220-z

Zhang, X. G., Zhang, Y. J., Zhang, Z. L., Mahadevan, S., Adamatzky, A., & Deng, Y. (2014). Rapid physarum algorithm for shortest path problem. *Applied Soft Computing*, *23*, 19–26. doi:10.1016/j.asoc.2014.05.032

Zuazola, G I.J., Moreno, A., Landaluce, H., Angulo, I., Perallos, A., Hernández-Jayo, U., Sainz, N., Ziai, M.A., Batchelor, J.C., Elmirghani, J.M.H. (2013). Telematics system for the intelligent transport and distribution of medicines. *Intelligent Transport Systems, 7*(1), 131-137.

KEY TERMS AND DEFINITIONS

Algorithm: A list of procedures to solve certain problems in computer science.
Public Transport Alternatives: List of all available transport modes in a study area.
Shortest Path: The minimum distance between two points in a given space.
Traveler: Any individual who needs a support in his/her decisions among travel alternatives.

Chapter 7
Modeling the Change of Urban Spatial Structure: Use Interzonal Travel Data to Estimate Urban Growth and Expansion by Hierarchical Cluster Analyses

Darcin Akin
Gediz University, Turkey

Serdar Alasalvar
Istanbul Greater Metropolitan Municipality, Turkey

ABSTRACT

The Urban spatial structure is affected by spatial interactions among various activity locations, and land uses in the city over the transportation system. Each city has its unique circulation pattern of passengers and freight due to its unique geographic conditions and the distribution of locations of economic activities. In that sense, it is claimed in this chapter per the authors that urban spatial structure can be modeled using interzonal (O/D) travel data. Thus, the chapter presents a case study of modeling spatial structures developed by employing Hierarchical Cluster Analysis (HCA) using travel pattern data for current and future scenarios. As a result, urban growth and expansion were estimated based on the level of interaction (represented by distance or similarity modeled based on trip interchanges) over the transportation system in terms of population and/or employment increases. The interaction was described by a measure of distance or similarity, modeled with respect to trip interchanges.

INTRODUCTION

Elements of the urban transportation system, namely, types, capacity and characteristics of travel modes, infrastructures, and users, and the cost or time of travel, have a unique spatial imprint which shapes the urban form. Each city presents different socioeconomic and geographical characteristics; therefore, the spatial imprint of the transportation system differs correspondingly (Rodrigue et al., 2013). For instance,

DOI: 10.4018/978-1-4666-8648-9.ch007

while some cities have an urban form shaped by mainly highway modes, and others have different spatial forms compared to the former because various public transport modes, including non-motorized modes and infrastructure developments are taken place in a more balanced transportation system. Urban spatial structure is affected by spatial interactions among different activity locations and land uses over the city's transport system since each city has its unique circulation pattern of passengers and goods. In that sense, it is claimed in this chapter per the authors that urban spatial structure can be modeled using interzonal (O/D) travel data. OD data are the outputs of the trip distribution step from the travel demand modeling (TDM) process. Thus, by comparing the spatial structures obtained from the models created by employing Hierarchical Cluster Analysis (HCA) for a base year and future scenarios, urban growth and expansion can be estimated in terms of population and/or employment changes, based on the level of interaction, represented by distance or similarity calculated based on trip interchanges, over the transportation network. For example, a network with substantial highways projects and the other having substantial public transportation investments can be compared with respect to the spatial structure impacted by the dominant elements of the transportation system such as highway, public transit modes, and major transport terminals. This chapter presents a case study of the analyses of urban spatial structures of Sakarya Province of Turkey (Figure 1) for base and future scenarios. Base year and future urban structures were modeled and compared to estimate the urban growth and expansion for years 2012 and 2023. Results of the analyses can be used to confirm the growth and the expansion of urban city core as well as other urban and suburban developments and their emergence with their neighboring clusters. Decision makers will be able to develop their capacities to plan ahead in order to meet future needs better from the spatial restructuring in urban agglomerations, when they know the extent and location of the future growth ahead.

BACKGROUND

In urban settings, characteristics of transportation system and infrastructure, and the attractiveness of transport modes shape mobility of travelers and demographic growth. Urban form is the spatial imprint of an urban transportation network. Therefore, "*urban (spatial) structure can be defined as the set of relationships arising from urban form and the underlying movements of goods and people*" (Rodrigue et al., 2013).

Transportation systems have long been recognized as vital parts of city systems. They affect the quality of live (Vuchic, 1999) and have strong effects on shaping the form of settlements. Urban transport systems can mainly be grouped into public, private, and freight transport systems. In most cases, they can be complementary, but they may usually compete against each other under the lack of a comprehensive planning effort (Rodrigue et al., 2013).

- Public Transportation: The purpose is to provide publicly accessible mobility where it is demanded, especially in urban settings. Its efficiency is based on the cost per passenger or traveled km. Major public transport modes must be operated in a fully or semi-exclusive right-of-way in order to provide high capacity and reliability. It includes modes such as cable (aerial tram, funicular, and so on), rail (tram, light rail, metro, and so on), and rubber tired modes (bus and minibus). Livable

Figure 1. Map of Turkey and Sakarya Province

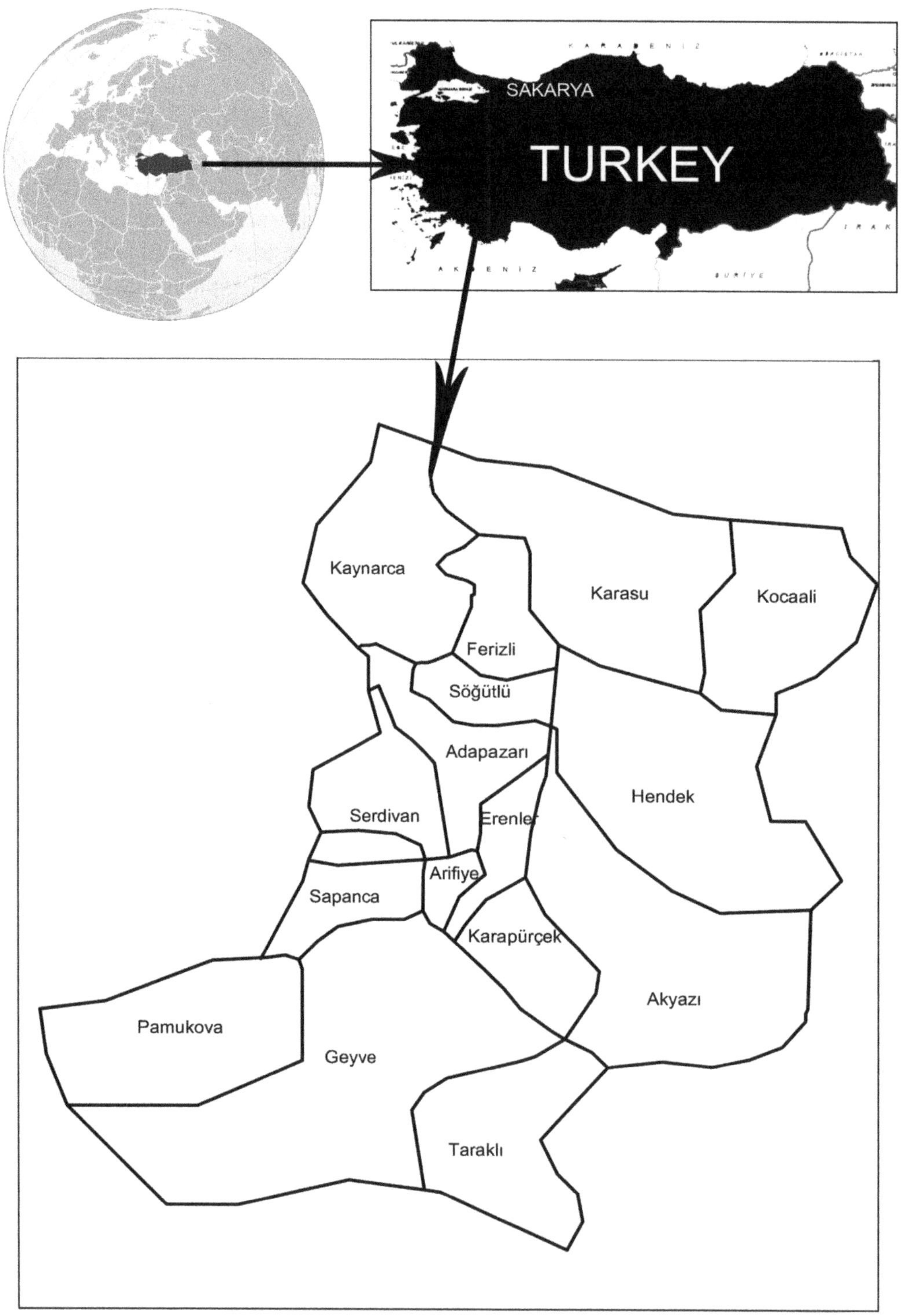

cities require intermodal transport systems in which public transit systems in addition to highway network must be provided in a balanced way for maximizing the benefits of the whole system. Moreover, provisions for pedestrians, bicyclists, and paratransit should not be excluded.

- Private Transportation: This includes any individual mode of transportation, including cars, bikes, motorcycles, and walking, of course. Consequences of excessive automobile dependence have been properly noted in the literature (Vuchic, 1999). On the other hand, fairly good number of people choose to walk, but the share of walking varies depends on many factors such as the comfort, availability and cost of transport modes, and other location-specific characteristics. While walking shares a great percent (88%) of all movements in Tokyo, this number is only 3% for Los Angeles. In major Turkish cities, walking constitutes up to 50% because of its convenience for some groups of people.
- Freight Transportation: Since cities have land uses which are dedicated as the origins of production and consumption, significant movements of freight are observed in urbanized areas as well. Thus, freights carried via trucks in urban areas create significant loads on the highway network between production and distribution centers, warehouses and retail activities as well as from/to primary terminals such as ports, rail yards, distribution centers and airports.

Cities are traditional places of economic activities, which stem from the interactions between people and infrastructures over the transport system. Traditional urban theories study how cities develop and grow through systematic interactions among production systems that benefit from co-location, have been identified as chief forces in the growth of cities. However, over the last decades the traditional understanding that the only outcomes of these forces should be an accelerating concentration of population, infrastructures and jobs has been challenged by the evidence of deconcentration, first in the United States, Canada and Australia, and later in Europe. The immigration of rural populations into the city known as a centuries old factor in rural depopulation and the dominant force in creating urban agglomerations is now giving way to a reverse migration into the countryside, at least in many western cities, as urban growth is experienced in terms of suburbanization and sprawl of urbanized areas (Besussi, Chin, Batty, & Longley, 2010).

Suburban or rural sprawling, or dispersion, has been observed in many urban agglemorations and in different types of cities, from dense, centralized European metropolitan cities such as Madrid, Paris, London and Dublin, to rapidly industrializing metropolitans such as Seoul, Shanghai, and Buenos Aires, to those experiencing recent, fast and uncontrolled urban growth, such as Istanbul, Bombay and Lagos. The evolution of transportation technology and modes has resulted in changes of urban form. The more radical the changes in transport technology and modes, the more the urban form has been altered. Vuchic (1999) documented the complexity of the relationship between the transport system and the character of cities and metropolitan regions. Among the most fundamental changes in urban form can be listed as the emergence of new clusters expressing new urban activities and new relationships between elements of the urban system due to faster and more economical transport modes (Rodrigue et al.,2013).

In many cities, downtown or the central business district (CBD), was once the primary destination of suburban commuters and served by public transportation. Later on this has been changed by new manufacturing, retailing, and management practices located in suburbs. Whereas traditional manufacturing depended on centralized workplaces and transportation, technological and transportation developments rendered modern industry more flexible. In many cases, manufacturing has been relocated in suburban areas, if not altogether to entirely new low-cost locations. Retail and office activities are also suburbanizing, which changes the urban form (Rodrigue et al., 2013).

There is considerable research in structural changes and the patterning of cities, but much of this has been focused on explaining urban structure and form at a single point in time, as if cities were in some sort of perpetual equilibrium. Clearly the absence of favoring data through time has been a major constraint on the ability to produce an appropriate science of urban dynamics and thus most of the thinking about urban change has been speculative and non-rigorous. However, this has already been changing. New data sets to control and manage urban sprawl, and new techniques such as urban remote sensing to process information from satellite and aerial photographic data, are becoming important tools for modeling the changes in urban form and structure (Besussi et al., 2010).

Thel lack of quantitative data related to urban (spatial) structure and city systems in developing countries sets barriers on the research with respect to spatial structure and urban changes (Ozus et al., 2012). In a research by Vliet et al. (2009), the constrained cellular automata (CA) was used for modeling changes in land use and urban growth in the Greater Vancouver Regional District (GVRD) between 1996 and 2001. Results showed that the model performed considerably better than a neutral allocation model. Cluster and radial analysis indicates that the model was capable of producing realistic urban growth patterns (Vliet et al., 2009). Various types of factors mainly affecting the geographical delineation of urban clusters can be named as the demographic characteristics of the cities, the number of public and private sector services located in them, as well as the total length of the road network in the study area. Furthermore, a comparative urban concentration indicator (UCI) was devised to reflect the role and the relative weight of every urbanized area in the study region as well as its spatial influence. Both the proposed methodological framework and the UCI were applied to the region of Thessaly, Greece (Tsompanoglou & Photis, 2013).

In recent years, there have been some efforts to model the urban growth and expansion using clustering techniques. For example, the research by Akin (2014), Ozus et al. (2012), and Akın (2011) employed hierarachical clustering (HC) in modeling the change of urban structure. Hierarchical clustering algorithms (HCA) and partitioning algorithms are two major clustering methods. Partitioning algorithms divide a data set into a single partition. However, hierarchical algorithms divide a data set into a sequence of nested partitions. In HCA, the clusters are represented hierarchically through a dendogram, at varying levels of granularity. Two major hierarchical algorithms can be named, such as agglomerative algorithms, and divisive algorithms. Depending upon whether this hierarchical representation is created in top-down or bottom-up fashion, these representations may be considered either agglomerative or divisive (Aggarwal & Reddy, 2013). Agglomerative algorithms start with every single object in a single cluster, while divisive ones start with all objects in one cluster and repeat splitting large clusters into small pieces (Gan et al., 2007). In general, clustering is an exploratory data mining, and a common technique for statistical data analysis, used in many fields, such as machine learning, pattern recognition, image processing, artificial intelligence, information technology, and biology, psychology, and marketing. Such analysis procedures help grouping a set of objects (zones in urban modeling) into clusters, where those objects are more similar based on defined measurements to each other than to those in other clusters (Jain et al., 1999). As such, researchers often find it very difficult to determine a proper algorithm for their applications and/or to compare novel ideas with existing results. The type of data is directly associated with data clustering, and it is a major factor to consider in choosing an appropriate clustering algorithm. There are many data types, and once researchers get familiar with data types, they should be able to adjust the algorithms accordingly. In many applications of cluster analysis, the raw data, or actual measurements, are not used directly unless a probabilistic model for pattern generation is available (Jain and Dubes, 1988). Preparing data for cluster analysis requires some sort of transformation, such as standardization or normalization. Data standardization makes data dimensionless. It is useful for defining standard indices. After standardization, all knowledge of the location and scale of the original data may be lost. It is necessary to standardize variables in cases where the dissimilarity measure, such as the Euclidean distance, is sensitive to the differences in the magnitudes or scales of the input variables (Milligan &Cooper, 1988; Aggarwal & Reddy, 2013). In many cases, raw data should be normalized and/or transformed before a cluster analysis is performed. A reason to do normalization is that objects in raw data may be described by variables measured with different scales, and another is to reduce the size of the data to improve the effectiveness of clustering algorithms (Gan et al., 2007).

A literature review revealed that studies with the subject of the multicenteric development course of cities were very few in developing countries due to the lack of quality data (Akin, 2014; Ozus et al., 2012; Akın, 2011; Zhao et al., 2011; Berkoz & Turk, 2010; Alpkokin et al., 2008; Bhandari et al., 2007; Bhandari et al., 2007; Alpkokin et al., 2005; Berkoz, 2000; Dokmeci & Berkoz, 1994). Thus, in this chapter the relationship between urban growth and travel patterns was studied as a case study of the province of Sakarya, Turkey, by using hierarchical cluster analyses.

DATA AND METHODOLOGY

Data that the analyses are based upon were borrowed from the case study of the metropolitan city of Sakarya. During 2012 and 2013, the Sakarya Transportation Master Plan for 2023 was prepared based on 2035 Landuse Development Plan. Population and employment projections were revised for years 2018 and 2023. A 4-step travel demand model (TDM) was developed, and calibrated for 2012 and validated for 2023.

Data

Interzonal travel data are the observed travel patterns of passengers for 2012, but the projected travel interactions resulted from the 4-step TDM in the Sakarya Transportation Master Plan. There are 16 districts in the province of Sakarya, and their population and employment details are given in Table 1. Growth of the urban area is expected due to population and employment increases due to the developments of industrial areas and logistics investments in the study area. Total population in the Province is projected to change from 888,556 to 1,156,227 by 30.1%. In overall employment, a change occurs from 176,756 to 244,861 by 38.5%. Furthermore, it is seen that districts far from the city center towards the north such as Kocaali and Kaynarca experience the least population increases 6 and 4.2% from 2012 to 2023, respectively. Karasu is also a northern district, but it has the advantage as a domestic tourism destination during summer months and therefore experiences 19.6% increase in population. On the other hand, Ferizli and Hendek districts experience great percent of employment increase by 121.8% and 73.7%, respectively due to the introduction of new organized industrial campuses. Pamukova also known for its vast agricultural lands has fairly good increase in population by 46.9%.

The districts in Table 1 are considered as the zones used in the analyses of travel patterns. Average daily weekday trip interchange matrix (T_{ij}) for 2012 (base year) is provided in Table 2 and for 2023 (fu-

Table 1. Populations and employment data for the 16 districts of Sakarya Province

No.	Districts	Population			Employment		
		2011 (SIS)	2023 (Projection)	% Change	2011 (Estimation)	2023 (Projection)	% Change
1	Adapazari	251,680	295,406	17.4	59,158	78,662	33.0
2	Akyazi	83,497	111,260	33.3	12,251	16,787	37.0
3	Arifiye	37,889	56,791	49.9	28,331	29,805	5.2
4	Erenler	75,682	96,667	27.7	23,094	36,578	58.4
5	Ferizli	23,654	32,805	38.7	1,685	3,737	121.8
6	Geyve	46,892	62,131	32.5	806	883	9.6
7	Hendek	73,918	98,050	32.6	17,078	29,662	73.7
8	Karapürcek	12,311	18,336	48.9	1,647	2,273	38.0
9	Karasu	53,928	64,496	19.6	N/A	N/A	N/A
10	Kocaali	22,203	23,524	6.0	N/A	N/A	N/A
11	Kaynarca	23,290	24,276	4.2	N/A	N/A	N/A
12	Pamukova	26,978	39,623	46.9	N/A	N/A	N/A
13	Sapanca	38,089	50,485	32.5	9,023	15,322	69.8
14	Serdivan	97,044	152,603	57.3	17,258	22,620	31.1
15	Söğütlü	14,229	18,252	28.3	6,425	8,532	32.8
16	Tarakli	7,272	11,521	58.4	N/A	N/A	N/A
17	Total	888,556	1,156,227	30.1	176,756	244,861	38.5

Note: N/A "not available"

ture plan year) in Table 3. A total of 1.85 million trips was observed within the Province of Sakarya in 2012. Total number of trips estimated for 2023 is around 1.9 million with the exclusion of five remote districts from the 4-step TDM due to a nonconformity in the calibration process.

Methodology

Hierarchical cluster analysis (HCA) was employed to analyze the data for 2012 (model calibration year) in Table 2 and for 2023 (future plan year) in Table 3. In the HCA, the analysis starts with an input matrix *D(i, j)* with i measurements (observations) of j variables, denoting similarity or distance among the objects. In data clustering, a similarity measure or distance (dissimilarity measure) is used to quantitatively describe the similarity or dissimilarity of two data points or two clusters. Similarity and distance measures are fundemental elements of a clustering algorithm, without which any meaningful cluster analysis cannot be obtained (Gan et al., 2007).

Distance-based methods are very popular in the literature, because they can be used with almost any data type, as long as an appropriate distance function is created for that data type. Thus, the problem of clustering can be reduced to the problem of finding a distance function for that data type. Therefore, distance function design has itself become an important area of research for data mining in itsown right (Aggarwal, 2003; Wang & Sun, 2012; Aggarwal & Reddy, 2013). The greater the similarity measure d_{ij} is, the more the resemblance is between the objects. The greater the distance measure d_{ij} is, then the

Table 2. Average weekday daily trip interchange matrix (T_{ij}) via all modes among the 16 districts of Sakarya for 2012

T_{ij}	1	2	3	4	5	6	7	8	9	10	11	12	13	14	15	16	Total
1- Adapazari	400,176	2,728	22,611	35,256	1,863	1,018	2,359	378	1,048	1,838	195	1,075	3,588	39,502	1,880	106	515,622
2-Akyazi	7,301	134,612	1,322	3,428	154	146	3,089	755	73	0	0	73	73	1,441	0	90	152,558
3-Arifiye	18,077	389	53,933	3,879	0	1,490	687	0	95	36	36	224	1,587	3,742	53	0	84,230
4-Erenler	48,033	1,145	6,477	94,327	202	542	1,217	263	196	159	0	498	1,450	6,236	190	53	160,984
5-Ferizli	3,316	73	795	294	31,409	73	0	0	526	47	0	109	248	669	2,772	0	40,331
6-Geyve	4,425	120	905	1,086	73	82,198	190	0	36	0	0	1,446	452	266	365	290	91,851
7-Hendek	6,144	2,341	2,021	1,041	0	47	149,791	53	95	106	95	146	184	1,353	53	0	163,471
8-Karapürcek	2,086	2,721	73	1,184	0	0	120	15,395	0	0	47	0	0	168	73	0	21,869
9-Karasu	3,111	0	529	305	464	81	176	0	88,612	126	918	0	0	758	564	0	95,642
10-Kocaali	2,907	53	378	378	332	0	53	0	316	60,152	0	0	36	941	142	0	65,687
11-Kaynarca	1,201	0	296	413	95	0	0	47	2,329	0	35,635	0	0	0	73	0	40,090
12-Pamukova	1,236	0	429	94	0	1,729	0	0	0	0	0	57,143	95	182	0	0	60,907
13-Sapanca	8,936	100	3,196	2,055	0	190	290	0	0	0	0	95	66,911	1,959	73	0	83,805
14-Serdivan	72,484	726	7,248	7,456	53	492	954	0	576	154	95	338	1,753	111,770	425	0	204,524
15-Söğütlü	10,808	36	1,105	1,197	1,461	168	53	95	316	0	47	0	146	1,289	36,075	0	52,796
16-Tarakli	310	53	0	100	0	243	0	0	0	0	290	0	0	95	0	15,870	16,961
Total	590,550	145,100	101,318	152,491	36,105	88,418	158,980	16,987	94,218	62,618	37,359	61,146	76,523	170,368	42,738	16,408	1,851,328

Table 3. Average weekday daily trip interchange matrix (T_{ij}) via all modes among the 16 Districts of Sakarya for 2023

T_{ij}	1	2	3	4	5	6	7	8	9	10	11	12	13	14	15	16	Total
1- Adapazari	363,836	25,397	32,328	67,845	9,419	2,363	32,732	5,565	N/A	N/A	N/A	N/A	20,028	69,142	22,118	N/A	650,773
2-Akyazi	22,181	48,153	5,392	9,638	974	556	15,856	3,899	N/A	N/A	N/A	N/A	4,141	6,300	2,085	N/A	119,174
3-Arifiye	41,871	7,676	41,853	17,584	1,455	1,879	9,666	1,800	N/A	N/A	N/A	N/A	9,765	15,151	3,153	N/A	151,853
4-Erenler	80,657	12,985	15,424	61,113	2,347	1,184	15,264	3,259	N/A	N/A	N/A	N/A	7,552	20,952	4,892	N/A	225,630
5-Ferizli	8,230	1,069	1,204	2,053	11,120	103	1,481	198	N/A	N/A	N/A	N/A	759	1,933	3,352	N/A	31,499
6-Geyve	3,568	962	1,176	1,243	160	1,762	1,117	218	N/A	N/A	N/A	N/A	1,186	1,176	391	N/A	12,959
7-Hendek	22,218	12,188	5,098	9,114	1,036	482	82,419	1,753	N/A	N/A	N/A	N/A	3,629	6,356	2,182	N/A	146,475
8-Karapürcek	8,048	6,735	2,014	3,797	329	227	3,617	6,696	N/A	N/A	N/A	N/A	1,322	2,366	725	N/A	35,874
9-Karasu	N/A	N/A	N/A	N/A	N/A	N/A	N/A	N/A	N/A	N/A	N/A	N/A	N/A	N/A	N/A	N/A	N/A
10-Kocaali	N/A	N/A	N/A	N/A	N/A	N/A	N/A	N/A	N/A	N/A	N/A	N/A	N/A	N/A	N/A	N/A	N/A
11-Kaynarca	N/A	N/A	N/A	N/A	N/A	N/A	N/A	N/A	N/A	N/A	N/A	N/A	N/A	N/A	N/A	N/A	N/A
12-Pamukova	N/A	N/A	N/A	N/A	N/A	N/A	N/A	N/A	N/A	N/A	N/A	N/A	N/A	N/A	N/A	N/A	N/A
13-Sapanca	25,463	6,010	8,598	9,247	978	1,483	6,856	1,243	N/A	N/A	N/A	N/A	47,709	9,593	2,208	N/A	119,389
14-Serdivan	148,705	13,567	20,136	37,212	4,029	1,785	17,807	3,080	N/A	N/A	N/A	N/A	14,240	98,323	9,559	N/A	368,442
15-Söğütlü	12,775	1,514	1,535	2,849	2,088	155	2,002	313	N/A	N/A	N/A	N/A	1,061	2,866	11,591	N/A	38,750
16-Tarakli	N/A	N/A	N/A	N/A	N/A	N/A	N/A	N/A	N/A	N/A	N/A	N/A	N/A	N/A	N/A	N/A	N/A
Total	737,551	136,255	134,759	221,694	33,935	11,979	188,816	28,025	N/A	N/A	N/A	N/A	111,393	234,157	62,254	N/A	1,900,818

Note: N/A "not available"

higher the dissimilarity is. In determining similarity measure to use in HCA, the nature of the observations plays a significant role (Hardle & Simar, 2007). In HCA, there are various measurements of distance. In this study, the "correlation distance" defined in terms of the correlation coefficient (r_{ij}) is seen in Eq. 1:

$$d(i,j) = 1 - r_{ij} \tag{1}$$

The correlation distance d_{ij} yields values between 0 and 1 for positive correlations, and between 1 and 2 for negative correlations. In this approach, positively correlated measurements are to be less apart than negatively correlated ones in the clustering analysis. Thus, in this regard the choice of the similarity measure looks just fine for the purpose of analyses of this study.

In agglomerative methods, a bottom-up approach is used, in which starting points are individual data points and successively merge clusters in order to create a tree-like structure. A variety of choices are possible in terms of how these clusters may be merged, which provide different tradeoffs between quality and efficiency. Some examples of these choices are single-linkage, all-pairs linkage, centroid-linkage, and sampled-linkage clustering. In single-linkage clustering, the shortest distance between any pair of points in two clusters is used. In all-pairs linkage, the average over all pairs is used, whereas in sampled linkage, a sampling of the data points in the two clusters is used for calculating the average distance. In centroid-linkage, the distance between the centroids is used. Some variations of these methods have the disadvantage of chaining, in which larger clusters are naturally biased toward having closer distances to other points and will therefore attract a successively larger number of points. Single linkage clustering is particularly susceptible to this phenomenon. A number of data domains such as network clustering are also more susceptible to this behavior (Aggarwal & Reedy, 2013). In this study, single, complete, and average linkage rules are employed as the commonly used methods to calculate the interzonal distances when there are multiple zones that may be clustered together. Detailed definitions of the linkage rules can be seen in Tan et al. (2006), Johnson & Wichern (1992), and Milligan (1980).

Single linkage rule is based on the cluster proximity between the closest two points that are in different clusters. When points x and y are in cluster C_1, and w and z are in C_2, then the distance (d) between two clusters, C_1 and C_2, is given as follows:

$$d(C_1, C_2) = \min(d(x,w),\ d(x,z),\ d(y,w),\ d(y,z)) \tag{2}$$

Complete linkage rule, however, alternatively takes the proximity between the farthest two points in different clusters as the cluster proximity. Therefore, the results can be sensitive to outliers. The distance between two clusters, C_1 and C_2, is then formulated in Eq. 3:

$$d(C_1, C_2) = \max(d(x,w),\ d(x,z),\ d(y,w),\ d(y,z)) \tag{3}$$

Another method as an alternative to the two techniques above, the group average rule defines the cluster proximity as the average pairwise proximities of all pairs of points among different clusters. In this case, the distance between clusters C_1 and C_2 with the number of observations n_1 and n_2, respectively, is defined in Eq. 4.

$$d(C_1, C_2) = \frac{\sum_{\substack{x \in C_1 \\ y \in C_2}} d(x, y)}{n_1 * n_2} \tag{4}$$

HCA Results for the Base year: 2012

Data visualization is the final step of data mining applications. The HCA was performed with both raw and normalized data, and it was observed that the results did not change. The travel data was normalized as follows:

$$t_{ij} = \frac{T_{ij}}{T_j} \tag{5}$$

where T_{ij}= number of average weekday daily travel volumes between districts i and j presented in Tables 2 and 3, T_j= total number of trips attracted to zones j, and t_{ij}=normalized travel volumes between zones i and j.

Thus, the HCA results for 2012 using the single, complete and average linkage rules are presented in Tables 4, 5, and 6, and visualized throrugh dendograms and maps in Figures 2, 3, and 4.

Table 4. Cluster Results for 2012 using the Single Linkage Method

Steps	No. of Clusters	Similarity	Distance	Clusters Joined	New Clusters	No. of Observations in New Clusters
1	15	71.2494	0.57501	1 - 14	1	2
2	14	69.0888	0.61822	1 - 4	1	3
3	13	67.4679	0.65064	1 - 3	1	4
4	12	52.2583	0.95483	5 - 15	5	2
5	11	49.7659	1.00468	2 - 8	2	2
6	10	49.0583	1.01883	9 - 11	9	2
7	9	48.8641	1.02272	1 - 13	1	5
8	8	48.6441	1.02712	6 - 12	6	2
9	7	48.567	1.02866	1 - 5	1	7
10	6	48.2231	1.03554	2 - 7	2	3
11	5	47.4142	1.05172	6 - 16	6	3
12	4	47.1029	1.05794	1 - 2	1	10
13	3	47.0503	1.05899	1 - 9	1	12
14	2	46.8978	1.06204	1 - 10	1	13
15	1	46.8072	1.06386	1 - 6	1	16

(C)luster 1: 1- Adapazarı 3-Arifiye 4-Erenler 5-Ferizli 13-Sapanca 14-Serdivan 15-Söğütlü; **(C)luster 2:** 2-Akyazı 7-Hendek 8-Karapürçek; **(C)luster 3:** 6-Geyve 12-Pamukova; **(C)luster 4:** 9-Karasu 11-Kaynarca; **(C)luster 5:** 10-Kocaali; **(C)luster 6:** 16-Taraklı.

Table 5. Cluster Results for 2012 using the Complete Linkage Method

Steps	No. of Clusters	Similarity	Distance	Clusters Joined	New Clusters	No. of Observations in New Clusters
1	15	71.2494	0.57501	1 - 14	1	2
2	14	57.3807	0.85239	1 - 3	1	3
3	13	55.8608	0.88278	1 - 4	1	4
4	12	52.2583	0.95483	5 - 15	5	2
5	11	49.7659	1.00468	2 - 8	2	2
6	10	49.0583	1.01883	9 - 11	9	2
7	9	48.6441	1.02712	6 - 12	6	2
8	8	47.4416	1.05117	1 - 13	1	5
9	7	46.4364	1.07127	7 - 10	7	2
10	6	46.3447	1.07311	7 - 9	7	4
11	5	46.3222	1.07356	6 - 16	6	3
12	4	46.1933	1.07613	6 - 7	6	7
13	3	46.0312	1.07938	2 - 6	2	9
14	2	45.7703	1.08459	2 - 5	2	11
15	1	43.8895	1.12221	1 - 2	1	16

(C)luster 1: 1- Adapazarı 3-Arifiye 4-Erenler 13-Sapanca 14-Serdivan; **(C)luster 2:** 2-Akyazı 8-Karapürçek; **(C)luster 3:** 5-Ferizli 15-Söğütlü; **(C)luster 4:** 6-Geyve 12-Pamukova; **(C)luster 5:** 7-Hendek 9-Karasu 10-Kocaali, 11-Kaynarca; **(C)luster 6:** 16-Taraklı.

Table 6. Cluster Results for 2012 using the Average Linkage Method

Steps	No. of Clusters	Similarity	Distance	Clusters Joined	New Clusters	No. of Observations in New Clusters
1	15	71.2494	0.57501	1 - 14	1	2
2	14	62.4748	0.75050	1 - 4	1	3
3	13	60.6710	0.78658	1 - 3	1	4
4	12	52.2583	0.95483	5 - 15	5	2
5	11	49.7659	1.00468	2 - 8	2	2
6	10	49.0583	1.01883	9 - 11	9	2
7	9	48.6441	1.02712	6 - 12	6	2
8	8	48.2057	1.03589	1 - 13	1	5
9	7	47.2974	1.05405	2 - 7	2	3
10	6	46.8682	1.06264	6 - 16	6	3
11	5	46.5197	1.06961	5 - 9	5	4
12	4	46.4506	1.07099	5 - 10	5	5
13	3	46.2590	1.07482	2 - 6	2	6
14	2	46.1821	1.07636	2 - 5	2	11
15	1	45.7265	1.08547	1 - 2	1	16

(C)luster 1: 1- Adapazarı 3-Arifiye 4-Erenler 13-Sapanca 14-Serdivan; **(C)luster 2:** 2-Akyazı 7-Hendek 8-Karapürçek; **(C)luster 3:** 5-Ferizli 15-Söğütlü; **(C)luster 4:** 6-Geyve 12-Pamukova 16-Taraklı; **(C)luster 5:** 9-Karasu 11-Kaynarca; **(C)luster 6:** 10-Kocaali.

Figure 2. Clustering results and the map of single linkage method for 2012

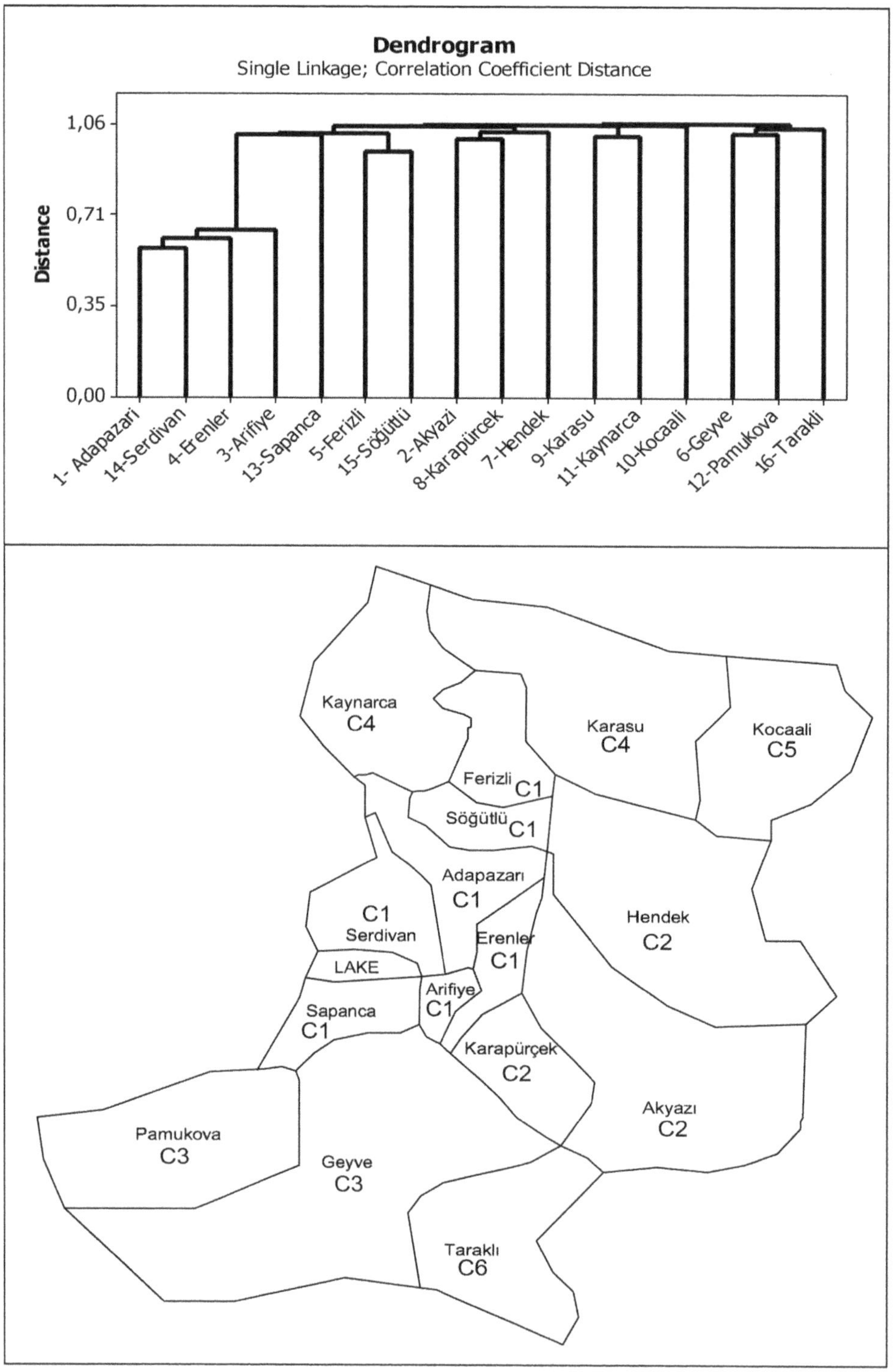

Figure 3. Clustering results and the map of complete linkage method for 2012

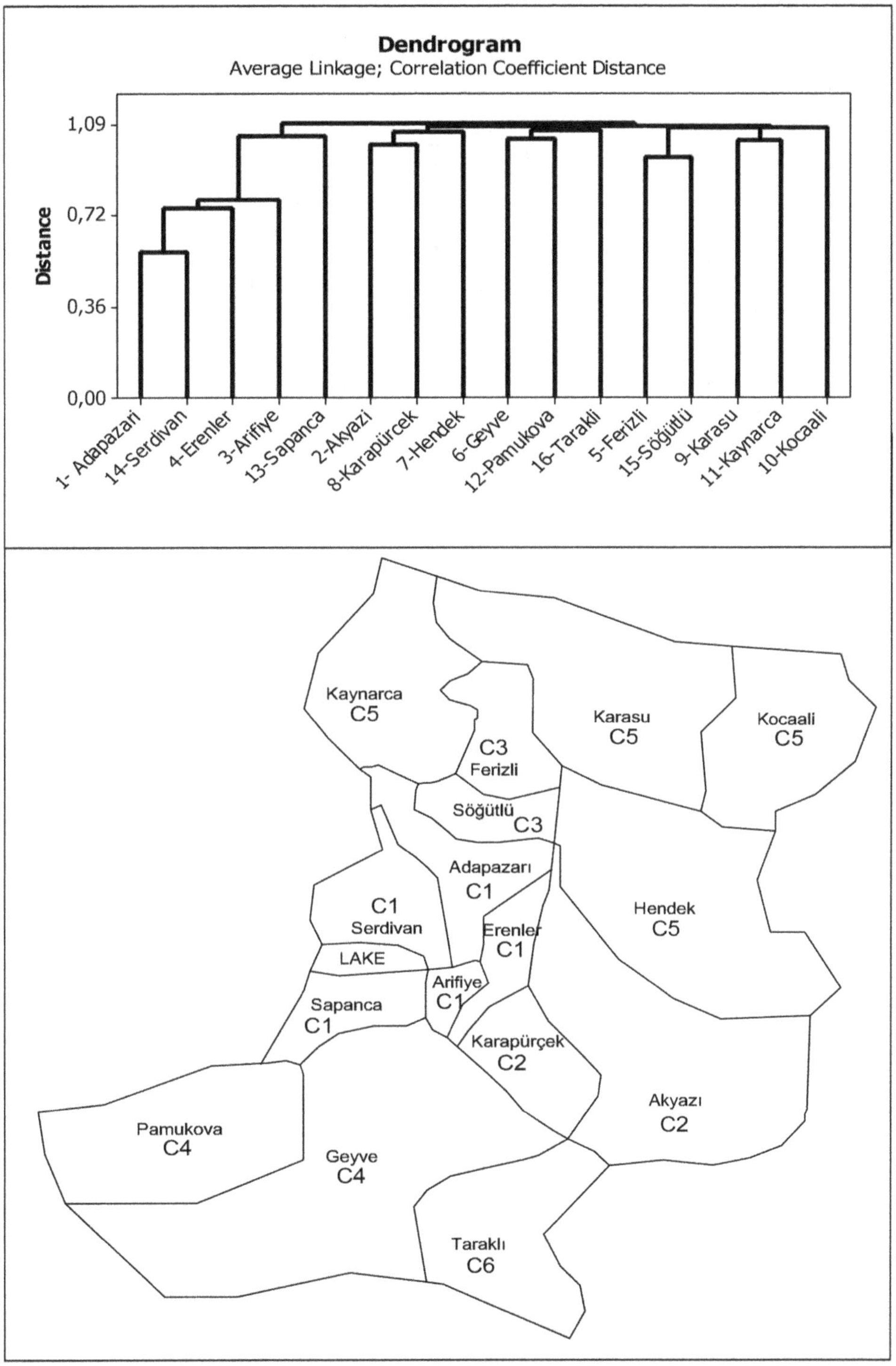

Figure 4. Clustering results and the map of average linkage method for 2012

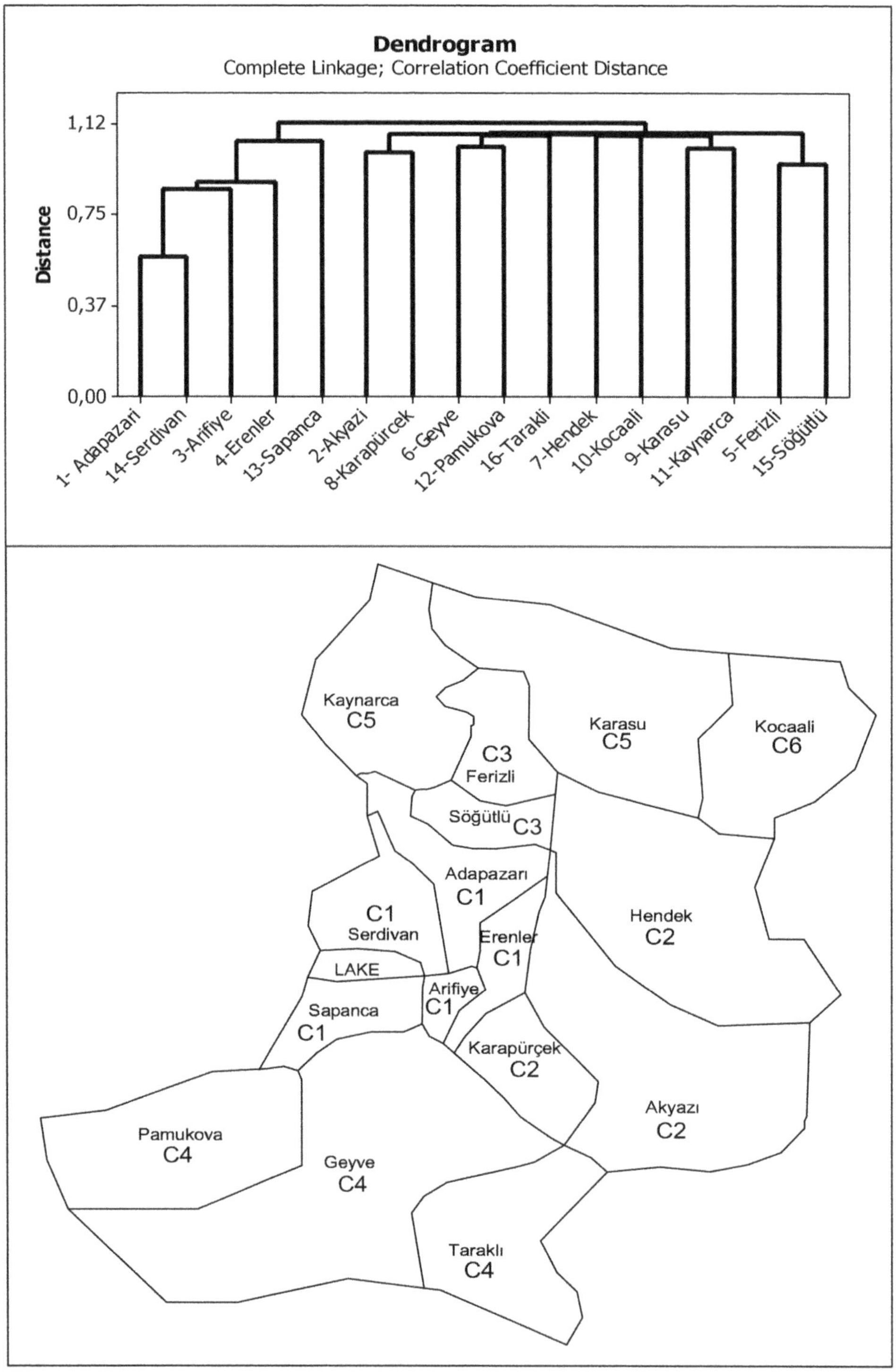

HCA Results for the Future year: 2023

The 2023 scenario includes no additional investment in the urban transport infrastructure (do-nothing scenario) after 2012, except the ones that have been already committed by departments of the local, as well as central government agencies. HCA results using the single, complete and average linkage clustering methods for 2023 are presented in Tables 7, 8, and 9, and the dendograms and maps are shown in Figures 5, 6, and 7.

Table 7. Cluster Results for 2023 using the Single Linkage Method

Steps	No. of Clusters	Similarity	Distance	Clusters Joined	New Clusters	No. of Observations in New Clusters
1	10	94.1124	0.117753	1 - 11	1	2
2	9	91.8893	0.162214	1 - 4	1	3
3	8	89.1444	0.217112	1 - 10	1	4
4	7	87.0824	0.258352	3 - 6	3	2
5	6	82.5869	0.348262	1 - 3	1	6
6	5	78.6998	0.426003	1 - 5	1	7
7	4	75.5295	0.489411	2 - 8	2	2
8	3	74.8432	0.503136	1 - 9	1	8
9	2	74.7073	0.505853	1 - 2	1	10
10	1	62.9061	0.741878	1 - 7	1	11

(C)luster 1: 1- Adapazarı 3-Arifiye 4-Erenler 6-Geyve 10-Serdivan 11-Söğütlü; **(C)luster 2:** 2-Akyazı; **(C)luster 3:** 5-Ferizli; **(C)luster 4:** 7-Hendek; **(C)luster 5:** 8-Karapürçek; **(C)luster 6:** 9-Sapanca.

Table 8. Cluster Results for 2023 using the Complete Linkage Method

Steps	No. of Clusters	Similarity	Distance	Clusters Joined	New Clusters	No. of Observations in New Clusters
1	10	94.1124	0.11775	1 - 11	1	2
2	9	87.0824	0.25835	3 - 6	3	2
3	8	84.7807	0.30439	4 - 10	4	2
4	7	83.1466	0.33707	1 - 4	1	4
5	6	75.5295	0.48941	2 - 8	2	2
6	5	70.5937	0.58813	1 - 3	1	6
7	4	58.5391	0.82922	1 - 9	1	7
8	3	56.1307	0.87739	2 - 7	2	3
9	2	50.0054	0.99989	1 - 5	1	8
10	1	48.6015	1.02797	1 - 2	1	11

(C)luster 1: 1- Adapazarı 4-Erenler 10-Serdivan 11-Söğütlü; **(C)luster 2:** 2-Akyazı 8-Karapürçek; **(C)luster 3:** 3-Arifiye 6-Geyve; **(C)luster 4:** 5-Ferizli; **(C)luster 5:** 7-Hendek; **(C)luster 6:** 9-Sapanca.

Table 9. Cluster Results for 2023 using the Average Linkage Method

Steps	No. of Clusters	Similarity	Distance	Clusters Joined	New Clusters	No. of Observations in New Clusters
1	10	94.1124	0.11775	1 - 11	1	2
2	9	87.7918	0.24416	1 - 4	1	3
3	8	87.0824	0.25835	3 - 6	3	2
4	7	85.6906	0.28619	1 - 10	1	4
5	6	78.3795	0.43241	1 - 3	1	6
6	5	75.5295	0.48941	2 - 8	2	2
7	4	67.8286	0.64343	1 - 5	1	7
8	3	62.5741	0.74852	1 - 9	1	8
9	2	60.5884	0.78823	1 - 2	1	10
10	1	55.7560	0.88488	1 - 7	1	11

(C)luster 1: 1- Adapazarı 3-Arifiye 4-Erenler 6-Geyve 10-Serdivan 11-Söğütlü; **(C)luster 2:** 2-Akyazı; **(C)luster 3:** 5-Ferizli; **(C)luster 4:** 7-Hendek; **(C)luster 5:** 8-Karapürçek; **(C)luster 6:** 9-Sapanca.

Cluster Validations

Clustering is an unsupervised process, i.e., there are no predefined classes and no examples that can show that the clusters found by the clustering algorithms are valid (Halkidi et al., 2002). Thus, cluster validation is a vital part of a cluster analysis to measure the performance of the linkage rules to model the data. These measures to judge the cluster validity are traditionally classified into three types, namely internal or unsupervised, external or supervised, and relative (supervised or unsupervised) (Tan et al., 2006; Jain & Dubes, 1988; Theodoridis & Koutroubas, 1999; Halkidi et al., 2002). The first two approaches involve statistical tests, which are computationally expensive. The third (relative) criterion does not involve statistical testing. Detailed explanations of some popular validation criteria can be found in Gan et al. (2007).

Internal criteria are used to evaluate the clustering structure produced by an algorithm using only quantities and features inherited from the data set. To apply the internal criteria, there are two situations: (a) hierarchy of clustering schemes (such as the hierarchical clustering algorithm) and (b) single clustering scheme (such as the partitional algorithm) (Gan et al., 2007). In this study, the cophenetic correlation (CC) is used to perform internal cluster validation. CC is a traditional evaluation measure for choosing the best hierarchical clustering among the alternatives created by using different linkage rules. The cophenetic distance between two objects is the proximity at which an agglomerative hierarchical clustering technique puts the districts in the same cluster for the first time (Tan et al., 2006).

The idea to validate the hierarchy of clustering schemes is to use the so-called cophenetic distance matrix *(CPDM)* and then to use the cophenetic correlation coefficient (*CPCC*) to measure the degree of similarity between *CPDM* and the proximity matrix $D(i, j)$. The cophenetic matrix *CPDM* is defined in such a way that the element $CPDM(i, j)$ represents the proximity level at which the two data points x_i and x_j are found in the same cluster for the first time (note the difference between agglomerative hierarchy and divisive hierarchy). The cophenetic correlation coefficient index (CPCC) is defined by Gan et al. (2007):

Figure 5. Clustering results and the map of single linkage method for 2023

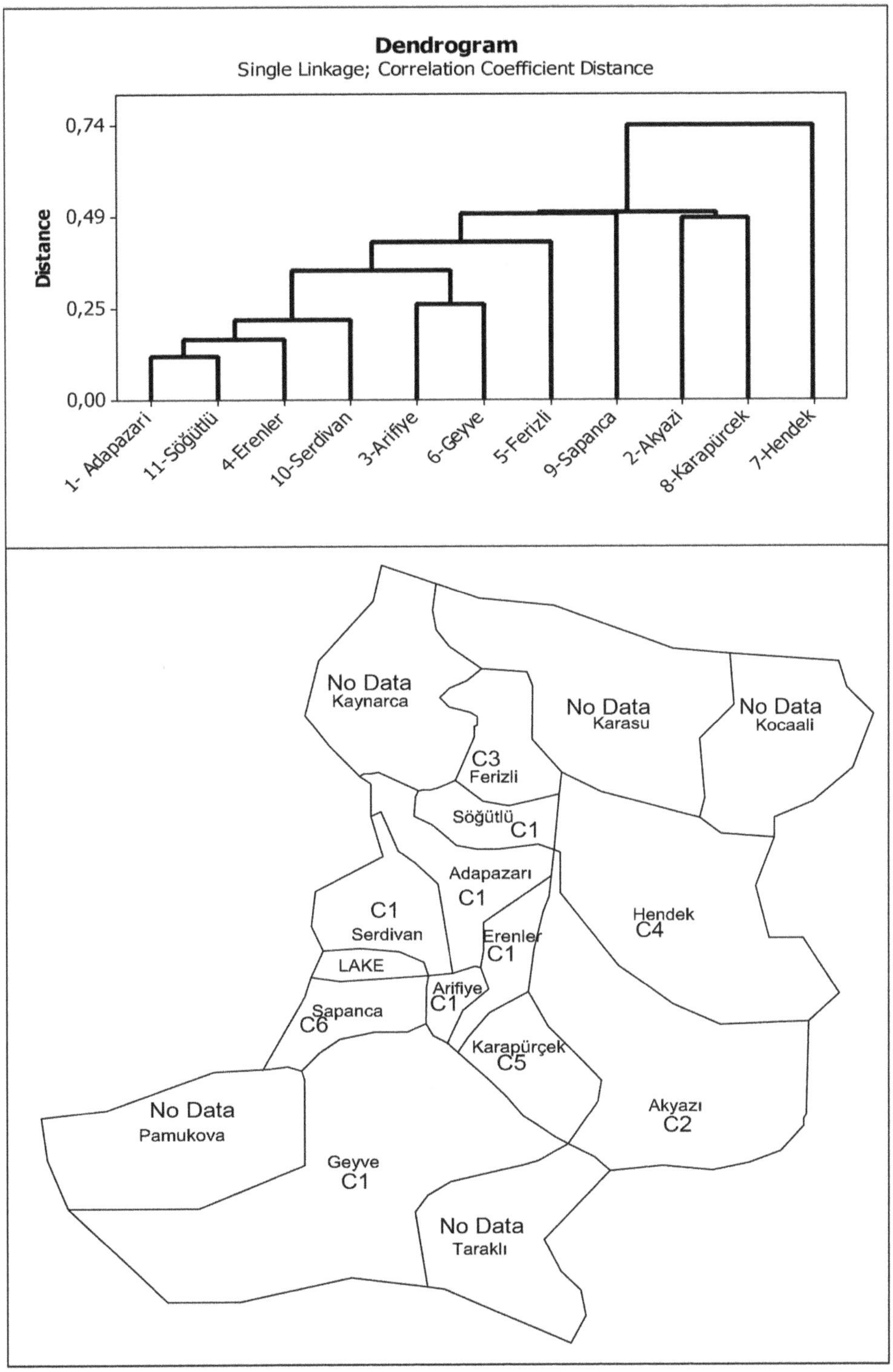

Figure 6. Clustering results and the map of complete linkage method for 2023

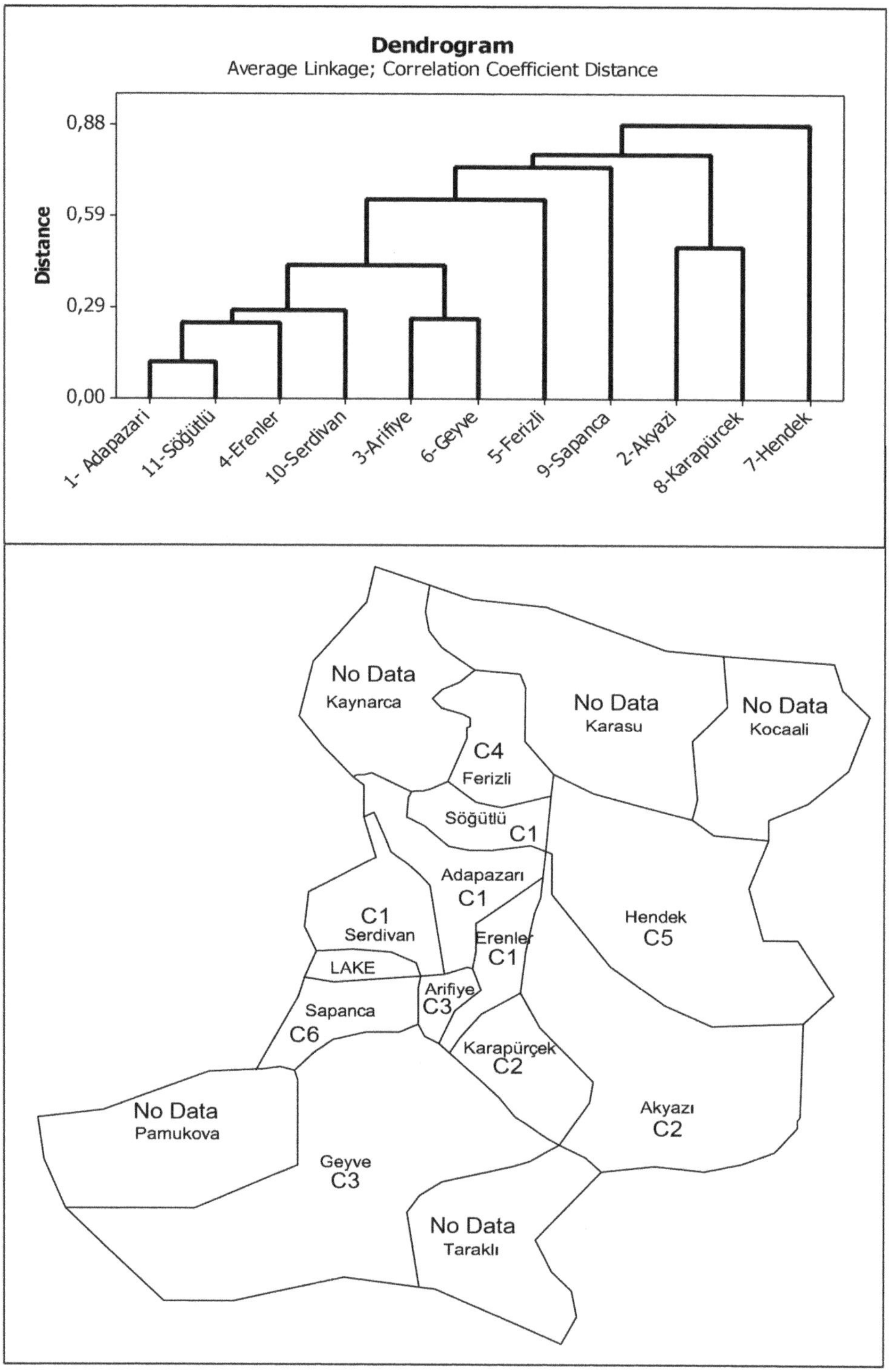

Figure 7. Clustering results and the map of average linkage method for 2023

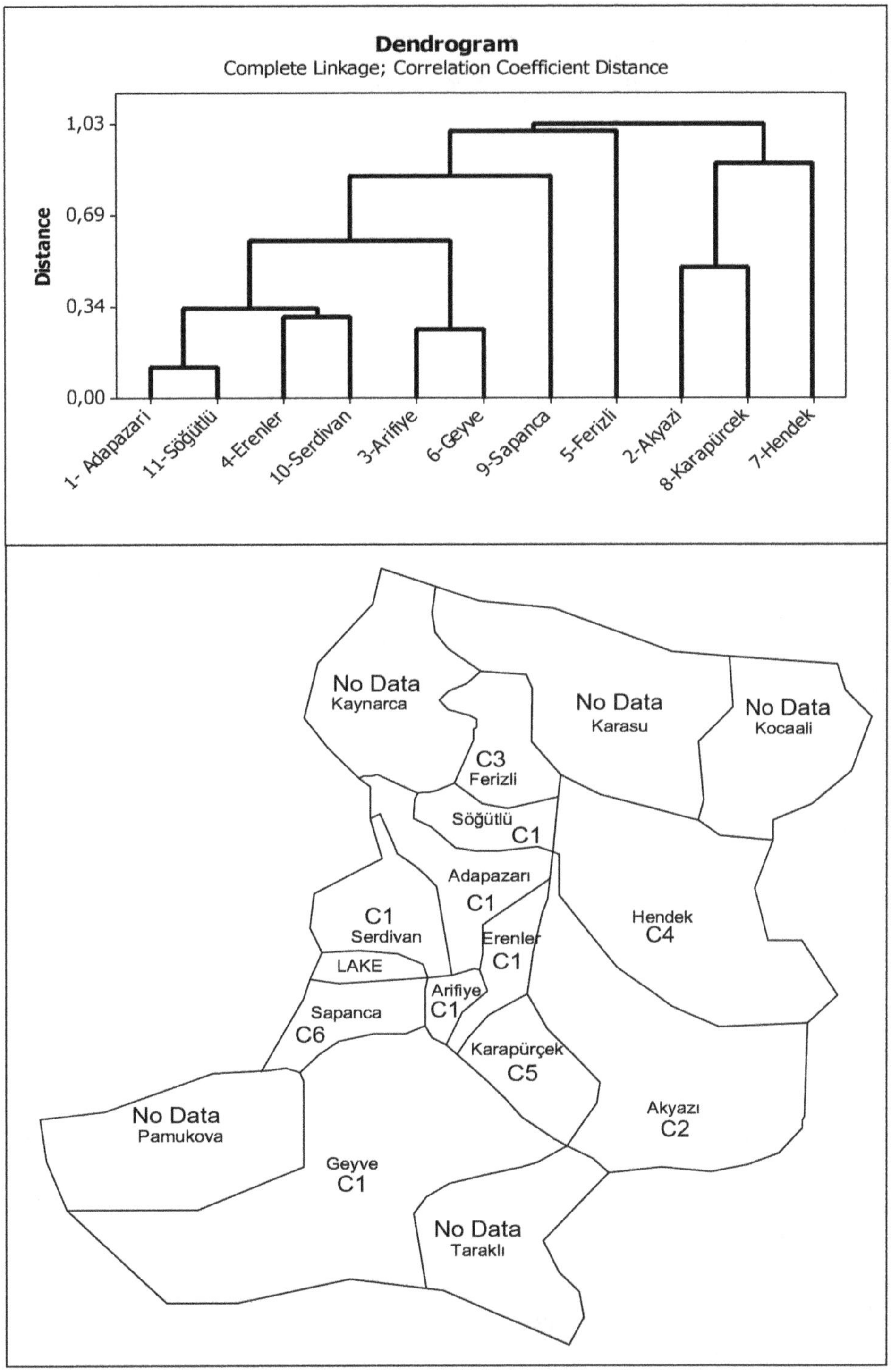

$$CPCC = \frac{\frac{1}{M}\sum_{i=1}^{n-1}\sum_{j=i+1}^{n} d_{ij}c_{ij} - \mu p\,\mu c}{\sqrt{(\frac{1}{M}\sum_{i=1}^{n-1}\sum_{j=i+1}^{n} d_{ij}^2 - \mu p^2)(\frac{1}{M}\sum_{i=1}^{n-1}\sum_{j=i+1}^{n} c_{ij}^2 - \mu p^2)}} \quad (6)$$

where $M = \frac{n(n-1)}{2}$ and μp and μc are defined as

$$\mu p = \frac{1}{M}\sum_{i=1}^{n-1}\sum_{j=i+1}^{n} d_{ij} \quad (6.1)$$

$$\mu c = \frac{1}{M}\sum_{i=1}^{n-1}\sum_{j=i+1}^{n} c_{ij} \quad (6.2)$$

where d_{ij} and c_{ij} are the elements of matrices *D (i, j)* and *CPDM (i, j)*, respectively. The range of *CPCC* is [−1, 1] and the high value indicates great similarity between the matrices *D (i, j)* and *CPDM (i, j)*.

The 2012 cophenetic correlation coefficients (CPCCs) were calculated in Table 10 from the correlation between the correlation distance matrix ($d_{ij}=1-r_{ij}$) in Table 11 and the symmetrical cophenetic distance matrices (CPDMs) for the linkage rules used in the analyses presented in Tables 12, 13, and 14. All methods seemed to fit the 2012 data well with high CPCCs; however, the best clustering was yielded by the group average method (the third row).

The 2023 cophenetic correlation coefficients (CPCCs) were calculated in Table 15 considering the correlation distance matrix ($d_{ij}=1-r_{ij}$) in Table 16 and the symmetrical cophenetic distance matrices (CPDMs) for the linkage rules employed shown in in Tables 17, 18, and 19. Reviewing the CPCC values, all methods seemed to fit the 2023 data well; however, the single or nearest neighbor method (the first row) yielded the best clustering.

Results

As seen there are six distinct clusters for 2012 as seen in Figure 8, but four in 2023 results as shown in Figure 9. Comparing the cluster maps, in 2023 scenario the districts in Cluster 1 (central city) grows

Table 10. 2012 Cophenetic Correlation Coefficients (CPCCs)

Linkage Method	N ((n-1)*m)/2	CPCCs	Significance (2-Tailed)
Single	120	0.689	< 0.000
Complete	120	0.927	< 0.000
Average	**120**	**0.948**	**< 0.000**

n=no. of rows and m=no. of columns of the matrices, which are equal to 16.

Table 11. 2012 Correlation Distance Matrix ($d_{ij} = 1 - r_{ij}$)

D (i,j)	1	2	3	4	5	6	7	8	9	10	11	12	13	14	15	16
1- Adapazari	0															
2-Akyazi	1.063	0														
3-Arifiye	0.651	1.092	0													
4-Erenler	0.618	1.058	0.857	0												
5-Ferizli	1.039	1.076	1.097	1.087	0											
6-Geyve	1.079	1.075	1.088	1.088	1.079	0										
7-Hendek	1.069	1.036	1.076	1.084	1.081	1.074	0									
8-Karapürcek	1.075	1.005	1.119	1.074	1.083	1.079	1.073	0								
9-Karasu	1.083	1.076	1.112	1.101	1.059	1.075	1.073	1.078	0							
10-Kocaali	1.062	1.074	1.105	1.091	1.066	1.075	1.071	1.076	1.068	0						
11-Kaynarca	1.095	1.076	1.118	1.103	1.077	1.076	1.072	1.073	1.019	1.073	0					
12-Pamukova	1.080	1.076	1.108	1.904	1.080	1.027	1.073	1.079	1.076	1.074	1.075	0				
13-Sapanca	1.023	1.080	1.023	1.051	1.081	1.072	1.075	1.083	1.081	1.078	1.080	1.077	0			
14-Serdivan	0.575	1.080	0.852	0.883	1.084	1.093	1.08	1.098	1.087	1.081	1.099	1.091	1.046	0		
15-Söğütlü	1.029	1.085	1.095	1.085	0.955	1.072	1.082	1.078	1.063	1.077	1.079	1.084	1.083	1.074	0	
16-Tarakli	1.094	1.068	1.122	1.101	1.079	1.052	1.073	1.076	1.074	1.072	1.064	1.074	1.079	1.099	1.082	0

Table 12. 2012 Cophenetic Distance Matrix ($CPDM_{ij}$) by the Single Linkage Method

CPDM (i,j)	1	2	3	4	5	6	7	8	9	10	11	12	13	14	15	16
1- Adapazari	0															
2-Akyazi	1.05794	0														
3-Arifiye	0.65064	1.05794	0													
4-Erenler	0.61822	1.05794	0.65064	0												
5-Ferizli	1.02866	1.05794	1.02866	1.02866	0											
6-Geyve	1.06386	1.06386	1.06386	1.06386	1.06386	0										
7-Hendek	1.05794	1.03554	1.05794	1.05794	1.05794	1.06386	0									
8-Karapürcek	1.05794	1.00468	1.05794	1.05794	1.05794	1.06386	1.03554	0								
9-Karasu	1.05899	1.05899	1.05899	1.05899	1.05899	1.06386	1.05899	1.05899	0							
10-Kocaali	1.06204	1.06204	1.06204	1.06204	1.06204	1.06386	1.06204	1.06204	1.06204	0						
11-Kaynarca	1.05899	1.05899	1.05899	1.05899	1.05899	1.06386	1.05899	1.05899	1.01883	1.05794	0					
12-Pamukova	1.06386	1.06386	1.06386	1.06386	1.06386	1.05172	1.06386	1.06386	1.06386	1.06386	1.06386	0				
13-Sapanca	1.02272	1.05794	1.02272	1.02272	1.02866	1.06386	1.05794	1.05794	1.05899	1.05794	1.05899	1.06386	0			
14-Serdivan	0.57501	1.05794	0.65064	0.61822	1.02866	1.06386	1.05794	1.05794	1.05899	1.05794	1.05899	1.06386	1.02272	0		
15-Söğütlü	1.02866	1.05794	1.02866	1.02866	1.02866	1.06386	1.05794	1.05794	1.05899	1.05794	1.05899	1.06386	1.02866	1.02866	0	
16-Tarakli	1.06386	1.06386	1.06386	1.06386	1.06386	1.05172	1.06386	1.06386	1.06386	1.06386	1.06386	1.05172	1.06386	1.06386	1.06386	0

Table 13. 2012 Cophenetic Distance Matrix ($CPDM_{ij}$) by the Complete Linkage Method

CPDM (i,j)	1	2	3	4	5	6	7	8	9	10	11	12	13	14	15	16
1- Adapazari	0															
2-Akyazi	1.12221	0														
3-Arifiye	0.85239	1.12221	0													
4-Erenler	0.88278	1.12221	0.88278	0												
5-Ferizli	1.12221	1.08459	1.12221	1.12221	0											
6-Geyve	1.12221	1.07938	1.12221	1.12221	1.07311	0										
7-Hendek	1.12221	1.07311	1.12221	1.12221	1.07311	1.07613	0									
8-Karapürcek	1.12221	1.00468	1.12221	1.12221	1.07311	1.07938	1.07311	0								
9-Karasu	1.12221	1.07127	1.12221	1.12221	1.07311	1.07311	1.07311	1.07311	0							
10-Kocaali	1.12221	1.07127	1.12221	1.12221	1.07311	1.07311	1.07127	1.07311	1.07311	0						
11-Kaynarca	1.12221	1.07127	1.12221	1.12221	1.07311	1.07311	1.07127	1.07311	1.01883	1.07311	0					
12-Pamukova	1.12221	1.07938	1.12221	1.12221	1.07311	1.02712	1.07311	1.07938	1.07311	1.07311	1.07311	0				
13-Sapanca	1.05117	1.12221	1.12221	1.12221	1.12221	1.12221	1.12221	1.12221	1.12221	1.12221	1.12221	1.12221	0			
14-Serdivan	0.57501	1.12221	0.85239	0.88278	1.12221	1.12221	1.12221	1.12221	1.12221	1.12221	1.12221	1.12221	1.05117	0		
15-Söğütlü	1.12221	1.07311	1.12221	1.12221	0.95483	1.07311	1.07311	1.07311	1.07311	1.07311	1.07311	1.07311	1.12221	1.12221	0	
16-Tarakli	1.12221	1.07356	1.12221	1.12221	1.07356	1.07356	1.07356	1.07356	1.07356	1.07356	1.07356	1.07356	1.12221	1.12221	1.07356	0

Table 14. 2012 Cophenetic Distance Matrix ($CPDM_{ij}$) by the Average Linkage Method

CPDM (i,j)	1	2	3	4	5	6	7	8	9	10	11	12	13	14	15	16
1- Adapazari	0															
2-Akyazi	1.08547	0														
3-Arifiye	0.78658	1.08547	0													
4-Erenler	0.7505	1.08547	0.78658	0												
5-Ferizli	1.08547	1.07636	1.08547	1.08547	0											
6-Geyve	1.08547	1.07482	1.08547	1.08547	1.07636	0										
7-Hendek	1.08547	1.05405	1.08547	1.08547	1.07636	1.07482	0									
8-Karapürcek	1.08547	1.00468	1.08547	1.08547	1.07636	1.07482	1.05405	0								
9-Karasu	1.08547	1.07636	1.08547	1.08547	1.06961	1.07482	1.07636	1.07636	0							
10-Kocaali	1.08547	1.07636	1.08547	1.08547	1.07099	1.07482	1.07636	1.07636	1.07099	0						
11-Kaynarca	1.08547	1.07636	1.08547	1.08547	1.06961	1.07482	1.07636	1.07636	1.01883	1.06961	0					
12-Pamukova	1.08547	1.07482	1.08547	1.08547	1.07636	1.02712	1.07636	1.07482	1.07482	1.07482	1.07482	0				
13-Sapanca	1.03589	1.08547	1.03589	1.03589	1.08547	1.08547	1.08547	1.08547	1.08547	1.08547	1.08547	1.08547	0			
14-Serdivan	0.57501	1.08547	0.78658	0.7505	1.08547	1.08547	1.08547	1.08547	1.08547	1.08547	1.08547	1.08547	1.03589	0		
15-Söğütlü	1.08547	1.07636	1.08547	1.08547	0.95483	1.07482	1.07636	1.07636	1.06961	1.07099	1.06961	1.07636	1.08547	1.08547	0	
16-Tarakli	1.08547	1.07636	1.08547	1.08547	1.07482	1.06264	1.07636	1.07636	1.07482	1.07482	1.07482	1.06264	1.08547	1.08547	1.07636	0

Table 15. 2023 Cophenetic Correlation Coefficients (CPCCs)

Linkage Method	N ((n-1)*m)/2	CPCCs	Significance (2-tailed)
Single	**55**	**0.859**	**< 0.000**
Complete	55	0.805	< 0.000
Average	55	0.789	< 0.000

n=no. of rows and m=no. of columns of the matrices, which are equal to 11.

Table 16. 2023 Correlation Distance Matrix ($d_{ij} = 1 - r_{ij}$)

D (i, j)	1	2	3	4	5	6	7	8	13	14	15
1- Adapazari	0										
2-Akyazi	0.659	0									
3-Arifiye	0.381	0.824	0								
4-Erenler	0.162	0.680	0.375	0							
5-Ferizli	0.452	1.004	0.800	0.632	0						
6-Geyve	0.348	0.904	0.258	0.400	0.923	0					
7-Hendek	0.765	0.742	0.902	0.795	1.026	0.991	0				
8-Karapürcek	0.506	0.489	0.739	0.527	0.993	0.876	0.877	0			
13-Sapanca	0.707	0.98	0.714	0.762	1.000	0.503	1.028	0.983	0		
14-Serdivan	0.217	0.767	0.447	0.304	0.627	0.382	0.846	0.642	0.724	0	
15-Söğütlü	0.118	0.806	0.537	0.326	0.426	0.588	0.876	0.722	0.829	0.337	0

Table 17. 2023 Cophenetic Distance Matrix ($CPDM_{ij}$) by the Single Linkage Method

CPDM (i, j)	1	2	3	4	5	6	7	8	13	14	15
1- Adapazari	0										
2-Akyazi	0.788232	0									
3-Arifiye	0.432410	0.788232	0								
4-Erenler	0.244164	0.788232	0.43241	0							
5-Ferizli	0.643429	0.788232	0.643429	0.643429	0						
6-Geyve	0.432410	0.788232	0.258352	0.432410	0.643429	0					
7-Hendek	0.884879	0.884879	0.884879	0.884879	0.884879	0.884879	0				
8-Karapürcek	0.788232	0.489411	0.788232	0.788232	0.788232	0.788232	0.88488	0			
13-Sapanca	0.748517	0.788232	0.748517	0.748517	0.748517	0.748517	0.88488	0.788232	0		
14-Serdivan	0.286186	0.788232	0.43241	0.286186	0.643429	0.432410	0.88488	0.788232	0.748517	0	
15-Söğütlü	0.117753	0.788232	0.43241	0.244164	0.643429	0.432410	0.88488	0.788232	0.748517	0.286186	0

Table 18. 2023 Cophenetic Distance Matrix ($CPDM_{ij}$) by the Complete Linkage Method

CPDM (i, j)	1	2	3	4	5	6	7	8	13	14	15
1- Adapazari	0										
2-Akyazi	1.02797	0									
3-Arifiye	0.58813	1.02797	0								
4-Erenler	0.33707	1.02797	0.58813	0							
5-Ferizli	0.99989	1.02797	0.99989	0.99989	0						
6-Geyve	0.58813	1.02797	0.25835	0.58813	0.99989	0					
7-Hendek	1.02797	0.87739	1.02797	1.02797	1.02797	1.02797	0				
8-Karapürcek	1.02797	0.48941	1.02797	1.02797	1.02797	1.02797	0.87739	0			
13-Sapanca	0.82922	1.02797	0.82922	0.82922	0.99989	0.82922	1.02797	1.02797	0		
14-Serdivan	0.33707	1.02797	0.58813	0.30439	0.99989	0.58813	1.02797	1.02797	0.82922	0	
15-Söğütlü	0.11775	1.02797	0.58813	0.11775	0.99989	0.58813	1.02797	1.02797	0.82922	0.11775	0

Table 19. 2023 Cophenetic Distance Matrix ($CPDM_{ij}$) by the Average Linkage Method

CPDM (i, j)	1	2	3	4	5	6	7	8	13	14	15
1- Adapazari	0										
2-Akyazi	1.02797	0									
3-Arifiye	0.58813	1.02797	0								
4-Erenler	0.33707	1.02797	0.58813	0							
5-Ferizli	0.99989	1.02797	0.99989	0.99989	0						
6-Geyve	0.58813	1.02797	0.25835	0.58813	0.99989	0					
7-Hendek	1.02797	0.87739	1.02797	1.02797	1.02797	1.02797	0				
8-Karapürcek	1.02797	0.48941	1.02797	1.02797	1.02797	1.02797	0.87739	0			
13-Sapanca	0.82922	1.02797	0.82922	0.82922	0.99989	0.82922	1.02797	1.02797	0		
14-Serdivan	0.33707	1.02797	0.58813	0.30439	0.99989	0.58813	1.02797	1.02797	0.82922	0	
15-Söğütlü	0.11775	1.02797	0.58813	0.11775	0.99989	0.58813	1.02797	1.02797	0.82922	0.11775	0

in population and expands to encompass other districts in Clusters 2, 3 and 4. Population of Cluster 1 was estimated to increase by 42.8%, from 500,384 in 2012 to 714,655 in 2023. Employment in Cluster 1 (central city) rose from 136,864 to 18,818 by 32.1% (Table 20).

SUMMARY AND CONCLUSION

The purpose of this chapter is to model urban growth and expansion based on observed and modeled travel patterns in the study area. In the analysis of interzonal trips using hierarchical cluster analysis (HCA) for 2012 and 2023 as a case study of the province of Sakarya, Turkey, the following conclusions were reached. Such analyses provided a tool to compare urban structure models representing possible snapshots of human settlements during the course of growth and expansion:

Figure 8. Cluster maps for 2012 HCA results

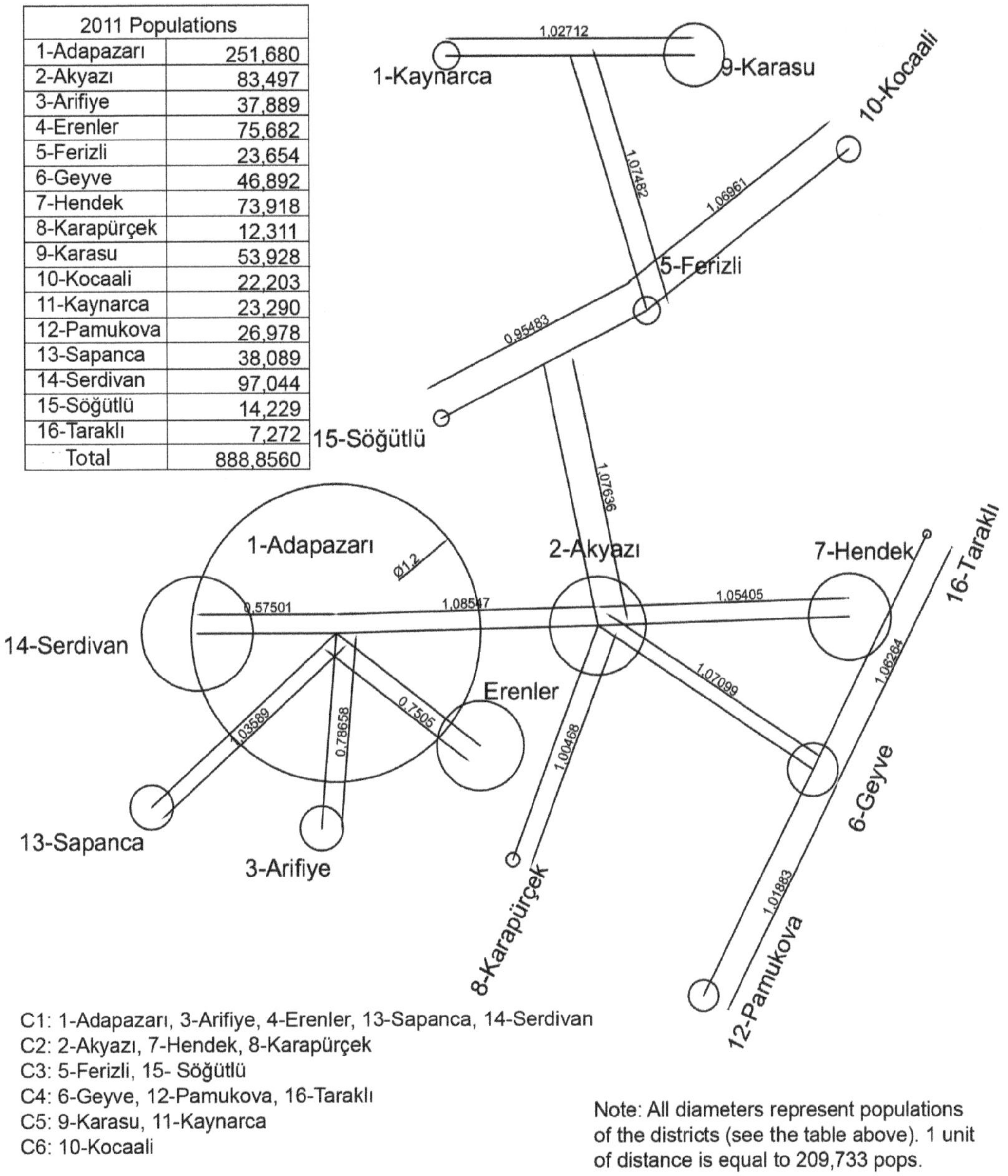

1. Modeling interzonal trips by HCA made possible to analyze the urban structure for different time horizons when O/D travel data are available to understand the interactions over the spatial structure of an urbanized area.
2. The results of the hierarchical clustering for years 2012 and 2023 seem to be highly consistent with the interzonal O/D trip matrices and the trip productions and attractions. In addition, projected changes in population, number of workers, and employment data between 2012 and 2023 are accounted for in the hierarchical clustering for 2023, the future plan year.

Figure 9. Cluster maps for 2023 HCA results

2023 Populations	
1-Adapazarı	295,406
2-Akyazı	111,260
3-Arifiye	56,791
4-Erenler	96,667
5-Ferizli	32,805
6-Geyve	62,131
7-Hendek	98,050
8-Karapürçek	18,336
9-Karasu	64,496
10-Kocaali	24,276
11-Kaynarca	23,524
12-Pamukova	39,623
13-Sapanca	50,485
14-Serdivan	152,603
15-Söğütlü	18,252
16-Taraklı	11,521
Total	1,156,227

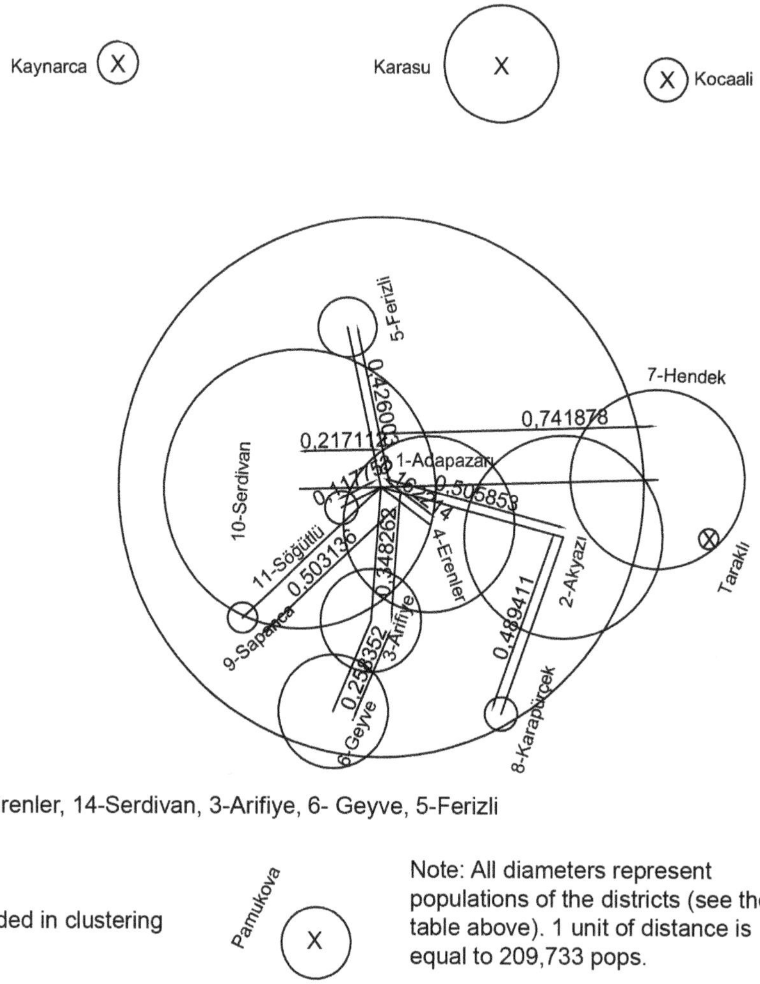

C1: 1-Adapazarı, 15- Söğütlü, 4-Erenler, 14-Serdivan, 3-Arifiye, 6- Geyve, 5-Ferizli
C2: 13-Sapanca
C3: 2-Akyazı, 8-Karapürçek
C4: 7-Hendek
[X] These districts were not included in clustering due to no data.

Note: All diameters represent populations of the districts (see the table above). 1 unit of distance is equal to 209,733 pops.

Table 20. Population and employment of the clusters based on 2012 and 2023 HCA

Clusters	2011		2023	
	Population	Employment	Population	Employment
0	0	0	163,440	N/A
1	500,384	136,864	714,655	180,818
2	169,726	30,976	50,485	15,322
3	37,883	8,110	129,596	19,060
4	81,142	806*	98,050	29,662
5	77,218	N/A	0	0
6	22,203	N/A	0	0

*: Missing data for Pamukova and Tarakli Districts; N/A: not available

3. According to the modeling results, the province of Sakarya is likely to be experiencing urban sprawl and growth during the next 11-year period because of the projected socio-economic changes in the 2030 Landuse and Development Plan and 2023 Transportation Master Plan of the Sakarya Metropolitan Municipality.

The results of this case study are expected be used by planners, administrators and policy makers to comprehend the interactions over a spatial structure, and to plan for creating efficient subcenters with more balanced distribution of travel patterns over urban agglomerations. Consequently, by this approach the impact of changes in land use, and population and employment densities on the hierarchical pattern of movements and hubs, which may indicate potential areas for future developments, can be modeled.

FUTURE RESEARCH DIRECTIONS

Modeling urban growth and expansion is very complex issue since it includes many factors such as socioeconomic factors, physical and environmental factors, and emerging market and development trends. Future research should include several of the following variables in modeling urban growth and expansion:

1. Socioeconomic factors.
    a. Population factors:
        i. Population and population density,
        ii. Birth rate and infant deaths,
        iii. Secondary school enrollment and graduation rates,
        iv. College enrollment and graduation rates.
    b. Economic factors:
        i. Employment and employment density,
        ii. Employment location: Downtown, urban, semi-urban or suburban.
        iii. Employment distribution: white and blue collar employment.
2. Physical and environmental factors.
    a. Physical factors:
        i. Buildings and vacant land,
        ii. Transportation network and level of service,
        iii. Park and open space,
        iv. Zoning,
        v. Land and building values,
        vi. Land use and distribution,
        vii. Previous plans.
    b. Environmental factors:
        i. Drinkable water resources,
        ii. Streams and waterways,
        iii. Watersheds,
        iv. Flood zones.

3. Emerging market and development trends.
    a. New demands for retirees and seniors,
    b. Multifamily development surge,
    c. Attractive large activity centers:
        i. Medical developments,
        ii. Educational complexes,
        iii. Shopping centers.
        iv. Suburban centers and suburban retrofits.
4. Availability of land to grow.
    a. Vacant/buildable land for new settlements:
        i. Low-density residential areas,
        ii. Middle-density mix land use areas.
    b. Other ways to accommodate growth:
        i. Suburban retrofitting,
        ii. Creating mixed-use centers,
        iii. Regeneration in downtown.

REFERENCES

Aggarwal, C. C. (2003). Towards systematic design of distance functions for data mining applications. *Proceedings of the ninth ACM SIGKDD international conference on Knowledge discovery and data mining*. ACM. doi:10.1145/956750.956756

Aggarwal, C. C., & Reddy, C. K. (Eds.). (2013). *Data Clustering: Algorithms and Applications*. CRC Press.

Akın, D. (2011). Kentsel Yapının Değişiminin Kentsel Yolculukların Hiyerarşik Kümeleme Analizi ile İncelenmesiyle Ortaya Konması: İstanbul Metropoliten Alanı 1996-2006 Yılları (*Studying the Change of Urban Structure by Hierarchical Cluster Analysis: Istanbul Metropolitan Region between 1996 and 2006*), KBAM 2. Sempozyumu (In *2nd Symposium of Urban and Regional Studies*). Ankara, Turkey.

Akin, D. (2014). Is Istanbul's growth monocentric or multicentric? Analyses of intraurban travel data between 1996 and 2006. In M.G. Karlaftis, N.D. Lagaros, M. Papadrakakis (Eds.), Proceedings of *OPT-i, 1st International Conference on Engineering and Applied Sciences Optimization* (pp. 2773-2799).

Alpkokin, P., Cheung, C., Black, J., & Hayashi, Y. (2008). Dynamics of clustered employment growth and its impacts on commuting patterns in rapidly developing cities. *Transportation Research Part A, Policy and Practice*, *42*(3), 427–444. doi:10.1016/j.tra.2007.11.002

Alpkokin, P., Hayashi, Y., Black, J., & Gercek, H. (2005). Polycentric employment growth and impacts on urban commuting patterns: Case study of Istanbul. *Journal of the Eastern Asia Society for Transportation Studies*, *6*(1), 3835–3851.

Berköz, L. (2000). Location of financial, insurance, and real estate firms in Istanbul. *Journal of Urban Planning and Development*, *126*(2), 75–88. doi:10.1061/(ASCE)0733-9488(2000)126:2(75)

Berkoz, L., & Turk, S. S. (2010). Intrametropolitan determinants of foreign investment firms in Istanbul. *Journal of Urban Planning and Development, 136*(3), 234–242. doi:10.1061/(ASCE)UP.1943-5444.0000025

Besussi, E., Chin, N., Batty, M., & Longley, P. (2010). The structure and form of urban settlements. In *Remote Sensing of Urban and Suburban Areas* (pp. 13–31). Netherlands: Springer. doi:10.1007/978-1-4020-4385-7_2

Bhandari, K., Alpkokin, P., Kachi, N., Jia, P., & Kato, H. (2007). *Dynamics of employment distribution and accessibility index: Case of Delhi.* JSCE.

Bhandari, K., Jia, P., Alpkokin, P., Mukhopadhyay, D., & Gangopadhyay, S. (2007). Policies, commuting patterns and accessibility in a non-monocentric city: Case study of Delhi. *Proceedings of the Eastern Asia Society for Transportation Studies* (Vol. 2007, pp. 192-192). Eastern Asia Society for Transportation Studies.

Dökmeci, V., & Berköz, L. (1994). Transformation of Istanbul from a monocentric to a polycentric city. *European Planning Studies, 2*(2), 193–205. doi:10.1080/09654319408720259

Gan, G., Ma, C., & Wu, J. (2007). *Data clustering: theory, algorithms, and applications* (Vol. 20). Siam. doi:10.1137/1.9780898718348

Halkidi, M., Batistakis, Y., & Vazirgiannis, M. (2002). Cluster validity methods: Part I. *SIGMOD Record, 31*(2), 40–45. doi:10.1145/565117.565124

Härdle, W., & Simar, L. (2007). *Applied multivariate statistical analysis* (Vol. 22007). Berlin: Springer.

Jain, A. K., & Dubes, R. C. (1988). *Algorithms for clustering data*. Prentice-Hall, Inc.

Jain, A. K., Murty, M. N., & Flynn, P. J. (1999). Data clustering: A review. [CSUR]. *ACM Computing Surveys, 31*(3), 264–323. doi:10.1145/331499.331504

Johnson, R. A., & Wichern, D. W. (1992). *Applied multivariate statistical analysis* (Vol. 4). Englewood Cliffs, NJ: Prentice hall.

Milligan, G. W. (1980). An examination of the effect of six types of error perturbation on fifteen clustering algorithms. *Psychometrika, 45*(3), 325–342. doi:10.1007/BF02293907

Ozus, E., Akın, D., & Çiftçi, M. (2012). Hierarchical cluster analysis of multicenter development and travel patterns in Istanbul. *Journal of Urban Planning and Development, 138*(4), 303–318. doi:10.1061/(ASCE)UP.1943-5444.0000121

Rodrigue, J. P., Comtois, C., & Slack, B. (2013). *The geography of transport systems*. Routledge.

Tan, P., Steinbach, M., & Kumar, V. (2006). Introduction to data mining. Addison-Wesley.

Theodoridis, S., & Koutroubas, K. (1999). *Pattern Recognition*. London: Academic Press.

Tsompanoglou, S., & Photis, Y. N. (2013). Measuring urban concentration: A spatial cluster typology based on public and private sector service patterns. *World Review of Science. Technology and Sustainable Development, 10*(4), 185–202.

Vliet, J. V., White, R., & Dragicevic, S. (2009). Modeling urban growth using a variable grid cellular automaton. *Computers, Environment and Urban Systems*, *33*(1), 35–43. doi:10.1016/j.compenvurbsys.2008.06.006

Vuchic, V. R. (1999). *Transportation for livable cities*. NJ, USA: Rutgers Center for Urban Policy Research.

Wang, F., & Sun, J. (2012). *Distance metric learning in data mining. SDM Conference (Tutorial).*

Zhao, P., Lu, B., & de Roo, G. (2011). The impact of urban growth on commuting patterns in a restructuring city: Evidence from Beijing. *Papers in Regional Science*, *90*(4), 735–754. doi:10.1111/j.1435-5957.2010.00343.x

KEY TERMS AND DEFINITIONS (some adopted from Rodrigue et al., 2013)

Cluster Analysis: Cluster analysis or clustering is the task of grouping (unsupervised classificaton) a set of objects or observed cases (without apriori information for classification) in such a way that the ones in the same group (called a cluster) are more similar (in some sense or another) to each other than to those in other groups (clusters). In clustering the problem is to group a given collection of unlabeled patterns into meaningful clusters. In a sense, labels are associated with clusters also, but these category labels are data driven; i.e., they are obtained solely from the data. Clustering is useful in several exploratory pattern analysis, grouping, decision making and machine learning situations. In such problems, in case there is little prior information (e.g., statistical models) available about the data and the decision-maker must make as few assumptions about the data as possible, clustering is particularly appropriate for the exploration of interrelationships among the data points to make a preliminary assessment on their structure.

Hierarchical Cluster Analysis: In data mining, hierarchical clustering (also called hierarchical cluster analysis or HCA) is a method of cluster analysis that seeks to build a hierarchy of clusters. Strategies for hierarchical clustering usually fall into two types: • Agglomerative: This is a "bottom up" approach: each observation starts in its cluster, and pairs of clusters are merged as one moves up the hierarchy. • Divisive: This is a "top-down" approach: all observations start in one cluster, and splits are performed recursively as one moves down the hierarchy.

Infrastructure: Capital goods that are not directly consumed and serve as support for the functions of a society (individuals and corporations). 1) In transport systems, all the fixed components, such as rights-of-way, tracks, signal equipment, terminals, parking lots, but stops, maintenance facilities, etc. 2) In transportation planning, all the relevant elements of the environment in which a transportation system operates.

Level of Service: 1) A set of characteristics that indicate the quality and quantity of transportation service provided, including characteristics that are quantifiable and those that are difficult to quantify. 2) For highway systems, a qualitative rating of the effectiveness of a highway or highway facility in serving traffic, in terms of operating conditions. A rating of traffic flow ranging from "A" (excellent) through "F" (heavily congested), and compares actual or projected traffic volume with the maximum capacity of the intersection or road in question. 3) For paratransit, a variety of measures meant to denote the quality of service provided, generally in terms of total travel time or a particular component of total travel time. 4) For pedestrians, sets of area occupancy classifications to connect the design of pedestrian facilities with levels of service.

Mobility: Refers to the movement of people or freight, denoted as travel activity by no. of trips or trip length or time spent per day or per capita, and so on. It can have different levels linked to the speed, capacity and efficiency of movements.

Mode of Transport: A vehicle by which the need for transportation is satisfied, including motorized and non-motorized modes.

(Mathematical) Model: Models are analytical tools (often mathematical) used by transportation modelers to assist in making forecasts of land use, economic activity, travel activity and their effects on the quality of resources such as land, air and water.

Physical Distribution: It is a collective term for the range of activities involved in the movement of goods from points of production to final points of sale and consumption. Physical distribution comprises all functions of movement and handling of goods, particularly transportation services (trucking, freight rail, air freight, inland waterways, marine shipping, and pipelines), transshipment and warehousing services (e.g. consignment, storage, inventory management), trade, wholesale and, in principle, retail. Conventionally, all these activities are assumed to be derived from materials management demands.

Planning: It refers to a process that allows people's needs, preferences and values to be reflected in decisions. Planning occurs at many different levels, from day-to-day decisions made by individuals or families to major decisions made by governments and businesses that have comprehensive and long-term impacts on society.

Public Transportation: Passenger transportation services, usually local in scope, that is available to any person who pays a prescribed fare. It operates on fixed or sometimes flexible schedules along designated routes or lines with specific stops and is designed to move relatively large numbers of people at one time. Thus, cost or energy consumptions decrease as the number of patrons is increased.

Spatial Interaction: A realized movement of people, freight or information between an origin and a destination. It is a transport demand / supply relationship expressed over a geographical space. Spatial interactions cover a wide variety of movements such as journeys for various purposes, such as work, migrations, tourism, the usage of public facilities, the transmission of information or capital, the market areas of retailing activities, international trade and freight distribution.

Sustainable Development: It is such a development that meets the needs of the present without compromising the ability of future generations to meet their needs.

Transportation Network.: It is a framework of transportation routes within a system of locations, identified as links and nodes. A route is a single link between two nodes that are part of a larger network such as roads and rails or air and sea corridors.

Urban Agglomeration: An urban agglomeration is an extended city or town area comprising the built-up area of a central place (usually a municipality) and any suburbs linked by continuous urban area. The term of agglomeration is also linked to conurbation, which is a more specific term for large urban clusters where the built-up zones of influence of distinct cities or towns are connected by continuous built-up development.

Urban Form: The spatial imprint of an urban transport system as well as the adjacent physical infrastructures and socioeconomic activities. Jointly, they confer a level of spatial arrangement to cities.

Urban or Suburban Sprawl: Described as the expansion of human populations away from central urban areas into previously remote and rural areas. Therefore, sprawl often results in communities reliant upon heavy automobile usage and extensive road infrastructure, mostly observed in developed nations (and primarily the United States, Canada, and Australia). It usually occurs toward the outward of a city to its suburbs, low-density and auto-dependent development on rural land. Urban sprawl can cause an expansion of the daily urban system, a high segregation between residential and commercial uses, and may encourage car dependency under the lack of fast, high-quality and dependable public transportation.

Urban (Spatial) Structure: The set of relationships arising from urban form and the underlying movements of people and goods.

Chapter 8
Analytical Hierarchy Process (AHP) based Decision Support System for Urban Intersections in Transportation Planning

Yetis Sazi Murat
Pamukkale University, Turkey

Ziya Cakici
Pamukkale University, Turkey

Turan Arslan
Uludag University, Turkey

Cengiz Akçam
General Highway Directorate, Turkey

ABSTRACT

In transportation planning, design of intersections is required some concrete decisions. It is mostly related to capacity, safety and economy. In many applications, capacity is considered as the main factor for design purpose and type of intersection is determined regarding the capacity. But it is known that safety and economy are as important as the capacity. In order to handle this problem, a multi-criteria decision making approach (AHP) is used. Four intersection types are used. Fuel consumption, emission rates, delay, cost of intersection type and safety parameters are taken into account in the decision process. These intersection types are tested by SIDRA Intersection software using the parameters listed above. The data obtained by SIDRA software is used in the AHP model. Weights of the factors used in the AHP model are varied considering the views of capacity and economy. And results and applicability of the model are discussed.

INTRODUCTION

As traffic flows from different directions at intersections, it causes crossing, merging or diverging points. Therefore, intersections must be planned and constructed with a special care. Improper planning and design may cause some important problems such as traffic accidents causing property and casualty loses, excessive delays causing congestions and fuel wastages etc.

Hence, the decision for the selection of an appropriate intersection design type is critical and one of the most important matters that decision makers face for designing efficient urban road networks.

DOI: 10.4018/978-1-4666-8648-9.ch008

However, in many real world applications, particularly in developing countries, this decision has usually been taken without any analytical proof. Therefore, the outcome expected from the implementation of a particular intersection design type can be disappointing. For instance, if an intersection with low traffic volumes is designed as a signalized intersection instead of sign controlled one, then excessive delays can possibly occur and result in poor performances in other quality characteristics. On the other hand, even at intersections with high traffic volumes designed as signal controlled, still excessive delays can possibly occur, however signal controlled intersections should be preferred for considering traffic safety issues. Therefore, the decision for selecting the appropriate intersection design type involves several and sometimes conflicting criteria that decisions makers must consider. Traffic capacity, width of the lanes, number of lanes, average delay time and costs are examples of such criteria. In some cases, more than one intersection types can be suitable for a particular intersection, therefore the decision for selecting the best one is the main problem that must be tackled. Thus, an analytical approach that considers assessing multi-criteria is essential as well as the cost-benefit analysis that evaluates consequences of different intersection design types in monetary terms.

In this study, the Analytical Hierarchy Process (AHP) method is utilized as a decision support system on intersection design type selection process. A real intersection with average annual daily traffic (AADT) of 25000 vehicles in Denizli, Turkey is analyzed in this study. Four intersection design types are taken into consideration for improving the performance of the intersection; signalized without left turn bay, signalized with left turn bay, roundabout and grade-separated intersections. Fuel consumption, emission rates, delay, cost of intersection type (as construction and operation costs) and traffic safety (represented by the number of conflict points) criteria are taken into account in the decision process. The SIDRA intersection software package is utilized to obtain the performance results of each intersection design type with respect to the criteria except for the traffic safety. The AHP model is then used for assessing intersection design types with regard to all criteria including the traffic safety criterion as well. The AHP procedure, modeling stages and the SIDRA analyses are all explained in following sections.

THE ANALYTICAL HIERARCHY PROCESS (AHP)

The Analytical Hierarchy Process (AHP) developed and refined by Saaty (1977; 1980; 1988; 1994) has become one of the most effective multi-criteria evaluation methods that have found uses in a wide range of practical applications in a variety of decision making processes. Since its introduction, the AHP has shown a broad impact on the field of multi-criteria or multi-attribute decision making, and recognized with its numerous applications such as choices, prioritization/evaluation, resource allocation, benchmarking and quality management, to name a few (Vaidya & Kumar, 2006). Triantaphyllou & Mann (1995) used the AHP method in industrial engineering applications and addressed some of the practical and computational issues involved. At the end of the study, they proved that the Analytical Hierarchy Process is an effective approach for this kind of decision problems. Al-Harbi (2001) studied the application of the AHP on the project management field. In the study, contractor prequalification problem was used as an example. A hierarchical structure was constructed for the prequalification criteria and the contractors wishing to prequalify for a project. By applying AHP method, contractors were prioritized. Piantanakulchai & Saengkhao (2003) applied the AHP for the evaluation of alternatives in highway route planning and proposed it to determine the best alignment which is socially the most preferable and has the least cost. Kahraman et al., (2004) investigated the multi-attribute comparison of catering service companies in Turkey using fuzzy AHP. In their studies, the customers of three Turkish catering firms were interviewed and the most important criteria taken into account by the customers were determined by the questionnaires. At the end of the study, these catering firms were compared with each other using fuzzy AHP. Tezcan (2007) applied Analytical Hierarchy Process on ready mixed concrete plant area selection problem. Pogarcic et al. (2008) focused on the application of AHP method in traffic planning as seeking the possibilities of utilizing the AHP method for making decisions regarding the traffic planning and the implementations of plans in traffic and ensuring the qualitative business logistics. Balli & Korukoglu (2009) used Fuzzy AHP method for the selection of computer systems of the firms. Fazlollahtabaret al. (2010) studied designing a new fuzzy expert system to evaluate alternatives in fuzzy analytic hierarchy process to solve the problems that otherwise could not be solved easily by the AHP method. Gulenc & Bilgin (2010) used AHP method in an Industrial facility manufacturer to make an investment decision. The analysis and solution of the application problem were completed using Expert Choice and then the most suitable alternative was selected. Ozdemirel et al. (2012) used the AHP method for the assignment and routing problem for determining the importance of the criteria affecting the utility of an assignment. Chatterjee & Mukherjee (2013) used AHP method for the selection of one of three potential hospital locations in Rural India. They considered three major factors and eleven sub factors in the evaluation process. At the end of the study, they concluded that cost of land, population density and proximity to public transport are three most significant sub factors. Effat & Hassan (2013) studied on designing and evaluation of three alternatives highway routes using the Analytical Hierarchy Process and least-cost path analysis. They used Geographic Information System (GIS) tools to develop a least-cost path for a corridor to link three cities in Sinai Peninsula desert environment. In their study, environmental and economic factors were integrated using Analytical Hierarchy Process. Michael et al., (2013) focused on urban sustainability indicators prioritization and they used AHP method for this issue. In their study, they produced list of sustainability indicators based on sorting criteria including accessibility, variability in spatial and temporal directions and measurability. They used AHP method to rank and prioritize urban sustainability indicators for Malaysia. Omurbek et al. (2013) used AHP method for the location decision. 7 districts of Isparta, Turkey was compared to each other considering 5 different criteria including location, environmental factors, labor and capital costs and laws. Arabameri (2014) used AHP method for the selection of fire station locations in Maku, Iran. In the study, classification and qualification of the numerous types of regions were made using AHP method. Mahendran et al., (2014) applied a fuzzy AHP approach for selection of measuring instrument for engineering college selection.

Decision making is one of the main tasks and sometimes a challenging process. Even regular people have to make decisions several times in their daily lives. They vary from routine decisions such as drinking, eating etc., which do not require any analytical approach to satisfy the personal needs, to more complicated ones such as the place to live in, to work etc. They carry greater importance for the professionals such as engineers or politicians who are responsible for making decisions on big projects that may directly affect community lives and well-beings. Therefore, an effective and efficient analytical approach that supports decision making process such as the AHP is an essential tool to lessen the burden on decision makers. In this section, the AHP modeling stages are introduced in general and the model developed for intersection type selection is presented in detail.

Basically, the AHP is a multi-attribute decision-making (MADM) method having its principal feature from its pairwise comparisons. In the AHP, the decision is first defined in a hierarchy which simplifies a complex decision making process. This stage also refers to the structuring complexity function of the AHP. The defined hierarchy for the studied case, which is the decision on selecting appropriate intersection design type for a particular intersection, is given in Figure 1. In this way, the complex problem of selecting appropriate intersection design type for a particular intersection is decomposed into a hierarchical structure that can be better coped with. After structuring the hierarchy, pairwise comparison matrices are prepared for all levels of the hierarchy such that for each level the pairwise comparisons are prepared with regard to each item at one upper level of the hierarchy, one at a time. For instance, for the lowest tier of the hierarchy, the pairwise comparison matrices are simply prepared with respect to each criterion one at a time among the alternatives which are signalized intersection with left turn bay, signalized intersection without left turn bay, roundabout and grade-separated intersection. The judgments are provided in ratios specific to the AHP. The most commonly used ones are suggested by Saaty (1980) as given in Table 1. This stage refers to measuring ratio scale function of the AHP. Such that every alternative is paired with every other alternative in turn for being evaluated in terms of performances with respect to each criterion specified. The judgments are provided in ratios as a_{ij}'s taken from Table 1, which indicate the strengths of preferences of alternative *i* over alternative *j* with regard to each criterion. Similar pairwise comparisons for the specified criteria, $C_1, \ldots C_K$, which are average vehicle delay, cost, fuel consumption, CO emission and traffic safety with respect to the overall goal, selecting appropriate intersection type in our case, can be prepared. As formulated by Saaty (1980), a pairwise matrix, **A**, among the alternatives with respect to criterion *k* can be indicated in the form of a matrix as:

Figure 1. Hierarchical structure of selecting intersection design type

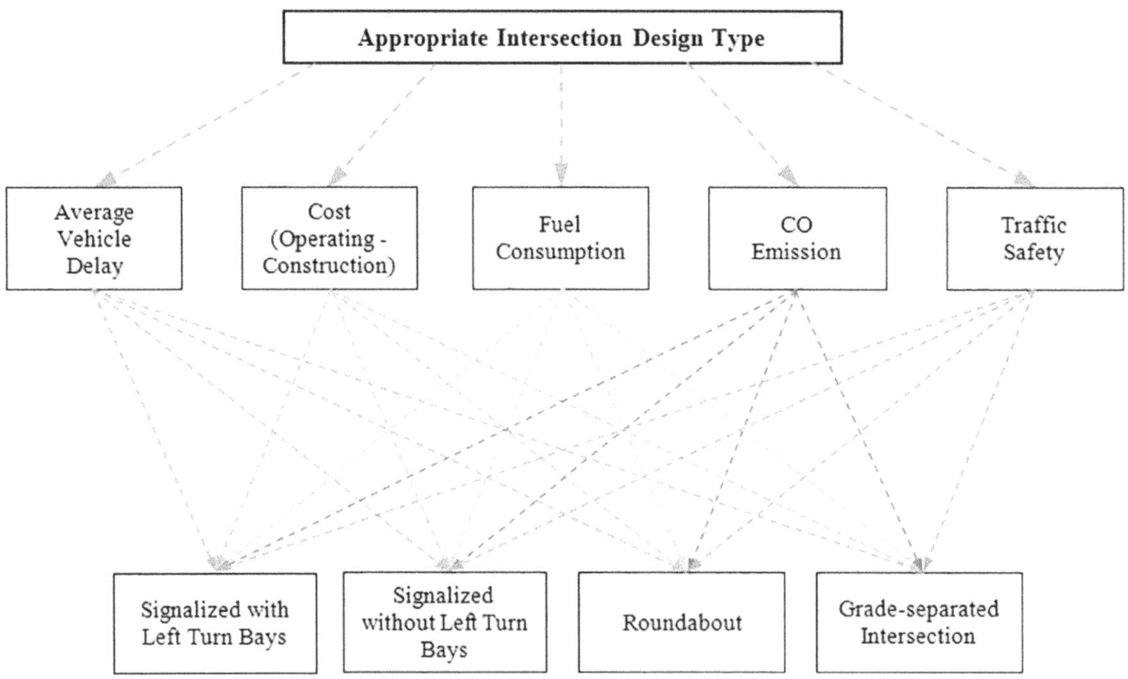

Table 1. Scale for pairwise comparisons

Intensity of importance	Definition	Explanation
1	Equal importance	Two activities contribute equally to objective
3	Weak importance	Experience and judgment slightly favor one activity over another
5	Essential or strong importance	Experience and judgment strongly favor one activity over another
7	Demonstrated importance	One activity is strongly favored and demonstrated in practice
9	Absolute importance	The evidence favoring one activity over another is of highest possible order of affirmation
2,4,6,8	Intermediate values between two adjacent judgments	When compromise is needed
Reciprocals of above	If activity *i* has one of the nonzero numbers assigned to it when compared with activity *j*, then *j* has the reciprocal value when compared with *i*.	

$$\mathbf{A}^k = \begin{array}{c} \\ A_1 \\ \vdots \\ A_n \end{array} \begin{array}{c} \begin{array}{ccc} A_1 & \cdots & A_n \end{array} \\ \begin{bmatrix} a^k_{11} & \cdots & a^k_{1n} \\ \vdots & \ddots & \vdots \\ a^k_{n1} & \cdots & a^k_{nn} \end{bmatrix} \end{array}$$

where n is the number of alternatives. Each entry a_{ij} in the matrix, **A**, represents the strengths of preferences that the decision maker believes that it exists for alternative *i* over alternative *j*.

As stated by Vargas (1990), the model is based on the following axioms:

1. Reciprocal comparison axiom; if A_i dominates A_j χ times then A_j dominates A_i $1/\chi$ times.
2. Homogeneity axiom; the dominance judgments are represented by means of a bounded scale.
3. Independence axiom; the weights of criteria are independent of the stimuli in a multi-criteria situation.
4. Expectation axiom; the hierarchic structure is assumed to be complete regarding the purpose of decision-making.

Starting from the last; expectation axiom states that in order to get accurate results the hierarchy defined for the decision making process, which is here selecting the appropriate intersection design type, should be correct. The independence axiom states that the importance of higher level objectives should not depend on the weights of lower level ones considered. The homogeneity axiom refers to the bounded scale as provided in Table 1. The first axiom corresponds to the reciprocal structure of the decision matrix. Accordingly, a pairwise matrix, whose elements are in form of a_{ij}'s, is actually a square positive reciprocal matrix by n x n as:

$$a_{ij} = \frac{1}{a_{ji}} \text{ and } a_{ii} = 1.00 \text{ for } \forall\ i,\ j = 1,\ldots,n$$

As each a_{ij} characterizes the estimate of the weight of the alternative i, w_i, to alternative j, w_j. Then, the priority weights among the alternatives, **W**, actually correspond to the normalized Eigen vector associated with the largest Eigen value, λ_{max} or λ_A of **A**. For a particular criterion it can be found from the following equation:

$$\mathbf{W} = \frac{AW}{\lambda_A} \text{ or } w_i = \frac{\sum_{j=1}^{n} a_{ij} w_j}{\lambda_A}$$

Naturally, decision makers cannot produce these pairwise priority weights accurately; therefore, some inconsistencies indeed exist in these matrices. Saaty (1977) shown that λ_A is always greater than or equal to n., and related to that he also stated that the closer the value of computed λ_A is to n, the more consistent is the decision matrix. Based on this premise, he suggested a measure of consistency, called consistency index, *CI*, indicating the deviation or degree of consistency calculated from the following equation:

$$CI = \frac{\lambda_A - n}{n - 1}$$

In order to make it more meaningful, this value is compared with another index, which is called random consistency index, *RI* that is actually the mean consistency index of randomly generated reciprocal matrices prepared using the same scale:

$$RI = \frac{\lambda_\theta - n}{n - 1}$$

where; λ_θ is the mean principle right eigen value of created pure random matrices. Therefore, whether all these rankings are realistic or not can be found from another equation called consistency ratio:

$$CR = \frac{CI}{RI} = \frac{\lambda_A - n}{\lambda_\theta - n}$$

Saaty (1977) provided a table for *RI* that was obtained from randomly generated pairwise matrices as given in Table 2.

If the consistency ratio is lower than a threshold (in general fixed to 0.10), then the normalized eigen vector assumed to be at an acceptable level for providing relative importance values among the alternatives compared. This implies that there is a 10% chance that the relative importance values would be random. In case of inconsistent results, however, the pairwise comparisons are usually repeated.

Table 2. Random consistency index, RI

n	1	2	3	4	5	6	7	8	9	10
RI	0.00	0.00	0.58	0.90	1.12	1.24	1.32	1.41	1.45	1.49

It should be noted that there are also some other methods such as geometric mean, arithmetic mean, the least squares and logarithmic least squares method other than Saaty's eigenvector method for estimating weights or priorities from the pairwise matrices but they are not explained here.

The final step of AHP which refers to synthesizing function of the AHP is essential for composing local priorities in the hierarchy to derive final priorities among the alternatives compared. This can usually be achieved by using additive weighting method in accordance with the hierarchy in which the problem is decomposed. The additive weightings are found from the following equation:

$$P_{im} = \sum_{k=1}^{K} w_k A_n \qquad \text{for all } n = 1, \ldots, N$$

Simply, the priorities obtained for each alternative with respect to each criterion are multiplied by the corresponding criterion weights and then they are summed up for each alternative for obtaining final priorities among the alternatives. Since both weights of criteria and priority values of alternatives are normalized in the range of 0-1, the final results are naturally also in the range of 0-1, summing up to one.

The Analysis

After constructing the decision hierarchy (Figure 1) as explained above, the following analysis are carried out accordingly:

- Preparing pairwise comparison matrices for criteria and for alternatives with respect to each criterion
- Obtaining weights (criteria) and prioritizations among alternatives via the eigenvector method
- Checking consistency ratios (if $CR < 0.10$ the final prioritization among alternatives are calculated, if not the process is repeated for the pairwise comparisons until consistent pairwise matrices are obtained)

Alternatives (Intersection Types)

Due to high level of activity and shared space among the road users, intersection design types vary substantially. They can be classified considering their various properties but in general depending on the treatment of crossing conflicts they can be classified as:

- At- grade intersections
- Grade- separated intersections

At-grade intersections, traffic flows approaching from different directions intersect at the same grade. Many traffic problems and traffic accidents occur on these types of intersections. At- grade intersections can be uncontrolled; signalized; roundabouts and signalized roundabouts. Uncontrolled at-grade intersections represent intersections where there is no control mechanism on any intersection approach (traffic signal or any other type of control devices). The movements of traffic flows are achieved by the drivers' behaviors, initiatives and understandings. On the other hand, on signalized at-grade intersections and signalized roundabouts, the movements of traffic are controlled by traffic signals; therefore, an improper setting or calculation of signal times can cause serious problems such as excessive delay, queuing, fuel consumption etc. Moreover, for regularity and continuity of traffic flows, more attention should be given to providing adequate traffic signal times (Murat et al., 2014). On the other hand, for the unsignalized roundabouts, although there is no control mechanism on any approach, the regularity and continuity of movement of the traffic flows are provided by a router island placed at the center of the intersection. On this type of intersections, traffic flows circulating around the center island have the priority over approaching ones.

In general, as the traffic volumes on crossing intersection legs reaches to a certain level, the capacity of at-grade intersections becomes inadequate. Then, this situation causes excessive queuing and delay on intersection legs causing severe traffic problems that result in serious accident rates at intersections. Therefore, for instance in Turkey, if the traffic volumes on intersection approaches become more than 1800 car/hour/lane, then the grade-separated intersection must be considered (Tunc, 2003).

On grade-separated intersections, crossing roads are connected at different levels. Although, on this type of intersections, there does not exist any traffic conflict point at the intersection area, merging and diverging (weaving) points can occur at associated entrance and exit ramps of intersections. Therefore, drivers have to be more careful at these conflicting point areas for avoiding the probable collisions. The grade-separated intersections include many different types of intersections such as; stack cloverleaf, diamond and trumpet type intersection etc.

In this study, four different intersection design types are considered for a particular intersection in Denizli, Turkey, with respect to existing traffic volumes of the specified intersection. The SIDRA Intersection software package is used for analyzing performances of each intersection design type for the existing traffic volumes. The existing traffic volumes used in the study are given in Table 3 with respect to the intersection as illustrated in Figure 2. All approaches at the intersection have different names as Pekdemir, Tali, Mugla and Pau. Table 3 shows the existing traffic volumes in matrix form, therefore, the value in a cross cell basically indicates how many vehicles moves to from the approach given on left to

Table 3. Existing traffic volumes of studied intersection

Intersection Approaches	Tali		Mugla		Pekdemir		Pau	
	HV	LV	HV	LV	HV	LV	HV	LV
Tali	-	-	2	71	3	94	3	70
Mugla	3	70	-	-	68	1033	10	284
Pekdemir	8	106	80	662	-	-	22	264
Pau	1	28	3	81	29	421	-	-
Total Number of Vehicle at Intersection: 3416 veh/hr HV: Heavy Vehicle / LV: Light Vehicle								

Figure 2. Studied intersection and its layout in Denizli, Turkey

the approach given at top. For instance, from Tali to Mugla, 2 heavy vehicles and 71 light vehicles make left turn; from Tali to Pekdemir 3 heavy vehicle and 94 light vehicles makes a right turn; and from Tali to Pau 3 heavy vehicles and 70 light vehicles go through. The average vehicle delay, operating cost, fuel consumption and CO emissions are adopted as the legitimate indicators of intersection performances. The analysis results obtained by the SIDRA Intersection software package for four different intersection design types are summarized in Table 4.

As the results all seen on Table 4 are scrutinized carefully, roundabouts and grade-separated intersections provide better performance values, as expected, than the others because both intersection types are actually non-signalized intersections, causing less or no interruption to moving traffic compare to signalized intersections with low traffic volumes. Therefore, either of them can be selected as the desired intersection type due to the performance based assessment. However, these criteria as well as some other factors such as and most importantly the traffic safety, which cannot be handled in the SIDRA analysis, may differ in the degree of preferability that can be assigned to the factors by policy makers or decision makers or even by the users of the system. In that case, the AHP method can be utilized in the decision-making process to efficiently deal with these sometimes conflicting needs.

Table 4. The analysis results obtained by SIDRA for intersection design types

Parameters	Signalized with Left Turn Bays	Signalized without Left Turn Bays	Roundabout	Grade-separated Intersections
Average Vehicle Delay (sec)	196.5	219.2	8.8	15
Operating Cost ($/hr)	7021.9	7646.56	1739.02	1771
Fuel Consumption (liter/hr)	469.2	501.1	215.4	319
CO Emission (kg/hr)	46.84	48.8	37.58	45.44

It should be clarified that the cost parameter is constituted of two different cost items considered for the assessment of the intersection, operating and construction costs. While the fuel cost represents the main part of the intersection operating cost, the design cost, which contains the cost of expropriation, construction and signalization system, represents the intersection construction cost. Because of being independent from each other, the intersection operating cost and intersection construction cost are examined separately in this study. In the first part, the analysis is carried out by considering only operating costs, then in the second part, construction costs are taken into consideration and results are analyzed.

The Providers/Decision Makers

The AHP procedure can be applicable to both individual and group decision settings. Eight traffic engineers from the Denizli Branch of 2nd Turkish Highways Izmir Regional Directorate are participated for this study. Their views and judgments are taken into consideration by reached consensus among them. Consensus refers here to the achieved agreement on constructing the hierarchy and on judgments among the participants. Therefore, in other words, pairwise comparison matrices among the criteria and pairwise comparison matrices among the alternatives with respect to each criterion are prepared based on the consensus of those participated people. It should be mentioned that there exist also other ways to set the priorities, if consensuses are not met: vote or compromise, geometric mean of the decision maker's judgments, separate models or players (Dyer & Forman, 1992; Vincent et al., 2002).

The AHP Results Considering Operating Cost

Firstly, the comparison matrices are prepared for determining the importance levels or weights of criteria considered at design stages. Eigen vector method as explained above is used for obtaining criteria weights. The pairwise matrices and resulting criteria weights are presented in Table 5.

Traffic accidents that incur losses in property and human lives cause immense losses on the social and economic structure of a region or a country. Therefore, traffic safety concern is essential for reducing the chance of collisions and accidents between vehicles, pedestrians and cyclists approaching to intersection from different directions. In other words, providing a safer environment for drivers as well as for other road-users at intersections is critical. The safety, in this study, is considered as the conflict points that are formed by conflicting movements in the traffic flow. The conflict points can be crossing,

Table 5. Defining criteria weights (n=5)

Criteria	Average Delay	Cost	Fuel Consumption	CO Emission	Traffic Safety	Eigen Vector	Priority Weights
Average Delay	1	1/2	5	4	1/4	0.304	0.167
Cost	2	1	6	4	2	0.659	0.362
Fuel Cons.	1/5	1/6	1	1/2	1/8	0.073	0.040
CO Emission	1/4	1/4	2	1	1/7	0.112	0.061
Traffic Safety	4	1/2	8	7	1	0.675	0.370
λ_{max}=5.305; CI=0.076 and RI=1.12 CR=0.07 (< 0.10)							

merging or diverging. In general, the number and distribution of conflicting points are used for assessing the intersection access and safety performance (Pan et al., 2013). It is evident and also shown by many studies that reducing the number of traffic conflict points significantly reduces also the accident rates (Davis, Hourdos, Xiong, & Chatterjee, 2011). The number of traffic conflict points can differ for each intersection design types as shown only for roundabouts and uncontrolled at-grade intersections with four legs in Figure 3. In case of high traffic volumes on the legs, grade-separated intersections should be preferred, since this type of intersection eliminates the crossing conflicts at intersection by a vertical separation of roadways in space (Tunc, 2003). Although, grade-separated intersections have no traffic conflict points occurring on intersection area, diverging or merging type conflict points can exist on associated entrance or exit lanes.

As a result, the traffic safety itself is considered to be the most important design criteria for determining the appropriate intersection design type. As seen in Table 5, therefore, the highest priority weight of 0.370 is assigned to this criterion. Effectively managing a budget is a very big concern for traffic engineers, decision makers or policy makers; therefore, the cost criterion follows the traffic safety very closely as the second most important criterion having the second highest priority weight of 0.361. Although, the delay is one of the most important performance criteria for intersections, it follows the safety and cost factors as the third one with a priority weight of 0.167. As seen, the weight assigned to the delay is almost half of the weights assigned to the cost and traffic safety, separately. The fuel consumption and CO emission factors are derivatives of each other. Hence, the importance levels or weights of these factors are quite close to each other's weight and less than the other criteria's weights. The fuel consumption

Figure 3. Traffic conflict points for roundabouts and at-grade intersection with four legs
Source: FHWA (2000)

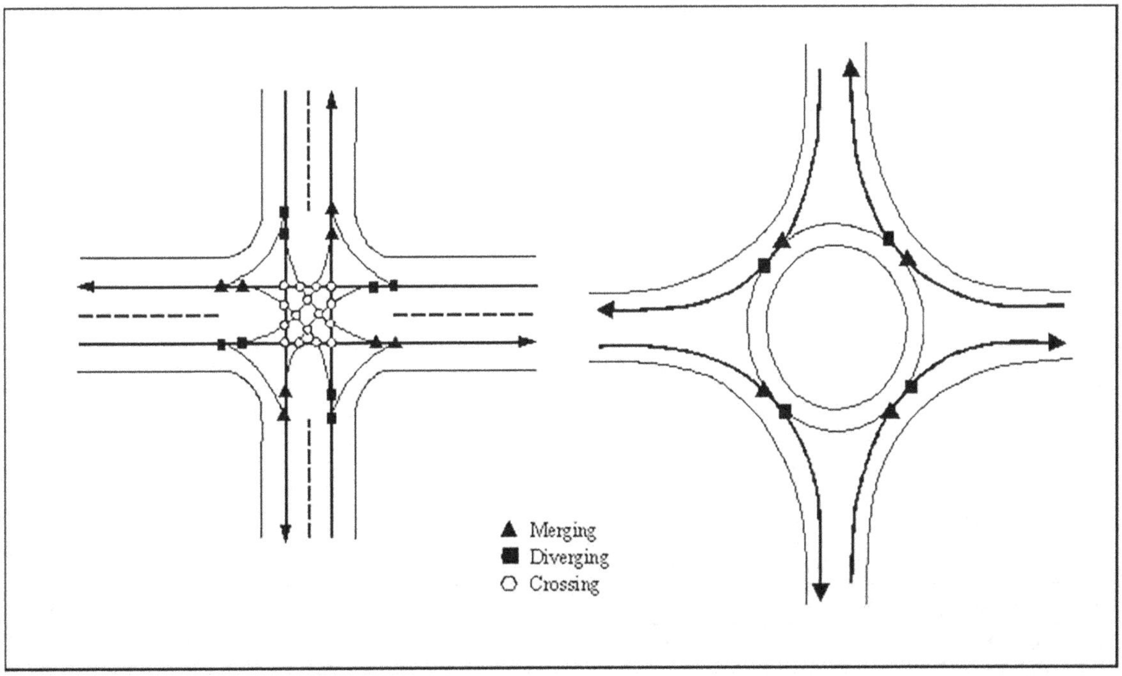

and CO emission together, which are 0.040 and 0.061, respectively, hold almost 10% of all the weights added together. This shows that environmental concerns do not play a significant role for determining the appropriate intersection design type yet, which is arguable.

After obtaining the criteria weights, all intersection design types proposed are evaluated with respect to each criterion. In this way, it would be possible to see the preferred intersection design type for the specified criterion in form of normalized priority. The AHP pairwise comparison matrices created for each criterion are presented in Table 6 through Table 10.

As seen in Table 6, with respect to the average delays, the roundabout type design receives the highest priority value of 0.534. The grade-separated type design follows it with a priority value of 0.310. These calculated values are also in accord with the simulation results obtained from the SIDRA, since both design types provide the lowest average vehicle delay in seconds, 8.8 and 15 seconds, respectively. Although, the SIDRA results suggest that the roundabout design type culminates in almost half of the average delay that the grade-separated design can cause, the priorities are not in that rate since that difference in actual practices is negligible. Moreover, it should be emphasized here that the results obtained from the AHP do not have to be in perfect harmony with the actual, or here simulation based results. However, an agreement to some extent is, of course, expected. The low *CR* value, 0.03, also indicates that providers of those pairwise comparison matrices are well-aware of or well anticipate the performances of intersection design types with respect to the average delays under existing traffic volumes.

Table 6. Pairwise comparisons of intersection types with respect to average delay (n=4)

Intersection Type	Signalized Intersection w/o LT	Roundabout	Grade-separated	Signalized w/ LT	Eigen Vector	Priorities
Signalized Intersection w/o LT	1	1/6	1/5	1/2	0.103	0.065
Roundabout	6	1	2	7	0.851	0.534
Grade-separated	5	1/2	1	4	0.494	0.310
Signalized w/ LT	2	1/7	1/4	1	0.145	0.091
λ_{max}=4.080; *CI*=0.027 and *RI*=0.90 *CR*=0.03 (< 0.10)						

Table 7. Pairwise comparisons of intersection types with respect to operating cost (n=4)

Intersection Type	Signalized Intersection w/o LT	Roundabout	Grade-separated	Signalized w/ LT	Eigen Vector	Priorities
Signalized Intersection w/o LT	1	1/8	1/7	1/2	0.081	0.051
Roundabout	8	1	2	6	0.827	0.523
Grade-separated	7	1/2	1	5	0.539	0.341
Signalized w/ LT	2	1/6	1/5	1	0.134	0.085
λ_{max}=4.055; *CI*=0.018 and *RI*=0.90 *CR*=0.02 (< 0.10)						

Table 8. Pairwise comparisons of intersection types with respect to fuel consumption (n=4)

Intersection Type	Signalized Intersection w/o LT	Roundabout	Grade-separated	Signalized w/ LT	Eigen Vector	Priorities
Signalized Intersection w/o LT	1	1/9	1/6	1/4	0.069	0.044
Roundabout	9	1	2	6	0.825	0.522
Grade-separated	6	1/2	1	5	0.521	0.330
Signalized w/ LT	4	1/6	1/5	1	0.164	0.104
λ_{max}=4.177; *CI*=0.059 and *RI*=0.90 *CR*=0.07 (< 0.10)						

Table 9. Pairwise comparisons of intersection types with respect to CO emission (n=4)

Intersection Type	Signalized Intersection w/o LT	Roundabout	Grade-separated	Signalized w/ LT	Eigen Vector	Priorities
Signalized Intersection w/o LT	1	1/8	1/6	1/3	0.076	0.048
Roundabout	8	1	2	7	0.844	0.536
Grade-separated	6	1/2	1	5	0.511	0.324
Signalized w/ LT	3	1/7	1/5	1	0.144	0.091
λ_{max}=4.132; *CI*=0.044 and *RI*=0.90 *CR*=0.05 (< 0.10)						

Table 10. Pairwise comparisons of intersection types with respect to traffic safety (n=4)

Intersection Type	Signalized Intersection w/o LT	Roundabout	Grade-separated	Signalized w/ LT	Eigen Vector	Priorities
Signalized Intersection w/o LT	1	9	7	1/2	0.618	0.386
Roundabout	1/9	1	1/4	1/8	0.062	0.039
Grade-separated	1/7	4	1	1/5	0.152	0.095
Signalized w/ LT	2	8	5	1	0.769	0.480
λ_{max}=4.245; *CI*=0.082 and *RI*=0.90 *CR*=0.09 (< 0.10)						

In a similar fashion, as Table 7 is scrutinized, one can see that with respect to the operating cost, again the roundabout and grade-separated design types earn the highest priority values of 0.523 and 0.341, respectively. These values are also in accord with the results obtained from the SIDRA, which are approximately 1,739 and 1,771 dollar per hour, respectively. Furthermore, yet the values obtained from the SIDRA are very close, the priorities are significantly different. The *CR* value, 0.02, is also very low indicating that participants are consistent about their agreed pairwise ratings.

As seen in Table 8, with respect to the fuel consumption, once again the roundabout and grade-separated design types take the highest priority values of 0.522 and 0.330, respectively, which are almost identical with the results given in Table 7, including priority values for signalized without left turn and signalized with left turn intersection design types. Somehow the *CR* value, 0.07, is a little higher than those obtained from ones with respect to average delay and operating cost, but still at acceptable level. The SIDRA results also follow the same pattern indicating that participants also well anticipate the performances that intersection design types display with respect to the fuel consumption.

Similarly, as seen in Table 9, with respect to the CO emission, the results are very close to the results obtained from the ones for the average operating cost and for the fuel consumption. Since they are highly related, the roundabout and grade-separated design types again receive the highest priority values of 0.536 and 0.324, respectively. The *CR* value, 0.05, is a little less than for the fuel consumption and a little higher than for the operating cost, but is is moderate and in acceptable level.

The pairwise comparison matrix with respect to the traffic safety criterion given in Table 10 is the only one that contains the results that, in any form, cannot be controlled or compared with the results obtained from the SIDRA. As seen clearly, the outcome of this matrix is totally different than previous ones. While the signalized with left turn bay type design receives the highest priority value of 0.480 which is followed by the signalized intersection without left turn bay type design with a priority value of 0.386, the grade-separated and roundabout design types receive lower priority values of 0.095 and 0.039, respectively. These results confirm that higher traffic volumes necessitate traffic signals for minimizing the probability of collisions between vehicles. Especially for the through traffic following the left turning vehicles, the probability of rear end collisions is usually high, therefore providing extra lane or lanes (it can sometimes be more than one lane) as left turn bays can substantially eliminate this type of accidents. It should be noticed that although grade-separated intersection design types eliminate the conflict points on intersection area, there still exists merging and diverging conflict points outside the intersection area but on associated entrance and exit lanes. Therefore, signalized intersection design types are safer than grade-separated intersection design types. The *CR* value, 0.09, is the highest of all the comparisons made for with respect to every criterion but still at the acceptable level. This can be due to the fact that the assessment of the traffic safety based on conflicting point approach is not an easy task for participants to well assess the safety issue. In other words, they can somehow anticipate which intersection design type is better than which, but not very sure about how much.

Since the weights of criteria and priorities among the alternatives with respect to each criterion are obtained, the overall composite weights of the alternatives, then, can be calculated using Equation.

The calculated overall composite weights are summarized in Table 11. The values in the parentheses given in the first row indicate the weights of each criterion obtained previously.

As seen in Table 11, the results suggest that the appropriate one should be the roundabout design type with a total weight of 0.35. The grade-separated and signalized with left turn bay design types follow it with 0.24 and 0.23, respectively. The signalized without left turn bay design is the last desired one with a total weight of 0.18. As expected, the traffic safety criterion played a very important role on defining those total weights. Since left turn volumes are very large, which are from Pau to Pekdemir: 450 veh/hr (80% of total traffic from Pau); from Mugla to Pau: 394 veh/hr (25% of total traffic from Mugla); from Pekdemir to Tali: 114 veh/hr (10% of total traffic from Pekdemir); and from Tali to Muğla: 73 veh/hr (30% of total traffic from Tali), it is not surprising that the intersection design types which allow an easier movement to the left turns receive always higher total weights than signalized without left turn bay design type which comparatively does not allow.

Table 11. Overall composite weights of intersection design types

	Average Delay (0.167)	Operating Cost (0.362)	Fuel Consumption (0.040)	CO Emission (0.061)	Traffic Safety (0.370)	Composite Weights
Signalized Intersection w/o LT	0.065	0.051	0.044	0.048	0.386	**0.18**
Roundabout	0.534	0.523	0.522	0.536	0.039	**0.35**
Grade-separated	0.310	0.341	0.330	0.324	0.095	**0.24**
Signalized w/ LT	0.091	0.085	0.104	0.091	0.480	**0.23**

The AHP Results Considering Construction Cost

Construction cost is one of the most important criteria that also play an important role in the selection of intersection design types, particularly if the budget is limited. Thus, the main deficiency of the grade-separated intersection design type rises from its relative higher construction cost compared to the others. It is, therefore, usually not preferred by decision makers except only for mandatory cases such as in areas where high traffic accidents rates are observed or expected. Thus, the construction cost is separately taken into account in this section following the same method explained previously. Since there will be no changes for the pairwise comparisons (with respect to the average delay, fuel consumption, CO emission and traffic safety) and priorities obtained from them, only the pairwise comparison matrix with respect to the construction cost is sought for all intersection design types. It should be emphasized that here the budget limitation is not totally considered, therefore the weight of the cost criterion is assumed to be same for the operating and construction cost. This is imperative since if the budget is limited than resulting weights may change drastically (for instance, if the budget is not enough to construct a grade-separated intersection design, then it can either totally leave it out from the alternatives or a very high weight for the construction cost can be obtained from a different pairwise comparison matrix among the criteria). The provided (agreed upon the consensus among the participants) pairwise comparison matrix with respect to the construction cost is given in Table 12.

Table 12. Pairwise comparisons of intersection types with respect to construction cost (n=4)

Intersection Type	Signalized Intersection w/o LT	Roundabout	Grade-separated	Signalized w/ LT	Eigen Vector	Priorities
Signalized Intersection w/o LT	1	1/4	7	4	0.395	0.261
Roundabout	4	1	9	5	0.904	0.598
Grade-separated	1/7	1/9	1	1/3	0.064	0.042
Signalized w/ LT	1/4	1/5	3	1	0.148	0.098
λ_{max}=4.204; CI=0.068 and RI=0.90 CR=0.08 (< 0.10)						

As seen in Table 12, the roundabout type design gets the highest priority value of 0.598 with respect to the construction cost. The signalized intersection without left turn bay type design follows it with a priority value of 0.261. The signalized intersection with left turn bay and grade-separated design type get priority values of 0.098 and 0.042, respectively. This order is in accord with the increasing construction costs that the design types can actually result in if constructed. Although, the *CR* value, 0.08, is a little high, it is still in acceptable range. As a result, the roundabout type design shows an obvious advantage over the others. In this case, the calculated overall composite weights change as given in Table 13.

In order to provide an easy comparison, the previous composite weights (where operation cost is considered) are also given in parentheses in the last column of Table 13. As seen, although the first preferred one (roundabout) is not changed (since it has also an obvious construction cost advantage) with the priority value of 0.37 (which was previously 0.35), the order of preferences somehow differs from the previous one. The second preferred one (grade-separated intersection), in this case losses its places to the signalized intersection either with left turn or without left turn bay design types and becomes the less preferred one (its priority value drops from 0.24 to 0.14). These results indicate how important role the construction cost can play on the intersection design type decision.

RESULTS AND CONCLUSION

In this study, a very well-known multi-criteria decision making procedure, the AHP, is used for the selection of an appropriate intersection design type among the four different intersection design types; roundabout, signalized with left turn bay, signalized without left turn bay and grade-separated, for an actual intersection at Denizli, Turkey. Two scenarios are examined in the analysis; in the first one the operating costs together with the other criteria are considered in the analysis, and in the second one the construction costs replaced for the operating costs in the analysis.

With regard to the traffic safety, the roundabout design type, which has eight conflict points, actually does not represents the best choice, however with respect to the other criteria; the average vehicle delay, operating cost, construction cost, fuel consumption and CO emission, it can offer more advantages than the other intersection design types can actually do as in the case studied here. Therefore, in overall, the roundabout design type received the highest priority values and was preferred as the first choice in both scenarios.

Table 13. Overall composite weights as construction cost considered

	Average Delay (0.167)	**Construction Cost (0.362)**	**Fuel Consumption (0.040)**	**CO Emission (0.061)**	**Traffic Safety (0.370)**	**Composite Weights**
Signalized Intersection w/o LT	0.065	0.261	0.069	0.076	0.618	**0.25 (0.18)**
Roundabout	0.534	0.598	0.825	0.844	0.062	**0.37 (0.35)**
Grade-separated	0.310	0.042	0.521	0.511	0.152	**0.14 (0.24)**
Signalized w/ LT	0.091	0.098	0.164	0.144	0.769	**0.24 (0.23)**

The grade-separated intersection design type offers the minimum interruptions to traffic flow, therefore this type of design can result in lower average delay, fuel consumption, operating cost and CO emission rates than the others, except for the roundabout design type, can result in. Therefore, it follows the roundabout design type as second with respect to the operating cost. However, since its construction cost is the highest among all, it becomes the last desired one in the second scenario. The signalized intersection with left turn and without left turn bays, in order, follow the grade-separated design in the first scenario, as the latter is the least desired one. However, in the second scenario, the signalized intersection without left turn bays follows the roundabout design type as second and the signalized intersection design type with left turn bays follows it as third. Although, the construction cost of providing bays can be higher, the left turn bays provide an extra space, thus the possibility of traffic accidents occurrence can be reduced and also the capacity of the adjacent lane can be improved as well. Therefore, considering the construction cost, the signalized intersection without left turn bays can offer a slight advantage over the one with left turn bays. The results obtained from the AHP, therefore, in harmony with all these findings.

It is also found that in the second scenario where the construction costs replaced for the operating costs, the order of preferences is also changed. Only the roundabout design type still remains the first preferred one. It affirms that the construction cost criterion is an important factor in the decision making of intersection design type selection. The results indicated that the grade-separated intersection is the least preferred one if the construction cost is taken into account. A similar finding is also observed for the signalized intersection with left turn bay. Due to the extra costs for adding turn bays (space for the bays, constructing layers and coating them by asphalt etc.), the signalized intersection without a left turn bay is preferred over it by a slim margin. It can be said that the sequence of preferences is totally defined by the construction costs.

In short, the AHP model can be used as a decision support tool for making a decision on the selection of the intersection design type for a specified intersection. The results obtained from the AHP are also partly supported by the SIDRA Intersection software package. It is partly since the SIDRA cannot handle all the criteria considered in the analysis. Such as the traffic safety criterion that can be better evaluated by the experts in the field. Furthermore, this AHP-based analytical approach can be used as a tool by traffic engineers and in transportation planning. It should, however, be emphasized that depending on the conditions such as traffic volumes and geometric conditions etc., the weights and priority values can differ for different locations and cases. Moreover, if some criteria can be assessed with any available method i.e. statistical models or simulations, here for instance the SIDRA Intersection software package, they should be assessed first by the available method or software. Then, the results should be conveyed to decision makers to let them gain more insights into the problem that can lead to more appropriate evaluation of the system as well as more concise results at the end.

REFERENCES

Al-Harbi, K. M. (2001). Application of AHP in project management. *International Journal of Project Management, 19*(1), 19–27. doi:10.1016/S0263-7863(99)00038-1

Arabameri, A. (2014). Application of Analytic Hierarchy Process (AHP) for locating fire stations: Case study Maku City. *Merit Research Journal of Art, Social Science and Humanities, 2*(1), 001-010.

Balli, S., & Korukoglu, S. (2009). Operating system selection using fuzzy AHP and TOPSIS methods. *Mathematical and Computational Applications*, *14*(2), 119–130.

Chatterjee, D., & Mukherjee, B. (2013). Potential hospital location selection using AHP: A study in Rural India. *International Journal of Computers and Applications*, *71*(17), 1–7. doi:10.5120/12447-9144

Davis, G. A., Hourdos, J., Xiong, H., & Chatterjee, I. (2011). Outline for a causal modal of traffic conflicts and crashes. *Accident; Analysis and Prevention*, *43*(6), 1907–1919. doi:10.1016/j.aap.2011.05.001 PMID:21819818

Dyer, R. F., & Forman, E. H. (1992). Group decision support with the analytic hierarchy process. *Decision Support Systems*, *8*(2), 99–124. doi:10.1016/0167-9236(92)90003-8

Effat, H. A., & Hassan, O. A. (2013). Designing and evaluation of three alternatives highway routes using the analytical hierarchy process and the least-cost path analysis, Application in Sinai, Peninsula, Egypt. *The Egyptian Journal of Remote Sensing and Space Sciences*, *16*(2), 141–151. doi:10.1016/j.ejrs.2013.08.001

Fazlollahtabar, H., Eslami, H., & Salmani, H. (2010). Designing a fuzzy expert system to evaluate alternatives in fuzzy analytic hierarchy process. *J. Software Engineering & Applications*, *3*(04), 409–418. doi:10.4236/jsea.2010.34046

Gulenc, I. F., & Bilgin, G. (2010). A model proposal for investment decisions: AHP method. *Journal of Proposal - Marmara University, 34*(17), 97-107.

Kahraman, C., Cebeci, U., & Ruan, D. (2004). Multi-attribute comparison of catering service companies using fuzzy AHP: The case of Turkey. *International Journal of Production Economics*, *87*(2), 171–184. doi:10.1016/S0925-5273(03)00099-9

Mahendran, P., Mohorty, M. B. K., & Saravanan, S. (2014). A fuzzy AHP approach for selection of measuring instrument for engineering college selection. *Applied Mathematical Sciences*, *8*, 2149–2161.

Michael, F. L., Noor, Z. Z., Zardari, N. H., & Meza, M. J. F. (2013). Analytical hierarchy process application in urban sustainability indicators prioritization. *Resources and Environment*, *3*, 1–5.

Murat, Y. S., Cakici, Z., & Yaslan, G. (2014). Use of Fuzzy Logic Traffic Signal Control Approach as Dual Lane Ramp Metering Model for Freeways. *Advances in Intelligent Systems and Computing*, *223*, 339–349. doi:10.1007/978-3-319-00930-8_30

Omurbek, N., Ustundag, S., & Helvacioglu, O. C. (2013). Use of analytical hierarchy process (AHP) in location decision: A study in Isparta Region. *Journal of Administrative Sciences*, *21*, 101–116.

Ozdemirel, A., Gokgur, B., & Eliiyi, D. T. (2012). An assignment and routing problem with time windows and capacity restriction. 15th Meeting of the Euro Working Group on Transportation. *Procedia: Social and Behavioral Sciences*, *54*, 149–158. doi:10.1016/j.sbspro.2012.09.734

Pan, F., Zhang, L., Lu, J., Zhao, J. J., & Wang, F. (2013). A method for determining the number of traffic conflict points between vehicles at major-minor highway intersections. *Traffic Injury Prevention*, *14*(4), 424–433. doi:10.1080/15389588.2012.713148 PMID:23531266

Piantanakulchai, M., & Saengkhao, N. (2003). Evaluation of alternatives in transportation planning using multi-stakeholders multi-objectives AHP modeling. *Proceedings of the Eastern Asia Society for Transportation Studies*, *4*, 1613–1628.

Pogarcic, I., Francic, M., & Davidovic, V. (2008). Application of AHP method in traffic planning. Proceedings of the *16th International Symposium on Electronics in Transport 2008.*

Roundabouts: an Informational Guide. (2000). United States Department of Transportation, Federal Highway Administration.

Saaty, T. L. (1977). A scaling method for priorities in hierarchical structures. *Journal of Mathematical Psychology*, *15*(3), 57–68. doi:10.1016/0022-2496(77)90033-5

Saaty, T. L. (1980). *The Analytical Hierarchy Process*. New York: McGraw Hill.

Saaty, T. L. (1988). *Multicriteria decision making: The analytic hierarchy process*. Pittsburgh: RWS Publications.

Saaty, T. L. (1994). *Fundamentals of decision making and priority theory with the AHP*. PA, U.S.A: RWS Publications.

Tezcan, O. (2007). Analytic hierarchy process method and application in area selection of ready mixed concrete plant. *National Journal of Ready Mixed Concrete, Nov.-Dec.,* 58-62.

Triantaphyllou, E., & Mann, S. H. (1995). Using the analytic hierarchy process for decision making in engineering applications: Some challenges. *International Journal of Industrial Engineering: Applications and Practice*, *2*, 35–44.

Tunc, A. (2003). *Traffic Engineering and Applications*. Ankara, Turkey.

Vaidya, O. S., & Kumar, S. (2006). Analytic hierarchy process: An overview of applications. *European Journal of Operational Research*, *169*(1), 1–29. doi:10.1016/j.ejor.2004.04.028

Vargas, L. G. (1990). An overview of the analytic hierarchy process and its applications. *European Journal of Operational Research*, *48*(1), 2–8. doi:10.1016/0377-2217(90)90056-H

Lai, V. SWong, B. K., & Cheung, W. (2002). Group decision making in a multiple criteria environment: A case using the AHP in software selection. *European Journal of Operational Research*, *137*(1), 134–144. doi:10.1016/S0377-2217(01)00084-4

KEY TERMS AND DEFINITIONS

AHP: A mathematical technique that can be used for multi-criteria decision making.

Intersection: Junction of two or more roads.

Multi-Criteria Decision Making: One of the operations research methods that considers multi-criteria for an efficient decision making.

Traffic Engineering: A branch of civil engineering that deals with efficient design of roads and traffic.

Transportation Planning: Rational evaluation approach of roads and transport facilities considering some analytical techniques.

Transportation: Movement of people and goods from one location to another.

Chapter 9
Understanding Accessibility:
Accessibility Modeling With Geographical Information Systems (GIS)

Kivanc Ertugay
Selcuk University, Turkey

Sebnem H. Duzgun
Middle East Technical University, Turkey

ABSTRACT

Accessibility measures are generally concerned with equity and a better distribution of services in a territory and can be accepted as key variables for supporting supply/demand, location/allocation and service/catchment area related planning policies and strategies at national, regional, and local levels. Since accessibility measures need organization of huge and complex spatial data sets, accessibility modeling often lends itself to Geographical Information Systems (GIS) for analysis and presentation. Since numerous accessibility measures and modeling techniques ranging from simple to sophisticated can be found in the literature, this work aim to provide an overview of the theoretical framework and relevant background about GIS-based accessibility modeling process. The results could provide a significant support for the decision makers who are supposed to deal with transportation planning, accessibility modeling, location/allocation and service/catchment area related issues.

INTRODUCTION

The term accessibility has long been used by geographers, economists, urban and transportation planners and reflects the relative ease of access to/from several urban/rural services by considering several travelling costs (Halden et al., 2000; Makri, 2002; McGrail & Humphreys, 2009). Accessibility measures are generally concerned with equity and a better distribution of services in a territory and help to evaluate the proximity/availability of several services like health, education, recreation, emergency or trade etc. by considering one or more modes of transportation such as such as walking, cycling, driving or public transport etc.

DOI: 10.4018/978-1-4666-8648-9.ch009

The accessibility measures basically help decision makers to

- identify regions that have inadequate or excessive service
- select appropriate sites for new or re-located services,
- test and improve the performance of the transportation system.

That is why, accessibility measures can be accepted as key variables for supporting supply/demand, location/allocation and service/catchment area related planning policies and strategies at national, regional, and local levels (Makri, 2002; Juliao, 1999; Kuntay, 1990; Halden et al., 2000; Radke & Mu, 2000).

Numerous accessibility measures, ranging from simple to sophisticated, can be found in the accessibility literature. While simple measures only consider proximity in terms of time and distance, sophisticated ones consider both proximity and availability including the size of supply and demand. Some of the most widely used accessibility measures in the literature are;

- Travel time/distance measures, service/catchment areas (travel time or distance to nearest supply/demand calculated from Euclidian/Network-based costs) (see Ghio et al., 2007; Joseph et al., 2006; Fortney et al., 2000; Sylvie, 2007; Brabyn, 2002; O'Sullivan et al., 2000; Juliao, 1999; Ebener et al., 2005),
- Cumulative opportunity measures (consider the total amount of demand/supply inside the catchment areas) (see Chapelet & Lefebvre, 2005; Boulos et al., 2001; Nadine et al., 2006; Black et al., 2004; Goulias, 2007),
- Supply to demand ratio measures (population to provider ratios, calculated inside the catchment areas) (see Luo, 2004; Scott et al., 2006; Bagheri et al., 2006),
- Kernel density measures (use the Gaussian kernel approach to calculate the density value of each demand/supply) (see Yang et al., 2006; Gibin et al., 2007; McGrail, & Humphreys, 2009),
- Gravity-based measures (a combined indicator of accessibility and gravity by considering the attractiveness of supply/demand) (see Kwan, 1998; Chen, 2000; Guagliardo, 2004),
- Two-step floating catchment area measures (2SFCA) (repeat the process of accessibility modeling considering the overlay areas (see Mitchel et al., 2008; Luo & Wang, 2003; Luo, 2004; Yang et al., 2006; Scott et al., 2006)

Since accessibility measures describe the characteristics of a location and need organization of huge and complex spatial data sets, accessibility modeling often lends itself to Geographical Information Systems (GIS) for analysis and presentation. GIS have unique capabilities to present spatially referenced information in a way, which aids decision-making and provides a powerful interface for handling, organizing, analyzing and presenting huge and complex spatial data sets. For example; data storage, management and manipulation capabilities for both graphical and attribute data, core data analyses capabilities such as buffer, overlay, proximity, shortest path, raster cost-distance etc., programming capabilities to handle current models or create new models and mapping and visualization capabilities to evaluate the results of the analyses (Black et al., 2004; Chen & Weng, 1999; Chen, 2000; Peters & Hall, 1999).

In a more specific way, GIS can handle important steps in accessibility modeling like;

- storing road networks and origin/destination-based geographical databases,
- calculating costs between origins and destinations on transportation networks,

- building regulations of streets such as one-way streets, closed streets, overpasses and underpasses,
- considering the delays in traffic network intersections,
- calculating travelling costs considering service area overlaps
- presenting results for a defined time or distance threshold (e.g. < 10 minutes or 10 kilometers) with several techniques such as zone, raster or isochronal technique with opportunity of different scales and various visualization alternatives etc. (MacFarlane, 2005).

In the light of the above mentioned greeting literature, this chapter aim to provide an overview of the theoretical framework and relevant background about physical accessibility in order to clarify the nature of GIS-based physical accessibility modeling and to support decision makers who are supposed to deal with transportation planning, accessibility modeling, location/allocation and service/catchment area related issues.

The chapter includes a detailed review of the literature about definitions of accessibility, usage areas of accessibility, components of accessibility (activity and cost elements), place accessibility measures (Travel time/distance, Cumulative opportunity, Population to provider ratio, Kernel density, Gravity measures, Two-step floating catchment area), individual accessibility measures (space-time), the linkage between accessibility measures and transportation planning, GIS-based accessibility modeling techniques (zone-based technique, isochrone-based, raster-based), the role of GIS in accessibility modeling (contribution of GIS technology into accessibility modeling), GIS-based accessibility modeling examples and applications, a detailed illustration of accessibility modeling process in GIS environment by using a hypothetical dataset. Finally it brings to a close with a conclusion part and future research directions part.

The basic framework of this research is based on a PhD thesis which is financially supported by the State Planning Organization (DPT) Grant No: BAP-09-11-DPT- 2002K125/510 (submitted to the Graduate School of Natural and Applied Sciences of Middle East Technical University by Ertugay K. and Duzgun S. "GIS-Based Stochastic Modeling of Physical Accessibility by Using Floating Car Data and Monte Carlo Simulations, 2011")

BACKGROUND

Definitions of Accessibility

The term accessibility is used by various disciplines and many different aspects, perspectives and hence definitions can be found in the accessibility literature.

Some of these different aspects are;

- Physical accessibility (sometimes considered as geographical accessibility) which is being able to reach a service/facility in spite of physical impedances
- Mental accessibility which is understanding and being able to use a given area and its facilities
- Social accessibility which is having friends and a job and being able to get to and from work, meet friends and participate in social activities
- Organizational accessibility which is having access to travel opportunities, information and service regarding a journey

- Financial accessibility which is being able to afford available public or private means of transport
- Virtual accessibility which is being able to access information and people without moving from a certain place, by using electronic facilities (Kwan, 1998; Makri, 2002).

The accessibility concept mentioned in this research is mainly physical accessibility. Several definitions related with physical accessibility can be found in the accessibility literature. Kuntay (1976b) defines physical accessibility as the ability to reach from one place to another securely and comfortably by shortest way, simple route, appropriate speed, and ability to reach the intended location for a specific aim. Dong et al. (1998) defines physical accessibility as the ease and convenience of access to spatially distributed opportunities with a choice of travel. Joly (1999) defines the physical accessibility as a geographical concept in transportation planning and as a capacity term to reach customers, or a service for evaluation of projects. Litman (2011) defines accessibility as people's overall ability to reach services and activities. Chen (2000) defines physical accessibility as a significant index that reflects the ease for travelers to achieve desired movements in urban areas. Although there are different definitions about physical accessibility in the literature, they mostly point out a common direction. Physical accessibility is a term that reflects the relative ease of access to/from several services by considering several costs of travelling.

However, it is important to indicate that; physical accessibility term used in the literature sometimes encapsulates more than a meaning of physical access considering physical impedances. Many factors may be possible to be considered within the term "physical accessibility" such as comfort, cost, frequency, security, connectivity, reliability, management etc issues (see more detailed information in Litman (2011).

Kwan (1998) also emphasizes that physical accessibility can be handled either for people (individual accessibility) or for places (place accessibility) according to the aim of the study. This means that physical accessibility can be handled as a property of people defining how easily an individual can reach activity locations, or can be handled as an attribute of locations indicating how easily certain places can be reached by the people or services.

Accessibility can also be viewed from various perspectives, such as a particular person, group, mode, location or activity and hence it is important to specify the considered perspective for evaluating accessibility. For example, accessibility of a particular location could be very high for some users, but be low for the other users. Similarly, the focus in accessibility studies could be very different specific to special disciplines. For example; the land use planners could focus on the time/distance cost between different land use and activities, on the other hand, the social service planners could focus on accessibility of disabled people to reach medical clinics or recreation centers etc. (Litman, 2011).

APPLICATION AREAS OF ACCESSIBILITY

The physical accessibility measures have long been used by geographers, economists and urban planners and directly or indirectly always been an important part of urban analyses. Accessibility measures are concerned with equity and a better distribution of services in the territory and can be accepted as key variables for the decision makers to test the accessibility level of several urban services and give vital clues for decision makers to define planning strategies. That is why accessibility measures, whether simple or sophisticated, are important variables that decision makers must consider in the early stages of their planning efforts (Makrí, 2001; Makrí & Folkesson, 1999; Juliao, 1999; Emelinda & Shashi, 1995).

Accessibility measures are widely used to check the benefits of urban plans as a planning control tool. They help to evaluate proximity and availability of several urban/rural services like health, education, recreation, emergency or trade etc. by considering several transportation types like pedestrian, bicycle, car, public transport etc. for a defined threshold of time or distance (e.g. 1 km, 5 minutes etc.) (Kuntay, 1976a, 1976b, 1990; Halden et al., 2000).

By the help of the accessibility measures, decision makers can;

- identify regions that have inadequate or excessive service
- select appropriate sites for new or re-located services,
- evaluate the performance of the transportation systems

That is why, accessibility measures can be accepted as key variables for the decision makers to support their supply/demand, location/allocation and service/catchment area related planning policies and strategies at national, regional, and local scales in different levels (Makri, 2002; Juliao, 1999; Halden et al., 2000; Radke & Mu, 2000).

Many factors affect accessibility, including mobility (physical movement), the quality and affordability of transport options, transport system connectivity, mobility substitutes, and land use patterns.

COMPONENTS OF ACCESSIBILITY

There are two fundamental components of accessibility in the literature, which are;

- Activity element and
- Cost element

The activity element of the accessibility usually includes the type of the traveler and distribution of various urban/rural services. All accessibility measures include representation of the activity and cost element, which need to be defined at a level of detail, according to the needs of the particular situation about accessibility.

The cost element of accessibility includes either un-constrained Euclidian distance-based costs (bird-flight distance-based costs) such as Buffer, Voronoi/Thiessen polygons etc. or constrained transportation network-based costs such as travel distance/travel time by considering transportation network and several transportation types (pedestrian, bicycle, car, public transport etc.) (Halden et al., 2000; Makrí, 2001).

Activity Element

The activity element of the accessibility generally consists of two elements, which are;

- Urban/rural services/facilities (supply points which are interested) and
- Type of the person/traveler (demand points)

Depending on the issue at hand, activity element in accessibility analyses is based on various urban/rural services which are interested, such as education/training facilities (like schools, colleges, universities, training centers), emergency facilities (like health centers, hospitals, police stations, fire brigades), or shopping/leisure facilities (like shops/shopping centers, cinemas/theatres, sports centers, outdoor activity opportunities, pubs/clubs) etc.

Type of person/traveler in accessibility analyses includes several factors such as employment status of the traveler (unemployed, retired, economically active etc.) or age of the traveler (adult, children etc.) or physical health of the traveler (healthy, disabled etc.) (Makrí & Folkesson, 1999; Halden et al., 2000). For example, economically active people and shopping centers can create the activity elements of an accessibility research, in which accessibility level of economically active people to shopping centers are investigated.

Cost Element

The cost element of accessibility usually comprises two basic elements, which are;

- Unconstrained Euclidian distance-based / bird-flight costs (such as Buffer, Voronoi/Thiessen polygons etc.) and
- Constrained transportation network-based costs (such as distance or time)

In representation of cost element in accessibility analyses, there are also several factors that must be considered in detail, which are;

- Time of the travel (rush hour, normal hour, etc.),
- Type of the travel (pedestrian, bicycle, car, public transport, etc.),
- Day of the travel (Sunday, Monday, etc.),
- Season of the travel (winter, summer, etc.),
- Characteristics of the travel (quality and capacity of the roads, the economy, comfort, cost and safety considerations)
- Type of the traveller (adult, children, normal, disabled etc.) or
- Mobility[1]of the traveler (Kuntay 1976a, 1976b; Halden et al., 2000).

For example, a 5-minute accessibility zone of a bicycle vehicle is different from a car vehicle. Similarly, a 5-minutes accessibility zone of a car vehicle in rush hour time traffic conditions will be different from the normal time traffic conditions.

Accessibility Measures

In general, accessibility measures can be handled either for places (place accessibility measures) or for people (individual accessibility measures).

Place Accessibility Measures

Place accessibility measures handle accessibility as an attribute of locations indicating how easily certain places can be reached by the people or services. At its simplest level, qualitative descriptions can be used to define the place accessibility of a location. Terms such as good accessibility, average accessibility or poor accessibility can be used as simple qualitative accessibility measures for describing the accessibility level of a location. These qualitative measures can be based on various factors such as;

- average time or distance cost between locations,
- the ease of walking or cycling to stations
- number of population/facilities within a defined time/distance threshold, number of stations within a region (e.g. 5 grocery within 1 minutes walk, 10 stations within 5 minutes walk)
- frequency of vehicles (e.g. 1 bus for every 15 minutes etc.),
- comfort, quality, convenience of transportation,
- cost, affordability, safety of transportation
- variety of transportation modes (such as rail, bus, bicycle, pedestrian etc), variety of lines passing from a location/station
- ease, quality, integrity of different transportation modes (e.g. transfer from metro to bus)
- connectivity of transportation network (e.g. the directness of travel between locations, the density of paths and intersections),
- proximity, density and mix of landuse/activities, distances between activities etc. (Halden et al., 2000; Litman, 2011)

The qualitative measures of accessibility (such as good accessibility, average accessibility or poor accessibility) are useful indicators and have been widely used to provide general accessibility level of locations. However, decision makers, who are supposed to deal with transportation planning, accessibility, location/allocation and service/catchment area related issues, usually need a more comparative and quantitative approach in order to support their planning policies and strategies. Numerous quantitative accessibility measures ranging from simple to sophisticated can be observed in the accessibility modeling literature. While the simple accessibility measures only consider proximity in terms of time and distance without considering the transportation network, sophisticated accessibility measures could consider both proximity and availability considering the size of the supply and demand and the transportation network. Some of the most widely observed quantitative accessibility measures in the literature are;

- Travel time/distance measures (travel time or distance to nearest supply/demand calculated from Euclidian/Network-based costs),
- Cumulative opportunity measures (consider the total amount of demand/supply inside the catchment areas),
- Supply to demand ratio measures (population to provider ratios, calculated inside the catchment area boundaries),
- Kernel density measures (uses the Gaussian kernel approach to calculate the density value of each demand/supply),
- Gravity-based measures (a combined indicator of accessibility and availability by considering the attractiveness of the supply/demand) and

- Two-step floating catchment area measures (FCA) (repeat the process of accessibility modeling process considering overlay areas), etc. (Luo & Wang, 2003; McGrail, & Humphreys, 2009; Guagliardo, 2004; Bagheri et al., 2006).

Although there are various accessibility measures ranging from simple to sophisticated, there is no best approach to measure accessibility. Different aims and situations can demand different measures and approaches (Makri, 2002).

Travel Time and Distance Measures

The travel time and distance measures are simple and commonly used measure of accessibility. They help accessibility related decision makers to understand the minimum, maximum or average travel cost between several opportunities (supply and demand points) and to determine the catchment/service area boundaries. The travel time and distance measures are widely used in the accessibility modeling literature and can be considered as the fundamental elements of all kind of accessibility measures, ranging from simple to sophisticated (Makri, 2002; Makri & Folkesson, 1999).

The travel time/distance costs can be measured as several ways such as;

- average travel time/distance to opportunities or
- minimum travel time/distances to opportunities
- average travel time/distance to nearest opportunity or
- minimum travel time/distance to nearest opportunity etc.

The shorter the travel time/distance mean the higher the accessibility. The estimation of these measures can be performed in two different ways. One is the simple Euclidian costs (known as straight-line costs bird-flight costs or unconstrained costs) and the other is more complicated transportation network-based costs (constrained costs). While the Euclidian costs are calculated from Buffer, or Voronoi/Thiessen polygons etc., the transportation network-based costs are generally calculated from transportation network based distances or time by using length of road segments and average speed on road segments such as 120 km/h for highways, 50 km/h for main streets and 30 km/h for local streets, etc. (See several examples; Ritsema van Eck & de Jong, 1999; Makri, 2002; Luo & Wang, 2003; Bixby, 2004; Lotfi &Koohsari, 2009; McGrail & Humphreys, 2009; Vahidnia et al., 2009).

Cumulative-Opportunity Measures

Cumulative-opportunity measures are evaluations of accessibility in terms of number or proportion of available opportunities within certain catchment area boundary (a threshold of travel distance or time). These measures provide an idea of the range of various choices available to supply/demand points in urban/rural environment.

The cumulative opportunity measures usually calculated from major facilities or centers of population such as cities, districts, central business districts or several public services such as hospitals, schools, recreation, emergency services etc. The cumulative opportunities can be the total number of jobs, floor spaces, people or employees etc. within a defined service/catchment area boundary threshold of distance or time. For example, the total number of schools within 500 meters of districts, the total number of customers within 30 minutes of shopping centers or the total number of people within 15 minutes of city center are good examples of this kind of accessibility measures (Makrí & Folkesson, 1999; Halden et al., 2000; Kwan, 1998).

The basic formulation for cumulative opportunity measure (A_t) is that;

$$A_t = \sum_t O_t \qquad (1)$$

in which t is the catchment threshold, and O_t is the cumulative opportunity that can be reached within threshold t.

As, all potential opportunities, whether closer or further, within the defined deterministic threshold are weighted equally and all potential opportunities beyond the defined deterministic threshold are not taken into consideration, defining a threshold is a critical factor in the calculation of cumulative opportunity measures and directly affect the results of the cumulative opportunity measures (Makri, 2002; Makri & Folkesson, 1999; Makrí, 2001).

Supply to Demand Ratio Measures

Supply to demand ratios, also known as provider to population ratios, are another type of accessibility measures, which are calculated within the bordered zones or geographical units such as states, countries, metropolitan statistical areas, districts, neighborhoods or catchment/service area boundaries. As advanced GIS tools and expertise is not needed to calculate and required data sources are relatively easier to obtain, they are widely observed in the accessibility modeling literature as simple accessibility measures.

The supply to demand ratios basically need two types of data source which are supply and demand sources. The supply sources are generally some service provider related indicators, such as number of schools, jobs, hospital beds or doctors and the demand sources are generally the population related indicators such as number of children, employees, economically active people etc., and mostly obtained from the census files.

Although supply to demand ratios are useful for making comparisons between several zones as indicators of availability, they have several limitations such as;

- they do not account for any measures of distance or travel impedance,
- results are blind to accessibility variations within bordered zones,
- results and interpretations obtained from deterministic bordered zones can vary greatly depending on the size of the zone, which is also well-known to geographers and spatial analysts as the modifiable areal unit problem (Guagliardo, 2004)

Kernel Density Measures

Kernel density measures are based on cells named pixels in raster environment. The value of each cell is represented by the help of a pre-defined kernel function which is generally a Gaussian kernel found in the GIS-based Spatial Analyst modules. The radius of the kernel reflects the catchment area boundary of the supply/demand locations.

There are mainly two types of data needed in the calculation of the kernel density measures, which are the location, and capacity of the demand and supply points. After the data is obtained, the density of the supply and demand is calculated separately according to the defined deterministic distance-based kernel size, which is a type of service/catchment area boundary. This calculation is performed in such a way that the cells near the kernel center receives higher values of supply or demand, and those near the kernel periphery receive lower values of supply or demand. A cell's value is inversely affected from its distance to the kernel's center. In the case of kernels overlap, either partially or fully, cells in these overlapping zones receive a higher score that is the sum of contributions from all overlying kernels. That is why the summed kernels can be quite peaked.

After the density calculation of supply and demand, the supply to demand ratio is computed in order to represent accessibility. The higher accessibility values represent higher supply and lower demand zones and lower accessibility values represent lower supply and higher demand zones in the final map (Guagliardo, 2004).

Although the kernel density measures give useful information about accessibility level to services, one of the biggest shortages of the measure is the usage of an unconstrained Euclidian distance-based kernel in determination of the catchment area boundary without considering the transportation network.

Gravity-Based Measures

Gravity-based measures represent accessibility of any location, by weighting the supply opportunities within a reasonable service/catchment area boundary according to their attraction (size, service capacity etc.) and evaluating each opportunity according to a measure of travel impedance (time or distance) (Kwan, 1998; Makrí, 2001; Guagliardo, 2004).

One of the widely used formulation for gravity model is that;

$$Ai = \sum_j \frac{S_j}{d_{ij}^{\beta}} \tag{2}$$

in which S_j represents the attraction factor of the supplies (service size or capacity), d_{ij}^{β} represents the impedance (time or distance) (Guagliardo, 2004). Although the simple gravity base measure formulations do not consider the size of the demand, the more sophisticated ones also take the demand V_j into consideration in such a way that;

$$Ai = \sum_j \frac{S_j}{d_{ij}^{\beta} V_j} \tag{3}$$

and

$$V_j = \sum_k \frac{P_k}{d_{kj}^{\beta}} \tag{4}$$

where P_k is population size at point k (the centroid of a census tract or block) d_{kj}^{β} is the distance between the demand point k and supply location j. The demand on provider location j is obtained by summing the gravity discounted demand influence of all population points within a reasonable catchment area boundary (Guagliardo, 2004).

Although the gravity-based measures can be considered as more realistic in terms of considering attraction of the opportunities (supply/provider), demand (population) and transportation characteristics, they have still limitations in terms of using deterministic catchment area boundaries.

Two-Step Floating Catchment Area (2SFCA) Measures

The two-step floating catchment area (2SFCA) measures, which are recently developed accessibility measures, are improved version of supply to demand ratio measures. There are two key differences between supply to demand ratio measures and the 2SFCA measures. Firstly, 2SFCA measures only consider time or distance-based catchment area boundaries rather than administrative zones such as districts or neighborhoods. Secondly, 2SFCA measures could consider accessibility differences in intersection zones by summing the population-to-provider ratios in the study area. 2SFCA measures consider the idea that the populations only use services within their catchment area. The size of the catchment area boundary is generally calculated from average or maximum traveling costs of time or distance, where all services within that boundary are considered accessible, and all other services out of the boundary are considered not accessible to the population.

The first step of calculating 2SFCA is to define a catchment area boundary for a defined time or distance threshold and determine the total demand that falls within the service/catchment area boundary for each of the service providers (supply). The division of the total potential populations (demand) within the defined catchment area boundary to the supply of the each service providers gives the service/catchment area-based population-to-provider ratios. The second step of calculating 2SFCA is to determine all available services (supply) for each of the populations (demand) that are within the catchment area boundary for a defined time or distance threshold. The final step of calculating 2SFCA is to sum all of the population-to-provider ratios, calculated in the second step, for each of the overlay areas.

2SFCA measures produce more realistic accessibility measures when compared with the supply to demand ratios (population-to-provider ratios), however they are still unrealistic in terms of using deterministic catchment area boundaries (McGrail & Humphreys, 2009).

Individual Accessibility Measures

Unlike place accessibility measures which handle accessibility as an attribute of locations, individual accessibility measures handles accessibility as a property of people, defining how easily an individual can reach locations considering the spatio-temporal constraints of the people and activities such as schedule, mobility, budget, time constraints etc. Individual accessibility measures are more sensitive to personal traveling abilities to reach activity locations considering space and time instead of assuming that all individuals in one place have the same level of accessibility (Hägerstrand, 1970; Lenntorp, 1976; Kwan, 1998; Pirie, 1979, Makri & Folkesson, 1999).

The most widely used individual accessibility measures in the literature are known as space-time measures. All types of space-time measures are developed based upon Hägerstrand's (1970) time-geographic framework which can be considered as an effective tool for understanding individual movement to reach service and activity locations in the environment. When compared with the research on place accessibility measures, the research on individual accessibility measures has slowly growed, mainly because of the lack of strong geocomputational platforms and georeferenced individual-level traveling data. Only from 1990s, GIS-based researches and technologies reprovide popularity in the field and several researchers have worked on individual accessibility measures (Neutens et al., 2007; Dong et al., 2006; Kwan, 1998; Miller, 2003; Kwan, 2004).

Space-Time Measures

The theory of individual space-time measures was first introduced by Hägerstrand (1970). Space-time measures model the accessibility of individuals by using the volume and projected area of the space-time prism as indicators of physical accessibility. Hägerstrand (1970) defines three types of constraints, which could shape of an individual's space-time prism:

- Capability constraints
- Coupling constraints
- Authority or steering constraints

Capability constraints limit the activities of individuals through their own biological necessities such as eating or sleeping and physical capabilities such as the resources they can command. For example, individuals eating and sleeping characteristics can be different from each other or individuals with private automobiles can generally travel faster through space than individuals who walk or rely on public transportation. Coupling constraints relate to where, when, and for how long individuals have to join other people, service and activity locations in space and time. Authority or steering constraints relate to the institutional context and refer to laws and regulations, which defines that specific locations are only accessible at specific times for specific people and for specific activities (Hägerstrand, 1970; Neutens et al., 2007; Kwan, 1998; Miller, 2003).

The basic conceptual item in the space- time framework is the space-time path, which traces the movement of an individual in space and time. In addition to tracing movement in geographic space from location to location, it also traces simultaneous movement from time to time. The path is vertical when the individual is stationary in space (but always moving in time) and the path is horizontal when the individual is moving in space and in time. The slope in the path indicates how fast individual is moving by using a potential transportation mode such as pedestrian, bicycle, car or public transport etc (Hagerstrand, 1970; Lenntorp, 1976; Miller, 2003)

A space-time prism (STP) is an extension of the space-time path that defines a potential accessibility space considering individual constraints. In this regard, a person must be at a given location (e.g. work) until time t1 and then must return to work again at time t2. If an average travel velocity is assigned for the individual's free time budget between t1 and t2, a potential path space (PPS) showing all locations in space and time that the person can occupy can be calculated. If the person wants to visit an activity location in his or her free time budget, its space-time path must intersect with the potential path space. Projecting the PPS to the two-dimensional geographic plane forms the potential path area (PPA). This area defines the set of geographic locations that the person can occupy.

Hägerstrand's time-geographic framework has inspired a great deal of researchers in their studies such as Lenntorp, 1976, 1999; Kwan, 1998; Miller, 2003; Kwan, 2004; Dong et al., 2006; Neutens et al., 2007. Although the framework provides as an effective infrastructure for understanding individual movement to reach service and activity locations, the major problem with space-time measures is that they depend on large amounts of individual information about completed trips and activities, which makes it difficult to use space-time measures in large-scale projects (Pirie, 1979; Kwan, 1998).

THE LINKAGE BETWEEN ACCESSIBILITY MEASURES AND TRANSPORTATION PLANNING

Litman (2011) is one of the best examples that explain the relation of accessibility measures with transportation planning.

According to Litman (2011);

- Although accessibility is the final goal of most of the transportation planning efforts, conventional transportation planning efforts generally pay less attention to integrate accessibility related measures and models into transportation planning process.
- Conventional Transportation planning generally considers "mobility/automobile" instead of "accessibility/people", evaluates transportation system performance with a car or public transportation based approach, takes traveling demand, road level of service, speed, volume, capacity and operating costs of vehicles and average type of users/traveller into account. During this process, different types of trips such as short, non-commute, non-motorized trips etc., different modes of transportation such as pedestrian, bicycle etc., different types of users/travellers such as enabled, elderly or children etc., are often neglected.
- Transportation planning process such as calculating travelling demands, allocating demands into supplies like vehicles and transportation infrastructures etc are generally very technical and uniform processes. Integration of accessibility measures into transportation planning could be a new/challenging research area for the researchers. More creative transportation planning approaches that could integrate accessibility measures into transportation modeling process could provide better understanding of transportation related problems, could provide more effective, accurate and realistic decision support to transportation related problems and could expand the scope of possible solutions to transportation related problems.
- Transportation planning efforts which does not take accessibility into consideration could result in misleading planning decisions. For example; "road design features that maximize motor vehicle traffic speeds may reduce walking and cycling accessibility etc.", or "locations convenient for automobile access, tends to be difficult to access by other modes of transportation etc.", or "central locations tend to be easier to access by walking, cycling and public transit tend to have lower traffic speeds, more congestion and more expensive parking".
- Accessibility measures could be used to evaluate the performance of the transportation systems by the decision makers. For example; by the help of the accessibility measures, the transportation planner could examine "how increased mobility effect overall accessibility?" or "how different services are accessible by different groups by different modes?" The results could help selecting from different plan alternatives or making improvements.

GIS-BASED ACCESSIBILITY MODELING TECHNIQUES

The literature on GIS-based accessibility modeling techniques can generally be divided into three, which are;

- Zone-based technique
- Isochronal (isochrone-based) technique
- Raster-based technique (see Makrí & Folkesson, 1999; Juliao, 1999; Chen, 2000).

The techniques are slightly different from each other and they have similar running steps, which are;

- data acquisition and integration phase,
- traveling cost calculation phase and
- visualization phase

Data acquisition and integration phase contain preparation of geographical information, which are mainly socio-economic, transportation and land use information. Traveling cost calculation phase contains calculation of cell crossing time in raster environment or calculation of the Euclidian/transportation network-based impedances in vector environment. Visualization phase is the last step in accessibility modeling and contains presentation of the calculated accessibility measures.

Zone-Based Technique

In zone-based technique, calculated accessibility measures are represented inside the defined bordered zones such as states, countries, metropolitan areas, districts, neighborhoods or any catchment/service areas. Determination of the size of the bordered zones is generally determined by the aim, the obtained data and the detail needs of the study. While a national or regional scale accessibility study generally requires a coarse zone representation such as state, country or district boundaries, a local scale accessibility study can require a smaller zone representation such as neighborhood or parcel boundaries. However, it must also be taken into consideration that the data is more difficult to obtain for the smaller zones such as parcels and neighborhoods when compared to coarse zones such as districts and countries (Halden et al., 2000).

In zone-based accessibility modeling technique, travelling cost calculation between supply and demand points are usually based on the zone centroids, which are geometric center of zones. In GIS environment, zonal centroids are generally used as representatives of the bordered zones and help to calculate traveling costs between supply and demand points. Although zone-based technique has an advantage of easier comparison of accessibility scores between the bordered zones, two main disadvantages of the technique are that the whole area inside the zones are represented with the same accessibility value and a constant catchment area boundary (average or maximum catchment area boundary) based on deterministic traveling costs such as Euclidian distance costs or constant transportation network-based costs are used to model accessibility.

Chen (2000)'s study is one of the examples of the zone-based technique for modeling of accessibility by GIS. By using the data of shopping opportunities (number of retail employment in each neighborhood and travel time data between neighborhoods, Chen (2000)'s study model shopping accessibility of Dallas/Fort Worth region by car and public transit during off-peak hours. The retail employment distribution boundaries are used as zone elements and different accessibility measures such as travel time and distance, cumulative opportunity and gravity are measured and compared.

The shopping accessibility as travel distance/time measure is calculated with the following equation;

$$A_i = \sum_j t_{ij} \tag{5}$$

where t_{ij} is the minimum total travel time between the centroids of zones i and j by auto and by transit in off peak hours. The calculated minimum travel cost values are used to model accessibility as a travel time distance measure in zone-based technique.

The shopping accessibility as cumulative opportunity measure is calculated with the following equation;

$$A_i = \sum_{t_{ij}<T} R_j \tag{6}$$

where i is the origin zone, j is the destination zone, R_j is the retail employments in destination zone, t_{ij} is the travel time between zones i and j, T is the time threshold which is set as 30 minutes for auto and 45 minutes for transit in the study. The cumulative retail employments, with traveling time less than the time threshold of 30 and 45 minutes, is used to represent shopping accessibility for each zone as cumulative opportunity measure

The shopping accessibility as gravity-based measure is calculated with the following equation;

$$Ai = \left[\frac{1}{j}\sum_{j=1}^{j}\frac{\log R_j}{\log H_{ij}}\right] \tag{7}$$

where R_j is the retail employment in zone j, J is the total number of zones in the area, H_{ij} is the transportation impedance element. The cumulative retail employments, divided by the transportation impedance without any defined threshold, is used to represent shopping accessibility for each zone as gravity-based measure.

Kwan (1998)'s study for Franklin County, Ohio is another example of GIS-based accessibility modeling in zone-based technique. In Kwan (1998)'s study, zone-based accessibility is modeled as a cumulative opportunity and gravity measure for 20, 30, and 40 minutes of time thresholds by using digital transportation network data of Franklin County and parcel boundaries data including various kinds of shopping and retail facilities such as restaurants, personal business establishments, banks, entertainment, outdoor activities, educational institutions and office buildings.

In calculation of travel time and distances between land parcels, the parcel boundaries are converted into point-based centroids and all point-to-point distances between land parcels are measured in terms of shortest travel time (minutes) by considering the Franklin County transportation network. Seven road classes in the digital street network are classified into three major categories in order to simplify the computational process in the study. Travel impedances are assigned to the transportation network as constant average traveling speeds such as;

- 55 miles per hour for controlled access freeways,
- 25 miles per hour for state highways and municipal arterials without access control and
- 15 miles per hour for other city streets

The travel time is further adjusted upward 25 percent to take the delays at traffic lights and turns into account.

In calculation of the weighted sum of shopping opportunities in a particular parcel for the gravity measure, each of the parcel area is multiplied by a building height factor. The building height factor is set to 1 except that; a value of 0.5 is assigned to multi-storey retail structures, a value of 2 is assigned to walk up commercial buildings with three or more stories and elevator commercial buildings with three or more stories, a value of 4 is assigned to non downtown locations and a value of 10 is assigned to downtown locations.

Isochronal (Isochrone-Based) Technique

In isochronal technique, accessibility measures are represented in terms of isochronal polygons, which are also known as the catchment or service area polygon boundaries. Isochronal polygon boundaries connect equal travel time or distance points away from one or more reference points (e.g. supply or demand).

Isochrone-based accessibility polygon boundaries are calculated from either constant average transportation network-based traveling costs such as 120 km/h for highways, 50 km/h for main streets and 30 km/h for local streets, etc. or unconstrained Euclidian distance based costs (straight-line/bird-flight based distances) such as buffer, voronoi (thiessen) polygons without considering the transportation network.

When an origin is defined as a reference point such as a demand or supply location, isochronal polygon boundaries can be drawn by connecting all points in all directions for an equal threshold of time or distance.

Buffer and voronoi polygons both define an unconstrained area of influence (proximity) around their origin points (supply/demand point). However, it is important to indicate that while a buffer polygon generally defines an equal distance zone around its origin point, a Voronoi/Thiessen polygon does not required to define an equal distance zone around its origin point. A Voronoi/Thiessen polygon defines a closer distance zone around its origin zone which means that any location inside a Voronoi/Thiessen polygon is closer to the origin in distance than any of the other origin points.

The buffer and voronoi-based isochrononal polygon boundaries have regular shape because of their unconstrained structure. However, transportation network-based isochronal polygon boundaries are constrained by the transportation network and can have irregular shape as the costs in a transportation network can provide traveling faster in some directions and traveling slower in other directions (Bureau of Transportation Statistics, 1997).

Isochronal technique can be used in calculation of several accessibility measures ranging from simple to sophisticated. For example,

- 10 minutes catchment area polygon boundary of supply/demand points can be calculated as a travel time/distance type of measure or
- Total number of cumulated supply/demand points within 10 minutes catchment area boundary can be calculated as a cumulative opportunity type of measure or
- Total number of weighted supply/demand points within 10 minutes catchment area boundary can be calculated as a gravity type of measure etc.

Dodge & White (1995)'s study in is an example of isochrone-based representation of accessibility for public services. The study calculates isochronal accessibility in terms of an unconstrained/bird flight distance measure and tries to understand how far people have to travel to reach a healthcare service in Wales. For this reason, 5 km buffer is created as a catchment area boundary around every supply points of healthcare services and critical zones are found where accessibility to the clinics could be a problematic.

Murad (2004)'s study also investigates catchment area boundaries of a medical center in Makkah City, Saudi Arabia in isochronal technique. By using a road network that shows all types of roads in the study area and a medical center location in point format, a transportation network-based service/catchment area is defined in ArcGIS Network analyst software environment and supply/demand opportunities inside the service/catchment area are analyzed. Then the outputs are used to define the priorities in health care plans, shortages in actual catchment area and the need for additional healthcare resources.

Although isochronal-based technique is widely used in accessibility modeling literature, one of the weaknesses of the isochronal technique is that accessibility measures are highly sensitive to traveling time/distance based costs and user defined thresholds. Slight changes in traveling costs and user-defined thresholds can create significant changes in catchment area polygon boundaries and hence directly affect the amount of supply and demand opportunities. Considering several costs and thresholds can provide more realistic decision support for decision makers who are supposed to deal with accessibility, location/allocation and service/catchment area related issues.

Raster-Based Technique

Pixel, which is also called cell, can be defined as the smallest unit in raster environment. In raster-based technique, accessibility measures are represented by raster-based pixels instead of vector-based polylines or polygons. The supply and demand locations and the transportation network are the main inputs of raster-based technique. By considering traveling costs in the transportation network, each pixel in raster environment generally gets an accessibility score, which is based on its proximity to nearest supply or demand opportunity.

There are three main phase in raster-based modeling of accessibility, which are;

- data acquisition and integration
- cost surface preparation
- accessibility modeling and visualization

Data acquisition and integration phase includes preparation of data in which supply/demand locations and transportation network data are obtained and converted into a common raster format. Cost surface preparation phase includes determination of traveling costs for each of the individual pixel on the transportation network and calculation of cell crossing costs. Finally, accessibility modeling and visualization phase includes measurement and representation of accessibility scores in raster-based environment by considering, supply/demand locations, traveling costs and cell crossing time.

Raster-based technique is generally preferred in regional studies, which does not necessitate high spatial accuracy. Because of pixel-based structure of the raster-based technique, working in raster environment reduces the geometrical accuracy of accessibility measures. However, it enables continuous representation of accessibility scores and opens a wide range of new raster analysis capabilities.

Juliao (1999)'s study is an example of raster-based representation of accessibility. The study calculates accessibility scores of municipality towns, highway nodes and city centre in Portugal and gives a detailed explanation of covered steps in terms of raster-based representation of accessibility.

In the data acquisition and integration phase, the basic data of supply and demand points and the transportation network are all converted into raster format. First, transportation network is classified according to the road types, which are Main highway (IP), Main lane (IP2), Complementary highway (IC), Complementary lane (IC2), National Road, Regional Road, Municipal Road etc. Then, all of the data are converted into raster format with a pixel dimension of 100 meters.

In the cost surface preparation phase, cell-crossing time for each of the pixel is calculated by using the following equation;

$$CCT = \frac{P * 60}{TS * 1000} \tag{8}$$

where CCT is the cell crossing time in minutes, P is the pixel size and TS is the average traveling speed in kilometers per hour (km/h) according to classified road segments in the transportation network. For example if a vehicle is travelling in road type Main Lane (IP2), cell-crossing time (CCT) is calculated as in below equation;

$$CCT = \frac{P * 60}{TS * 1000} = \frac{100 * 60}{80 * 1000} = \frac{6}{80} = 0.0750 \tag{9}$$

With the same logic, entire cell crossing times according to average speeds are calculated by considering different road categories in the transportation network.

To fill the gaps between road infrastructures and to have accessibility scores for the whole territory, an average walking speed of 6 Km/h is used as an average pedestrian speed outside of the transportation network.

In the accessibility modeling and visualization phase, supply/demand locations and cell crossing time are used to calculate several accessibility maps in Portugal for different time thresholds. These are accessibility map of the municipality towns, accessibility map of the highway nodes and accessibility map of the city centre Lisbon.

Ebener et al. (2005)'s study is another research example for raster-based technique that models healthcare accessibility for pedestrians. According to the study, most of the accessibility modeling research involve vector approach, which relies on high quality road network and supply/demand information. However, an advantage of raster-based technique is that they do not restrict traveling by physical road network and provide a continuous accessibility environment with a free travel across the terrain. With this in mind, healthcare accessibility for pedestrians is analyzed for 180 minutes of threshold time by using two different types of traveling costs, which are called isotropic (do not consider effect of slope and land cover in calculation of traveling costs) and anisotropic (consider the effect of slope and land cover in calculation of traveling costs). In isotropic approach, a fixed average travelling speed is used. In anisotropic approach several average travelling speeds considering different land use and slope type are assigned and the accessibility map results are compared.

THE ROLE OF GIS IN PHYSICAL ACCESSIBILITY MODELING

The Contribution of GIS Into Physical Accessibility Modeling

Since physical accessibility measures describe the spatial characteristics of a location and need large amount of computation and organization between huge and complex spatial data sets, accessibility modeling often and unavoidably lends itself to Geographical Information Systems technologies in terms of data collection, manipulation, programming, topology, analysis and presentation related issues.

As GIS have unique capabilities to handle spatial data and operations related to positions on the Earth's surface, with an integrated database of basic transportation, land-use and socio-economical data, GIS could provide a powerful interface and infrastructure for the decision makers who are supposed to deal with accessibility, location/allocation and service/catchment area related issues. As accessibility measures such as transportation, land use and/or socio-economical data, accessibility modeling needs a GIS environment.

The general support of GIS in accessibility modeling can be summarized as below;

- Data collection related supports (GIS have capabilities to capture, store, integrate and convert spatial and attribute data). For example,
  - GIS can support storing complex transportation network, supply/demand or origin/destination related datasets in a common raster or vector environment
- Data manipulation related supports (GIS have capabilities to select, query, calculate, update, classify spatial and attribute data). For example,
  - GIS can support classification of transportation network data according to road type or road capacity and classification of supply/demand or origin/destination locations data according to their weight, importance etc.)
  - GIS can support calculation of new information from classified datasets by using attribute table data (e.g. calculation of traveling time on main road segments by using road length and average speed data)
- Spatial analysis related supports (GIS have capabilities to operate vital raster and vector based spatial analysis functions such as proximity (buffer, voronoi, density etc.), spatial overlay (union, intersect, zonal statistics etc.) and network analysis (shortest path, service area, cost distance, etc.) For example;
  - GIS can support calculating proximity between origin/destination locations or performing network analyses, (e.g. creating 1 kilometers Eucledian distance buffer or creating 5 minutes transportation network based service area boundary

- Topology related supports (GIS can handle accessibility related topological relationships between nodes, arcs, polygons, centroids etc. and help to understand how transportation network or supply/ demand segments connect and relate to each other). For example;
  - GIS can support building traffic regulations of streets such as one-way streets, closed streets, overpasses, underpasses, delays in intersections
  - GIS can support detecting cumulative number of opportunities inside catchment area boundaries
- Programming related supports (GIS have capabilities to create, edit, and manage current models or create new models by providing a specific programming environment). For example,
  - GIS can support developing a new toolbox / user-interface that handles stochastic modeling of accessibility)
- Mapping and presentation related supports (GIS have capabilities to present accessibility maps with opportunity of different scales (1/1000, 1/5000 etc.), different data classification methods (natural breaks, quantile, equal interval, standard deviation etc.), different dimension (in 2D or 3D dimension), various color choices (red, yellow etc.) and various figure choices (square, circle etc.)). For example,
  - An accessibility map can be presented 5, 10, 15, 20 minutes of accessibility of fire brigades in 3D with different colors) (Black et al., 2004; Peters & Hall, 1999; MacFarlane, 2005; Ebener et al., 2005)

GIS-Based Accessibility Modeling Examples

As physical accessibility modeling unavoidably lends itself to a GIS platform, many GIS-based accessibility modeling research can be observed in the literature, especially in the last decades. This part of the research summarizes some basic GIS-based accessibility modeling research examples in the literature.

Boulos et al. (2001), Ghio et al. (2007), Chapelet & Lefebvre (2005) and Black et al. (2004) are some of the GIS-based examples that model accessibility as Euclidian-distance based travel distance measure and cumulative opportunity measure.

Boulos et al. (2001)'s study calculates healthcare accessibility for educational facilities in London, as Euclidian-distance based travel distance and cumulative opportunity measure by using isochronal technique. 100, 200 and 300 meters of constant deterministic Euclidian-based buffers are created around each educational facility as a travel distance measure and number of health services inside each of the buffer area is calculated as a cumulative opportunity measure in ArcView GIS software environment. Ghio et al. (2007)'s study is another example that measure healthcare service accessibility as Euclidian-distance based travel distance and cumulative opportunity measure by using isochronal technique. In order to find out the average distance to a healthcare facility from a populated area and the critical areas that should be targeted for a new facility (a mobile clinic), a GIS-based buffer analysis is performed and un-served or limitedly served populations in Yemen that currently fall outside of a suitable buffer distance to healthcare services are analyzed. Chapelet & Lefebvre (2005)'s study calculates accessibility to general practitioners as Euclidian-distance based travel time/distance measure. By using a raster-based cost distance function in GIS environment, the study develop an analytical methodology for showing the attractiveness of different land use types by considering their total area and their distance to a general practitioner in Gurgaon (e.g. 75% of commercial areas are at less than 600 meters from a general practitioner or 10% of industrial areas are at less than 1 km from a doctor etc). Black et al. (2004)'s study calculates healthcare accessibility in Honduras as Euclidian-distance based travel time/distance and cumulative opportunity measure. By using Thiessen/Voronoi polygons as a catchment area boundary, accessibility to healthcare facilities (CESAMO's) are calculated by considering several input data which are; population and populated places (towns and villages), land cover, road network, digital elevation model (DEM), municipality boundaries, healthcare facilities (CESAMO's) and number of physicians. In order to detect critical settlements in terms of healthcare accessibility, several travel time/distance and cumulative opportunity measures are calculated for each populated place (towns and villages). The calculated measures in the study are;

- Available physicians inside each Thiessen polygon catchment area
- Distance (kilometers) to the nearest CESAMO
- Traveling time (minutes) to the closest road segment by walking (In calculation of traveling costs by walking, average speed of 5 kilometer per hour, weighted by the slope, is used)
- Traveling time (hours) to the closest CESAMO by car (In calculation of traveling costs by car, the road types are considered. For primary roads, average speed of 100 kilometers per hour weighted by the slope, for secondary roads average speed of 70 kilometers per hour weighted by the slope, and for Rodera's roads average speed of 20 kilometers per hour, weighted by the slope are used.

Brabyn (2002), Brabyn & Skelly (2002), Nadine et al. (2006) and Messina et al. (2006) are some of the GIS-based accessibility modeling examples that model accessibility as a transportation network based travel time/distance and cumulative opportunity measure.

Brabyn (2002) models physical accessibility of General Practitioners (GPs) as a transportation network based travel time/distance measure in zone-based technique. By considering average deterministic transportation network costs, Brabyn (2002) calculates minimum travel time and distance from the centroids of the population blocks to the closest GP in New Zealand. In calculation of average constant transportation network costs, roads are classified according to types and an average traveling speed is assigned for each of the pre-determined road classes. Similarly, Brabyn & Skelly (2002) estimates physical accessibility of New Zealand public hospitals as a transportation network based travel time/distance measure in zone-based technique. By using average deterministic transportation network costs in the digital road network data, minimum traveling distance and time between census centroids and closest hospital are calculated as a travel time/distance measure and population more than 60 minutes from a hospital are calculated as a cumulative opportunity measure in a GIS environment. Nadine et al. (2006) is an example that measure healthcare service accessibility as transportation network based travel time/distance and cumulative opportunity measure by using isochronal technique. The hospital locations, populated census blocks and digital road network data of British Columbia are used as input data and alternate accessibility scenarios are modeled for different type of healthcare services (e.g. all hospitals, hospitals with critical care etc.) and for different travel time thresholds (e.g. 30 minutes, 1 hour) including the estimates of percentage of population that is served or not served. Messina et al. (2006) model accessibility to community hospitals as transportation network based travel time/distance and cumulative opportunity measure by using raster based technique. The aim of the study is to identify relatively remote locations from existing community hospitals and provide accessibility related decision support to policy makers in Michigan. For this aim, by using hospital locations data, under-served population locations, which are within maximum 30 minute travel time to suitable hospitals and total population inside the under-served locations are modeled for normal and rush hour traffic conditions. In calculation of time cost for normal traffic conditions, speed limits defined by classified road segments are used (range from 40.2 to 112.7 kilometers per hour). In calculation of time cost for rush hour traffic conditions, all speed limits are reduced by 25%.

Yang et al. (2006) is a GIS-based accessibility modeling example that model accessibility for dialysis centers, as a two-step floating catchment area (2SFCA) and kernel density measure in Cook county, Chicago, USA. In calculation of 2SFCA measure scores, isochrone and zone-based accessibility modeling techniques are used. For each census tract within a 30-minute service area, a supply to demand ratio is computed by dividing the number of stations in each dialysis center by the sum of all population that requires dialysis treatments. Demand is calculated for each census tract by multiplying the 2000-year census tract population by a constant factor, 0.12%, which is the percent of population that requires dialysis treatments observed for Chicago in 2000. Supply is calculated as the number of stations in each dialysis center. In calculation of Kernel density measure scores, raster and zone-based accessibility modeling techniques are used. First, by defining suitable kernel radii as a Euclidian-based service area boundary, kernel density surfaces are created for both demand and supply locations. Then, supply to demand ratio surface was created for each census tract by dividing the supply density surface by the demand density surface. Finally, the Yang et al. (2006)'s study compares accessibility scores of 2SFCA measures with the Kernel density based measures by using a zone-based technique.

CASE STUDY

A Hypothetical Example: Accessibility Modeling Process in GIS Environment

In order to able to provide a simple guide for the decision makers who are supposed to deal with transportation planning, accessibility modeling, location/allocation and service/catchment area related issues, the process of accessibility modeling in GIS environment is demonstrated by using a simple hypothetical dataset that is composed of a group of hospital locations as supply points, a group of population locations as demand points and a group of transportation network (see Figure 1).

When a "travel time or distance based accessibility measure" is needed to be calculated by the decision maker, a "buffer" function, a "voronoi" function or a "service area" function can be run in GIS environment.

The inputs for the "buffer" function are the origin points (supply/demand) and the Eucledian distance threshold (such as 500, 750, 1000 meters etc) decided by the decision maker. The output is a polygon data that represents equal distance buffer polygon(s). Buffer function does not consider the transportation network.

The inputs for the "voronoi" function are only the origin points (supply/demand). A threshold is not needed in voronoi function as the polygon boundaries are calculated automatically by GIS considering the Delaunay triangulation. The output is a polygon data that represents closer distance voronoi polygon(s). Voronoi function does not consider the transportation network.

The inputs for the "service area" function are the origin points (supply/demand) and the network based distance threshold (such as 500, 750, 1000 meters etc) decided by the decision maker and the transportation network (roads). The output is an equal distance service area polygon(s) that considers

Figure 1. Hypothetical dataset of supply points (hospitals), demand points (population locations) and the transportation network (roads) (the numbers represent capacity of the supply or size of the demand)

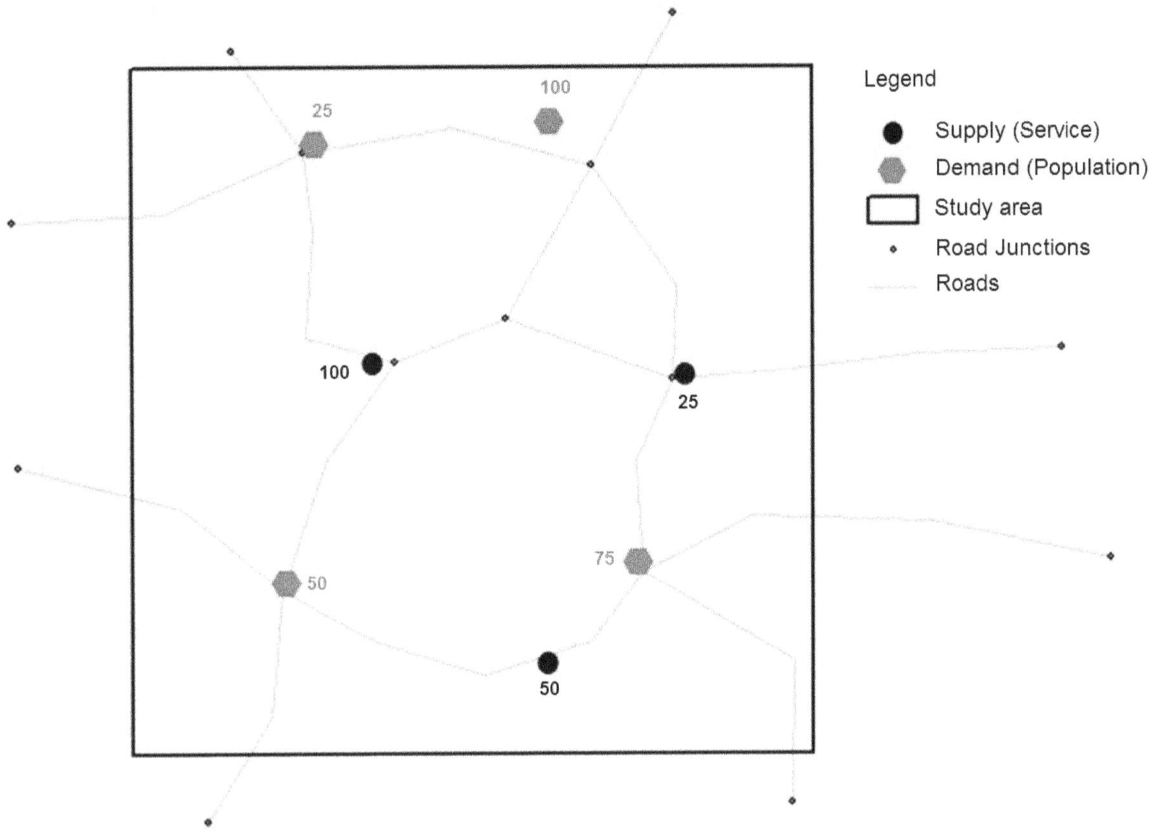

transportation network. If "time" based measure is needed instead of "distance" based measure, then an additional average speed data is needed (such as 90 km/h for the main streets 30 km/h for the inner streets etc.). Service area function is dissociated from buffer and voronoi functions as it considers the transportation network.

The "buffer", "voronoi" and "service area" functions are performed in GIS environment as a travel time or distance based measure. As the voronoi does not require a threshold determination because of its modeling structure, a 750 meters threshold is only considered for the "buffer" and "service area" operations (see Figure 2). The outputs are examples of isochronal accessibility modeling technique.

Both buffer, voronoi and service area functions are widely used by the decision makers to understand proximity of demand to supply (or vice versa) in the territory, however voronoi function can better help decision makers to understand which demand locations will be served by which supply locations.

As the calculation steps are similar, the rest of the accessibility modeling process is illustrated only by using the results of buffer operation. However the decision makers can also perform similar analysis by using the results of voronoi or network based service area operations.

When the cumulative opportunities inside the boundaries are calculated, then the travel time/distance based measures turns into a "cumulative-opportunity measure". The cumulative opportunity measures can be calculated in GIS environment by "spatial join" function of GIS softwares.

Figure 2. The example outputs of travel time/distance based accessibility measures

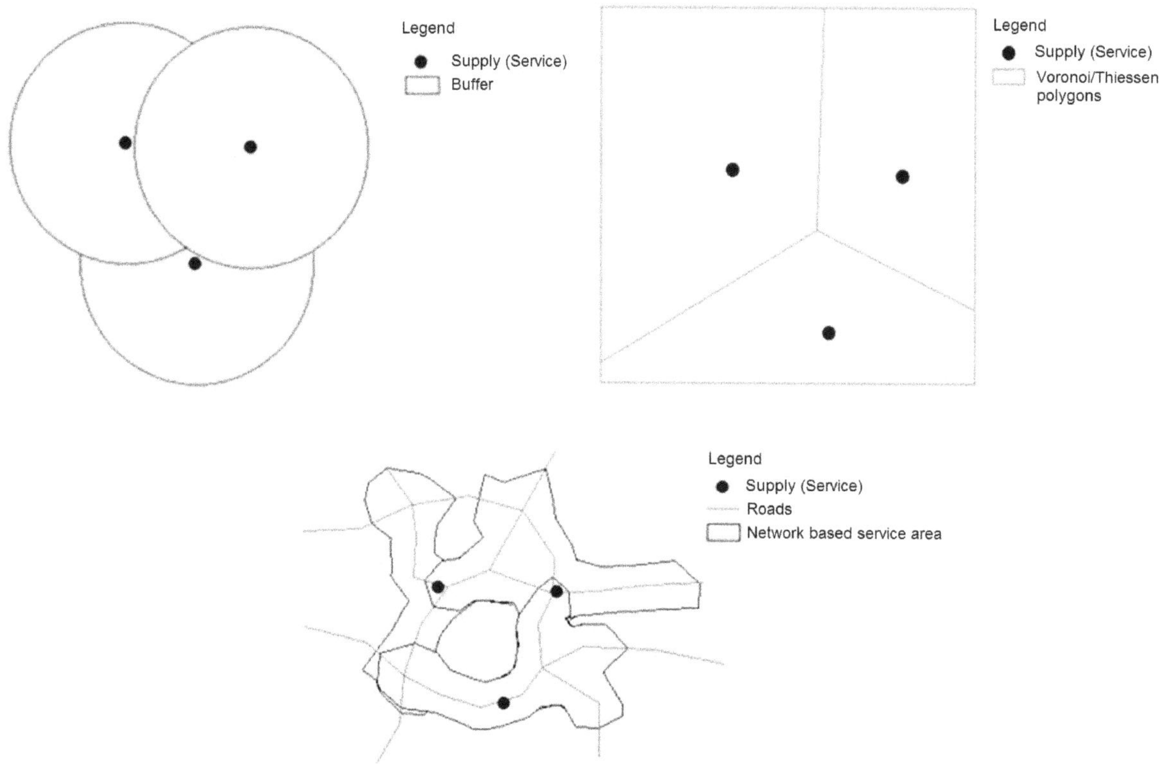

The inputs for the "spatial join" function are generally the polygon output of buffer, voronoi or service area operations and the point opportunities data. The output is a polygon data that have summed opportunities in the database such as total number of jobs, floor spaces, people or employees.

As an illustration, cumulative population (demand) inside 750 meters buffered health services (supply) is calculated by using "spatial join" function of GIS as a cumulative opportunity measure (see Figure 3). The outputs are examples of zone based accessibility modeling technique. Cumulative opportunity measures could enable decision makers to analyze the cumulative demand inside service area polygons.

When the decision maker analyze provider to population ratio inside service areas, then cumulative opportunity measures turns into a supply to demand ratio measures. The modeling process of supply to demand ratio measures is mostly the same with a cumulative opportunity measure. However, the only difference is that by using "field calculator" function of GIS, the ratio of cumulative supply (e.g.: capacity indicator of health services such as number of beds, doctors, etc) to cumulative demand (e.g.: total population) is calculated in GIS environment. The illustrated outputs of the supply to demand ratio measures are presented below (see Figure 4). In the illustration, the ratio of "cumulative population living within 750 meters to health services" to "cumulative number of beds in health services" are calculated by using "spatial join" and "field calculator" capabilities of GIS as an example to supply to demand ratio measure.

The kernel density measures are similar to supply to demand ratio measures. They both represent the ratio of cumulative supply to cumulative demand in the territory. However modeling of kernel density

Figure 3. The example output of a cumulative opportunity based accessibility measure

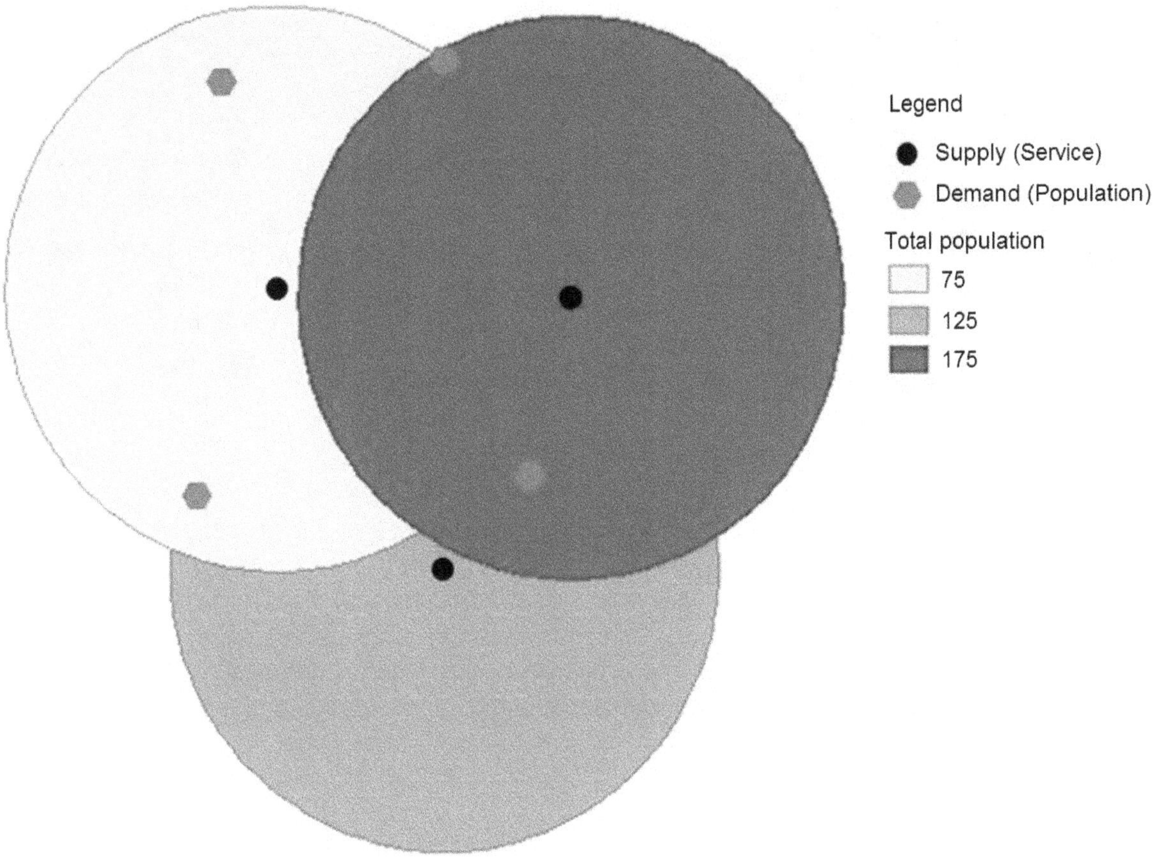

measures in GIS environment handled in raster GIS environment (see accessibility measures part of the chapter). As an illustration, kernel density of both supply and demand points are performed in raster GIS environment by using the kernel density function and raster calculation functions of GIS.

The inputs of kernel density function are the location and capacity of the demand and supply points and an user defined distance-based kernel size. The density of the supply and demand is calculated separately according to the kernel size, which is a type of service/catchment area boundary. In the illustration the kernel size is defined as 750 meters in radius. The outputs are density of supply and density of demand in the territory. After this step, the supply to demand ratio is computed by using "raster calculate" function of GIS. The output is a map that represents kernel density measure of accessibility. The higher accessibility scores in the supply to demand ratio map represent higher supply and lower demand zones and lower accessibility scores represent lower supply and higher demand zones in the territory (see Figure 5).

When the gravitational effects of services are considered by the decision maker in measuring accessibility, then the supply to demand measures turn into a gravity based measure. Sometimes services can be more attractive for the demand/users by the reason of gravity forces and the decision maker need to consider these effects in accessibility modeling. This time, supply/services are needed to be weighted according to their gravity/attraction factor considering the demand/users. The gravity weights are gener-

Figure 4. The example output of a supply to demand ratio based accessibility measure

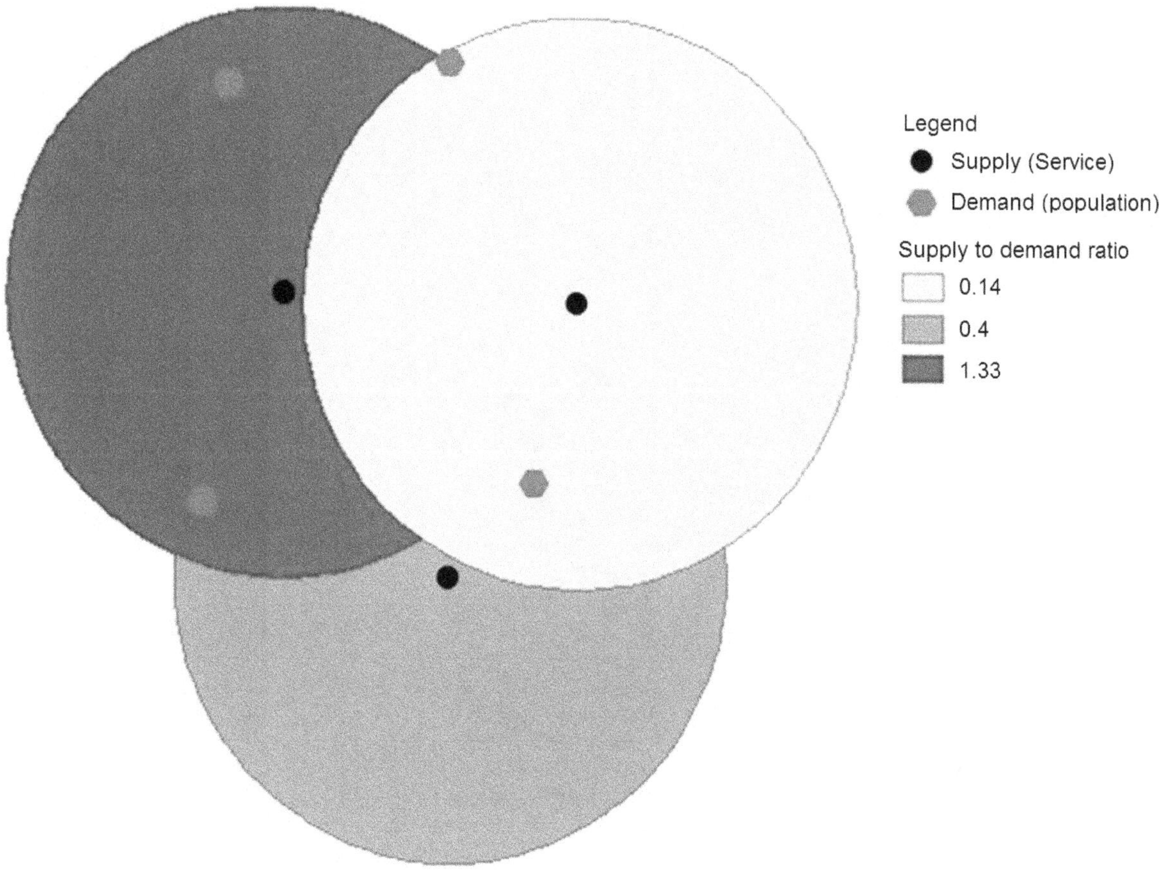

ally determined according to several criterias such as service quality, service popularity, service diversity, service size, etc.)". The gravity based measures are generally calculated the same as the supply to demand ratio measures. However the difference is that; the ratio of supply to demand is multiplied by the gravity factor of supply/services by using the "field calculator" capabilities of GIS. As an illustration, a hypothetical gravity factor for the supply locations (health services) is determined as in Figure 6.

The output of the gravity based accessibility measure is presented in Figure 7.

The result could demonstrate the impact of gravity factor on the results of accessibility scores. The lower accessibility zones in previous examples such as supply to demand ratios (see Figure 4) could have higher accessibility scores in gravity based accessibility measures (see Figure 7) or vice versa.

When the accessibility differences in intersection/overlay zones of service area buffers are considered, then supply to demand ratio measures turns into a 2SFCA measures. Actually this is an improved version of supply to demand ratio measures and hence can be considered as more realistic (see 2SFCA measures part of the chapter). By using "spatial join" capability of GIS, the supply to demand ratios in the intersection/overlay areas can easily be summed up in GIS environment. The 2SFCA outputs can be considered as more realistic when compared with the previous models.

The illustrated outputs of the 2SFCA measures are presented below (see Figure 8).

Figure 5. The example outputs of kernel density based accessibility measures

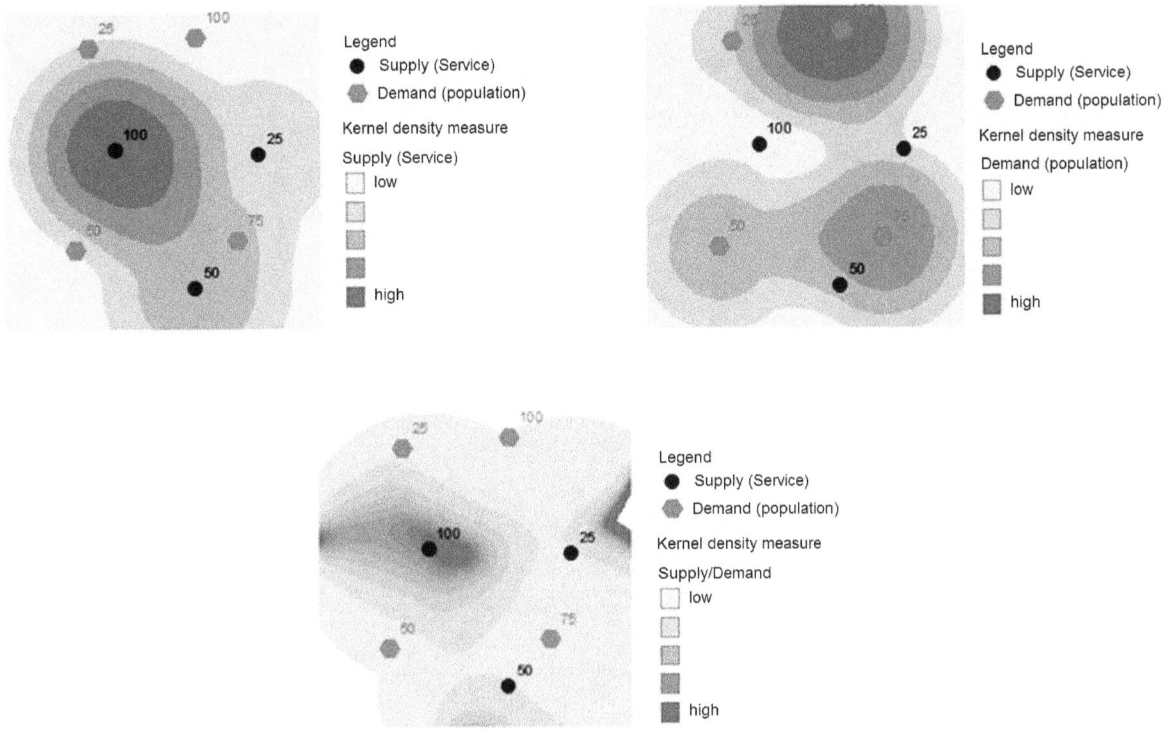

Figure 6. Hypothetical example of gravity factor for supply locations (health services)

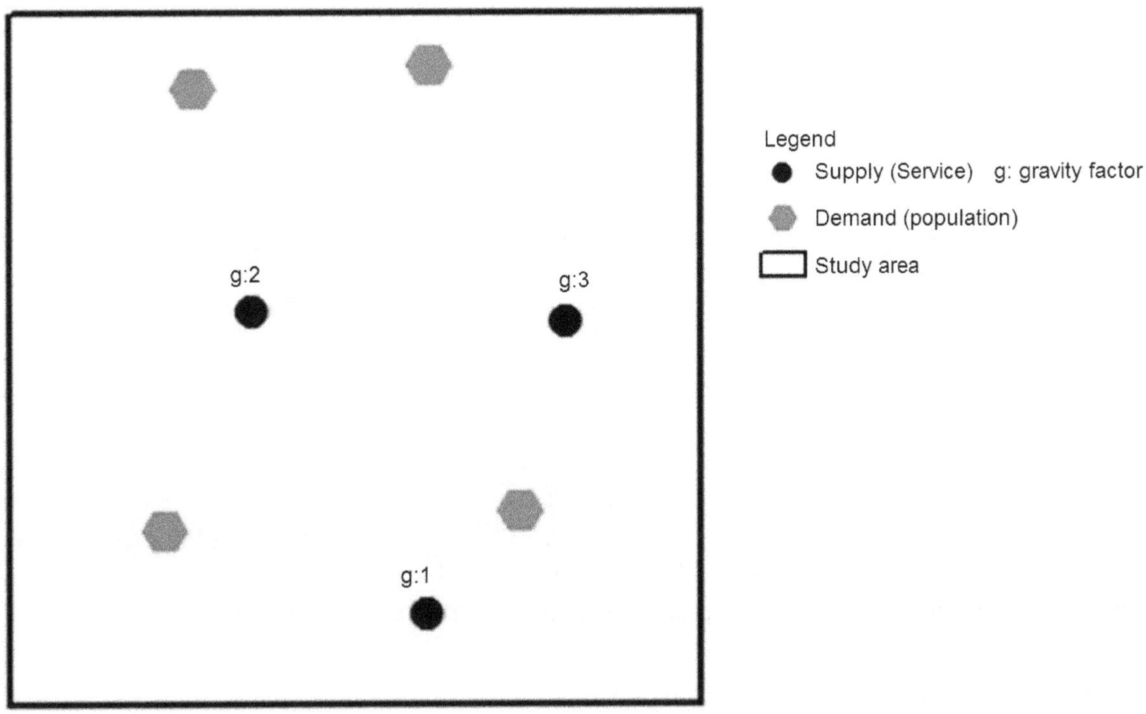

Figure 7. The output example of a gravity based accessibility measure

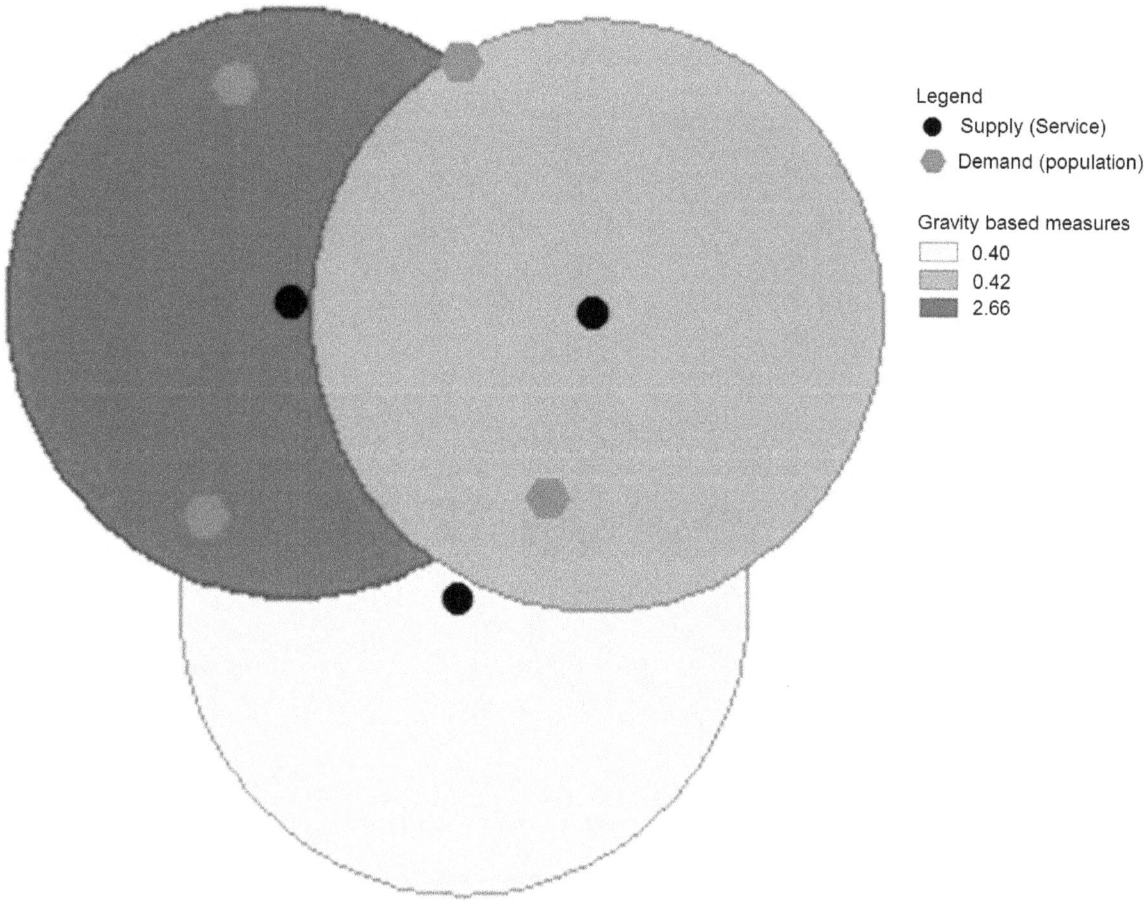

CONCLUSION

This chapter aim to provide an overview of the theoretical framework and relevant background about physical accessibility in order to clarify the nature of GIS-based physical accessibility modeling and to support decision makers working in local and metropolitan municipalities, ministries (e.g. health, education, agriculture, culture and tourism, energy and natural resources, environment and forest, transport and communication, industry and commerce, etc.), and public and private sector (e.g. real estate, industry, trade companies, shopping center administrations, etc.) and dealing with transportation planning, accessibility modeling, location/allocation and service/catchment area related research areas.

The provided framework could help decision makers to model accessibility, ranging from simple travel time measures to more sophisticated cumulative opportunity, gravity, two-step floating catchment area measures, etc which could be adapted to several accessibility related research such as central business district accessibility, job accessibility, recreational accessibility, trade center accessibility or educational accessibility etc. by considering several transportation modes such as pedestrian, bicycle, car or public transport etc.

Figure 8. The output example of a 2SFCA based accessibility measure

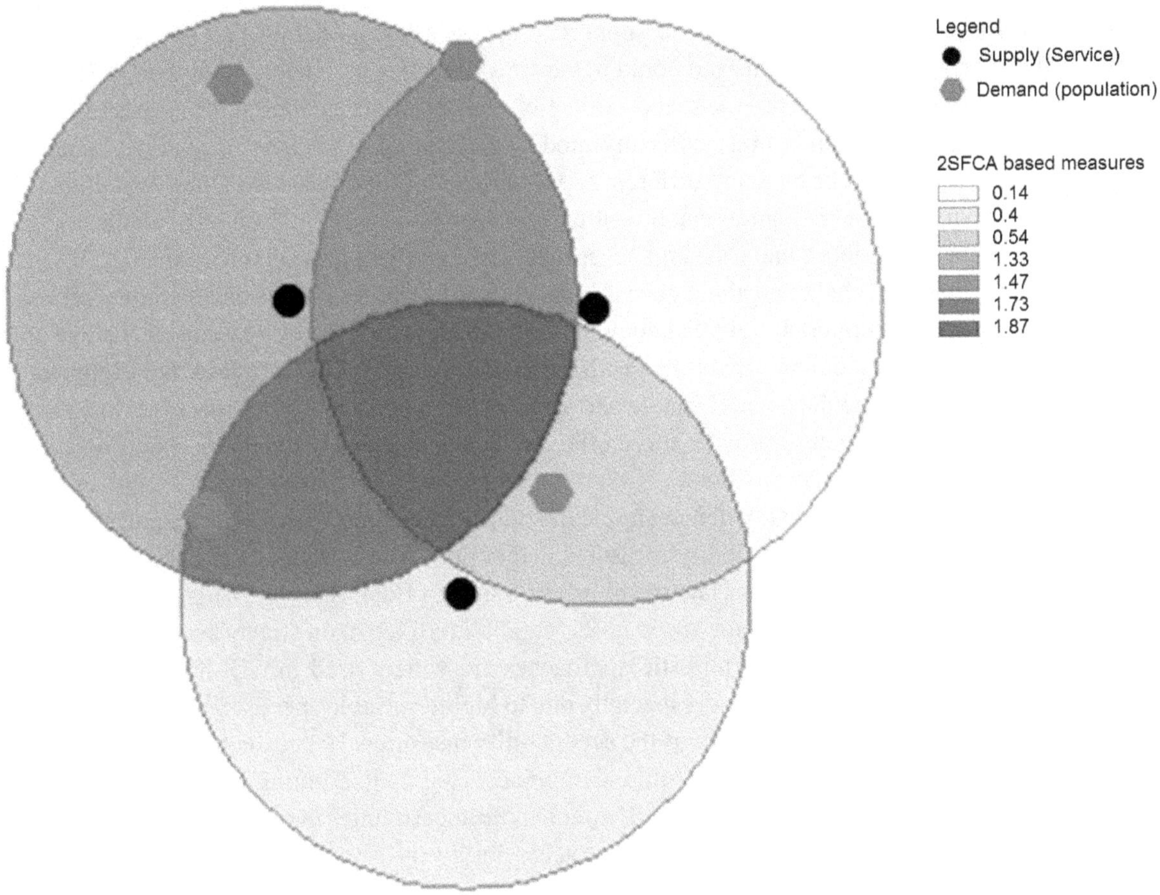

Although accessibility is the final goal of most of the transportation planning efforts, conventional transportation planning efforts generally pay less attention to integrate accessibility related measures and models into transportation planning process. However, accessibility measures could be used to evaluate the performance of the transportation systems by the decision makers. For example; by the help of the accessibility measures, the transportation planner could examine "how increased mobility effect overall accessibility?" or "how different services are accessible by different groups by different modes?" The results could help selecting from different plan alternatives or making improvements.

Because of the similarity in the accessibility modeling steps, "the hypothetical example" part of the chapter is only illustrated by using the "buffer" based service area boundaries. However it is also possible to use other kinds of service area boundaries (such as district, voronoi or network based boundaries) in accessibility modeling process by the decision maker according to aim, data availability and specific needs of the study.

FUTURE RESEARCH DIRECTIONS

Although GIS environment provides a vital support to accessibility modeling process, standard GIS software are general-purpose systems and could not always provide a user friendly support to the decision makers in accessibility measurement and evaluation. For example, modeling of accessibility in a GIS environment by using different accessibility modeling techniques (such as; isochrone, raster and zone-based techniques) and/or by using different types of accessibility measures (such as; cumulative opportunity, gravity, two-step floating catchment area measures etc.) are not directly applicable and requires huge effort, experience and time and necessitates many other types of software.

Consequently, the real challenge about accessibility modeling process is a broader framework such as a GIS-based decision support system that could incorporate decision maker's expert knowledge with specialized accessibility modeling capabilities with a user-friendly graphical interface. Development of such a user friendly system which could enable performing different accessibility measures in a simple user interface environment could provide more effective decision support for the decision makers in accessibility measurement and evaluation.

It is also important to stress that; whether simple or sophisticated, one of the fundamental shortcomings of the current GIS-based accessibility measures is that they are generally calculated from a fixed catchment area boundary based on constant traveling costs such as Euclidian (bird-flight) distance costs or transportation network-based average speed costs (e.g. 50 km/h for main streets and 30 km/h for local streets, etc.). Although such deterministic approaches are widely used in GIS-based accessibility modeling literature, they are not realistic, especially due to highly variable speeds in road segments and uncertainty in the accuracy and reliability of the accessibility measures. Hence another challenge about accessibility modeling process in a close future can be accepted as integrating stochastic approaches into accessibility modeling process which could better handle uncertainties in transportation costs, could create improvement on accuracy and reliability of accessibility measures in terms of catchment area boundaries (see Ertugay & Duzgun, 2011).

Transportation planning process such as calculating travelling demands, allocating demands into supplies like vehicles and transportation infrastructures etc are generally very technical and uniform processes. Integration of accessibility measures into transportation planning could be a new/challenging research area for the researchers. More creative transportation planning approaches that could integrate accessibility measures into transportation modeling process could provide better understanding of transportation related problems, could provide more effective, accurate and realistic decision support to transportation related problems and could expand the scope of possible solutions to transportation related problems.

REFERENCES

Bagheri, N., Benwell, G. L., & Holt, A. (2006). *Primary health care accessibility for rural Otago: ''A spatial analysis''. Health Care & Informatics*. Review Online.

Bixby, L. R. (2004). Spatial access to health care in Costa Rica and its equity: A GIS-based study. *Social Science & Medicine*, *58*(April), 1271–1284. doi:10.1016/S0277-9536(03)00322-8 PMID:14759675

Black, M., Ebener, S., Vidaurre, M., Aguilar, P. N., & El Morjani, Z. (2004). Using GIS to Measure Physical Accessibility to Health Care. *Proceedings of Health GIS Conference*. San Diego.

Boulos, M. N., Roudsari, A. V., & Carson, E. R. (2001). Health Geomatics: An Enabling Suite of Technologies in Health and Healthcare. *Journal of Biomedical Informatics, 34*(3), 195–219. doi:10.1006/jbin.2001.1015 PMID:11723701

Brabyn, L. (2002). Modelling population access to New Zealand's general practitioners. Proceedings of *International ESRI User Conference. San Diego.*

Brabyn, L., & Skelly, C. (2002). Modeling population access to New Zealand public hospitals. *International Journal of Health Geography, 1*, 1, 3.

Mobility, Access and Transportation. (1997). Bureau of Transportation Statistics. Retrieved from http://www.bts.gov/transtu/tsar/tsar97/chap06.pdf

Chapelet, P., & Lefebvre, B. (2005). Contextualizing the urban healthcare system. Methodology for developing a geodatabase of Delhi's healthcare system. *CSH Occasional Paper, N 11* (p.135). New Delhi: French Research Institutes in India, Rajdhani Art Press.

Chen, Q. (2000). *Measuring Accessibility in GIS.* Retrieved from http://mather.ar.utexas.edu/students/cadlab/chen/measuring_accessibility_in_gis.htm

Chen, Q., & Weng, W. (1999). *Urban Accessibility Evaluation in GIS.* Retrieved from http://mather.ar.utexas.edu/Courses/parmenter/gis/fall/index.html

Dodge, M., & White, S. (1995). GIS and public services in Wales: Analyzing supply and demand for healthcare and education. *Mapping Awareness Journal, 9*(7).

Dong, B., Bowman, J. L., & Walker, J. (1998). *Analysis of Activity-Based Accessibility.* Retrieved from http://web.mit.edu/jwalker/www/JW_ABA.pdf

Ebener, S., El Morjani, Z., Ray, N., & Black, M. (2005). Physical accessibility to health care: from isotropy to anisotropy. *GIS@ development, 9*(6).

Ertugay, K., & Duzgun, S. (2011). GIS-based stochastic modeling of physical accessibility using GPS-based floating car data and Monte Carlo simulation. *International Journal of Geographical Information Science, 25*(9), 1491–1506. doi:10.1080/13658816.2010.528419

Fortney, J., Rost, K., & Warren, J. (2000). Comparing Alternative Methods of Measuring Geographic Access to Health Services. *Journal Health Services and Outcomes Research Methodology, 1*(2), 173–184. doi:10.1023/A:1012545106828

Ghio, C., Mark, L., Abdulkadir, N., & Ahmed, A. (2007). *Health GIS Tools and applications informing decisions in Yemen.* Geospatial Application Papers, Retrieved from [REMOVED HYPERLINK FIELD] http://www.gisdevelopment.net/application/health/overview/me05_012a.htm

Gibin, M., Longley, P., & Atkinson, P. (2007). Kernel density estimation and percent volume contours in general practice catchment area analysis in urban areas. *Proceedings of the GIScience research UK conference (GISRUK).* Ireland. Maynooth.

Goulias, K. G. (2007). *An Optimal Resource Allocation Tool for Urban Development Using GIS-based Accessibility Measures and Stochastic Frontier Analysis. California PATH Program*. Institute of Transportation Studies, University of California at Berkeley.

Guagliardo, M. F. (2004). Spatial accessibility of primary care: Concepts, methods and Challenges. *International Journal of Health Geographics*, *3*(3). PMID:14987337

Hagerstrand, T. (1970). What about people in regional science? *Papers / Regional Science Association. Regional Science Association. Meeting*, *24*(1), 7–21. doi:10.1111/j.1435-5597.1970.tb01464.x

Halden, D., Mcguigan, D., Nisbet, A., & Mckinnon, A. (2000). *Guidance on accessibility measuring techniques and their application*. Scottish Executive Central Research Unit.

Joly, O. (1999). Recent Overview of Spatial Accessibility Indicators Applied. In *France: 1st Synthesis of the French Research Network Contributions. Anexe 5, Part II: Study Program on European Spatial Development (SPESD)*. France: SDEC.

Juliao, R. P. (1999). *Measuring Accessibility Using Gis*. Geo-Computation-99, Retrieved 03.09.2008, from http://www.Geovista.Psu.Edu/Sites/Geocomp99/Gc99/010/ Gc010.Ppt.

Kuntay, O. (1976a). Planlama kontrol araci olarak fiziksel erişilebilirlik ve çekim gücü (Planning tool as physical accessibility and traction control.). *Şehircilik Enstitüsü Dergisi* (Urban Development Institute Journal), 33-47.

Kuntay, O. (1976b). *Mekanda Fiziksel Erişilebilirlik ve Çekim Gücü*. (Physical accessibility and gravity in space.) (Unpublished dissertation). Black Sea Technical University, Trabzon, Turkey.

Kuntay, O. (1990). Erişilebilirlik Kesin Bir Öncelik. *Planlama Dergisi*, *90*(7), 1–2.

Kwan, M. P. (1998). Space-time and integral measures of individual accessibility: A comparative analysis using a point-based framework. *Geographical Analysis*, *30*(3), 191–216. doi:10.1111/j.1538-4632.1998.tb00396.x

Kwan, M. P. (2004). GIS Methods in Time-Geographic Research: Geocomputation and Geovisualization of Human Activity Patterns. *Geografiska Annaler. Series B, Human Geography*, *86*(4), 267–280. doi:10.1111/j.0435-3684.2004.00167.x

Lenntorp, B. (1976). *Paths in space-time environments: A time geographic study of movement possibilities of individuals*. Lund, Sweden: Gleerup.

Lenntorp, B. (1999). Time-geography-At the end of its beginning. *GeoJournal*, *48*(3), 155–158. doi:10.1023/A:1007067322523

Litman, T. (2011). *Measuring Transportation; Traffic, Mobility and Accessibility*. Victoria Transport Policy Institute.

Litman, T. (2013). The New Transportation Planning Paradigm. *ITE Journal*, *83*(6), 20–28.

Lotfi, S., & Koohsari, M. J. (2009). Measuring objective accessibility to neighborhood facilities in the city (A case study: Zone 6 in Tehran, Iran). *Cities (London, England)*, *26*(3), 133–140. doi:10.1016/j.cities.2009.02.006

Luo, W. (2004). Using a GIS-based floating catchment method to assess areas with shortage of physicians. *Health & Place*, *10*(1), 1–11. doi:10.1016/S1353-8292(02)00067-9 PMID:14637284

Luo, W., & Wang, F. (2003). Measures of spatial accessibility to health care in a GIS environment: Synthesis and a case study in the Chicago region. *Environment and Planning. B, Planning & Design*, *30*(6), 865–884. doi:10.1068/b29120

MacFarlane, R. (2005). A guide to GIS applications in integrated emergency management. *Emergency planning College.*

Makrí, M. B. (2001). *Accessibility Indices. A Tool for Comprehensive Land-Use Planning*. Retrieved from [REMOVED HYPERLINK FIELD]http://www.infra.kth.se/tlenet/meet5/papers/Makri.pdf

Makrí, M. B. (2002). Accessibility indices and planning theory. *Proceedings of Eighth International Conference on Urban Transport and the Environment for the 21st Century, Urban Transport VIII*. Seville, UK. Witt Press.

Makri, M. C., & Folkesson, C. (1999). *Accessibility Measures for Analyses of Land-Use and Travelling with Geographical Information Systems*. Retrieved from http://www.trafikdage.dk/td/papers/papers99/papers/paper/bpot/makri/ makri.pdf

Matthew, R., McGrail, J., & Humphreys, S. (2009). Measuring spatial accessibility to primary care in rural areas: Improving the effectiveness of the two-step floating catchment area method. *Applied Geography (Sevenoaks, England)*, *29*(4), 533–541. doi:10.1016/j.apgeog.2008.12.003

Messina, J., Shortridge, A. M., Groop, R. E., Varnakovida, P., & Finn, M. J. (2006). Evaluating michigan's community hospital access: Spatial methods for decision support. *International Journal of Health Geographics*, *5*(42), 1–18. PMID:16995948

Messina, J. P., Shortridge, A. M., Groop, R. E., Varnakovida, P., & Finn, M. J. (2006). Evaluating Michigan's community hospital access: Spatial methods for decision support. *International Journal of Health Geographics*, *5*(1), 42. doi:10.1186/1476-072X-5-42 PMID:16995948

Miller, H. J. (2003). What about People in Geographic Information Science. *Computers, Environment and Urban Systems*, *27*(5), 447–453. doi:10.1016/S0198-9715(03)00059-0

Mitchel, L., Gary, H., Jonathan, R., & Sean, W. (2008). Urban Population Distribution Models and Service Accessibility Estimation. *Computers, Environment and Urban Systems*, *32*(1), 66–80. doi:10.1016/j.compenvurbsys.2007.06.001

Dong, X., Ben-Akiva, M. E., Bowman, J. L., & Walker, J. L.(2006). Moving from trip-based to activity-based measures of accessibility. *Transportation Research Part A, Policy and Practice*, *40*(2), 163–180. doi:10.1016/j.tra.2005.05.002

Murad, A. A. (2004). Creating a GIS application for local health care planning in Saudi Arabia. *International Journal of Environmental Health Research*, *14*(3), 185–199. doi:10.1080/09603120420002186060 PMID:15203450

Neutens, T., Witlox, F., & Demaeyer, P. (2007). Individual accessibility and travel possibilities: A literature review on time geography. *European Journal of Transport and Infrastructure Research*, *7*(4), 335–352.

O'Sullivan, D., Morrison, A., & Shearer, J. (2000). Using desktop GIS for the investigation of accessibility by public transport: An isochrone approach. *International Journal of Geographical Information Science*, *14*(1), 85–104. doi:10.1080/136588100240976

Parentela, E. M., & Sathisan, S. K. (1995). *GIS-based allocation of emergency response units along a transportation route. In 1995*. San Diego: ESRI User Conference.

Peters, J., & Hall, G. B. (1999). Assessment of ambulance response performance using a geographic information system. *Social Science & Medicine*, *49*(11), 1551–1566. doi:10.1016/S0277-9536(99)00248-8 PMID:10515636

Pirie, G. H. (1979). Measuring accessibility: A review and proposal. *Environment & Planning A*, *11*(3), 299–312. doi:10.1068/a110299

Radke, J., & Mu, L. (2000). Spatial decompositions, modeling and mapping service regions to predict access to social programs. *Geographic Information Sciences*, *6*(2), 105–112.

Ritsema van Eck, J. R., & de Jong, T. (1999). Accessibility analysis and spatial competition effects in the context of GIS-supported service location planning. *Computers, Environment and Urban Systems*, *23*(2), 75–89. doi:10.1016/S0198-9715(99)00016-2

Schuurman, N., Fiedler, R.S., Grzybowski, S.C.W., & Grund, D., (2006). Defining rational hospital catchments for non-urban areas based on travel-time. *International Journal of Health Geographics*, *5*(43), 1–11. doi:10.1186/1476-072X-5-43 PMID:16390549

Scott, J., Larson, A., Jefferies, F., & Veenendaal, B. (2006). Small-area estimates of general practice workforce shortage in rural and remote Western Australia. *The Australian Journal of Rural Health*, *14*(5), 209–213. doi:10.1111/j.1440-1584.2006.00811.x PMID:17032297

Sylvie, D. (2007). Gis-based accessibility analysis for network optimal location model. *Cybergeo, Systems, Modelisation Geostatistiques,* Article 407.

Vahidnia, M. H., Alesheikh, A. A., & Alimohammadi, A. (2009). Hospital site selection using fuzzy AHP and its derivatives. *Journal of Environmental Management*, *90*(10), 3048–3056. doi:10.1016/j.jenvman.2009.04.010 PMID:19477577

Wu, Y. H., & Miller, H. J. (2002). Computational tools for measuring space-time accessibility within transportation networks with dynamic flow. *Journal of Transportation and Statistics*, *4*(1), 1–14.

Yang, D. H., Goerge, R., & Mullner, R.(2006). Comparing GIS-Based Methods of Measuring Spatial Accessibility to Health Services. *Journal of Medical Systems*, *30*(1), 23–32. doi:10.1007/s10916-006-7400-5 PMID:16548411

KEY TERMS AND DEFINITIONS

Geographical Information Systems (GIS): Geographical Information Systems (GIS) is a system that provides a powerful interface for storing, editing, analyzing and presenting spatially referenced data.

Physical Accessibility: Physical accessibility is a term that reflects the relative ease of access to/from several services by considering several costs of travelling.

ENDNOTES

Mobility is a critical component of accessibility. The term mobility refers to the potential for movement. A number of factors affect mobility including the availability and cost of transportation infrastructure. For example, if two people have the same residential location, but one person has a car and the other does not, each person's access to employment and shopping activities may be very different (Bureau of Transportation Statistics, 1997).

Section 3

Sustainability/ Environmental Risks of Transport

Chapter 10
Sustainable Transport System: Transport On-Demand

Maria Spichkova
RMIT University, Australia

Margaret Hamilton
RMIT University, Australia

ABSTRACT

Transport systems are major emitters of greenhouse gases, which makes environmental sustainability of any transport a crucial issue. Another issue is the lack of a systematic approach to the modeling and implementation of public transport systems. Finally, there are problems with the human interfaces to public transport systems, which do not encourage, and many do not allow, comfortable and simple interaction with the system. In this chapter, the authors discuss their solutions for these problems, explaining how to cover the existing gaps in a methodological and systematic way. The main contribution of this chapter is a model of an on-demand transport system that covers all the points mentioned above and focuses on spatial planning and optimizations including environmental issues in transport planning.

INTRODUCTION

One of the powerful drivers for sustainable transportation is public transport. The more people who share the one means of transport, the less the cost and damage to the environment. However, there are many specific problems and challenges, which make the use of public transport inaccessible and inconvenient for potential new passengers. Some of these issues, which we will consider in this chapter, are:

- *Time* - Long waiting times. A passenger may be required to wait indefinitely for the next bus/train/tram, and any time longer than 20 minutes is perceived as being too long;
- *Distance* - Long walking distances to the next nearest stop/station. Passengers may live a long way away from the nearest public transport route, or the connecting multimodal options may be far separated;

DOI: 10.4018/978-1-4666-8648-9.ch010

- *Timetables* - Non-punctuality. If the public transport vehicle is not following the timetable, e.g., is running late due to disruption, delays, congestion, passengers may miss their connections and be late for work.

It might appear that increasing the number of services and so making the timetable denser could solve these problems, but this solution would mean

- higher costs for the transport company, and
- further emission of greenhouse gases.

Another factor which could make this solution even more questionable would be if the bus were to be empty (idle drive) which could happen for some parts of any route during some times of the day. Hence quick and obvious options are not always the best solutions, and a systematic approach is needed to deal with these problems. A crucial question here also is, how to model this idea in such a way that it

- is applicable to any kind of transport system; and
- allows extensions for the special cases of transport systems having different flexibility in their operation possibilities; and
- covers modelling of exceptions (problems with the traffic due to some disruptions).

In this book chapter we will develop and explain our model for making public transport sustainable, *transport on-demand,* through the use of various definitions, examples and scenarios. Our methodology is aimed at providing smart solutions for sustainable outcomes, so that a transport system becomes a part of a 'smart city', cf. (Ercoskun, 2011; Nam & Pardo, 2011). We conform with the desire to develop sustainable software and systems, i.e., such software and systems that "meet the needs of the present without compromising the ability of future generations to meet their own needs" (World Commission on Environment and Development, 1987). The sustainability of software and systems development becomes increasingly important every year for many reasons, including expanding the software usage as well as growing demand for building further applications based on it. The three pillars of sustainability, namely the economy, society, and environment need are not balanced, and to deal this problem, a "green software" solution is required, cf. also (Hilty & Aebischer, 2014; Easterbrook, 2010). We also work within the classification of green software according to the Green Software Engineering project where the aim is to apply "green" principles to "software products, software development processes and their underlying software process models", (GreenSoft, 2009). However, we mainly focus not on the green development process, but on the development of software and systems that can have a positive impact on the environment and human beings.

The main idea behind our approach is to identify whether a bus, tram, or train would normally be empty during the day/night for some parts of particular routes. It could be empty for some cases if the timetable were denser or if the route were extended to outlying districts. We aim to have these parts of the route provided as optional choices in the timetable, during these hours. This would imply both lower costs for the transport company with less emission of greenhouse gases, as well as enabling the transport system to be more attractive for actual and potential passengers due its flexibility. To make this idea realistic, a proper model of the system is essential.

We present our vision of a decision support system for the drivers of public transport. *Transport on-demand* means dynamic timetables, which are influenced by passengers in real time. For this case drivers need additional assistance from the system to avoid manual analysis to decide which parts of the route should be served/skipped in each particular case. Thus, a *decision support system* (DSS) should be a key component of the on-demand transport system, to ensure human-oriented development of the driver interface to the system and as a result safety on the road.

The rest of the chapter is organised as follows: In the next section, background information on the related research is given. Then our model for transport on-demand and the corresponding optimisation methods is introduced. The formal model description and the presentation of the system architecture are followed by two on-demand bus scenarios, for a simple general example and for the case study, which is based on the spatial information of an existing real bus route 445 operating in the Melbourne area of Victoria, in Australia. The case study includes a prototype of the system, implemented as a web application. After this, the future research directions are discussed. Finally, the main contributions of the presented work are evaluated in the conclusion section.

BACKGROUND

Our aim is to design a sustainable on-demand transport system model, which is good for passengers and drivers. Unfortunately, we know that the most unreliable component of any computer system is the people who use it. This idea, applied to formal methods, is termed the *Engineering Error Paradigm* (Redmill & Rajan, 1997). Human factors targeted by the engineering error paradigm typically include the design of the user interfaces as well as any corresponding automatization. In this paradigm people are seen as being equivalent to software and hardware components in the sense of operations with data and other components, but at the same time humans are seen as being "the most unreliable component" of the whole system because they are inconsistent and unreliable. This implies that designing humans out of the main system actions through automatization of some system design steps is considered a good proposal for reducing risk. A *decision support system (DSS)* can be viewed as a special way of partially designing humans out of the main system actions, particularly if we assume that the human will follow the decision recommended by the support system, cf. e.g., (Marakas, 2003; Bonczek et al., 1981; Fick & Sprague, 1980). DSSs have been designed for industrial and organisational users, but as web-based technologies have progressed, the applications area has become more broad (Bhargava et al., 2007). Large parts of DSSs are now developed for casual users, e.g., for passengers interested in taking the fastest connection from some address to a desired destination point. Moreover, from the development of traditional "desktop applications based" DSSs have evolved to the development of distributed, web-based applications (Power, 2013; Mitra & Valente, 2007). In this chapter we propose a web-based DSS for the public transport driver, where the request information for the route is collected from the passengers dynamically in realtime.

We recognise that another problematic area for transport on-demand systems is the user interface for these systems, both for passengers and for drivers. The research field of Human Computer Interaction (HCI) is concerned with the development of interfaces for every-day applications, and there are many approaches to integrating human interface engineering with software engineering, e.g., (Volpert, 1991; Heumann, 2002; Constantine & Lockwood, 2002; Constantine & Lockwood, 1999). Many early approaches in this area were initiated because of mistakes in usage and the development of safety-critical systems. One of the widely cited examples of the HCI-related accidents in safety-critical systems is the case of massive radiation overdoses of the Therac-25 machine used in treating cancer over 1985-1987. The machine, produced by Atomic Energy of Canada Limited (AECL), delivered overdoses leading to deaths and serious injuries of patients, some of whom were given over one hundred times the normal dose of radiation (Miller, 1987; Leveson & Turner, 1993). The causes of these accidents have been attributed to software failures as well as to problems with the system interface and the lack of decision support for the operators of this machine. Operators of the Therac-25 were not able to recognise dangerous error situations and continued the treatment even after the machine showed warning messages. These types of accidents have shown that studying the human errors and their causation should be a significant part of software and system engineering at least in the case of safety-critical systems.

An appropriate system interface allowing correct human computer interaction as well as decision support is equally as important as the correct, error-free behaviour of the developed system. Even if the system we develop behaves in the correct way, this would not help much for a case where the system interface is unclear to the user or is too complicated to be used correctly. According to statistics presented in (Dhillon, 2004), the human is responsible for 30% to 60% the total errors which directly or indirectly lead to the accidents, and in the case of aviation and traffic accidents, 80% to 90% of the errors can be attributed to human error. Thus, it is essential to take human factors into account when developing computer systems, especially decision support and safety-critical systems.

The fundamental goal of human factor engineering, as claimed in (Wickens & Hollands, 2000), is to reduce errors, and increase productivity and safety when the human interacts with a system. Engineering psychology applies psychological perspectives to the problems of system design and focuses on the information-processing capacities of humans. The goals of modelling (including formal methods) are almost the same: to reduce errors, and increase the productivity and safety of the developed systems.

Another important point of the engineering error paradigm is that human errors often occur as a result of mismatches in the HCI and overestimation of the physical capabilities of a person or of some devices. A decision support system can provide the required estimation automatically, consistently and accurately using predefined schemas.

There are several applications of formal methods to HCI analysis, e.g., (Shackel & Richardson, 1991; Følstad et al., 2012), as well as many approaches to system modelling, e.g., (Broy et al., 2008; Berger, 2002; Donath et al., 2008). In our previous work (references have been removed for double-blind review), we focused on human factors in modelling processes as well on readability and usability of formal languages and interfaces, with the aim of combining the achievements in the fields of HCI and modelling approaches.

In our earlier research on the Melbourne public transport system, we identified that 90% of commuters consider the time spent travelling as their "in-between" time and use it to catch up with their family and work colleagues on mobile phone or email (references have been removed for double-blind review). This demonstrates a willingness of the majority of commuters to connect, and indicates there could be a willingness to engage with the public transport system to inform their journey, if such an opportunity were available. Also, earlier work on an automatic passenger counting system installed on Yarra Trams (tram network in Melbourne, Australia), highlighted that if commuters knew the numbers of other people travelling on their route in advance, many would consider alternative options (Fihn & Finndahl, 2011).

The approach we present in this chapter is based on the experiences from several projects of software and systems engineering in mobility, modelling, formalization and testing, especially within existing transport systems and automotive fields in Europe and Australia, cf. e.g., (Hamilton & Berry, 2010; Hamilton et al., 2011; Spichkova et al., 2012; Spichkova, 2011; Feilikas et al., 2011). In the rest of the chapter, we will emphasize the development of a formal model for incorporating the various personalized transport solutions for on-demand public transportation planning systems.

DEVELOPING A MODEL FOR TRANSPORT ON-DEMAND

In this section we introduce the idea of load modes for the transport system, based on passenger numbers at particular places for various times. We introduce our model and develop some corresponding optimizations, which will save not only on fuel costs, but also reduce greenhouse gas emission and noise pollution.

In this chapter, we define a vehicle to be a public transport vehicle, such as a bus, a tram, or a train. The general ideas of our approach can be applicable to any kind of public transport vehicle, including ferries and cars or taxis, but it is easiest to see that bus transport systems can benefit the most from our innovation, due the flexibility of this kind of vehicle. Bus delays are less critical than for the rest of the transport system because buses can easily deviate from their route, if necessary. This gives us an opportunity for further extensions of the system. For example, the bus stops on the optional parts of the route can be defined as "floating" stops, so that the bus driver can pick up passengers between the regular stops (assuming that the vehicle stopping is allowed on these streets) thus increasing the overall quality of the service. Another example could be deviations from the route, for hours out of the peak times, to pick up disabled passengers with restricted physical abilities.

We model a public transport route R from place A to place B by a sequence of stops/stations $[S_0, \ldots, S_n]$. Thus, the set of route stops is defined by $Stops = \{S_0, \ldots, S_n\}$, where each stop/station S_i has special attributes:

1. $departT_i$ – time, when the vehicle should depart on the stop/station,
2. $location_i$ – spatial information about the stop/station ($location_0$ corresponds to A, and $location_n$ corresponds to B),
3. $option_i$ – Boolean marker whether the vehicle should operate on this stop/station of the route ($option_i = true$ means that the vehicle should stop at S_i).

While modeling a simple route, which operates static timetables and does not allow any additional requests from passengers or dynamic optimization of the route, it is enough to use the first two attributes, *arrivalT* and *location*. For modelling an on-demand transport system, the option attribute is essential.

In general, we can say that a public transport system has two modes of operation on any route:

- *irregularLoad* mode, when the vehicle could be empty. This means there could be times when no passengers are around to travel over parts of the route, and
- *normalLoad* mode, when there is at least one passenger in the vehicle.

Our goal is to avoid the situations where the bus has an idle trip for all or parts of its route. In the ideal case, the system always runs in the *normalLoad* mode, but if not, we can recommend a system optimization in the following manner:

Case1. End of route. When the *irregular connections* are at either end of the route, the optimization is implemented in the real transport system in a straightforward way discussed below, in the next section.
Case 2. Along route. Optimization for the *irregular connections* in the middle of the route (i.e., anywhere between "regular" connections) is practical for the cases when the bus can take a shortcut along the route, taking a shorter trip to the beginning of the "regular" connections.

We call the timetable for the optimized parts of the route *optional*, because the vehicle follows this timetable and approaches the corresponding stops/stations only on demand, when requested by a passenger.

For each particular run of the route *R*, we also define $RegularStops^R$ to be the sequence of stops/stations, on which the vehicle should stop during this run, i.e. for all i, $0 \leq i \leq n$,

$$S_i \in RegularStops^R \Leftrightarrow option_i = true$$

This sequence is statically specified for the *normalLoad* mode, but in the case of the *irregularLoad* mode it should be specified dynamically, according to the requests of passengers.

Route Optimisation

End of Route Optimisation

For Case 1, when the *irregular connections* are at either end of the route, these parts of the route become *optional* (the *option* attribute for the corresponding stops is set to *false*), and we introduce so-called *provisional ends* of the route (stops/stations, where the regular connections end) if

1a. at the provisional end there are no passengers left in the vehicle, and
1b. the driver has no message that the vehicle is requested between the provisional and the actual end of the line (in either direction).

In this case the vehicle need not go to the actual end of the line and can stay at the provisional end until according to the timetable it would need to proceed in the opposite direction. Thus, the driver's decision, about whether to stop at the provisional end or continue to the actual end, leads to an automatic system operating on passengers' requests and also on historical statistical data collected about each route.

Along Route Optimisation

For Case 2, when the *irregular connections* occur during the route and not at either end. If there is a possible shortcut, we introduce so-called *provisional shortcut stops* of the route if

2a. the vehicle can shortcut to miss at least a few of the irregular connections:
   i. the stops/stations between which the shortcut can be taken and become the *provisional shortcut ends*, and
   ii. the part of the route to shortcut becomes the *optional part of the route* (the *option* attribute for the corresponding stops is set to *false*), and

2b. at the provisional shortcut end there are no passengers left in the vehicle, and
2c. the driver has no message that the vehicle is requested at any stops in the optional part of the route.

In both cases, the request for the vehicle should be sent before some predefined time (which we call the *on-demand request time*, denoted by *requestT*) before the vehicle arrives at the stop/station after which it will proceed according to the optional timetable. For usability reasons, we recommend setting this time to the maximum of all required request times within the route, rounding this maximum upward to 5 minutes steps. For example, if the calculated maximum over all required request times were 12 minutes, we would set the *on-demand request time* to 15 minutes to make it easier for passengers to remember. We demonstrate this in the bus scenario below.

The *on-demand request time* could also be set not for a single route but for the whole transport system, which would make it easier to remember. However, if the differences in between on-demand request times of several routes exceed 5-10 minutes, it may be more convenient for passengers to have different request times for different bus routes, where for some routes these times are much shorter.

Each optional stop/station S_i obtains an additional parameter, $statistics_i$ to collect information about how often and when the system has obtained requests to serve this stop/station.

Architecture of the System

To have a sustainable and human-oriented solution, we need three views on the transport system (cf. Figure 1 for the architecture of the system):

Figure 1. Architecture of the on-demand transport system

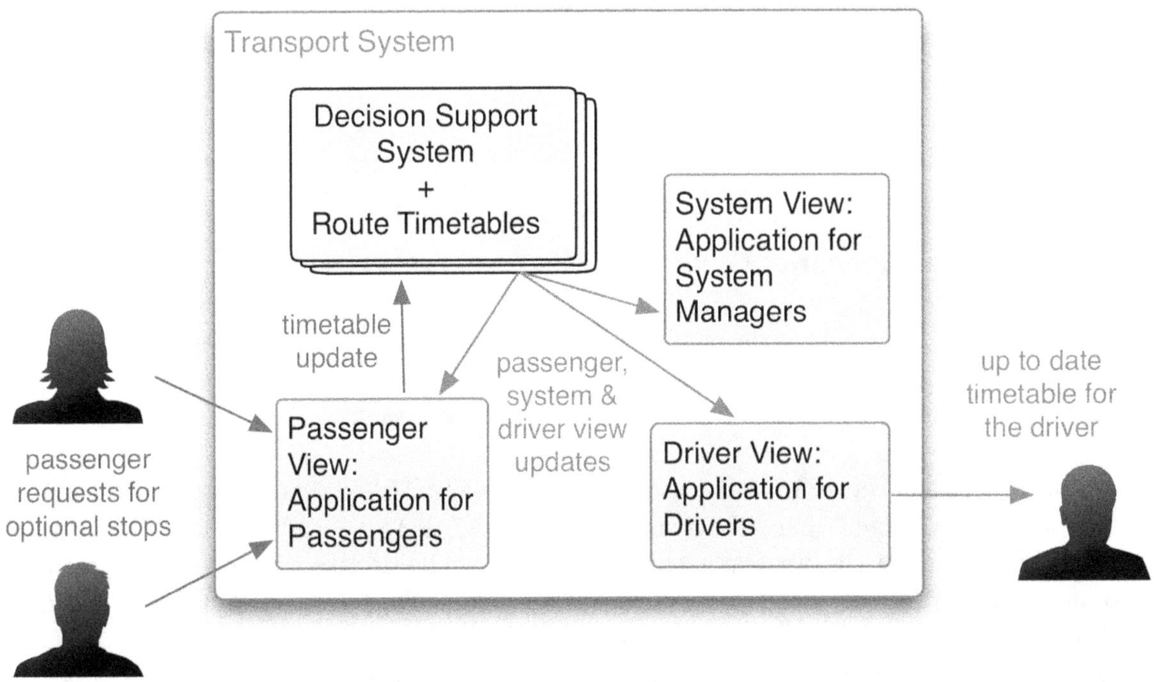

- a *system view*, which comprises the complete information about the route and its runs, and also includes the statistics of the requests;
- a *passenger view*, which is focused on the possibilities of requesting the vehicle to travel to a certain stop/station. It would make no sense for a passenger to send a request for a vehicle to service a particular stop at the time *t*, if there were another earlier request for a vehicle to serve this same stop at this same time. Thus, before sending their own request, passengers need to know the current state of the route – which stops/stations are still optional and which are already "regular" due some previous request(s);
- a *driver view*, which contains only the information that is currently required to complete the run in the most efficient and safest way – the sequence $RegularStops^R$, where the beginning of the sequence is truncated before the current stop. When the driver leaves a stop, it disappears from his plan (but not from the system view) on this route. The driver view can be represented
    - in a *general mode*, containing the complete current route plan, or
    - in a *simplified mode*, containing the information about 4 stops/stations only: the current stop, which the vehicle is/should be approaching right now, and three next consecutive stops.

In the next sections we present a discussion and comparison of these modes. The main advantage of the simplified mode is better readability of the plan: if there are no disruptions or problems with traffic along the route, it is not important for drivers to know what the distant stops will be or the route actualisation for them, at this point in time. Moreover, dynamically updating information about these stops can distract from safe driving, where a compact view in a simplified mode provides only the information which is currently required. The driver and the system views comprise information on

- the $departT_i$ values (i.e. times when the vehicle should depart the stop/station) for each stop/station along the route plan for this run, as well as
- the actual and planned locations (in relation to the stops/stations) of the vehicle at the current moment in time.

In this chapter, we focus on the driver and the system views, as they are important for the decision support system.

On-Demand Bus Scenario: General View

Assume an abstract simple bus route consists of 15 stops, cf. Figure 2(a). For this case, we specify a set of $Stops = \{S_0, \ldots, S_{14}\}$, and represent the route by the sequence $[S_0, \ldots, S_{14}]$. The timetable density of this route is assumed to be reasonable for all hours of the bus operation, and well synchronized with other routes.

For the connections between S_1 and S_5 as well as between S_9 and S_{12}, the bus is never empty (denoted by the solid lines in Figure 2), but at the connections between S_0 and S_1,, S_5 and S_9, as well as between S_{12} and S_{14} for some hours *H*, e.g., very early/late hours or middle of the day, the bus is often empty (denoted by dashed lines). We can say that the bus system has two modes of operation on this route, *normalLoad* mode (out of the hours *H*) and *irregularLoad* mode (during the hours *H*).

In addition, we assume that for the bus travelling along this route, a direct shortcut between stops S_5 and S_9 is impossible from either direction, and:

Figure 2. Example of a simple transport route. (a) Identification of the optional parts; (b) Optimization of the route; (c) Route plan at the beginning of the run. We assume there were no requests to stop at S_0, S_7, S_8, S_{13}, and S_{14}. We also assume that the vehicle waited too long at stop S_1 and according to the system plan should be at S_3. Hence the actual and the planned locations of the vehicle are different on the plan. (d) Route plan for the Case 1.1; (e) Route plan for the Case 1.2.

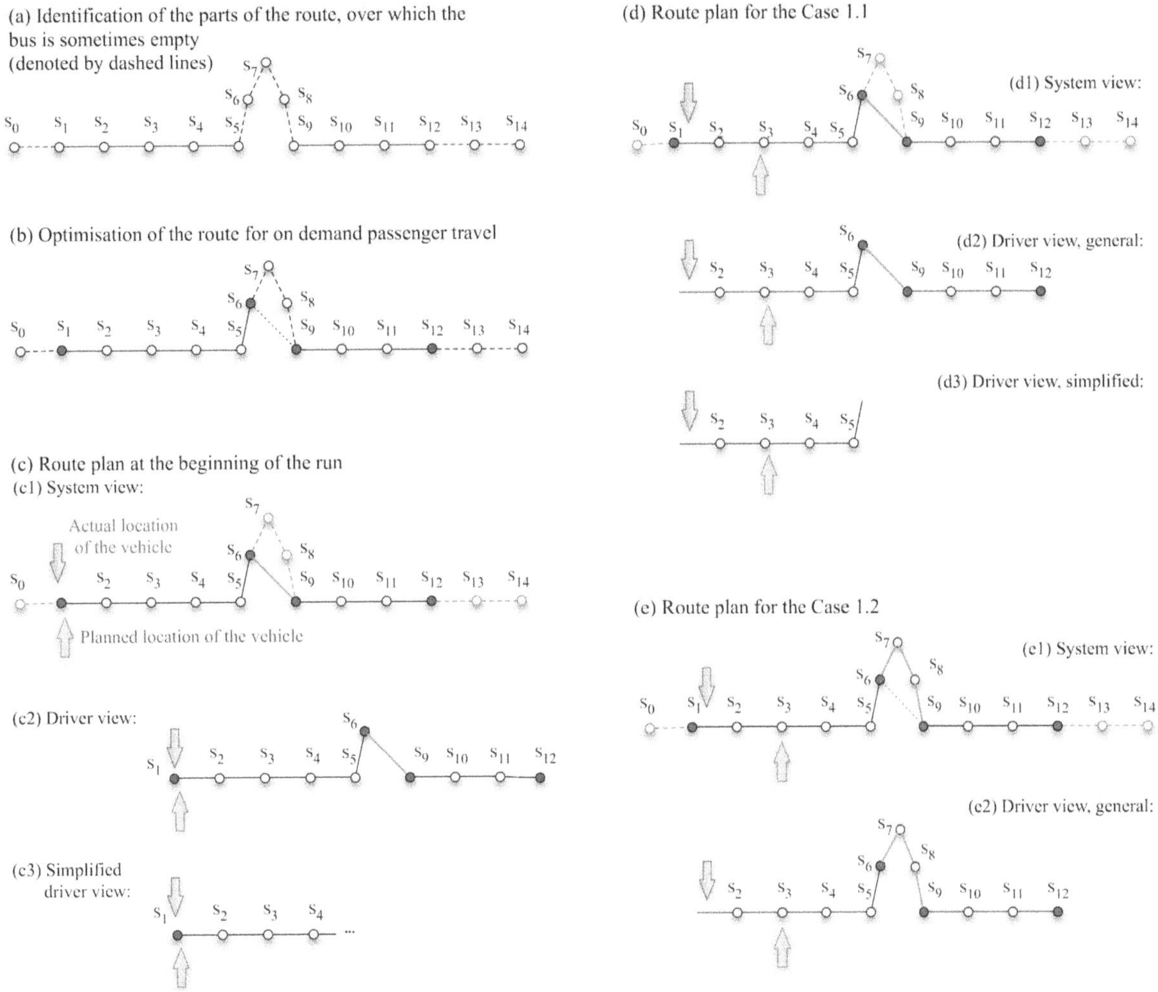

- for the direction $S_0 \rightarrow S_{14}$, after the stop $S_{5,}$ the only efficient way for the bus to go is to travel directly to the stop S_{6},. However the bus can shortcut the path S_6 - S_8 - S_8 - S_9 by travelling from S_6 directly to S_9;
- for the reverse direction $S_{14} \rightarrow S_0$, after the stop S_9, the bus can shortcut the route to the stop S_6, but cannot go directly from S_9 to S_5. This could happen, e.g., in the case of a one-way street.

This would mean that the *provisional shortcut stops* become not S_5 and S_9, but S_6 and S_9, see Figure 2(b).

Therefore, we optimize the route by dividing it into 5 parts: two regular parts between S_1 and S_6 as well as between S_9 and S_{12} (the corresponding connections are denoted by solid lines), and three optional parts between S_0 and S_1, S_6 and S_9, and between S_{12} and S_{14} (denoted by dashed lines). The dotted lines in Figure 2 denote possible shortcuts of the route. S_1 and S_{12} become *provisional ends* of the route; S_6 and S_9 become *provisional shortcut stops.*

The required request time for the optional parts of the route can be calculated as follows:

- The part $S_0 - S_1$: this optional part of the route consists of a single stop only. If, according to the optional timetable, the bus should take t_{1A} minutes to go from stop S_1 to stop S_0 and t_{1B} minutes to go from S_0 to S_1, then we can define $t_1 = \text{maximum}(t_{1A}, t_{1B})$ as the maximum request time required for this optional part of the route.
- The part $S_6 - S_9$: When moving from S_6 to S_9, the bus should take t_{2A} minutes to go from the stop S_6 to the stop S_9 and t_{2B} minutes to go from the stop S_9 to the stop S_6, then $t_2 = \text{maximum}(t_{2A}, t_{2B})$ becomes the maximum required request time for this optional part of the route.
- In the same way we can specify the maximum required request time t_3 for the part $S_{12} - S_{14}$.

Thus, the *on-demand request time* for this route should be the maximum of t_1, t_2, and t_3, rounded up to the nearest 5 minutes.

In the above scenario, the bus operates on parts of the route only on demand, which increases both profit for the transport company and overall sustainability of the public transport system, without any disadvantages for the passengers. To implement this optimization we need to identify which parts of the route are "optional" and during which hours, and provide an interface to enable comfortable and easy requesting of the bus at the optional stops according to the timetable. As technology improves, the numbers of passengers on board vehicles is becoming more accurately calculated and available. Yarra Trams (providers of the tram network in Melbourne, Australia) have been trialling automatic passenger counting systems, which count the number of passengers who board and exit the tram at each stop (Fihn & Finndahl, 2011). This information can inform the transport providers so they know which stops along the route to identify as optional, and at which times of the day.

In general, the optional section of the route is quite long and has many provisional shortcut stops, because we could have many possibilities for optimizing the route according to the concrete route load and passengers' requests (an example of such a route is presented in the next section). The choices regarding where to drive and which shortcut to choose should be made very quickly to avoid additional delays in the traffic. This means tighter constraints on the driver if he/she needs to do it manually, using a map, or from memory. Thus, on the one hand this optimization can help to save the environment and make public transport more sustainable in the environmental sense, but on the other hand, if we do not have a decision support system for the driver, the decision overload (caused by the optimization) could lead to human error: driving mistakes on the road, wrong shortcut selection, lack of alertness because of stress and tiredness, not seeing passengers, etc. One of the most common causes of human error in using safety-critical systems is an overestimation of human ability to stay alert for long periods of time or to make decisions in parallel with performing some safety-critical action (as in the case of driving on the road). Following the *Engineering Error Paradigm* (Redmill & Rajan, 1997), we recommend taking the optimisation decision function from the driver to make the system more reliable and road-safe: we suggest to accompany the optimization of the public transport routes with an appropriate decision support system, which can also be seen as a dynamically extended/shortened route during the times when the route is in the *irregularLoad* mode.

Let us discuss the dynamic plan for the route shown in Figure 2 (assuming that the route is in the *irregularLoad* mode, and while the vehicle goes in the direction $S_0 - S_{14}$). We represent this as a general version using notation S_0, …, S_{14} and listing all the stops along one line, where in the implementation we would be employing a vertical listing (more readable for the small screens) and using the names of the stops as specified in the timetable available for passengers. The stops where some route planning extensions can occur (i.e. the provisional ends and shortcut stops) are marked bold and use orange colour to highlight and draw attention to the possible changes in the connections. In the Case 1, we assume the vehicle was empty at the stop S_1 when travelling previously in the opposite direction and there were no requests from the stop S_0, before starting to drive in the direction the $S_0 - S_{14}$, so the route plan would start from S_1. We have six options for updating the current plan of the route, depending on further requests:

Case 1. There were no requests from the stops S_7, S_8, S_{13}, and S_{14}: The sequence *RegularStops*^{R} for this case is $[S_1, S_2, S_3, S_4, S_5, S_6, S_9, S_{10}, S_{11}, S_{12}]$, and the route plan is defined as follows (cf. also Figure 2(c), which shows the system view as well as general and simplified driver views for this scenario):

$$\mathbf{S_1} - S_2 - S_3 - S_4 - S_5 - \mathbf{S_6} - \mathbf{S_9} - S_{10} - S_{11} - \mathbf{S_{12}}$$

In this case, when leaving S_1, once more we have six options for the updating of the current plan of the route, depending on the requests, cf. Figure 2(d). We also assume that the vehicle waited too long at S_1 and according to the system plan should already be at S_3, meaning that the actual and the planned locations of the vehicle are different for this case, as shown in Figure 2(c). Please note, that it is impossible to receive a request for this stop on this run, because this stop already belongs to the "traversed" part of the route. The important point here is that the simplified driver view is the same for all six subcases cf. Figure 2(d3).

Case 1.1. There were no requests from the stops S_7, S_8, S_{13}, and S_{14}. The sequence *RegularStops*^{R} remains unchanged, and the current route is defined as follows, cf. Figure 2(d):

$$S_2 - S_3 - S_4 - S_5 - \mathbf{S_6} - \mathbf{S_9} - S_{10} - S_{11} - \mathbf{S_{12}}$$

Case 1.2. There was a request from at least one of the stops S_7 and S_8 (a request from one of this stops makes both of them "regular" for this run), but there were no request from S_{13}, and S_{14}. The sequence *RegularStops*^{R} becomes $[S_1, S_2, S_3, S_4, S_5, S_6, S_7, S_8, S_9, S_{10}, S_{11}, S_{12}]$, and the current route is updated, cf. Figure 2(e):

$$S_2 - S_3 - S_4 - S_5 - \mathbf{S_6} - \mathbf{S_7} - \mathbf{S_8} - \mathbf{S_9} - S_{10} - S_{11} - \mathbf{S_{12}}$$

Case 1.3. There was a request from the stop S_{13}, but there were no request from S_7, S_8 and S_{14}. The sequence *RegularStops*^{R} becomes $[S_1, S_2, S_3, S_4, S_5, S_6, S_9, S_{10}, S_{11}, S_{12}, S_{13}]$, the current route is updated as shown in Figure 3(a):

$$S_2 - S_3 - S_4 - S_5 - \mathbf{S_6} - \mathbf{S_9} - S_{10} - S_{11} - \mathbf{S_{12}} - \mathbf{S_{13}}$$

Case 1.4. There was a request from the stop S_{14}, but there were no request from S_7, and S_8. The sequence *RegularStops*^{R} becomes $[S_1, S_2, S_3, S_4, S_5, S_6, S_9, S_{10}, S_{11}, S_{12}, S_{13}, S_{14}]$, the current route is updated as shown in Figure 3(b):

Figure 3. Route plan for the Cases 1.3 – 1.6

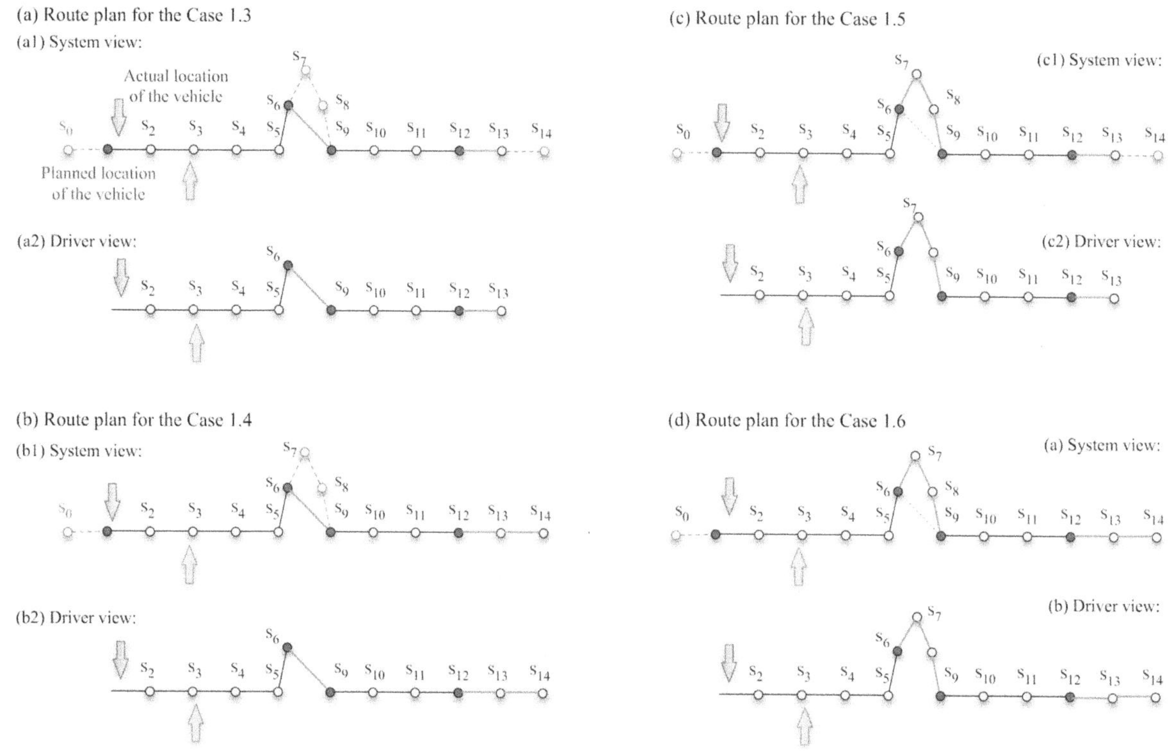

$$S_2 - S_3 - S_4 - S_5 - \mathbf{S_6} - \mathbf{S_9} - S_{10} - S_{11} - \mathbf{S_{12}} - \mathbf{S_{13}} - \mathbf{S_{14}}$$

Case 1.5. There was a request from at least one of the stops S_7 and S_8, as well as request from S_{13} (but no request from S_{14}). The sequence *RegularStops*^{R} becomes [S_1, S_2, S_3, S_4, S_5, S_6, S_7, S_8, S_9, S_{10}, S_{11}, S_{12}, S_{13}], the current route is updated as shown in Figure 3(c):

$$S_2 - S_3 - S_4 - S_5 - \mathbf{S_6} - \mathbf{S_7} - \mathbf{S_8} - \mathbf{S_9} - S_{10} - S_{11} - \mathbf{S_{12}} - \mathbf{S_{13}}$$

Case 1.6. There was a request from at least one of the stops S_7 and S_8, as well as request from S_{14} (in this case it does not matter whether there was a request from S_{13}, because a request from S_{14} makes both S_{13} and S_{14} "regular" for this run). The sequence *RegularStops*^{R} becomes [S_1, S_2, S_3, S_4, S_5, S_6, S_7, S_8, S_9, S_{10}, S_{11}, S_{12}, S_{13}, S_{14}], which also means that no further optional stop can be added to the sequence, because it already contains all the optional stops in the driving direction. The sequence remains unchanged until the end of the run. The current route is updated as shown in Figure 3(d):

$$S_2 - S_3 - S_4 - S_5 - \mathbf{S_6} - \mathbf{S_7} - \mathbf{S_8} - \mathbf{S_9} - S_{10} - S_{11} - \mathbf{S_{12}} - \mathbf{S_{13}} - \mathbf{S_{14}}$$

Case 2. There was a request from the stop S_7, or S_8, but there were no requests from S_{13}, and S_{14}. The sequence *RegularStops*^{R} for this case is [S_1, S_2, S_3, S_4, S_5, S_6, S_7, S_8, S_9, S_{10}, S_{11}, S_{12}], and the route plan becomes

$$\boldsymbol{S_1} - S_2 - S_3 - S_4 - S_5 - \boldsymbol{S_6} - \boldsymbol{S_7} - \boldsymbol{S_8} - \boldsymbol{S_9} - S_{10} - S_{11} - \boldsymbol{S_{12}}$$

In this case, stops S_7 and S_8 were already marked as "regular" for this run (the parameters $option_7$ and $option_8$ are set to be true), therefore we should not receive any further requests from these stops. Thus, when leaving S_1, we have three options for the updating of the current plan of the route.

Case 2.1. There were no requests from the stops S_{13} and S_{14}. The sequence $RegularStops^R$ and the route plan remain unchanged.

Case 2.2. There was a request from the stop S_{13}, but there were no requests from S_{14}. The sequence $RegularStops^R$ becomes $[S_1, S_2, S_3, S_4, S_5, S_6, S_7, S_8, S_9, S_{10}, S_{11}, S_{12}, S_{13}]$, and the route plan becomes

$$S_2 - S_3 - S_4 - S_5 - \boldsymbol{S_6} - \boldsymbol{S_7} - \boldsymbol{S_8} - \boldsymbol{S_9} - S_{10} - S_{11} - \boldsymbol{S_{12}} - \boldsymbol{S_{13}}$$

Case 2.3. There was a request from the stop S_{14}. The sequence $RegularStops^R$ becomes $[S_1, S_2, S_3, S_4, S_5, S_6, S_7, S_8, S_9, S_{10}, S_{11}, S_{12}, S_{13}, S_{14}]$, the route plan becomes

$$S_2 - S_3 - S_4 - S_5 - \boldsymbol{S_6} - \boldsymbol{S_7} - \boldsymbol{S_8} - \boldsymbol{S_9} - S_{10} - S_{11} - \boldsymbol{S_{12}} - \boldsymbol{S_{13}} - \boldsymbol{S_{14}}$$

Case 3. There was a request from the stop S_{13}, but there were no requests from S_7, S_8, or S_{14}. The sequence $RegularStops^R$ for this case is $[S_1, S_2, S_3, S_4, S_5, S_6, S_9, S_{10}, S_{11}, S_{12}, S_{13}]$, and the plan becomes

$$\boldsymbol{S_1} - S_2 - S_3 - S_4 - S_5 - \boldsymbol{S_6} - \boldsymbol{S_9} - S_{10} - S_{11} - \boldsymbol{S_{12}} - \boldsymbol{S_{13}}$$

In this case, stop S_{13} is already marked as "regular" for this run (the parameters $option_{13}$ is set to be true), therefore we cannot receive any further requests from this stop. When leaving S_1, we can have four options for updating of the route plan.

Case 3.1. There were no requests from the stops S_7, S_8, and S_{14}. The sequence $RegularStops^R$ and the route plan remains unchanged.

Case 3.2. There was a request from S_7 or S_8, but no requests from S_{14}. The sequence $RegularStops^R$ becomes $[S_1, S_2, S_3, S_4, S_5, S_6, S_7, S_8, S_9, S_{10}, S_{11}, S_{12}, S_{13}]$, the plan becomes

$$S_2 - S_3 - S_4 - S_5 - \boldsymbol{S_6} - \boldsymbol{S_7} - \boldsymbol{S_8} - \boldsymbol{S_9} - S_{10} - S_{11} - \boldsymbol{S_{12}} - \boldsymbol{S_{13}}$$

Case 3.3. There was a request from S_{14}, but no requests from S_7, or S_8. The sequence $RegularStops^R$ becomes $[S_1, S_2, S_3, S_4, S_5, S_6, S_9, S_{10}, S_{11}, S_{12}, S_{13}, S_{14}]$, the plan becomes

$$S_2 - S_3 - S_4 - S_5 - \boldsymbol{S_6} - \boldsymbol{S_9} - S_{10} - S_{11} - \boldsymbol{S_{12}} - \boldsymbol{S_{13}} - \boldsymbol{S_{14}}$$

Case 3.4. There was a request from S_7 or S_8, as well as a request from S_{14}. The sequence $RegularStops^R$ becomes $[S_1, S_2, S_3, S_4, S_5, S_6, S_7, S_8, S_9, S_{10}, S_{11}, S_{12}, S_{13}, S_{14}]$, which also means that no further optional stop can be added to the sequence. The plan becomes

$$S_2 - S_3 - S_4 - S_5 - \boldsymbol{S_6} - \boldsymbol{S_7} - \boldsymbol{S_8} - \boldsymbol{S_9} - S_{10} - S_{11} - \boldsymbol{S_{12}} - \boldsymbol{S_{13}} - \boldsymbol{S_{14}}$$

Case 4. There was a request from the stop S_{14}, but there were no requests from S_7 or $S_{8.}$ The sequence *RegularStops*^{R} for this case is [S_1, S_2, S_3, S_4, S_5, S_6, S_9, S_{10}, S_{11}, S_{12}, S_{13}, S_{14}], and the plan becomes

$$\boldsymbol{S_1} - S_2 - S_3 - S_4 - S_5 - \boldsymbol{S_6} - \boldsymbol{S_9} - S_{10} - S_{11} - \boldsymbol{S_{12}} - \boldsymbol{S_{13}} - \boldsymbol{S_{14}}$$

In this case, stops S_{13} and S_{14} were already marked as "regular" for this run (the parameters $option_{13}$ and $option_{14}$ are set to be true), therefore we cannot receive any further requests from these stops. Thus, when leaving S_1, we can have two options for the updating of the current plan of the route.

Case 4.1. There were no requests from S_7 and S_8. The sequence *RegularStops*^{R} and the route plan remains unchanged.

Case 4.2. There was a request from S_7 or S_8. The sequence *RegularStops*^{R} becomes [S_1, S_2, S_3, S_4, S_5, S_6, S_7, S_8, S_9, S_{10}, S_{11}, S_{12}, S_{13}, S_{14}], which also means that no further optional stops can be added to the sequence. The plan becomes

$$S_2 - S_3 - S_4 - S_5 - \boldsymbol{S_6} - \boldsymbol{S_7} - \boldsymbol{S_8} - \boldsymbol{S_9} - S_{10} - S_{11} - \boldsymbol{S_{12}} - \boldsymbol{S_{13}} - \boldsymbol{S_{14}}$$

Case 5. There was a request from S_7 or S_8, as well as request S_{13} (but not from S_{14}). The sequence *RegularStops*^{R} becomes [S_1, S_2, S_3, S_4, S_5, S_6, S_7, S_8, S_9, S_{10}, S_{11}, S_{12}, S_{13}], and the route plan becomes

$$\boldsymbol{S_1} - S_2 - S_3 - S_4 - S_5 - \boldsymbol{S_6} - \boldsymbol{S_7} - \boldsymbol{S_8} - \boldsymbol{S_9} - S_{10} - S_{11} - \boldsymbol{S_{12}} - \boldsymbol{S_{13}}$$

In this case, stops S_7, S_8 and S_{13} were already marked as "regular" for this run (the parameters $option_7$, $option_8$ and $option_{13}$ are set to be true), therefore we cannot receive any further requests from these stops. Thus, when leaving S_1, we can have two options for the updating of the current plan of the route.

Case 5.1. There were no requests from the stop S_{14}. The sequence *RegularStops*^{R} and the route plan remains unchanged.

Case 5.2. There was a request from the stop S_{14}. The sequence *RegularStops*^{R} becomes [S_1, S_2, S_3, S_4, S_5, S_6, S_7, S_8, S_9, S_{10}, S_{11}, S_{12}, S_{13}, S_{14}], no further optional stop can be added to the sequence, and the route plan becomes

$$S_2 - S_3 - S_4 - S_5 - \boldsymbol{S_6} - \boldsymbol{S_7} - \boldsymbol{S_8} - \boldsymbol{S_9} - S_{10} - S_{11} - \boldsymbol{S_{12}} - \boldsymbol{S_{13}} - \boldsymbol{S_{14}}$$

Case 6. There was a request from S_7, or S_8, as well as from S_{14}. The sequence *RegularStops*^{R} becomes [S_1, S_2, S_3, S_4, S_5, S_6, S_7, S_8, S_9, S_{10}, S_{11}, S_{12}, S_{13}, S_{14}], no further optional stops can be added to the sequence, and the route plan becomes

$$\boldsymbol{S_1} - S_2 - S_3 - S_4 - S_5 - \boldsymbol{S_6} - \boldsymbol{S_7} - \boldsymbol{S_8} - \boldsymbol{S_9} - S_{10} - S_{11} - \boldsymbol{S_{12}} - \boldsymbol{S_{13}} - \boldsymbol{S_{14}}$$

In this case, no further optional stops can be added to the sequence *RegularStops*^{R}, the plan already contains all optional parts of the route (like in *normalLoad* mode).

It is easy to see that even for a very small route with not very complicated optional parts there are many possibilities for options for the updating of the current plan of the route. Thus, it would be problematic for a driver to check for the timetable updates continuously and to analyse the current plan manually. It becomes especially complicated when a request to serve a particular stop implies that not only one, but many stops become “regular” for this run. This implies that a driver decision support system representing the updated plan of the route is required. A simplified driver view, as shown in Figure 2, improves the ease of system use by the driver and allows for focussing on the necessary current driving goals.

On-Demand Bus Scenario: Analysis of a Specific Bus Route

This example is based on the spatial information and the timetable of an existing real bus route 445, which operates in Melbourne area of Victoria, in Australia. Bus route 445 takes passengers between two railway stations at Hoppers Crossing and Werribee Station, cf. Figure 4. Depending on the particular time of the day, the whole route can take in the non-optimised case approximately 34-38 minutes in the direction of “Werribee Station to Hoppers Crossing” and 32-38 minutes in the direction of “Hoppers Crossing to Werribee Station” (according to the timetables on the webpage of the Public Transport Victoria (PTV), dated 27 July 2014).

Figure 4. Bus route 445 in Melbourne area, Australia (2014): Identification of the optional sections of the route and optimization of the route.

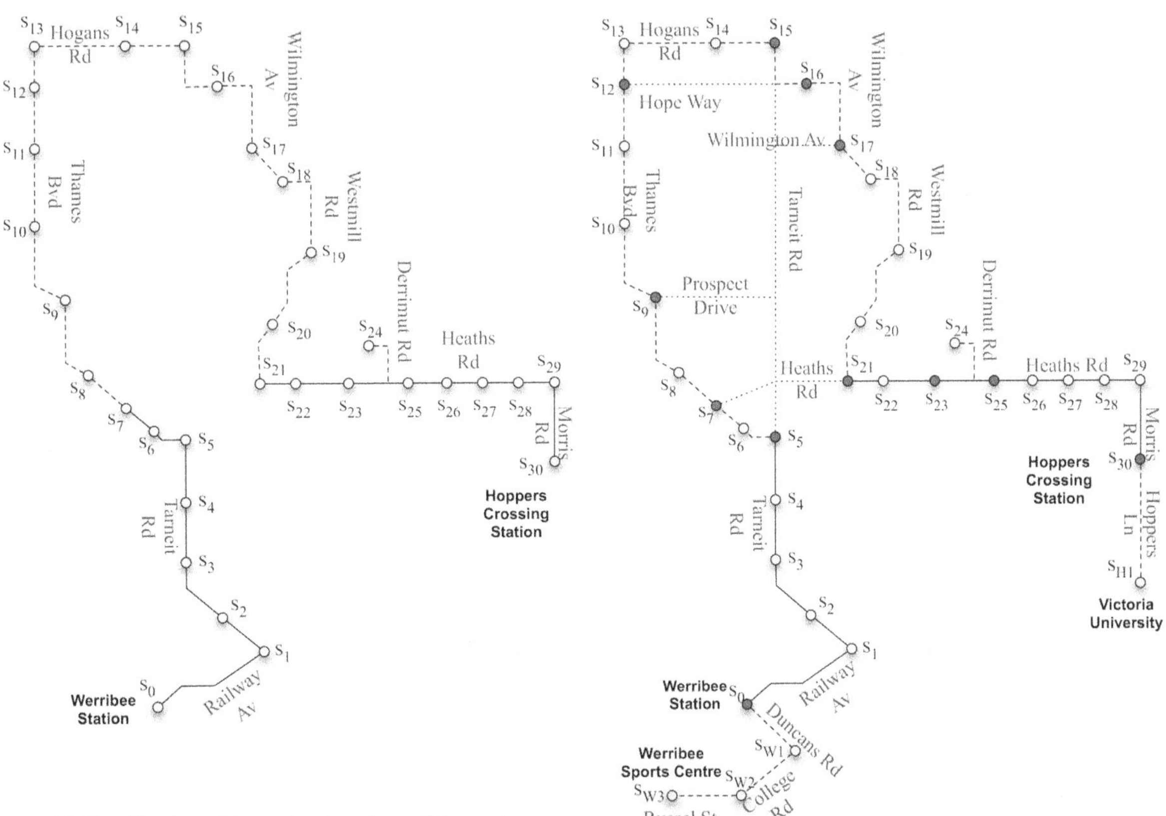

(a) Identification of the optional sections of the route (marked by dashed lines)

(b) Optimized route

For the case when the bus is travelling in the direction to Hoppers Crossing Station, the bus departs from Werribee Station and runs via Manly St, Market Rd, Railway Av, Tarneit Rd, Thames Bvd, Hogans Rd, Tarneit Rd, Wilmington Av, Westmill Dr, Heaths Rd and Derrimut Rd to Werribee Plaza Shopping Centre; then via Derrimut Rd, Heaths Rd and Morris Rd to Hoppers Crossing Station.

For the case when the bus is travelling in the direction to Werribee Station, the bus departs from Hoppers Crossing Station and runs via Morris Rd, Heaths Rd and Derrimut Rd to Werribee Plaza Shopping Centre; then via Derrimut Rd, Heaths Rd, Westmill Dr, Wilmington Av, Tarneit Rd, Hogans Rd, Thames Bvd, Tarneit Rd, Railway Av, Market Rd and Manly St to Werribee Station, see Figure 4. In our representation, we use a short notation S_0, …, S_{30} for the stops.

All buses on this route provide wheelchair accessible services, which makes this route a good example for deviations along the route, for hours out of the peak times, to pick up disabled passengers with restricted capabilities.

We have identified optional sections of the route when the bus may be empty, by the *dashed lines* in Figure 4(a). To collect more precise and real time data about the route load, one can use approaches like (Fihn & Finndahl, 2011) which publishes information about the number of passengers at each stop along the tram route. For this scenario, we have not built the statistics by collecting real data, but by analysing the spatial information about the route 445 and other public transport routes in this area, as well as Google map information for this suburb. A larger case study using PTV statistics is planned for future work. People who live far from a railway station may choose to travel on public transport but not every day (housewives, school children or elderly people), or prefer to drive in their own private cars or taxis, if they need to go to another part of the city during the middle of the day or late evenings, i.e., times outside of peak hours. In general, the closer the bus stop is to the railway station, the more people catch the bus from this stop. Thus, the bus load on the route 445 is higher at the ends of the route but could be zero during some hours in the middle of the route, e.g., between stops S_7 and S_{21} (which is the mainly residential part of the suburb) as well as between stops S_{23} and S_{25} (the stop S_{24} is an appendix to the main road, to bring passengers to the Werribee Plaza Shopping Centre). This means, we do not need to optimize the ends of the route, but between stops S_7 and S_{21} and between stops S_{23} and S_{25} an optimisation could be required. However, after analysis of the city map as well as the spatial information on other bus routes operating in this area, we recommend extending the ends of the route by additional optional stops at each end, S_0 (Werribee Station) and S_{30} (Hoppers Crossing Station), namely S_{W1}, S_{W2}, S_{W3} *and* S_{H1}. Thus, the stops S_0 and S_{30} become the *provisional ends* of the route, where the new actual ends of the route are S_{W3} (Werribee Sport Centre) *and* S_{H1} (Victoria University).

We present the extended and optimized route in Figure 4(b). The dotted lines represent possible shortcuts of the route.

In the rest of this section we discuss particular scenarios that are possible for this route in the *irregularLoad* mode. For simplicity, we are focussing only on the direction $S_{W3} - S_{H1}$, i.e. from the Werribee Sport Centre to the Victoria University. Here we have four parts of the route that could be optimized:

1. S_{W3} to S_0
2. S_5 to S_{21}
3. S_{23} to S_{25}
4. S_{30} to S_{H1}

For this scenario, we can deal with the parts at the ends of the route in the same way as for the general scenario. The case of the stop S_{24}, which is an appendix part of the route, is trivial. Thus, we focus here on the most interesting optional part of the route, between stops S_5 and S_{21}.

When the bus is approaching the stop S_5, we could have seven possible cases for the bus driving scenarios, cf. also Figures 5 – 7 where we present the parts of the route between stops S_5 and S_{21} for these scenarios. Please note that we present here scenarios for system views – as we do not present system views for a concrete moment of time, we omit here the information about departure times and actual/planned location of the bus. However, the simplified driver views are presented assuming a concrete moment of time during the route run.

Figure 5. Bus route 445: Possible scenarios after the stop S_5 (Cases 1, 2 and 3.1)

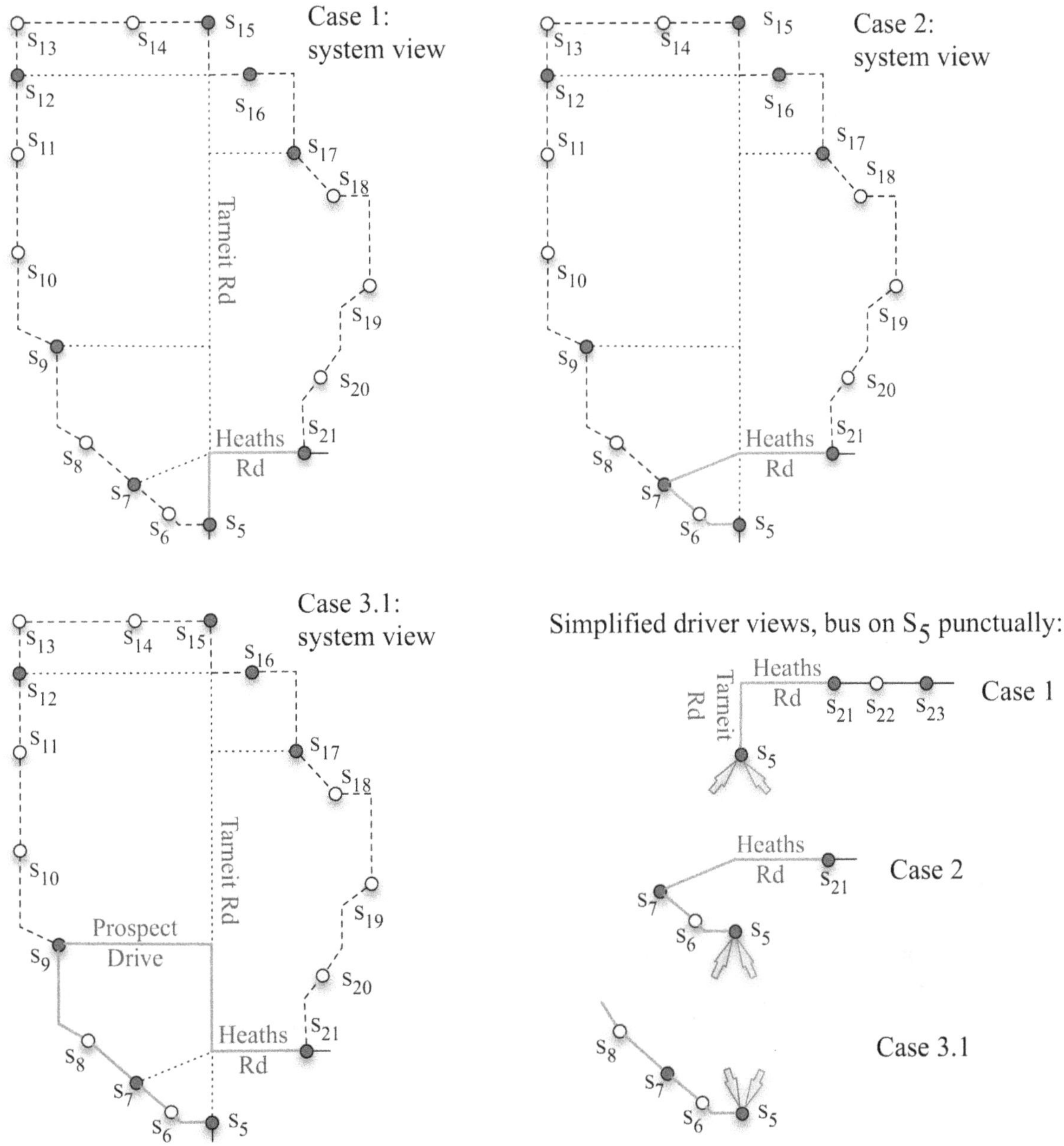

Case 1. There are no requests for the whole part of the route. In this case, the bus should go directly to the stop S_{21}, following Tarneit Rd until the intersection with Heaths Rd, then turning on to Heaths Rd. For the graphical representation cf. the route in Figure 5, Case 1.

Case 2. There are requests only from the stop S_6, and/or S_7. The bus proceeds to the stop S_7 (via S_6), and then goes directly to the stop S_{21} via Heaths Rd (cf. the route in Figure 5, Case 2).

Case 3. There are requests from the stop S_8, and/or S_9. The bus proceeds to the stop S_9 (via S_6, S_7, and S_8), and the rest of the route plan depends on the following five options (for all these options, the simplified driver view is the same, cf. the simplified driver view for the Case 3.1 in Figure 5):

Case 3.1. There are no further requests from the stops S_{10},…, S_{20}. The bus proceeds directly to the stop S_{21} following the Prospect Drive until the intersection with Tarneit Rd, then via Tarneit Rd until the intersection with Heaths Rd, then turning on to Heaths Rd, (cf. the route in Figure 5).

Case 3.2. There are some requests from the stops S_{10}, S_{11}, and/or S_{12}, but no requests from S_{13}, S_{14}, and S_{15}. From S_9, the bus proceeds to the stop S_{12} (via S_{10}, and S_{11}), and then we have three additional options:

1. There were no requests from S_{16}, …, S_{20}. From S_{12}, the bus proceeds to S_{21} following the Hope Way until the intersection with Tarneit Rd, then via Tarneit Rd until the intersection with Heaths Rd, then turning on to Heaths Rd, as shown in Figure 6, Case 3.2.1.
2. There is a request from S_{16}. From S_{12}, the bus proceeds to S_{16} following the Hope Way. After S_{16} it proceeds to S_{21} like in *normalLoad* mode, as shown in Figure 6, Case 3.2.2.
3. There is no request from S_{16}, but there is a request at least from one of the stops S_{17}, …, S_{20}. From S_{12}, the bus proceeds to S_{17} following the Hope Way, until the intersection with Tarneit Rd, then via Tarneit Rd until the intersection with Wilmington Av, and then turning on to Wilmington Av. After S_{17} it proceeds to S_{21} as in *normalLoad* mode, as shown in Figure 6, Case 3.2.3.

Case 3.3. There are some requests from the stops S_{13}, S_{14}, and/or S_{15} (for this case, it is not important whether there are requests from S_{10}, S_{11}, and/or S_{12}). From S_9, the bus proceeds to S_{15} (via S_{10}, …, S_{14}, as in *normalLoad* mode). The rest of the route plan until S_{21} depends on three following options:

1. There is no request from S_{16},…, S_{20}. From S_{15}, the bus proceeds to S_{21} following Tarneit Rd, then turning to Heaths Rd Av, cf. Figure 6, Case 3.3.1.
2. There is a request from S_{16}. From S_{15}, the bus proceeds to S_{21} (via S_{17}, …, S_{20}) as in *normalLoad* mode, as shown in Figure 7, Case 3.3.2.
3. There is no request from S_{16}, but there are requests from S_{17},…, S_{20}. From S_{15}, the bus proceeds to S_{17} following Tarneit Rd until the intersection with Wilmington Av, then turning on to Wilmington Av. After S_{17} it proceeds to S_{21} as in *normalLoad* mode, cf. Figure 7, Case 3.3.3.

Case 3.4. There are no requests from the stops S_{10}, S_{11}, …, S_{15}, but there is a request from S_{16}. From S_9, the bus proceeds directly to the stop S_{16}, following Prospect Drive, then turning on to Tarneit Rd, then via Tarneit Rd until the intersection with Hope Way, then turning on to Hope Way. After S_{16} it proceeds to S_{21} as in *normalLoad* mode. The system view on the route is shown in Figure 7, Case 3.4.

Case 3.5. There are no requests from the stops S_{10}, S_{11}, …, S_{16}, but there is a request from at least one of the stops S_{17},…, S_{20}. From S_9, the bus proceeds directly to the stop S_{17}, following Prospect Drive, then

Figure 6. Bus route 445: Possible scenarios after the stop S_5 (Cases 3.2.1 – 3.3.1, system views)

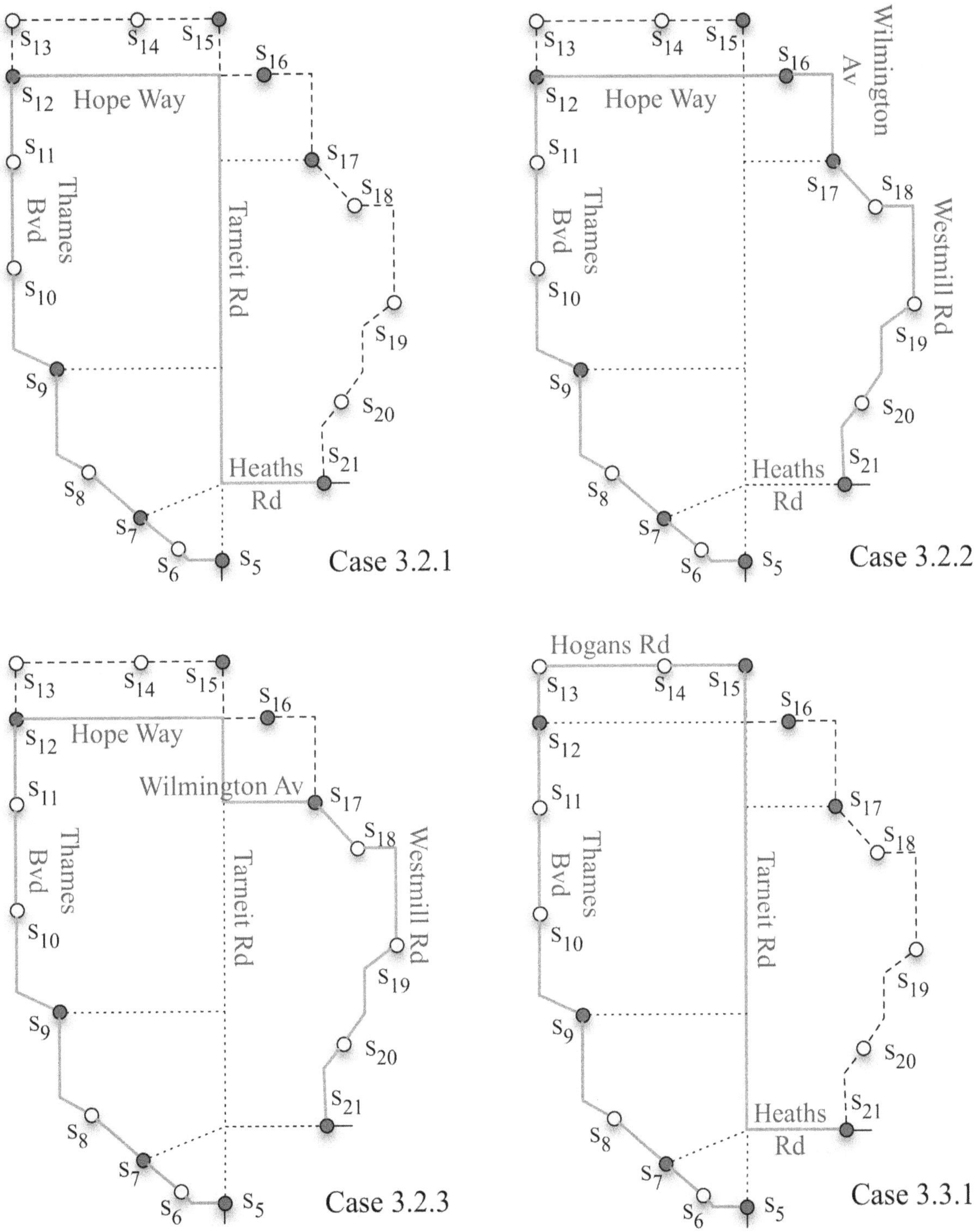

Figure 7. Bus route 445: Possible scenarios after the stop S_5 (Cases 3.3.2 – 3.5, system views)

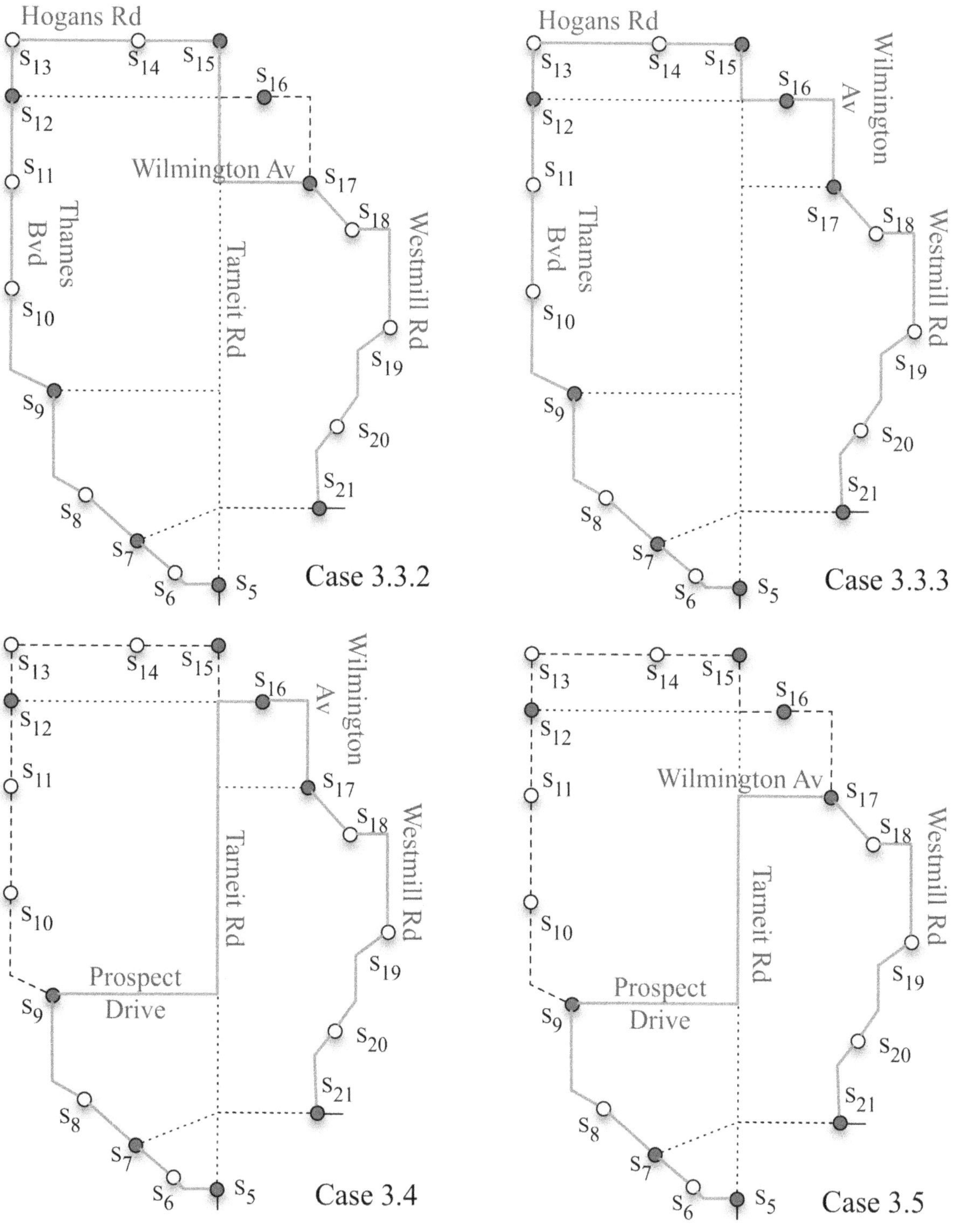

turning on to Tarneit Rd, then via Tarneit Rd until the intersection with Wilmington Av, then turning on to Wilmington Av. After S_{17} it proceeds to S_{21} as in *normalLoad* mode. The system view on the route is shown in Figure 7, Case 3.5.

Case 4. There are no requests from the stop S_8, and/or S_9, but there is a request from at least one of the stops S_{10}, S_{11}, S_{12}. The bus goes to the stop S_{12} (via S_6, …, S_{11}), and after that we have four additional options:

Case 4.1. There are no requests from S_{13}, …, S_{20}. From S_{12}, the bus proceeds to the stop S_{21} following Hope Way until the intersection with Tarneit Rd, then via Tarneit Rd until the intersection with Heaths Rd, then turning on to Heaths Rd. This scenario is equal to the Case 3.2.1, cf. Figure 6.

Case 4.2. There is a request from at least one of the stops S_{13},…, S_{15}. From S_{12} the bus proceeds to the stop S_{15} as in *normalLoad* mode. The scenarios for the rest of the route until S_{21} are as described for the Cases 3.3.1 – 3.3.3, cf. Figures 6 and 7 for the corresponding system views.

Case 4.3. There are no requests from S_{13}, …, S_{15}, but there is a request from S_{16}. From S_{12} the bus proceeds to the stop S_{16} following Hope Way. After S_{16} it proceeds to S_{21} as in *normalLoad* mode. This scenario is equal to the Case 3.2.2, cf. the corresponding system view in Figure 6.

Case 4.4. There are no requests from S_{13},…, S_{16}, but there is a request at least from one of the stops S_{17},…, S_{20}. From S_{12} the bus proceeds to S_{17} following Hope Way, until the intersection with Tarneit Rd, then via Tarneit Rd until the intersection with Wilmington Av, then turning on to Wilmington Av. After S_{17} it proceeds to S_{21} as in *normalLoad* mode. This scenario is equal to the Case 3.2.3, cf. the corresponding system view in Figure 6.

Case 5. There are no requests from the stops S_8, …, S_{12}, but there is a request from at least one of the stops S_{13},…, S_{15}. From S_5 the bus proceeds to the stop S_{15} as in *normalLoad* mode. The scenarios for the rest of the route until S_{21} are as described for the Cases 3.3.1, 3.3.2, 3.3.3, and 4.2 (cf. Figures 6 and 7).

Case 6. There are no requests from the stops S_8, …, S_{15}, but there is a request from S_{16}. From S_5 the bus proceeds directly to the stop S_{16} following Tarneit Rd until the intersection with Hope Way. After S_{16} it proceeds to S_{21} as in *normalLoad* mode. This route in this case is equal to the route described for the Case 3.4 (cf. Figure 7 for the system view).

Case 7. There are no requests from the stops S_8, …, S_{16}, but there is a request from at least one of the stops S_{17},…, S_{20}. From S_5 the bus proceeds directly to the stop S_{17} following Tarneit Rd until the intersection with Wilmington Av, then turning on to Wilmington Av. After S_{17} it proceeds to S_{21} as in *normalLoad* mode. This route in this case is equal to the route described for the Case 3.5 (cf. Figure 7 for the system view).

We are interested in the required *on-demand request time* for this route. We assume that the bus should take 3 minutes to go from stop S_{30} to stop S_{H1} or from S_{H1} to S_{30}, as well as 5 minutes from stop S_0 to stop S_{W3} or from S_{W3} to S_0. According to the current timetables, the bus needs 11 minutes to go from S_5 to S_{21} in *normalLoad* mode. Thus, the required request time for this route would be 11 minutes. We round this time up to the nearest 5 minutes and obtain the *on-demand request time* for this route: 15 minutes. This means that during the *irregularLoad* mode the request for the bus to come to a stop on an optional part of the route should come at least 15 minutes earlier than the time specified in the bus timetable.

As an example of this irregular route, let us consider the stop S_{14} (Harmony Dr/Hogans Rd) and analyse the timetable for Monday to Friday, assuming the *irregularLoad* mode is between 10am and 2:30pm, as well as after 8pm (marked bold). We find the following for the given times:

6:22, 7:16, 8:11, 9:01,

10:01, 11:01, 12:01, 1:01, 2:01,

3:01, 4:01, 5:01, 5:58, 7:04,

8:08, 9:08.

Thus, a passenger, wishing to catch the bus from this stop, at 11:01am, should send a request to the on-demand bus system no later than at 10:46am. To make it easier for passengers, we could add the on-demand request times to the timetable in the following manner:

6:22, 7:16, 8:11, 9:01,

10:01$^{\textbf{R. until 9:46}}$, **11:01**$^{\textbf{R. until 10:46}}$, **12:01**$^{\textbf{R. until 11:46}}$, **1:01**$^{\textbf{R. until 12:46}}$, **2:01**$^{\textbf{R. until 11:46}}$,

3:01, 4:01, 5:01, 5:58, 7:04,

8:08$^{\textbf{R. until 7:53}}$, **9:08** $^{\textbf{R. until 8:53}}$.

These ideas could be implemented as a mobile application that allows passengers to send their on-demand requests to the driver, as well as a decision support application for the driver.

By serving the optional parts of this route only on-demand, we can optimise the timetables making them denser, i.e., enabling the bus to pass by, not every 60 minutes approximately, but every 30 minutes instead.

Implementation

To make this model really useful for the passengers and for the drivers, the system must be implemented as a smartphone or web application. The first prototype of the system was implemented within the "Buses on demand" student project at our School of Computer Science and IT in RMIT University, Australia.

The prototype allows passengers to send requests to the transport system any time and from any location, choosing for the *irregularLoad* mode which "optional" stops should be operated, as well as see which stops have already been "requested" by other passengers, i.e., which stops are currently marked to be "regular" (cf. Figure 8 for the passenger view of the system and for the graphical representation of the system view). To increase the safety on the road, an automated decision support system was implemented: the prototype also provides the driver with instructions, on how the route should be shortcut if optimisation is required. Thus, the prototype provides the driver with the latest version of the timetable and the optimized route: the driver's view of the route comprises of only the "regular" stops, and when an "optional" stop becomes "regular", the driver's view of the route is immediately updated. The system view provides additional information to the system manager on the requests statistics. The prototype supports two kinds of representations: graphical, using Google maps, as presented in Figure 8, and textual, using tabular representation.

Figure 8. Prototype of the bus on-demand system: passenger view (the "print screen" image is cut in the middle) and the graphical representation of the route using Google maps (red markers denote "regular" stops, blue markers denote "optional" stops).

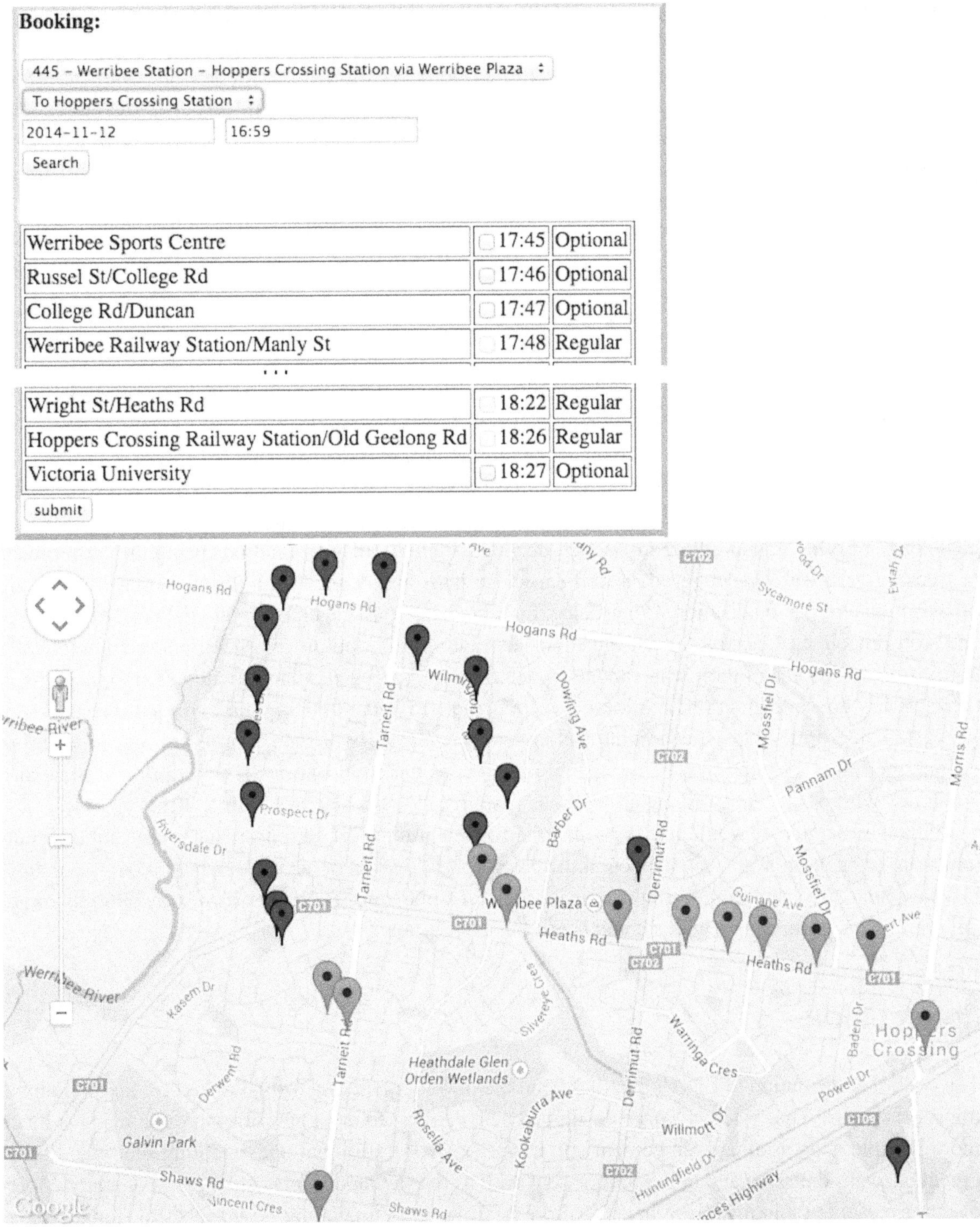

FUTURE RESEARCH DIRECTIONS

Currently the buses in the public transport system in Melbourne do not have GPS devices, or if the buses are new and GPS has been fitted, it is turned off. There is the possibility that the positions of buses fitted with GPS tracking technology can be integrated into the wider public transport system, cf. (Covan, 2014). This would be a tremendous advantage for our proposed on-demand model, and would allow for integrating the modelling of exceptions on the road (such as problems with traffic due to road works, car accidents, etc.) into the model directly. In further research on the modelling of on-demand systems, we plan a number of case studies based on analysis of actual train and tram routes in Melbourne, Australia.

There are further opportunities to develop mobile applications to notify commuters if their trains or buses are late. Such systems might rely on the vehicle with the GPS tracking device to arrive at a particular stop and notify people using the mobile application that it has arrived. This reminder could be used to predict when the bus would arrive at their stop and remind them within 15 minutes of reaching it. The time could be flexible and set by the passenger to enable them to reach their stop in whatever time is required to catch the bus.

The GPS could also be used to remind people travelling in the bus about when they are nearing their destination. For disabled people, or commuters engrossed in their reading material or games, they may otherwise have missed the stop, or not managed to get to the exit door in time.

There are opportunities for more complex multimodal journey planning, which would notify passengers that they are running late and will miss their connecting train. Such a system would inform the passengers of other transport options, which may take them to their destinations using different modes of transport, if they were to get off a stop earlier perhaps and connect in a different place. There are other options for personalisation, for instance, for wheelchair passengers, to let them know which buses and stops enable easy access for wheelchairs. There are safety options for encouraging commuters to notify selected people of their whereabouts, when they are approaching a particular stop to meet them, or when they are setting out on their journey to inform them of the time of their expected arrival. There are sustainable options for encouraging passengers to walk parts of their journey for exercise and to encourage fitness, or to by-pass congested areas and consider other options. These options can be also combined with our recent work on autonomous systems, cf. (Spichkova & Simic, 2015).

It may be that people would like to be informed of other transport options, which exist near their current stop, or position. This may happen in the case of a traffic jam or accident, where the bus is caught in a particular street and cannot continue its journey for some time. There may be nearby optional routes on other forms of transport, which could avoid this congestion.

CONCLUSION

The overall contribution of this chapter has been the development of a formal model for an on-demand transport system. The system we have developed here can be seen as a part of a smart city, and is much more flexible and sustainable in comparison to the existing traditional one, enabling the reduction of greenhouse gas emission and noise pollution, as well as the saving of fuel costs for the company. Moreover, the presented solution allows having a more flexible/dense timetable and also possible longer routes as well as subjective shorter travelling times.

The key features of the defined on-demand transport system include the optionality of particular stops along the route and at either end of the route, with the specification based on the actual load modes of the transport system. However, the development of the on-demand transport system is more complicated, requiring a large number of additional aspects to be taken into account, and a DSS to be built to enable the driver to react promptly and effectively to the dynamic changes in the timetables. In this case, a proper formal model is required to build the system correctly.

Our model covers several aspects of sustainable transport and focuses on spatial planning and optimization including environmental issues in transport planning. The model incorporates three distinct views of operation:

1. The driver view: This provides a list of all the stops, optional and regular, where the bus is expected to travel at the current moment in time. The route plan is updated dynamically, but there is a DSS embedded in this view, so the driver will not be distracted while driving, but keep a regular update on whether some parts of the route need to be skipped or changed during that particular run. From the DSS, the driver always has the latest version of the timetable for his route and the information that is currently required to complete the run in the most efficient and safest way. To increase the usability, we have introduced two modes of this view: the general mode where the driver can overview the complete current route plan, where the simplified mode contains the information about 4 stops/stations only: the current stop, where the vehicle is/should be approaching, and three next consecutive stops. This allows the driver to focus on the current part of the route, where the DSS takes control over the further updates of the plan.
2. The passenger view: This describes an interface for passengers to provide real-time information about the transport system. It enables them to select a particular route and to request the system to book the bus to stop at any optional parts of the route.
3. The system view: this provides a map of the overall route where the bus is travelling. It has identified the regular stops and includes any optional stops, which have been selected by passengers for the requested time slot.

The model, various scenarios and optimizations have been evaluated as case studies based on the analysis of an actual bus route 445 in Melbourne area, Australia (Public Transport Victoria, 2014). The current integrated system described above exists as a prototype only, and requires incorporation and testing on a real time bus system. If fully implemented, across all modes of transport, this system could provide for a much more sustainable, flexible and user-friendly public transport system. If passengers knew such a system existed, they might be encouraged to travel, especially the optional routes, and this might stimulate potential passengers to become actual ones and to switch from private transport to public.

REFERENCES

Berger, A. (2002). *Embedded Systems Design: An introduction to processes, tools, and techniques.* Lawrence, KS: CMP Books.

Berry, M., & Hamilton, M. (2010). Changing urban spaces: Mobile phones on trains. *Mobilities*, *5*(1), 111–129. doi:10.1080/17450100903435078

Bhargava, H. K., Power, D. J., & Sun, D. (2007). Progress in web-based decision support technologies. *Decision Support Systems*, *43*(4), 1083–1095. doi:10.1016/j.dss.2005.07.002

Bonczek, R. H., Holsapple, C. W., & Whinston, A. B. (1981). *Foundations of decision support systems*. San Diego, CA: Academic Press.

Broy, M., Fox, J., Hölzl, F., Koss, D., Kuhrmann, M., & Meisinger, M. et al. (2008). Service-oriented modeling of CoCoME with Focus and AutoFocus. In A. Rausch, R. Reussner, R. Mirandola, & F. Plasil (Eds.), Lecture Notes in Computer Science: Vol. 5153. *The Common Component Modeling Example* (pp. 177–206). Berlin Heidelberg, Germany: Springer. doi:10.1007/978-3-540-85289-6_8

Constantine, L. L., & Lockwood, L. A. (1999). *Software for use: A practical guide to the models and methods of usage-centered design*. New York, NY: Addison-Wesley.

Constantine, L. L., & Lockwood, L. A. (2002). Usage-centered engineering for Web applications. *IEEE Software*, *19*(2), 42–50. doi:10.1109/52.991331

Covan, P. (2014, June 11). Auditor General slams Victorian bus GPS tender. *ITnews*. Retrieved from http://www.itnews.com.au/News/387872,auditor-general-slams-victorian-bus-gps-tender.aspx

Dhillon, B. (2004). *Engineering usability: fundamentals, applications, human factors, and human error*. Stevenson Ranch, CA: American Scientific Publishers.

Donath, U., Haufe, J., Blochwitz, T., & Neidhold, T. (2008). A new approach for modeling and verification of discrete control components within a Modelica Environment. *Proceedings of the 6th International Modelica Conference*. Retrieved from http://publica.fraunhofer.de/documents/N-70395.html

Easterbrook, S. M. (2010). Climate change: A software grand challenge. *Proceedings of the FSE/SDP Workshop on the Future of Software Engineering Research*. New York, NY: ACM. doi:10.1145/1882362.1882383

Ercoskun, O. Y. (Ed.). (2011). *Green and ecological technologies for urban planning: Creating smart cities*. Hershey, PA: IGI Global; doi:10.4018/978-1-61350-453-6

Feilkas, M., Hölzl, F., Pfaller, C., Rittmann, S., Schätz, B., Schwitzer, W., et al. (2011). A refined top-down methodology for the development of automotive software systems – the Keyless Entry System case study [Technical Report TUM-I1103]. *Technische Universität München*. Retrieved from http://mediatum.ub.tum.de/?id=1094230

Fick, G., & Sprague, R. H. (1980). *Decision support systems: issues and challenges*. Oxford, United Kingdom: Pergamon Press.

Fihn, J., & Finndahl, J. (2011). *A framework for how to make use of an automatic passenger counting system* [Masters Thesis]. RMIT University. Retrieved from http://www.essays.se/essay/de25b1166d/

Følstad, A., Law, E., & Hornbæk, K. (2012). Analysis in practical usability evaluation: A survey study. *Proceedings of the SIGCHI Conference on Human Factors in Computing Systems*. New York, NY: ACM. doi:10.1145/2207676.2208365

GreenSoft. (2009). *Green Software Engineering*. Retrieved from http://www.green-software-engineering.de/en/project.html

Hamilton, M., & Berry, M. (2010). Transmesh: Creating a mobile technological urban space. *Proceedings of the Mobile Technology in the Urban Field Symposium*. Melbourne, Australia: RMIT University. Design Research Institute.

Hamilton, M., Salim, F., Cheng, E., & Choy, S. L. (2011). Transafe: A crowdsourced mobile platform for crime and safety perception management. *ACM SIGCAS Computers and Society*, *41*(2), 32–37. doi:10.1145/2095272.2095275

Heumann, J. (2002). Use cases, usability requirements, and user interfaces [Tutorial Notes]. In *Proceedings of Object-Oriented Programming, Systems, Languages, and Applications (OOPSLA 2002)*. New York, NY: ACM.

Hilty, L. M., & Aebischer, B. (Eds.). (2015). *ICT innovations for sustainability. Advances in Intelligent Systems and Computing, 310*. Zurich, Switzerland: Springer; doi:10.1007/978-3-319-09228-7

Leveson, N. G., & Turner, C. S. (1993). An investigation of the Therac-25 accidents. *IEEE Computer*, *26*(7), 18–41. doi:10.1109/MC.1993.274940

Marakas, G. M. (2003). *Decision Support Systems* (2nd ed.). Upper Saddle River, NJ: Prentice-Hall.

Miller, E. (1987). The Therac-25 Experience. *Proceedings of Conference State Radiation Control Program Directors*. Washington, DC: Division of Engineering Safety, Office of Nuclear Regulatory Research, U.S. Nuclear Regulatory Commission.

Mitra, G., & Valente, P. (2007). The evolution of Web-based optimisation from ASP to e-Services. *Decision Support Systems*, *43*(4), 1096–1116. doi:10.1016/j.dss.2005.07.003

Nam, T., & Pardo, T. A. (2011). Conceptualizing smart city with dimensions of technology, people, and institutions. *Proceedings of the 12th Annual International Digital Government Research Conference: Digital Government Innovation in Challenging Times*. New York, NY: ACM. doi:10.1145/2037556.2037602

Power, D. J. (Ed.) (2013). *Engineering Effective Decision Support Technologies: New Models and Applications*. Hershey, PA: IGI Global. doi:10.4018/978-1-4666-4002-3

Public Transport Victoria. (2014). *Route timetables. Route 445 - Werribee Station - Hoppers Crossing Station via Werribee Plaza*. Retrieved from http://ptv.vic.gov.au/route/view/8266

Redmill, F., & Rajan, J. (1997). *Human factors in safety-critical systems*. Newton, MA: Butterworth-Heinemann.

Shackel, B., & Richardson, S. J. (1991). *Human factors for informatics usability*. Cambridge University Press.

Spichkova, M. (2011). Architecture: Requirements + Decomposition + Refinement. *In Softwaretechnik-Trends, 32(4)*. Retrieved from http://pi.informatik.uni-siegen.de/gi/stt/31_4/index.html

Spichkova, M. (2012). Human Factors of Formal Methods. *Proceedings of the IADIS Interfaces and Human Computer Interaction 2012 (IHCI 2012)*. Lisbon, Portugal: IADIS Press.

Spichkova, M. (2014). Design of formal languages and interfaces: "Formal" does not mean "unreadable". In K. Blashki & P. Isaias (Eds.), *Emerging Research and Trends in Interactivity and the Human-Computer Interface* (pp. 301–314). Hershey, PA: IGI Global; doi:10.4018/978-1-4666-4623-0.ch015

Spichkova, M., Hölzl, F., & Trachtenherz, D. (2012). Verified System Development with the AutoFocus Tool Chain. In *Proceedings of the 2nd Workshop on Formal Methods in the Development of Software*. Electronic Proceedings in Theoretical Computer Science 86. Retrieved from http://arxiv.org/abs/1207.1841

Spichkova, M., & Simic, M. (2015). Towards formal modelling of autonomous systems. In E. Damiani, R. J. Howlett, L. C. Jain, L. Gallo, & G. De Pietro (Eds.), *Intelligent Interactive Multimedia Systems and Services* (pp. 279–288). Springer International Publishing; doi:10.1007/978-3-319-19830-9

Spichkova, M., Zhu, X., & Mou, D. (2013). *Do we really need to write documentation for a system? CASE tool add-ons: generator+editor for a precise documentation*. Proceedings of the International Conference on Model-Driven Engineering and Software Development. Retrieved from http://arxiv.org/abs/1404.7265

Volpert, W. (1991). Work Design for Human Development. In C. Floyd et al. (Eds.), *Software Development and Reality Construction*. Berlin, Germany: Springer.

Wickens, C. D., & Hollands, J. G. (2000). *Engineering psychology and human performance*. Upper Saddle River, NJ: Prentice Hall.

World Commission on Environment and Development. (1987). *Our Common Future*. Oxford, United Kingdom: Oxford University Press.

KEY TERMS AND DEFINITIONS

(Public Transport) Vehicle: A bus, a tram, a train, a ferry, or a car used as a public transport proposes. In this chapter we restrict this definition to buses, trams and trains.

Irregular Connection: A part of the public transport route, which can be empty over some period of time during the hours of public transport operation.

Irregular Load Mode of a Route: A mode of operation when the vehicle could be empty, i.e. have idle drive, for part of the route.

Mode: A set of behavioural properties, which is valid in a particular context.

Normal Load Mode of a Route: A mode of operation when the vehicle is not empty for any part of the route.

On-Demand Request Time: a predefined time, which is specified by the public transport system as a deadline for sending requests (during the irregular load mode of the route operation) for attending the vehicle on an optional part of the route.

Optional Part of the Route: a part of the public transport route (more precisely, of its irregular connections, cf. above), which can be shortcut between two provisional shortcut stops (if it is in the middle of the route) or skipped from the provisional to the actual end of the route (if it's at the end of the route).

Provisional End of the Route: A stop/station, where the regular connection ends for this direction of the route.

Provisional Shortcut Stop of the Route: A stop/station, from which the public transport vehicle can shortcut an irregular stop when the public transport route is in irregular load mode.

Chapter 11
Smart Technologies for Sustainable Mobility

Ozge Yalciner Ercoskun
Gazi University, Turkey

ABSTRACT

The emergence of easy mapping tools, widespread cellular network, declining costs of smartphones and increasing internet use by public agencies provide new opportunities towards the betterment of public transport management. Applying decision support systems such as Geographic Information Systems (GIS) and Information and Communication Technologies (ICT) to public transportation tailors transit services to urban dwellers' needs, reducing eco-footprint while making them more accessible. This chapter aims to explore how GIS and ICT can be used for more eco-friendly public transport and improved sense of community. The chapter posits the positive impact of mobile phones and social media towards sustainable mobility. Multi-modal journey planners, mobile bus ticketing, demand responsive transit systems, eco-taxis, electric car and bicycle sharing are explained with examples from the world. The chapter concludes with a discussion of challenges and future options for using smartphones, social networking and the position of disadvantaged groups in sustainable mobility.

INTRODUCTION

Decision support systems (DSS) are knowledge-based information systems designed to capture, handle and analyze information that in the decision making processes of professionals. DSS refer to applications that are designed to support, not replace, decision making. The classifications of DSS are based on ultimate technologies that determine the characteristics of the decision-making:

- *Spatial DSS:* Spatial decision support is the computational assistance for making better decisions about problems that involve geographic components. Spatial decision support may be employed to facilitate the development, evaluation or selection of proper policies, plans, scenarios, interventions, or solution strategies of a settlement. The spatial DSS process solves large, disparate data sets that include numerical data, maps, images, texts, and other forms; multi-dimensional and heterogeneous geo-data or generates decision alternatives (The Redlands Institute website),

DOI: 10.4018/978-1-4666-8648-9.ch011

- *Communications-driven DSS* use network and communications technologies to facilitate collaboration and communication.

Data-driven DSS, document-driven DSS, knowledge-driven DSS and model-driven DSS are the other types of DSS (Bresfelean & Ghisoiu, 2010) which are not considered in this chapter.

Some of the attributes and benefits of spatial and communication-driven DSS can be given as follows (Bresfelean & Ghisoiu, 2010, Marques da Costa et al. 2013):

- Easy access (to real-time data),
- Rich communication (of results and new ideas in a practical form),
- Facilitated analysis (of data through use of computational intelligence),
- Time-saving and cost-saving strategic advantage,
- Greater consistency and accuracy,
- Smarter response (to changes and failures),
- Worker empowerment,
- Greater user satisfaction,
- Organizational enhancement,
- Increased innovation and productivity.

Smart technologies provide tools such as DSS for transport planning which enables data summarization and comparison, new location options, evaluation of measures and scenario development possibilities at local levels. GIS enhances spatial DSS; ICT tools and social networking utilizes communications-driven DSS. GIS as a part of spatial DSS is designed to support decision research process making for solving complex spatial problems. Spatial DSS integrates database management systems with analytical models, graphical display and tabular reporting capabilities, and the expert knowledge of decision makers (Densham, 1991). GIS can collect, store, and retrieve information based on its spatial location, identify locations within a targeted environment which meet queries, facilitate selecting analytical models and give alternatives on the selected environment (Crossland et al. 1995).

"From computers to data to information to communication to democracy" (Saco, 2002, p.xiii). New GIS and ICT platforms have evolved rapidly and transformed vastly during the last 30 years. Some of the concepts used to define the transformation in the Digital Age put forward by Castells (2009), can be considered beyond the scope of this chapter since they relate to media and business processes, and the cultural change in the globalized world. However, the popularization of the internet and the wireless systems have caused changes in today's communication platform. The internet was invented in 1969, but its commercialization expanded in the 1990s. Since then, users grew from 40 million in 1995 to 1.4 billion in 2008 (Castells, 2009). After the convergence of internet and wireless systems in 2000s, the meaning of 'being connected' has shifted (Pinzon, 2013); as Castells points "the key feature of wireless communication is not mobility but perpetual connectivity" (2009, p.69).

Recent developments in geo-information technologies on wireless platforms to link spatial and relational data provide new and smart insights into patterns of urban life, such as traffic flows within cities, urban mobility and proximity of social relations. Mobile phones allow for large scale monitoring of people's movements and physical proximities over time, giving the possibility of understanding cognitive relationships and social ties. Popular social network sites (SNS), like *Swarm, Facebook check-in* and *Quora*, extend this real-time and location-based geo-information across highly connected networks of users. More than 235 million mobile users in the U.S. and Europe consumed mobile media in 2011, a 30% increase over 2010. Globally, through rapid adoption of smart phone technologies and rise of geo-social networking, the number of mobile social media users is expected to reach 1.3 billion by 2016 (Suh et al., 2012).

GIS and ICT utilize mobile devices that have changed our lives. They shape how we interact with each other and how we organize friends and family as well as broader networks. They have become the 'digital fabric of our lives'. The opening question for the usual telephone conversation has changed from "Whom?" to "Where?" since today we know who we are talking to before answering the phone. Mobile communication coordinates our daily mobility activities, for example, who is going to do the shopping, or who is going to pick up the children that day. These are not planned but talked and decided in real-time. However, phones also limit our intimate space. There is always the possibility of someone entering into our private sphere uninvited. While keeping in touch with social networks has its advantages, constant contact might prevent us from interacting with our real life environment (Mobile Tech, 2013).

Social networking refers to a group of web-based applications that encourage users to interact with one another virtually. Many public transport agencies have begun to adopt these networking tools and their reasons for using these can be listed as (TCRP Synthesis 99, 2012):

- Real-time updates: Social media enable agencies to share real-time service information and recommendations with their users,
- Public information: Many public transport agencies use social media to provide the public with information about services, fares, and long-range planning projects,
- User connectivity: Transportation agencies take advantage of the interactive aspects of social media to connect with their user informally,
- Employee recognition: Social networking can be an effective tool for recognizing current workers and finding new employees,
- Entertainment: Agencies often use social media to entertain their users through songs, videos, campaigns and contests.

The cultural revolution of social media is a central pillar of the new Digital Age in the 21st century (Castells, 2009). Social media content can reveal urban dwellers' preferences regarding mobility decisions. The important role that information plays in transport decision making ranges from understanding preferences of routes and modes, to participatory scheme evaluation. Surveys are often used to capture information such as travel times, waiting times, vehicle occupancies and perceived service quality (Gal-Tzur et al., 2014).

GIS uses captured information and integrates this information into the urban transport planning process, which in turn facilitates stakeholder information, discussion and decision making. GIS can simulate different scenarios; integrate into cellular automata, location-allocation models which find optimal routes, multi-criteria decision analysis which considers multiple criteria in deriving different transport planning options (Ocalir et al., 2010). However, there are some constraints which can be highlighted as increase of working volume, lack of data, high cost of data, and the need of specialized experts etc. (Marques da Costa et al. 2013).

Distance played a major role in the formation and the structure of cities. Planners started to pay more attention towards technologies that affect the costs of traversing distances. The costs of distance include energy, time and money, while causing air and noise pollution (Cohen et al. 2002). GIS and ICT are the prominent technologies here; since both can reduce the costs associated with distance. The increase of mobile technologies and wireless connection overcomes time and distance limitations, which leads to increased positive communications and relationships.

The other dimension of urban transportation is the effect on city's ecological footprint. Urban areas are responsible for a significant portion of carbon dioxide emissions, with transport related activities taking the lead in emission within cities. Applying GIS and ICT to public transportation and traffic management tailors transit services to urban dwellers' needs, significantly reducing settlements' ecological footprint while at the same time making them more accessible. This chapter aims to explore how the application of GIS and ICT can promote innovative practices on public transport for reducing carbon emissions while fostering sense of community. The chapter's scope extends from the impact of mobile phones and social media on sustainable mobility to environmental dimension to address innovative, sustainable examples for urban transport planning. Information was gathered through a literature review, and case examples. Because the field of social media is still evolving, the literature review featured information from online sources, includes blog posts, websites, conference presentations, electronic journals and publications covering technology and urban public transport management.

This chapter proposes the need for sustainable urban mobility to analyze the potential of mobile technologies and social media; and considers the possibilities of GIS and ICT in enriching discourses around cities of tomorrow. The chapter is motivated by questions that go beyond the emergence of new mobile technologies and social media use in public transport, such as: What are the types of DSS? What profound changes can new communication mechanisms support in urban mobility network? What are the different dimensions of sustainable mobility at a local scale? How ultimate mobile technologies affect communities and contribute to eco-friendly use public of transport? Why is social media important? How is social networking being used in transport? Where can the best practices of GIS and ICT in the creation of sustainable mobility be found from around the world? What are the motivations behind using the internet and online maps for collective organization? How can ICT help disadvantaged people with accessing public transport services?

This chapter provides an overview of ultimate GIS and ICT used in urban public transport. It begins with DSS types and focus on the impact of mobile phones and social media on urban services which enhance sustainable mobility. The characteristics of social media networking and their use in transport agencies are reviewed. Multi-modal journey planners, traffic maps, mobile bus ticketing systems, mobile applications of alternative public bus services, demand responsive transit systems, commuter shuttles, eco-taxis, electric car sharing and bicycle sharing, sustainable urban mobility cases for disadvantaged groups are explained as best practices from Europe, North America and Turkey. The chapter concludes with a discussion of current challenges and future options for using smartphones, social networking and disadvantaged groups in sustainable mobility.

ULTIMATE TECHNOLOGIES FOR ECO-FRIENDLY USE OF PUBLIC TRANSPORT

Mobility is a hot topic in the transportation sector especially together in combination with the term sustainability. Mobility is all about getting from A to B as fast as possible and in a convenient way. For sustainable mobility, the resources needed to create and fuel all modes of transport, and absorb the waste from their creation and use should save the planet. According to Banister (2008), the principles of sustainable mobility can be listed as:

- Making the best use of technology in transport modes, ICT and in the transport system itself, and in giving industry priorities on vehicles with alternative fuels. For the general public, there would also need to be behavioral change, with increased use of public transport as well as with ecological driving.
- Regulation and pricing should be reflected in the actual costs of travel through higher fuel prices or road user charging. Demand management reduces congestion and improves environmental quality, decreases the numbers of trips, trip distance, change modal shares and promotes public transport.
- Physical land-use planning and development patterns are used to support shorter travel distances. Improved levels of proximity would contribute to trip reduction and modal split changes.
- Clearly targeted personal information, awareness raising, demonstration, acceptability is also essential for sustainable mobility.

The goal of sustainable mobility is addressed in connected sustainable cities. Connected sustainable cities employ networked intelligence for the efficient use of the limited resources such as energy and water, together with the effective management of waste that a city produces, such as carbon emissions to the atmosphere. ICT can be employed to help reduce carbon emissions. Broadband-based applications and services can create links that make buildings and infrastructure, mobility, work, and energy functions all more efficient for the environment. A city that adopts ICT solutions is referred to as a Smart City. Smart Cities have the potential to become the cities of tomorrow. For developing a Smart City concept, the city has to use numerous technologies, including short range wireless technologies, NFC (Near Field Communication) technology and its close range transmission ability (Ronay & Egger, 2013). Connectivity can support new ways of planning, working, and living that make social connections tight and lead to collective sustainable behavior. ICT allows for sustainable urban policies across neighborhoods, and the entire social fabric of a city (Mitchell & Casalegno, 2008).

In many cities, public transportation systems are not effective, they are confusing and unpleasant to use. Therefore, urban dwellers often choose their private automobiles for transportation. However, GIS and ICT can greatly reduce this barrier towards more sustainable mobility. There are three levels to applying ICT to create enhanced public transportation systems. First, the system must cover the whole city sufficiently. Second, GPS must be integrated to public vehicles so that trip planning and real time personal scheduling can be made possible. Finally, users should be able to easily find user-friendly software in transit systems. This means not only guidance and scheduling but includes online social networking, as a resource to help organize trips (Mitchell & Casalegno, 2008). For instance, The MIT Mobile Experience Lab is interested in improving social and economic factors within sustainable public transportation. By focusing on the social characteristics of ICT (such as empathy and customization, as well as interaction and social networking), the Lab aims to make these ultimate solutions more attractive to urban dwellers. The most elegant solution is useless if nobody uses it (Casalegno & Chiu, 2008).

The user-friendly solutions of ICT have changed our ways of exchanging information, interactions among city people, movements, and use of urban space. These interactions generate mobility in the cellular network, based on location and route. These are useful for observing the movement of people regularly, in large scales and with reduced costs. The handover data was collected from cellular towers in Lisbon, Portugal to understand the city dynamics through GIS visualization and statistical analysis. The visualizations provided a qualitative explanation of how the movement of calls is useful in the flow of people in the city (Demissie et al., 2013). A significant obstacle to personalizing the public transport experience has been the lack of data about individual traveler preferences. However, the widespread adoption of smartphones offers a potential channel to provide this missing data, as well the ability to observe and collect passenger behavior which was until very recently inaccessible. Following the trend of utilizing people as sensors, these powerful devices can be used to sense how people travel (when, from where, to where, with what mode, etc.) as well as allow them to communicate with route planning services dynamically. The important challenges of a mobile travel information system are multi-modal routing, user preferences, alerts, and responds to the user's query. Routing systems allow travelers to choose a most economic, most environment friendly or a most scenic route (Lathia et al., 2012). The recent ICT offers the opportunity of a new route for dynamic information flow that captures the views, needs and experiences of the travelling public through social media text posts (Gal-Tzur et al., 2014).

Social media is defined by Dourish & Satchell (2011, p.21) as the "digital systems through which people engage in social relations of many sorts social dynamics are supported by technologies such as social networking sites (e.g. Facebook, LinkedIn, Bebo), social virtual realities (e.g. Second Life, Friendster), blogging and micro-blogging (e.g. Twitter), always-on mobile communications (e.g. Smart-phones) and systems for using media and user-generated content (e.g. from SMSs to posts on YouTube or Wikipedia)". Top social media platforms are Facebook, Youtube, and Wikipedia (Pinzon, 2013). Castells refers to social media as mass self-communication and defines it as: "Self-generated in content, self-directed in emission, and self-selected in reception by many who communicate with many. This is a new communication realm, whose backbone is made of computer networks, whose language is digital, and whose senders are globally distributed and globally interactive" (Castells, 2009, p.70).

Why is the social media so important? Social media use is exploding day by day (over 800 million Facebook users in the world). Mobile devices bring connectivity everywhere and all the time. Social media drives users to various other virtual and real products. Many applications are being developed to provide information via social networks (e.g. *Facebook* pages) about companies, products, etc. (CIVITAS forum, Nash, 2011). The social networks can be existing, real-world networks such as families or coworkers, or virtual networks, such as online communities. Both types of social networks can be used to achieve a common goal of sustainable mobility and encouraging behavior change in users. Competition between virtual and real-world social networks can encourage increased rates of motivation for participation (Casalegno & Chiu, 2008).

Here is an example, a written scenario on using social networking. Without having a schedule; Lucy meets her friend at a cafe, The online social networking site *Swarm* makes it possible to let friends know where she is, locate where friends and friends of friends are within a 1 km radius of her location, and find venue and send messages to friends. *Loopt* alerts users when their friends are nearby and can show what they are doing by interactive maps on their phones. The system also allows users to share a variety of information with friends in their address book or through online social networks (Mitchell & Casalegno, 2008).

How is social networking being used in transport? Many agencies provide smart traveler services which bring access to real time information, internet applications, media reports and social networking sites. This affects transport system operations. Public transport agencies can gather and share information and make general marketing. Real time information in crisis or disruptions can be obtained; surveys can be made for feedbacks. Complaints can be taken, problems can be reported, and inputs can be provided for planning. Positive feelings can be created, fan groups can be built, and photo sharing activities can be made. Social networks can be used for controlling traffic signals, organizing carpools and any other peer to peer services (CIVITAS forum, Nash, 2011).

For instance, Cardiff Bus, a local bus company in Wales, UK, has embraced social media to improve customer satisfaction and to promote its local bus services. It carries about 100,000 passengers every day on 3,000 journeys. Since 2009, Cardiff Bus has been using both Twitter and Facebook daily to communicate with the public, primarily for receiving customer feedback with the aim of improving services. Twitter is used to communicate service changes, delays and cancellations to users on a real time basis. The timing of the launch in a period of heavy snow allowed the bus company to advise users of disruptions during this period (Eltis 1).

Case Studies

Multi-modal travel planning is a key element of intelligent transport systems in smart cities. It provides the traveler with comprehensive door-to-door information allowing for well-informed travel decisions. It seamlessly integrates information for different modes, based on a strong backbone of local public transport. In Planning and Design for Sustainable Urban Mobility Report (UN Habitat, 2013), examples of cities with multi-modal infrastructure of ICT and payment elements that promote multi-modal connectivity are given in Table 1. Many cities in the world serve multi-modal planner systems which can provide timetable, routing and other real time travel information about public transport services (bus, train, tram, metro) and about transportation networks (roads, footpaths, cycle routes). These systems integrate a number of different types of data about the planned services for the public transport system, including:

- Information to identify and navigate stops, stations and other transport modes,
- Information about origins and destinations that users want to travel to and their relation to stops and stations, for example points of interest and their access points,
- Information about the topography of a region and its relation to transport systems,
- Information about the network, the routes and lines,

Table 1. Different cities with multi-modal planner systems (UN Habitat, 2013)

City	Multi-mode	ICT	Integrated Payment
London	Metro, bus, bike-sharing, taxis, light rail, trams	iBus, Web and Mobile information systems	Oyster smart card
Paris	Metro, tram, bus, bike-sharing	IMAGE project (real time traffic info.)	Navigo pass
Singapore	Metro, light rail, bus, taxis	How2go, Web and Mobile information systems	EZ-Link, NETS FlashPay
Hong Kong	Metro, bus, minibus	Next Train mobile app, Passenger info. sys.	Octopus smart card
New York City	Metro, BRT, Local and express bus	MTA Bus Time	Metrocard
Mexico City	Metro, BRT, bike-sharing	Web based passenger info.sys.	Metrobus card
Guanzhou	Metro, BRT, bike-sharing	Web based and station based passenger info.sys.	Yang Cheng Tong
Budapest	All public transport modes	Centrally coordinated ticketing	Travel card 24h
Chicago	Bus metro, bike-sharing, car-sharing	Bus and Train Tracker (real time bus and train information)	Chicago Card

- Schedules for different modes of transport,
- Schematic maps of the transport network,
- Maps of the city and of the area around stops,
- Information about facilities at stops and stations,
- Information about fares, tariff zones and fare products,
- Information about the carbon footprint costs of usage of different modes etc.

Moreover, Scotty is a Belgian multimodal journey planner which offers real-time information about all modes of transport, launched in 2012. The aim of Scotty, a new mobility information website, is to provide real-time information about the current mobility situation and possible routes. It makes intelligent use of all information provided by its partner mobility providers but also uses real traffic data generated by cameras and traffic observation devices. The info is completed with GPS and cell phone data into reliable information.

The Scotty-platform offers routes for any transport mode (walking, cycling, bus/tram/subway, train and car) complete with travel times, exact location and extra information about the surroundings. Scotty also offers a traffic map which shows traffic flows, accidents or traffic jams, car sharing locations, electric charging stations, bicycle sharing stations etc. Users can customize their page and login with an email or a Facebook, Twitter or LinkedIn account. Every feature is also accessible via IPhone, IPad or Android-devices. The biggest advantage of Scotty is giving reliable information, directly from public transport agency partners such as the railways, bus operators, public transport providers, bike-sharing schemes, car-sharing and parking companies (Eltis 2).

The MIT Mobile Experience Lab developed a personalized public bus assistant for Amsterdam. This prototype includes updates on bus reservations, traffic reports, scheduled construction, city events, weather conditions, and other factors which affect a commuter in Amsterdam. The recommendations are made by the personal travel assistant in personal preferences (commute duration, route, and departure time) and real-time commute data (road conditions, weather, and current capacity). The personal travel assistant can learn a user's preferences, and automatically integrate them into its recommendations. Once the user selects a commuter route and confirms her reservation, the system uses GPS on the bus and the user's mobile phone to coordinate boarding. Strong optimization software organizes both the travel recommendation system and the routing system, including destinations and rate of travel, service pickup and delivery points, overall transit system status, and environmental factors such as weather or pollution (Casalegno & Chiu, 2008).

Again, The MIT Mobile Experience Lab developed another application for San Francisco named "The Passport". This application runs in smart phones and has different modules such as personalized route planning, the green gauge, wearable urbanism, location-based social networking, personalized interaction with city services, and dynamic, intelligent traffic management. The Passport application helps with making public transit a more attractive choice. The Passport application's main function is to connect the individual with the overall transit system. It provides real-time GPS information about bus arrival times and system delays. It can suggest several options, calculating both the time it will take to walk to the bus stop and when the bus will arrive. The user can then make a decision to catch the bus. These calculations consider factors such as the overall system status, performance along the bus line, weather, traffic, city events, and planned works (Casalegno & Chiu, 2008). The Passport can integrate personal information into its calculations, such as health conditions. Taking these conditions into consideration, the system can recommend maximizing or minimizing walking, for example, to a person with diabetes or disabilities.

The Green Gauge in the Passport acts as a kind of city health monitor and is located both inside and on the sides of buses. As a bus drives through the different districts of the city, it collects electricity usage and real-time environmental data about each district. The Green Gauge provides access to users about their cities and neighborhood's environmental status and raises awareness (Casalegno & Chiu, 2008). The Smart Trip Advisor in this application can generate a complete overview of a journey targeted towards a specific user. The advisor alerts the user to multi-modal options which may speed their journey or help reduce their environmental impact. For example, if there is traffic congestion, the system may recommend that a user stop at a public teleconference facility. These recommendations have an impact on a decision-maker.

WISETRIP is another innovative journey planner which provides multimodal trip information. Personalized services delivery is achieved; users select their trip, build personal schedule of notifications, configure alerts that assist trip execution and enable continuous validation of trip data. Enhanced WISETRIP considers real-time events, mobile and social media extensions, geographical and modal expansion, environmental criteria, personal preferences and trip strategies. Detailed maps for walking segments, relative distance in meters, information on stairs, lifts, escalators etc., is required particularly for disabled people and those carrying luggage. The inclusion of the CO_2 footprint metric within the presentation, the criteria selection and filtering process is provided (Spitadakis & Fostieri, 2012).

These ultimate multi-modal planners help passengers make informed travel decisions based on factors such as price, environmental impact, and time-to-destination. The user can match her priorities (time, money, CO_2) with the options listed by the system, or make changes. Systems estimate incorporate data such as scheduled events; street closures and construction; accidents; environmental information such as weather, pollution, or seasonal changes; and current traffic volume which facilitate decision-making (Casalegno & Chiu, 2008).

Smart technologies for sustainable mobility can be exampled in mobile ticketing systems. UK bus operator, Go North East is pioneer at the development of ticketing technologies. In 2010, they launched a smartcard scheme called 'the key'. Extending the 'key' brand to a mobile smart phone was the next project. The key mobile app, is free to download, provides up-to-date travel information fed by the company's social media activity and helps users to find their nearest bus stop and the departure time of the next bus using the GPRS technology in their Smartphone. Users can see the route and stops in a Google map and can talk to customer services via Facebook or a web chat, as well as download timetable information. The key mobile ticketing app was available with one day and seven day tickets 'Get Around' ticket range for young people and students. Tickets purchased using the key mobile app are cheaper when compared with on-bus ticket. Since key mobile app was launched it had been downloaded more than 11,000 times and is growing by around 2,000 downloads a month. The key mobile app's simple design and practical use seems to be popular in the UK (Eltis 3).

The UK bus services are pioneers at mobile ticketing systems. Another company, Arriva in the UK introduced a mobile phone ticket purchasing system on all its bus networks across the UK in 2009. M-ticketing is an application which allows customers to buy, receive and use bus tickets via mobile phone (Figure 1). M-ticketing is a free application which can be downloaded directly to mobile phones, via the internet or SMS. It can be used on any smartphones. Passengers register either online or by sending a text message. The ticket is displayed on the screen of the mobile phone and must be shown to the driver to be

Figure 1. Arriva Bus App (https://play.google.com/store/apps/details?id=de.hafas.android.arriva)

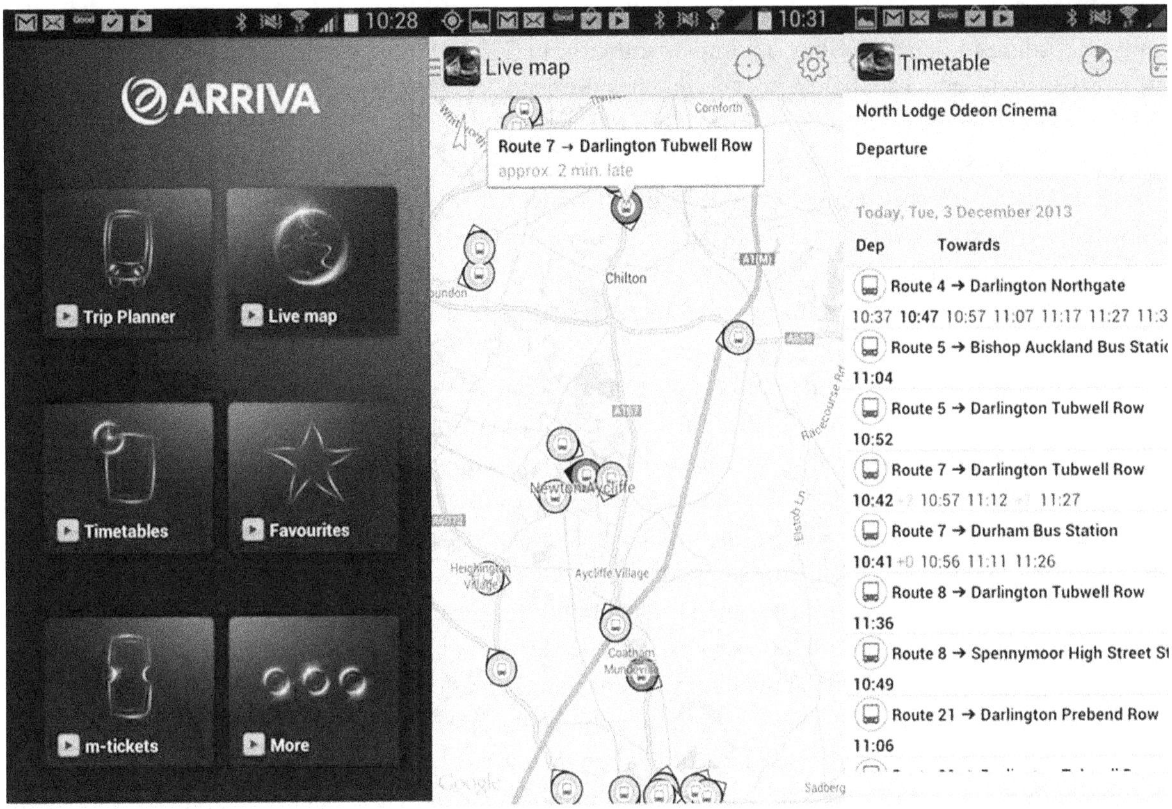

activated. The original aim was to have 20,000 customers sign up to m-ticketing by the end of the first year, but the figure was actually more than 30,000 after 12 months, and reached 40,000 after 18 months by 2012. 30% of visits to the website were made on a mobile device; in the UK only 20% of website visits are made from a mobile device. The number of journeys made with m-tickets has increased to 2.6 million. The company won the 2011 UK Bus Awards for Innovation (Eltis 4).

In Warsaw, Poland, it is possible to purchase tickets via mobile phone including 20, 40, 60-minute tickets and one day and 3-day passes, using the following applications: mPay, SkyCash, moBILET. The validated ticket is stored in the phone's memory and can be used offline. The system is easy because one can buy a ticket at any place and at any time. It saves time; there is no need to look for a kiosk or a ticket machine. One does not need cash or coins. An electronic ticket never gets wet or worn out and one can buy tickets in all cities covered by the system. In 2012, close to 5.5 million were encoded on city cards (Eltis 5).

Another form of mass transit is demand responsive systems which is defined as "an advanced, user-oriented form of public transport characterized by flexible routing and scheduling of small vehicles operating in shared-ride mode between pick-up and drop-off locations according to passengers needs" (Casalegno & Chiu, 2008, p.2). The flexible routing systems are energy-efficient and able to serve a wider geographic area, including low-density suburban neighborhoods (Spinak, Chiu & Casalegno, 2008).

An example of these demand responsive transit systems, Drin Bus is a flexible bus service that connects the hilly, low-density areas of Genoa through a many to many (pickup/drop-off points) operational model. Riders can reserve the bus up to 30 minutes prior to their desired departure time via telephone. GPS-GIS integrated bus-monitoring software allows a central call center to manage the bus fleet dynamically according to demand. Since its implementation in 2002, the percentage of residents in the Drin Bus served areas who use their cars everyday has dropped 4%. Drin Bus saves €34,500 per year due to decreased environmental impact (Spinak, Chiu & Casalegno, 2008).

In San Francisco, a pilot project was begun in 2007 to encourage people to use the city's Muni buses. Using advanced Internet technology, the hybrid Connected Bus with onboard touch screens provide information on bus arrivals and other Muni information, wireless internet access from laptop computers or mobile devices, and external displays about the environmental benefits of the vehicle. The objective is safe and more reliable rides for buses to spend less time in traffic and have lower carbon emissions (Mitchell & Casalegno, 2008).

The Google employee shuttle collects passengers from many pick-up points to bring them to one end-point (Spinak & Casalegno, 2012). The biodiesel-fueled Google shuttle buses make 142 trips each day to transport 1200 employees from and to 40 pick-up locations each business day. Google tracks and records real-time information about its bus fleet, employees can get updates on bus delays and schedules via their computers or smartphones. The buses, which are not marked as Google shuttles, are capable of holding up to 37 passengers, and are equipped with wireless internet, and conference tables (Casalegno & Chiu, 2008). The shuttle has been so popular that employees have started moving to new neighborhoods to better take advantage of the shuttle service. Shuttle service is not the only new corporate trend affecting housing choices; for example, the company Facebook gives employees who live within a km of the company campus in Palo Alto a $600/month subsidy (Helft, 2007). Commuter shuttles have an effect on communities and public space. Users have noticed an increased sense of community from residents walking to shuttle stops. Residential shifts make possible group telecommuting options (Spinak et al., 2008).

Taxi, is an indispensable component of urban transport. Compared to other modes of transport, taxi is a semi-public transport system which has a relative advantage with the comfort and convenience that it provides 24 hours a day to its users (Ocalir et al., 2010). Eco-taxis are spreading across Europe which is a very environment friendly application. Eco-taxi in Switzerland (esp. in Locarno, Bellinzona etc.) is a good example; it is dialed for free from anywhere and the passenger can directly contact the driver. This choice of taxi is ecological and cheap. In 2003, eco-taxi saved 40% of fuel and emissions compared to a traditional taxi (http://www.ecotaxi.ch/).

Sustainable mobility research focuses on understanding and developing new models and services that integrate smart technologies, including GIS, ICT and alternative fuels, to develop new transportation choices and services. Researches in this area include car-sharing and bike-sharing.

New transport alternatives can be classified on the basis of different aspects of daily mobility: some of them try to improve the average level of car occupancy by car-sharing; others discourage the use of private cars or try to make public transportation competitive such as The Dial-a-Ride or Park-and-Ride. Car-sharing is a collective transportation system based on a shared use of private cars with the objective to reduce the number of cars on the road. Car-sharing is most effective for daily commuters. It is necessary to collect some information about service users and to make them available to the service planner. Car-sharing can be operated in two main ways: Daily Car Sharing Problem (DCSP) or Long-term Car Sharing Problem (LCSP). In the case of DCSP, each day a number of users declare their availability for picking up and later bringing back colleagues on that particular day (Calvo et al., 2004). The problem is to assign clients to servers and to identify the routes to be driven by the servers in order to minimize service costs. This can be based on a credits mechanism to motivate the cooperation among users. Users can spend the accumulated credits on parking facilities (Ferreira et al., 2009). In the case of LCSP, each user is available both as a server and as a client and the objective is to define user pools where each user will pick up the remaining pool members, in turn, on different days. The objective is to maximize pool sizes and minimize the total distance travelled by all users when acting as server. ICT facilitates updating the database through remote connections for any user, while the GIS is used to build the web pages which display the optimal route (Calvo et al., 2004).

This chapter also highlights case studies of best practices in car and bike sharing programs. New and smart car and bike sharing programs are being developed in the North America and Europe. These systems offer transportation flexibility and can be linked with public transportation systems.

Autolib' is a system of electric car-sharing for the city of Paris. A fleet of 3000 cars are deployed to 1000 rental stations throughout Paris (Figure 2). A central computer system allows users to specify a return location when they pick up their car. Autolib' members have the option either to purchase a monthly pass or to pay as they drive. The monthly pass runs approximately €250 per month. Alternatively, for people interested in using Autolib' only occasionally, this option allows Autolib' members to pay €10 for a car by the hour (Grünig & Marcelino, 2009). Paris' mayor is part of a larger initiative to reduce air pollution in Paris by 32% before 2015.

Figure 2. Car sharing system modules (Ferreira et al., 2009)

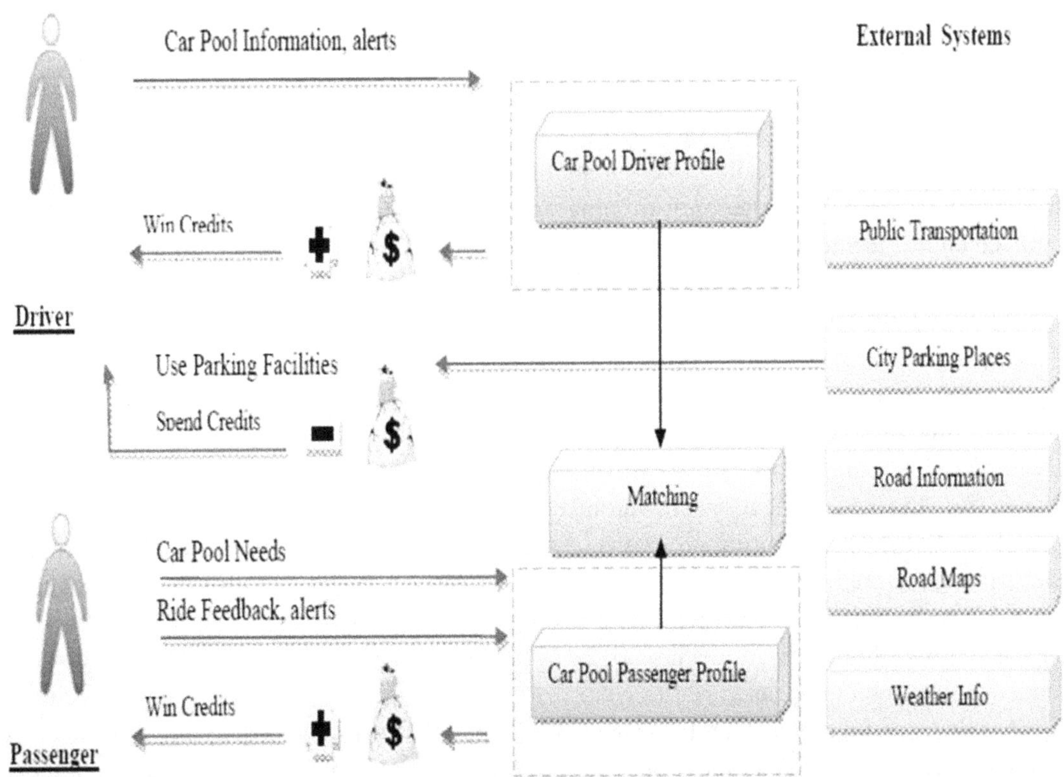

Figure 3. Autolib' App (https://play.google.com/store/apps/details?id=com.polyconseil.autolib)

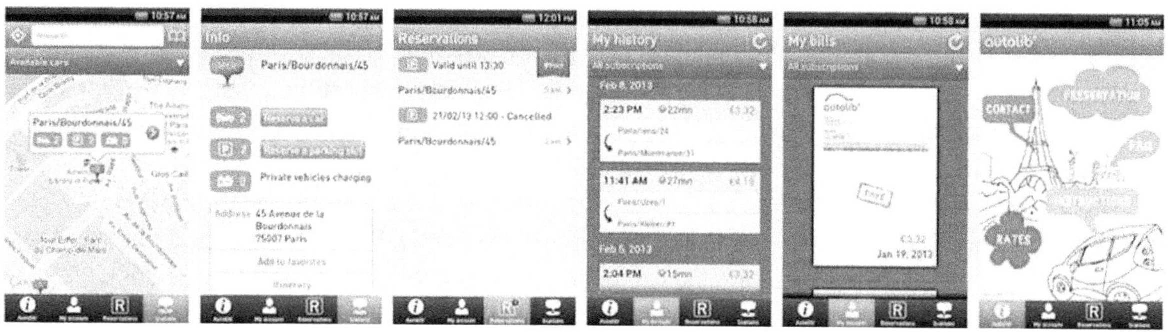

Here are some different car-sharing best practices:

- City Car Share, in San Francisco, offers lower night rates, in order to encourage members to drive at non-peak traffic times,
- The North American company Zipcar locates cars not only in dense urban areas, but also places where shoppers need a car (e.g. Ikea stores),
- Toronto's AutoShare partners already use an electric car fleet which utilizes green energy,
- The London City Car Club uses London's Oyster Card (Table 1) as its member card for accessing its cars,
- Switzerland's Mobility CarSharing is a cooperative; members can buy a subscription to the service for about the cost of four years of annual membership fees (Spinak, Chiu & Casalegno, 2008).
- Car2go users can simply find one of the specially-painted cars on the streets of Ulm, Germany or they can locate the next available car2go on the internet (Grünig & Marcelino, 2009).
- Sunfleet's car fleet in Gothenburg running on ethanol, biogas, biodiesel and/or having new advanced engine technologies such as battery electrical cars or hybrids has more than 11,000 members and over 7,000 trips per month. Sunfleet is used by municipalities, municipal housing companies, big companies, private persons and students (Eltis 6).

Some cars in sharing programs include a hardware unit which can provide numerous driving behavior alerts in real time. These are configurable to the operator's specifications such as speeding and sudden acceleration. The unit provides real time information on car status and location. It can also receive commands from the management system automatically. These can be used as a remote control. In summary, studies of several car-sharing programs have found the following behavior changes among their members: increased public transportation use, increased use of bicycles and walking for transportation purposes. Members tend to change their minds about purchasing a new car or sell their current car after joining car sharing (Spinak & Casalegno, 2012). Members drive less per year than their cities' average and reduce their carbon footprint.

Paris has a successful municipal bike-sharing program, Vélib' started in 2007 with over 18,000 rental bikes and over 1,000 of rental stations. Bicycles can be rented on a one-way basis; they do not need to be returned to the same rental station. A user can pay an annual membership fee ($35), but one day and one week passes are also available. For each rental, the first 30 minutes are free of charge, and fees are time dependent after that. Daily use averages 60,000 to 95,000 trips. Moreover, Vélib has contributed to changes in commuting behavior in Paris. P'tit Vélib', for children in Paris promotes soft mobility by allowing children to learn how to cycle safely. Through this new service, P'tit Vélib' allows families to rent 4 different types of bikes for children between the ages of 2 and 8 in different green or pedestrian areas in Paris. Because good habits develop early, Parisians wish to familiarize their children in the use of more eco-friendly transport means at the youngest possible age (http://blog.velib.paris.fr/en/ptit-velib/).

One of the popular European cities, Madrid launched e-bike sharing scheme in Spain. This system encourages people of all ages and abilities to get involved in cycling via electric bikes. Madrid's Bici-MAD scheme provides 123 self-service stands with 3126 docks for 1560 electric bicycles around the city centre. An annual membership to BiciMAD costs €25, or €15 for those already subscribing to the Community of Madrid's intermodal travelcard. It can be fully monitored and operated on the website of BiciMAD including maps and stations (http://www.bicimad.com).

The sustainable mobility is a strong factor in a city's achievement. The Slovenian capital city of Ljubljana has won the European Green Capital Award 2016. The city's popular bike-sharing scheme, BicikleLJ, has been used for 1.6 million trips since it was launched in 2011. In 2012, Ljubljana adopted goals that would see residents using public transport and non-motorised modes more than private cars by 2020 (EU Green Capital website).

After summarizing the best practices on sustainable public transport modes and their link to GIS and ICT, some examples of mobile applications and social networking can be given from the world.

66% of government agencies in North America used some form of social networking in 2009. Most government agencies report they use social media to stay engaged with citizens, for public information, and for employee learning and development. Transportation officials often take advantage of the real-time structure of Twitter and Facebook to post service updates and announcements. Based on statistics compiled for 19 social networking sites, the average user is 37 years old. Older users can adopt social media at a faster rate. The survey shows that individuals with more education and higher household incomes are more likely to use online government services than others (TCRP Synthesis 99, 2012).

Here are some interesting applications on internet about public transport (CIVITAS, Nash, 2011):

Cyclopath (www.cyclopath.org) includes an editable bicycling map where users can share notes about roads, put tags about special locations, and fix map problems,

Next Stop Design – (http://www.nextstopdesign.com/welcome) US application for designing a better bus stop, includes educational information and tools to help users develop their ideas,

FixMyTransport (www.fixmytransport.com) online community (crowd sourced) information on public transport problems,

GreenCityStreets (www.greencitystreets.com) crowd sourced information on public transport planning (rather than problems) includes educational component.

According to the Vodafone report, with the introduction of the 3G networks in 2009, mobile broadband and smartphones are rapidly replacing older generation devices in the Turkish market. There are already 4.8 million Smartphone users (2012) and the number of Facebook users stand at 32,408,540, the 7th largest usage base per country in the world (Mobile Tech, 2013). The largest cities such as Istanbul, Ankara and Izmir started to develop mobile applications of public transportation trip planner for smartphones.

Izmir public transport mobile application serves updated information about bus, ferry, metro, light rail and airport conditions. Real time information about buses and bus stops can be checked on the map and the route can be tracked online. Every vehicle has a GPS device attached that allow users to follow the route to their destinations or and check the remaining stops and measure the distances involved. Kentkart is the keycard for the public transport system where users can check the fare or choose the nearest kiosk where they can buy a Kentkart. The system also supports offline use. The Municipality website offers a multi-modal planner which is currently in its testing phase (Izmir Metropolitan Municipality website).

Bus and Metro Lines of the General Directorate of EGO in Ankara Metropolitan Municipality introduced its mobile application in 2013 for the purpose of providing real time bus information to smartphone users. The system displays bus features, speed, license number and the current estimated arrival time at the desired bus stop (EGO website). Using GIS, users can find the shortest path to their destinations. Users can select the best route or the route of least impedance between two or more points, the system takes connectivity and travel restrictions such as one-way streets and rush-hour traffic into account when displaying the best route. Traffic can be monitored through instant traffic cameras since The Municipality mobile application provides traffic density map, road information, construction studies and live camera demos. After launching the application in the beginning of 2013, the number of users increased to 266,000 in Ankara (Ankara Metropolitan Municipality Catalogue, 2013).

Istanbul is famous for its traffic jams. A 'Congestion Index' prepared by a navigation company; Tom-Tom, stated that Istanbul has the worst traffic congestions in Europe, in 2013 (TomTom 2013 Report). City authorities provide applications for mobile telephone users to help deal with this issue. IETT, the municipal body responsible for public transportation in Istanbul, provides live tracking on their buses. Their program can detect position data and suggest alternative bus routes. Meanwhile, Istanbul Metropolitan Municipality provides live traffic data through their mobile application. Users can watch live footage from cameras positioned along highways and other major routes. However, Stareveren (2014) stated that the trip planner of Istanbul is often incomplete, inaccurate or has no data. So the reliability is the important factor for public transportation system users.

DISCUSSION

The DSS provides a powerful tool for monitoring the transport service as well as for supporting the decision processes used in the provision of this service. The DSS offers its users both information analysis and scenario testing capabilities, so that they can study the effects of their decisions before implementation. The DSS uses state of the art technology such as GIS and ICT, it consist of a data warehousing system for the integrated management of a wide variety of data sources and online analytical processing tools for the efficient "what-if" scenario testing. Perhaps the most important function of the DSS is the monitoring and planning of the transport service supplied. To achieve this, the appropriate measures, analyses and categories through which the transport service should be defined. Having defined measures, the user can then query and get specific answers to questions spatially with GIS and real-time with ICT and GPS (Matzoros, 2002).

Smartphones use geo-location technology to enable mobile network operators to connect with users at the right time and in the right place. Integrating location-based social networking into the transit system enables the system to send alert to riders on nearby buses and their friends throughout the wider transportation network. This information provides a different sociality to public transportation. Passengers waiting at a bus stop can search a ride with friends on arriving buses, providing other criteria for riders to use when deciding which bus to catch. Social-networking information can also be used as a platform for urban gaming and as a platform for making place recommendations. For example, as your bus passes a cafe, you can take a copy of a critique written by a friend (Casalegno, Chiu, 2008).

However, there is a constant reliance on smartphones which have a negative impact on people's skills. The use of texting and Facebook or Twitter and other sites as a form of communication erode people's ability to write sentences that communicate real meaning and inhibit the art of dialogue. Smartphones limit our ability to communicate through face-to-face dialogue and reduce communication within families (Medlock, 2013). Smartphone addiction is one of the important research problems in this digital age. On the other hand, Castells (2009) agrees that this is a key social change brought about by the internet. Talking about Facebook, he maintains that this tool "does not invent social networking but it certainly facilitates and expands it" (2011, p.165). Without considering utilities like Facebook as communities in-themselves, Miller sees social media working as a possible counter-balance to community. For example, allowing a space for shy people to participate in group discussions without face-to-face interventions (Pinzon, 2013).

Smartphones enable people to connect to each other directly or via social network sites. These sites provide a platform to get to know each other and to interact. That means that anyone can easily disseminate texts, pictures or videos from their mobile phones. This connectedness has played an important role in the recent Arab Spring movement in North Africa and the Middle East (EPOMM, 2013) or Gezi events in Istanbul, Turkey. This is what makes social media an essential tool in the way social movements build their capacity for action in order to achieve their goals (Pinzon, 2013).

Despite the advantages to using social networking, a series of barriers and concerns are identified, including the following (TCRP Synthesis 99, 2012):

- Resource requirements: Although setting up social media sites is generally free, web pages require ongoing maintenance and monitoring. Staff availability is the greatest barrier to adopting social media.
- Managing employee access: Public and private sector organizations take actions to address employee use of social media.
- Accessibility: Internet accessibility for people with disabilities has improved but graphics, videos, and user-generated content has created some accessibility challenges.
- Security: ICT security experts concerned that social media can create cyber threats, from spam to malware.
- User privacy: Social media sites on third-party platforms are typically governed by the privacy policy of the application.
- Changing social media pattern: Expert opinions about the future of social media vary, but all agree that interactive media will stay. The challenge for transit organizations is to keep track of changes in this dynamic environment and to adapt accordingly.

Coming to the disadvantaged group usage, smartphones can help older people to live a comfortable life in a familiar environment. For example, doctors can easily monitor these people's well-being for the relevant day-to-day health services. Or you can transfer money without having to travel with mobile banking systems (Mobile Tech, 2013). Finally, smartphones are often only way of communication without having to travel under difficult circumstances (EPOMM, 2013).

In order to reach the sustainability goals, mobility management serves several target groups (EPOMM, 2013):

- Company staff,
- Students,
- Residents,
- The elderly,
- Tourists and tourism operators,
- Visitors to events.

The strategy for each target group can differ for providing the best possible service for various solutions. Gscheid Mobil in Munich is one of the best examples in Europe for the mobility of different groups. It is important to learn more about the various target groups that should be influenced: what are their characteristics, motivations and resistance? Emotional marketing strengthens the development of such a mobility culture; for example, campaigns to make cycling popular. Elderly people and immigrants also need extra support (EPOMM, 2013).

The Munich-based environmental organization Green City and the computer training centre Medi-encenter 50plus organized an internet course focused on mobility for older people. The main goal of the course was to enable older passengers to use the internet to get information about mobility services and even buy tickets, make reservations and participate in car sharing. The course included lessons about a CO_2 calculator of different transport modes, links for interesting mobility web pages in Munich, obtaining travel information via smartphones, route planning with Google Earth, a virtual ticketing machine exercise and how to rent a public bike and car. The internet courses were very popular among the participants. The average age of participants was 67 years. The majority hardly used the internet to plan their trips before but use it more often now, as a result of this training (Eltis 7).

Finally, online participatory planning encourages the urbanites to be more responsible for, involved in and aware of their role in local governance. Engaging citizens through social media in sustainable urban mobility planning has an important role in Aberdeen, UK. Aberdeen has final Sustainable Urban Mobility Plan (SUMP) in 2012. Its SUMP has already been successfully involving stakeholders and citizens as now recognized by the EC. Aberdeen's SUMP draws its key elements from the problems that the citizens themselves have identified. Web links and information were provided on several sites including Aberdeen City's Facebook and Twitter pages. The city council observed that utilizing social media has led to higher response rates than previously achieved for city-wide transport related surveys. It has reached a broader audience, engaging younger citizens. The use of social media also allows individuals to see what their friends have been looking at, which generates indirect promotional opportunities. The jury of the EC's first SUMP Award honored Aberdeen with the 10,000 Euro Award prize: "Its successful use of social media demonstrates the Council's ambition for innovation and connecting to citizens. Good response rates from citizens prove the appropriate application of the chosen tools" (Eltis 8).

CONCLUSION

Smart travel will contribute to a new mobility culture in cities. Green transport technologies and better infrastructure for cycling, walking and public transport and its networks as well as new and smarter mobility systems and intermodal services are necessary in today's world. There is an ever increasing need for raising awareness and promoting behavioral changes. To achieve sustainable mobility, awareness-raising, regulations and fiscal measures are needed, combined with infrastructure, technology and mobility management. Creating partnerships, sharing and transferring best practices at the national and European level are key factors needed for success. The stakeholders in sustainable mobility can be listed as:

- Companies: the business sector is interested in new ways such as tele-working,
- Schools: student travel plans is important,
- Energy companies: they are concerned about sustainable mobility for reducing carbon footprint,
- Healthcare insurers: they may finance bicycling or walking campaigns,
- Tourism operators: they may contribute to the city marketing.

Creating partnerships with the companies, regions and cities, but also with NGOs, is a key pillar of sustainable mobility. For instance, Austrian klima:aktiv mobil programme that has gained over 4,000 partners (more than 2,800 companies, 500 cities, 500 tourism operators, 200 schools and youth groups) implementing CO_2-reducing mobility projects, saving 540,000 tonnes of CO_2 per year (EPOMM, 2013).

New tools and applications can promote sustainable mobility and behavior. Smart technologies allow easier information-sharing. On these emerging sharing platforms, people can find and capitalize on lessons learned from best practices around the world. These types of tools can advance the rise of a new bottom-up culture of decision making and promote civic engagement. A new dynamic of participatory citizenship can be leveraged on the emerging technologies described above. "Connected sustainable cities will not emerge automatically. We will have to use our imaginations to invent them, and we will need to be determined and resourceful in pursuing implementation" (Mitchell and Casalegno, 2008, p. 112).

This chapter provided some points for the necessary goals and priorities for sustainable mobility, and for the evaluation of smart technology options for decision making. After the literature survey, some key strategies are summarized here for public transport agencies:

- Keep social media informed,
- Identify the real costs and staff requirements for smart technologies,
- Find the right voice and language to engage people,
- Listen to the users for feedback,
- Respect the strengths of social media for quick communication,
- Entertain interactively,
- Just get started using smart technologies and social networks.

A social media policy is critical. It should be followed by metrics, such as counting the followers. Additional indicators and cost-benefit analysis should be made. Internet security should be developed and maintained against viruses and other types of malware. All applications should be accessible to disabled persons. Social media platforms can bridge the digital divide, or the perceived gap between people who have access to information technology and those who do not. Further research could provide more data on the demographics of social media users and help determine whether public transportation agencies need to take additional actions to ensure that all riders can access online information and social networking sites. Location-based technology and social-buying services should be developed.

Furthermore, connectivity, social networking, and contact for geographically disparate households should be reinforced. Mobile services enable participants to act together more efficiently in pursuit of shared objectives by promoting cooperation among social networks; these improve social capital and social cohesion. Additionally, ICT assists in limiting the carbon footprint of the different sectors of the economy and this is the hot topic in the globalised world.

This chapter will increase our understanding of the relationship between communication and geo-informational trends and the public transportation issues in the city. Today, the spatial layout of modern cities often gives the impression that people are less important than cars. In such cities, public spaces are "traffic nodes" rather than meeting places (Ercoskun, 2010, p. 50). However, the key for a greener future in urban planning and traffic management is in the hands of the technology; technology that supports local applications. ICT-enabled developments of mixed-use, walkable, and sustainable neighborhoods can improve the environmental performance of cities and reduce the carbon footprint significantly.

The development of connected sustainable cities will require multi-partnerships between researchers, governmental bodies and technology firms to create state of the art devices and systems. The city planners will rethink the spatial organization of the cities. Smart technologies will be integrated to all infrastructures such as transport channels, energy systems, water, waste and recycling processes. To raise the quality of life, we should enhance smart technologies and connectivity to create patterns of city life which sustain the limited resources.

REFERENCES

Ankara Metropolitan Municipality Catalogue. (2013). *Ankara Buyuksehir Bld.* Ankara: Bilgilendirme ve Faaliyetlerden Yararlanma Kilavuzu.

Banister, D. (2008). The sustainable mobility paradigm. *Transport Policy*, *15*(2), 73–80. doi:10.1016/j.tranpol.2007.10.005

Bresfelean, V. P., & Ghisoiu, N. (2010). Higher education decision making and decision support systems. *WSEAS Transactions on Advances in Engineering Education*, *2*(7), 43–52.

Calvo, R. F., De Luigi, F., Haastrup, P., & Maniezzo, V. (2004). A distributed geographic information system for the daily car pooling problem. *Computers & Operations Research*, *31*(13), 2263–2278. doi:10.1016/S0305-0548(03)00186-2

Casalegno, M., & Chiu, D. (2008). *White Paper on Sustainable Transportation*, MIT Mobile Experience Laboratory, Cisco Systems Inc., 1-18. Retrieved from http://www.connectedurbandevelopment.org/pdf/sust_trans/sustainable_transportation.pdf

Castells, M., Fernandez-Ardevol, M., Linchuan Qiu, J., & Sey, A. (2007). *Mobile communication and society: A global perspective*. London: The MIT Press.

Castells, M. (2009). *Communication power*. Oxford: Oxford University Press.

CIVITAS forum, Nash, A. (2011) *Social media in transport*. Retrieved from http://www.andynash.com/nash-publications/2011-Nash-Social-Media-Trans13oct2011.pdf

Cohen, G, Salomon, I. & Nijkamp, P. (2002). Information–communications technologies (ICT) and

Cohen, G., Salomon, I., & Nijkamp, P. (2002, February). Transport: Does knowledge underpin policy? *Telecommunications Policy*, *26*(1-2), 31–52. doi:10.1016/S0308-5961(01)00052-0

Crossland, M. D., Wynne, B. E., & Perkins, W. C. (1995). Spatial decision support systems: An overview of technology and a test of efficacy. *Decision Support Systems*, *14*(3), 219–235. doi:10.1016/0167-9236(94)00018-N

Demissie, M. G., Correria, G. H. A., & Bento, C. (2013). Exploring cellular network handover information for urban mobility analysis. *Journal of Transport Geography*, *31*, 164–170. doi:10.1016/j.jtrangeo.2013.06.016

Densham, P. J. (1991). Spatial decision support systems. In D. J. Maguire, M. F. Goodchild, & D. W. Rhind (Eds.), *Geographic information systems: Principles and applications* (pp. 403–412). London: Langman.

Dourish, P., & Satchell, C. (2011). The moral economy of social media. In M. Foth etal. (Eds.), *From social butterfly to engaged citizen. Urban informatics, Social media, ubiquitous computing, and mobile technology to support citizen engagement* (pp. 21–37). London: The MIT Press.

EGO website, http://map.ego.gov.tr:8080/ego/index.aspx

EPOMM. (2013) Managing mobility for a better future: Mobility management: The smart way to sustainable mobility in European countries, regions and cities. Retrieved May, 6, 2014 from http://epomm.eu/index.php?id=2790

Carreno, M. (2012). Using social media to promote bus travel, Cardiff Bus, UK. *Eltis.org*. Retrieved from http://www.eltis.org/discover/case-studies/using-social-media-promote-bus-travel-cardiff-bus-uk

Christiaens, J. (2012). Scotty, the multimodal journey planner and information platform (Belgium). *Eltis.org*. Retrieved from http://www.eltis.org/discover/case-studies/scotty-multimodal-journey-planner-and-information-platform-belgium

Carreno, M. (2012). 'The Key', travel smartcard scheme, UK. *Eltis.org*. [REMOVED HYPERLINK FIELD]Retrieved from http://www.eltis.org/discover/case-studies/key-travel-smartcard-scheme-uk

McBrierty, C. (2012). M-Ticketing, Arriva Buses, UK. *Eltis.org*. Retrieved from http://www.eltis.org/discover/case-studies/m-ticketing-arriva-buses-uk[REMOVED HYPERLINK FIELD]

Wiszniewska, A. (2013). Public Transportation Ticket on a Mobile Phone in Warsaw (Poland). *Eltis.org*. Retrieved from [REMOVED HYPERLINK FIELD]http://www.eltis.org/discover/case-studies/public-transportation-ticket-mobile-phone-warsaw-poland

Grahn, M. (2009). Sunfleet Gothenburg - 37 carsharing pools using environmental friendly cars only (Sweden). *Eltis.org*. Retrieved from http://www.eltis.org/discover/case-studies/sunfleet-gothenburg-37-carsharing-pools-using-environmental-friendly-cars-only

Schuster, A. (2010). Internet courses for senior citizens in Munich, Germany. *Eltis.org*. Retrieved from [REMOVED HYPERLINK FIELD] http://www.eltis.org/discover/case-studies/internet-courses-senior-citizens-munich-germany#field_case_study_user_contact

Roeller, D. (2013). Engaging citizens through social media in sustainable urban mobility planning in Aberdeen (UK). *Eltis.org*. Retrieved from http://www.eltis.org/discover/case-studies/engaging-citizens-through-social-media-sustainable-urban-mobility-planning

Ercoskun, O. Y. (2010). Green Urban Planning and Design for Smarter Communities. In A. Hallin, & T. Karrbom Gustavsson (Eds.), *Organizational Communication and Sustainable Development* (pp. 41–59). Hershey, PA: Information Science Publishing. doi:10.4018/978-1-60566-822-2.ch003

EU Green Capital website. (n. d.). Retrieved from http://ec.europa.eu/environment/europeangreencapital/ljubljana-wins-european-green-capital-award-2016/index.html

Ferreira, J., Trigo, P., & Filipe, P. (2009). Collaborative Carpooling System. *World Academy of Science. Engineering and Technology*, *3*, 609–613.

Gal-Tzur, A., Grant-Muller, S. M., Kuflik, T., Minkov, E., Nocera, S., & Shoor, I. (2014). The potential of social media in delivering transport policy goals. *Transport Policy*, *32*, 115–123. doi:10.1016/j.tranpol.2014.01.007

Grünig, M., & Marcelino, D. (2009). A Missing Link to Sustainable Mobility: Connecting Public Transportation to Car and Bike Sharing Programs: Best Practice Examples from Europe and the U.S. Heinrich Böll Stiftung Washington, Retrieved August, 5, 2014 from http://www.ecologic.eu/3185

Harvey, D. (2008). The right to the city. *New Left Review*, *53*, 23–40.

Helft, M. (2007). Google's Buses Help Its Workers Beat the Rush. *The New York Times*.

Ecotaxsi. (n. d.). Retrieved from http://www.ecotaxi.ch/

The Velib blog. (n. d.). P'tit Vélib'. Retrieved from http://blog.velib.paris.fr/en/ptit-velib/

BiciMAD. (n. d.). BiciMAD.com. Retrieved from http://www.bicimad.com

Cyclopath. (n. d.). Cyclopath.org. Retrieved from http://www.cyclopath.org

Nextstopdesign (n. d.). Retrieved from http://www.nextstopdesign.com/welcome

Fixmytransport. (n. d.). Retrieved from http://www.fixmytransport.com

Green City Streets – Information technology for better public transportation. (n. d.). Greenscitystreets.com. Retrieved from http://www.greencitystreets.com

Arriva Bus App. (2015). Google Play Store. Retrieved from https://play.google.com/store/apps/details?id=de.hafas.android.arriva

Autolib' app. (2015). Google Play Store. Retrieved from https://play.google.com/store/apps/details?id=com.polyconseil.autolib

Izmir Metropolitan Municipality website. (n. d.). Retrieved from http://asps.unibel.com.tr/anasayfa

Lathia, N., Capra, L., Magliocchetti, D., De Vigili, F., Conti, G., & De Amicis, R. etal. (2012). Personalizing mobile travel information services. *Procedia: Social and Behavioral Sciences*, *48*, 1195–1204. doi:10.1016/j.sbspro.2012.06.1095

MacKinnon, R. (2011). *Let's take back the Internet!* TEDglobal 2011, filmed July 2011. Retrieved from http:// www.ted.com/talks/rebecca_mackinnon_let_s_take_back_ the_internet.html

Marques da Costa, E., Marques da Costa, N., Abrantes, P., & Louro, A. (2013). ICT *Tools to enhance the utility of a spatial decision support system-methodological essays.* Proceedings of AESOP-ACSP Joint Congress. Dublin.

Matzoros, A. (2002). Decision support systems for public transport management: The Athens public transport authority project. *Transportation Planning and Technology*, *25*(3), 215–237. doi:10.1080/0308106022000018973

Medlock, L. S. (2013). Mobile phones and society- how being constantly connected impacts our lives. *The Mobile Issue*, 29. Retrieved from http://source.southuniversity.edu/mobile-phones-and-society-how-being-constantly-connected-impacts-our-lives-137313.aspx

Mitchell, W. J., & Casalegno, F. (2008). *Connected sustainable cities.* Boston: MIT Mobile Experience Lab Publishing.

Mobile technologies – The digital fabric of our lives. (2013). Vodafone Institute for Society and Communications, Retrieved from http://www.vodafoneinstitut.de/uploads/media/Studies_komplett_FINAL_von_250913.pdf

Pinzon, L. (2013). *Beyond connectivity: The impacts of social media in urban development in Puerto Ayora, Ecuador* [DPU Working Paper No. 160]. London: The Bartlett Development Planning Unit.

Ronay, E., & Egger, R. (2013). NFC smart city: Cities of the future a scenario technique application (pp. 565–577). In Z. Xiang & I. Tussyadiah (Eds.), (2014). Information and Communication Technologies in Tourism. Springer.

Ocalir, E. V., Ercoskun, O. Y., & Tur, R. (2010). An integrated model of GIS and fuzzy logic (FMOTS) for location decisions of taxicab stands. *Expert Systems with Applications*, *37*(7), 4892–4901. doi:10.1016/j.eswa.2009.12.026

Saco, D. (2002). *Cybering democracy. Public space and the Internet.* Minneapolis, London: University of Minnesota Press.

Spinak, A., Chiu, D., & Casalegno, M. (2008). *SII-Sustainability Innovation Inventory*, Connected Urban Development, Retrieved from http://www.connectedurbandevelopment.org/mit/sustainability_innovation_inventory

Spinak, A., & Casalegno, F. (2012). Sustainable and equitable urbanism: The role of ICT in ecological culture change and poverty alleviation. In O. Y. Ercoskun (Ed.), *Green and Ecological Technologies for Urban Planning: Creating Smart Cities* (pp. 37–57). Hershey: Information Science Publishing. doi:10.4018/978-1-61350-453-6.ch003

Spitadakis, V., & Fostieri, M. (2012). WISETRIP-International multimodal journey planning and delivery of personalized trip information. *Procedia: Social and Behavioral Sciences*, *48*, 1294–1303. doi:10.1016/j.sbspro.2012.06.1105

Stareveren, R. V. (2014) *How open data access will revolutionize Istanbul's public transit systems*. Sustainablecitiescollective. Retrieved from http://sustainablecitiescollective.com/reneevanstaveren/290891/how-open-data-access-will-revolutionize-istanbul-s-public-transit-systems

Suh, K., Smith, T., & Linhoff, M. (2012). Leveraging socially networked mobile ICT platforms for the last-mile delivery problem. *Environmental Science & Technology*, *46*(17), 9481–9490. doi:10.1021/es301302k PMID:22877137

Uses of social media in public transportation. (2012). Transit Cooperative Research Program. Retrieved from http://onlinepubs.trb.org/onlinepubs/tcrp/tcrp_syn_99.pdf

Redlands Institute website. (n. d.). University of Redlands. Retrieved from http://www.spatial.redlands.edu/redlandsinstitute/giscience/sds/

TomTom Report. (2013). *TomTom European Congestion Index*. Retrieved from http://www.tomtom.com/lib/doc/trafficindex/2013-1101%20TomTomTrafficIndex2013Q2EUR-km.pdf

UN Habitat. (2013). *Planning and design for sustainable urban mobility*. New York: Routledge.

KEY TERMS AND DEFINITIONS

Car-Sharing: A collective transportation system based on a shared use or rent of cars for short periods of time, often by the hour with the objective to reduce the number of cars on the road.

Crowdsourcing: The process of obtaining needed services, ideas, or content by soliciting contributions from a large group of people, and especially from an online community.

Demand Responsive Transit Systems: An advanced, user-oriented form of public transport characterized by flexible routing and scheduling of small vehicles operating in shared-ride mode between pick-up and drop-off locations according to passengers' needs.

Mobile Ticketing: The process whereby customers can order, pay for, obtain and validate tickets from any location and at any time using mobile phones or other mobile handsets.

Multi-Modal Journey Planners: Online systems which provide timetable, routing and other real time travel information about public transport services and about transportation networks.

Smartphone: A mobile phone with more advanced computing capability and connectivity than basic feature phones. It includes the features such as a personal digital assistant, a media player, a digital camera, and a GPS navigation unit, also web browsing, Wi-Fi, 3rd-party apps, motion sensor, mobile payment and 3G.

Social Media: The digital systems through which people engage in social relations of many sorts social dynamics are supported by technologies such as social networking sites, social virtual realities, blogging and micro-blogging.

Chapter 12
Integrating Spatial Planning of Protected Areas and Transportation Infrastructures

Mihai Răzvan Niţă
University of Bucharest, Romania

Mihăiţă Iulian Niculae
University of Bucharest, Romania

Gabriel Ovidiu Vânău
University of Bucharest, Romania

ABSTRACT

Our study focuses on the position of protected areas in relation to the transport infrastructure planning and their specific requirements (fragmentation, artificial edge effects, buffer zones, etc.). The presence of transport infrastructure in protected areas is discouraged by the European Union legislation and although their avoidance should be easy, there are numerous cases in which knowledge of the protected areas lacked from the transport infrastructure planning. Our chapter presents the theoretical aspects of integrating protected areas with transportation infrastructures and the main software and methodologies which can be used in the planning process exemplified with case studies. The multitude of challenges found in transportation and conservation planning requires complex decision support systems (DSS), such are the tools based on Geographical Information Systems (GIS) we showcased in this chapter.

1. INTRODUCTION

The environment of our planet is dramatically changing under an ever increasing pressure on the natural resources by the large human population and its wide range of needs. One of the most visible and important direction of change is the destruction of natural habitats, affecting the very functionality that entire ecosystems rely on (Primack, Pătroescu, Rozylowicz, & Iojă, 2008). There are immediate and long term consequences, not only on the natural balance, but also on the human communities (M.R. Niă,

DOI: 10.4018/978-1-4666-8648-9.ch012

2011). As an example, clearing the forests for resource exploitation, agricultural uses or to accommodate expanding cities, sets in motion a chain of events including loss of habitats and biodiversity, increased erosion and loss of soil, increased water runoff and floods, climate change and desertification (Acevedo et al., 2008). To remove the natural habitats, which at the present seems entirely possible and the direction the human population is choosing, would imply the loss of ecosystem services and resources with catastrophic consequences (MEA, 2005).

Protected areas emerged as an important tool to oppose this process and safeguard natural elements vital for our present life and future existence, and are one of the most common land conservation policies worldwide. For the International Union for Conservation of Nature (IUCN) a protected area is a clearly defined geographical space, recognized, dedicated and managed, through legal or other effective means (Chape, Blyth, Fish, Fox, & Spalding, 2003), to achieve the long-term conservation of nature with associated ecosystem services and cultural values (Dudley, 2008). Considering this widely known definition, protected areas ensure not only the preservation of the natural aspects, but also of the coexistent cultural values (Antrop, 2005), acknowledging in this way on one hand the juxtaposition of human presence on the same space as the natural one and on the other hand the necessary public positive involvement (Primack et al., 2008). The needs of the population living in the protected areas and of the general population, including relevant stakeholders, are to be taken into consideration and adjusted with the conservation requirements if long term sustainability is to be achieved (Ontario Ministry of Natural Resources, 2009).

Successful protected areas should be established on principles such as the precise identification of values (species, habitats, biomes, ecosystems etc.) and pressures (del Saz-Salazar & Rausell-Köster, 2008), adequate legislation (including management plans) and objectives (Lichtenberg, Tra, & Hardie, 2007), appropriate zoning (M. E. Watts et al., 2009), balanced management of uses (prohibited and permitted activities), permanent public consultation and involvement, creative awareness and education programs (Sevenant & Antrop, 2010), sustainable financing, intelligent administration or effective monitoring and response, easy access to relevant information about the present situation and the possible effects of proposed policy changes (Audenaert, De Cleyn, & Vankerckhove, 2008; Handy & Mokhtarian, 1995).

In order to provide the much needed system equilibrium at regional and global scale, one of the most important principles when establishing protected areas is to ensure not only that they maintain the conservation of their components but also are covering in a network distribution all the key ecosystems. The European Union's Nature 2000 network of protected areas ensures the long-term survival of Europe's most valuable and threatened species and habitats (Ioja, Rozylowicz, Patroescu, Nita, & Onose, 2011). Worldwide, in the 2003 UN list of protected areas there were 100 000 protected areas in 227 countries or territories, occupying 11.5% of the planet's land surface area (Chape et al., 2003).

The activities that exists inside and outside with potentially negative side-effects needs to be carefully administrated. Every activity has an ecological footprint (M.R. Niă, 2008; Pătroescu, Niă, Iojă, & Vânău, 2009), from the most destructive ones, such as urban development, forest exploitation/deforestation, farming and mining to the traditional, less intensive ones. Not always activities that produce environmental externalities are addressed by legislative measures or incentives for reducing their impact. An element that is always there, no matter what activity is taken into consideration, is the transport infrastructure. The maintenance of ecosystem functions and structures increases in difficulty as they have to coexist with the transportation infrastructure (Zhu & Dale, 2000).

Transportation infrastructure is an essential component of our present way of life. It sustains the mobility of people and it is vital to the production and distribution of goods (Hensher & Button, 2003). The development of modern fast ways of transport, including aviation, rail, road, and ship transport led to the intensification of the economic activity (Calthrop & Proost, 1998). The modernization and development of transportation infrastructure represents a long-term activity aimed at producing economic and social positive effects (Hensher & Button, 2003), and providing a good circulation of people and goods (Zhu & Dale, 2000).

The consequences were far reaching, changing not only society, but also the natural landscapes. Places and resources considered not so long ago isolated and too remote to be reached with efficiency are now fully exploitable (Beinat & van Drunen, 1998). At the same time, better transport infrastructure means larger metropolitan areas, more urban development and generally more areas transformed to accommodate our needs (Taleai, Sharifi, Sliuzas, & Mesgari, 2007), as more infrastructure investment and energy consumption will be stimulated by the huge transportation demand (Cui & Li, 2014)

The extent and nature of the growing traffic demand in Europe pose several challenges for sustainable transportation, putting pressure on the decision-makers to provide new facilities for both passenger and freight transportation, and hence an ever increasing strain on infrastructure planning policy (Nocera, Maino, & Cavallaro, 2012). When planning for transport infrastructure, the main objectives are: to improve the accessibility by reducing travel/transport time or increasing the potential to travel/transport (the quantity of economic or social activities, including work, that can be reached using the transport system) and the efficiency of different economic activities (by time, energy and cost savings, reliability, safety).

The following objectives are suggested to be targeted by the transport system, corresponding to the three axes of sustainable development (Petersen et al., 2009):

- Economic: economic growth, generation of employment, reduction and avoidance of congestion;
- Environmental: reduction and avoidance of climate change effects by reducing greenhouse gases, harmful local pollutants, noise nuisance from transport, protection of environmentally-sensitive areas from transport encroachment;
- Social: reduction and avoidance of fatal and serious accidents, provision of accessibility to opportunities/services, enhancement of social cohesion, participatory approach to transport planning, good quality transport, high quality standards of employment within the transport sector.

The ***planning*** of transport systems in modern terms incorporates not only requirements meant to offer direct benefits to the users but also principles that ensure the reduction of undesirable effects on the economy, society and the environment. In theory, a set of mathematical equilibriums could be developed to describe all relations between all transport drives and to design an ideal transport network. This approach is called the "Network Design Problem (NDP)". It allows determining the optimal investment program, so that the social welfare is maximized (de Ceuster et al., 2010).

The most used in practice are the geographical and transport demand oriented approaches. The two are correlated, since the spatial structure determines the pattern of activities and the activities lead to a demand for transport. This demand subsequently manifests itself as traffic on the infrastructure. The transport demand approach has the tendency to concentrate on the development of the existing infrastructure, increasing its capacity and quality, and has solid economic premises, while the geographical approach it's more inclusive in terms of social cohesion, but also subject to poor resource allocation and prioritization. Another practice often employs a bottleneck approach, solving transportation problems as they manifest. This lacks vision and in the long term do not offer any durable solutions.

In order to develop a European Union optimum transport network, the Trans-European transport network planning methodology (de Ceuster et al., 2010) proposes the following steps:

- Selection and functional classification of nodes and defining a network on the connection level. This consists in the identification of nodes (cities, activities, points of interest, convergence points) at various territorial scales, from high to low level scale. An ease way is to use the settlement hierarchy by population, although the intensity of the economic activities, generating transport demand beside passenger traffic, is an essential factor.
- Quality criteria for the connections (links) in the network and assessment of missing links. To offer specific accessibility, the function and the desired quality of the connections needs to be clearly defined. Quality implies aspects such as speed, reliability and comfort. Intrinsically related with the connections' quality is the price of the infrastructure. In current practice the capacity approach the primary consideration, while the other quality aspects receive insufficient attention and could hinder the functionality of the transport connection.
- Network assessment strategy: ranking (MCA) and evaluating (CBA) projects. A cost-benefit analysis (CBA) is the most appropriate tool for evaluating the projected development of the transport infrastructure. A multi-criteria analysis offers the possibility to integrate numerous other aspects, non-easily and directly transferable into economic costs. A hierarchy in the access points is established (e.g. based on number of inhabitants) and used as a selection criterion. The size of the flows between the various access points can be used to check the selection.

The linkage between ***spatial planning of transport infrastructure and management of protected areas*** is present in both planning texts and regulation as in management plans of protected areas, as different categories of protected areas require specific environmental impact assessment for the transportation infrastructure (Dudley, 2008). The multitude of challenges found in transportation and conservation planning requires complex decision support systems (DSS) correlating large sets of multiple criteria to cope with the complexity of practical decisions in the case of conflicting goals (Matsatsinis, Ipsilantis, & Samaras, 2008). Environmental decision making is complicated as it involves multiple decision-makers, stakeholders, regulations and provisions, as well as competing objectives of different group of interests (Junier & Mostert, 2014). While transport infrastructure's planners aim for immediate, easier to advocate, economic and social benefits, protected areas managers have sometimes difficult to monetize goals, as it is the biodiversity conservation. Furthermore, transportation planning is better established and provided with better funding than protected areas in most countries, since it responds to more direct and immediate needs.

These support systems allow decision makers to navigate large amounts of information quickly (Matsatsinis et al., 2008); explore interrelationships between factors that may influence their decisions; facilitate understanding, learning, and negotiation and improve the basis of interactions among regional stakeholders (Zhu & Dale, 2000). Although quantitative modeling in transportation planning has been the target of criticism due to its rather inflexible character, and focus more on technical capabilities, while needing to correspond with the background and procedures of policy makers it still help decision makers efficiently and effectively utilize data, information, and knowledge available to solve their problems (Matsatsinis et al., 2008).

The decision support systems (DSS) usable in the planning and management of protected areas, targeting issues related to the adjacent transport infrastructure, are usually too complex to be operated directly by decision makers. The administration of every component of this category of DSS, from database to model and interface, has to be done by qualified personnel. The disadvantages (difficulty of finding or training operators, increase in the personnel costs) are outweighed by many advantages: better constructed and less error-prone databases, knowledge and use of state of the art models, improvement and adaptation of models if needed, adequate outputs for specific tasks. Beside or alongside statistical DSS, tools based on Geographical Information Systems (GIS) are extremely effective when dealing with data that has to be spatially-correlated. Since the planning and management of both protected areas and transport infrastructure have a strong territorial dimension, using DSS with spatial correlation capability is not only advisable, but a necessity. One other major advantage of GIS based DSS is that the results, while generated by skilled handling of the data, can be presented in forms (maps) that are user friendly and easy to understand even by the general population.

Our study tackles the problem of insufficient data concerning the protected areas position and specific requirements (avoidance of fragmentation, reducing of artificial edge effects, buffer zones, etc.) in relation to the transport infrastructure planning. The presence of transport infrastructure in the protected areas territory or proximity is discouraged by the European Union legislation. Although avoidance of the protected areas should be easy most of the times, there were many cases where even simple knowledge of protected areas position and spatial requirements lacked from the transport infrastructure planning. This led to plans that were rejected further on the approval process, increasing the monetary and time cost, new plans having to be drawn. Present trends in transport infrastructure planning are to be as inclusive as possible, when considering the interests involved. Protected areas management and objectives should be also considered.

Considering the potentially divergent goals of transport infrastructure and management of protected areas our paper aims at finding the right balance between the two of them from the planning perspective. How can we ensure accessibility, while at the same time avoiding sensible conservation areas? How can we build a transport infrastructure that has the lowest possible impact on protected areas' environment (fragmentation, edge effect, pollution etc.)? How can plan a transport infrastructure that helps conservation, by carefully controlling the flows of visitors? These are some of the questions that our study will help answer. Spatial planning is the preferred tool we use in this case, trying to accommodate some important objectives for both the transportation infrastructure and the conservation of protected areas. Integrating requirements from transportation infrastructure building, on one hand and protected areas management, on the other hand, by planning better in advance should provide benefits for both (better fulfillment of objectives, avoidance of legal, social and economic conflicts, lower costs).

The title of our chapter promotes the notion of integration between the planning of protected areas and transportation infrastructure. This integration in our opinion is bidirectional and should be included in all phases of a project (planning, design, construction or maintenance) from either category that can have a significant impact on the other. Planners and administrators have the difficult task of integrating the conservation objectives of the protected areas with the economic and social objectives of the transportation infrastructure. While this is a difficult task, the planning and management of protected areas states that any activity that is contrary to the conservation objectives should be discouraged. At least in the European Union, the legal framework gives considerable importance to environmental conservation and can be used to block any project that doesn't properly address this issue. DSS that integrates aspects related to both fields, such are the ones we used in our study, help to avoid some of the problems and offer easier ways to achieve solutions.

The chapter starts with an introductory part presenting the relation between protected areas and transportation infrastructure, with a focus on their components which are conflicting in the planning phase. We described in the methodology section the data sources and methods we used for our case studies, together with a brief presentation of other instruments that can be used in the process. Spatial planning between protected areas and transportation infrastructures is presented on two scales: at local level and at regional and national level. For each of the scales we present general theoretical notions, our case studies followed by their discussions. The chapter ends with a summarization of the main findings.

The ***general objectives*** of our chapter are: (1) to present the existing linkages between transportation infrastructure and protected area, such as the fragmentation of protected areas by the transport infrastructure, obstruction for building new transport routes, due to the existence of protected areas, competition for limited space for development; (2) to assess how spatial planning at local, regional and national level can help integrate the conservation objectives with the transportation infrastructure by avoiding sensible areas, directing the traffic flows to carefully chosen interest spots, including in the budgets the costs for the right mitigation measures and (3) to present case studies of using specific software packages and techniques in planning for conservation, as the ones helping to determine transport infrastructure densities, fragmentation and edge effect indicators, the size and distribution of a protected area network that fulfill the conservation needs, connectivity of protected areas.

2. METHODOLOGY

We identified several scientific methods and software used in protected areas planning. Among these the most encountered ones are Marxan, Marxan with Zones, Zonae Cogito, C- plan, ConsNet, Zonation, Conefor. Choosing the appropriate instrument for the analysis is highly depended both on the available data and the expected results. These software packages help at improving both the spatial planning and the environment policies and strategies by promoting innovative methods that will identify, prioritize and organize the relation between protected areas, species and forms of human pressure.

Marxan is a software package offering support in the projection of a protected areas network at national or regional level. The main purpose is to identify a network of protected areas, as defined by a set of planning units, aimed at fulfilling the biodiversity conservation objectives (or targets) with minimum costs (Game & Grantham, 2008; Possingham, Ball, & Andelman, 2000). The costs can represent land ownership or different other social and economic costs. The software uses as input data among others planning units that can be user-defined according to the various interests, and in which the distribution of conservation features can be determined. Marxan represents the most used software in the world for conservation planning (M. E. Watts et al., 2009). The presence of transportation infrastructure in a planning unit increases its cost and reduces the chances of being included in the resulting protected areas network. ***Marxan with Zones*** (Marxan Z) represents an expansion pack for the Marxan software, a decision support system aimed at presenting land use zonation option in different geographical regions (M. E. Watts et al., 2009), using Spatially Explicit Annealing. Its main objective is to improve the initial software by developing new functions increasing its efficiency in resolving problems specific to spatial conservation planning, both in terrestrial and marine environments. The software is capable of allocating the selected planning units in multiple areas and includes multiple costs in conservation planning.

Zonae Cogito (ZC) acts as a support instrument for the decision making and management of databases used in Marxan, with the advantage of directly visualizing results (M. Watts, Stewart, Dan, & Kircher, 2011) as maps presenting the best solution, selection frequency or each individual solution. It also allows modifying the solution and calibrating the input parameters. ***C-Plan*** is a research tool for conservation planning designed as a tool and support system in the decision process. It is based on GIS techniques and illustrates the option for achieving a conservation objective in a certain region, allowing the user to decide which sites should implement conservation. A specific element is the un-replacement value for selected areas as an indicator of the conservation targets (Pressey, Watts, Ridges, & Barrett, 2005).

ConsNet is a software package used in projecting and analyzing protected area networks using multiple spatial criteria, such as size, connectivity, compactness, alienation, etc., allowing users to introduce also socio-economic criteria (Ciarleglio, Sarkar, & Barnes, 2010). The analyzed region is divided into units containing data on the presence and abundance of species or the economic costs and benefits. ***Zonation*** is a decision support tool and represents a suitable framework for prioritizing spatial conservation at large scales. The main objective is to identify surface of habitats and different landscapes with important role in maintaining the quality of habitats and species connectivity. Using the Zonation algorithm the software removes cells from the area using the criteria minimization of marginal loss until the final solution is present. In our case study we used ***Conefor Sensinode*** 2.2 software (figure 1) developed for decision support regarding the planning and conservation process. The software package quantifies the importance of habitats in maintaining or improving landscape connectivity (Saura & Torne, 2009). The method used for analyzing connectivity is the graph structures and the metrics are based on habitat availability metrics (Pascual-Hortal & Saura, 2006; Urban & Keitt, 2001), both recent approaches in the connectivity analysis (Saura & Rubio, 2010).

The sources of data from our case studies are represented by a series of vector data available for Romania. For the protected areas we used a dataset realized by the Ministry of Environment and Climate Change, containing the limits of natural and national parks from Romania, the limits of Natura 2000 sites (SPA and SCI), actualized for the year 2011, and the limits of scientific reserves, nature reserves and nature monuments. All the shapes are available in Stereo 70 projection. From the transportation infrastructure we selected the road network, represented only by national and European categories, as we aimed at identifying the negative effects associated with traffic as well. For that purpose we used the dataset available online at www.geo-spatial.org (Nacu, 2013).

For the local scale we focused our case study on the Defileului Jiului – Natura 2000 Site of Community Importance, partially overlapping the National Park Defileul Jiului (in translation Jiu Gorge), thus with a double protection status. The protected area is situated south-west Romania (Gorj and Hunedoara counties), and characterized by a large land cover with broadleaf forests (*Fagus sp.* and *Quercus sp.*) and contains a number of species from the annexes of the Habitat Directive (*Ursus arctos, Lynx lynx, Lutra lutra* etc.). For the regional and national scale we focused our analysis on the Natura 2000 network from the alpine biogeographic region in Romania. It contains large surfaces of the Romanian Carpathians, a mountainous range characterized by high percent of forest surfaces.

3. SPATIAL PLANNING AT LOCAL LEVEL

Management plans of protected areas generally nominate the conservation action needed, and sometimes even the desired outcomes considering the local social and economic constraints. When regarding the transportation infrastructure the aspects that are of equal importance in their relation to protected areas, are their physical dimension and their traffic parameters. A series of negative effects of roads and railroads found at local level represent challenges to the planning process, such as habitat fragmentation and associated edge effects, road mortality, visual disturbance, noise, chemical pollution, increased predation and invasive species along edges (Barber et al., 2011).

Figure 1. Data input in the Conefor Software package

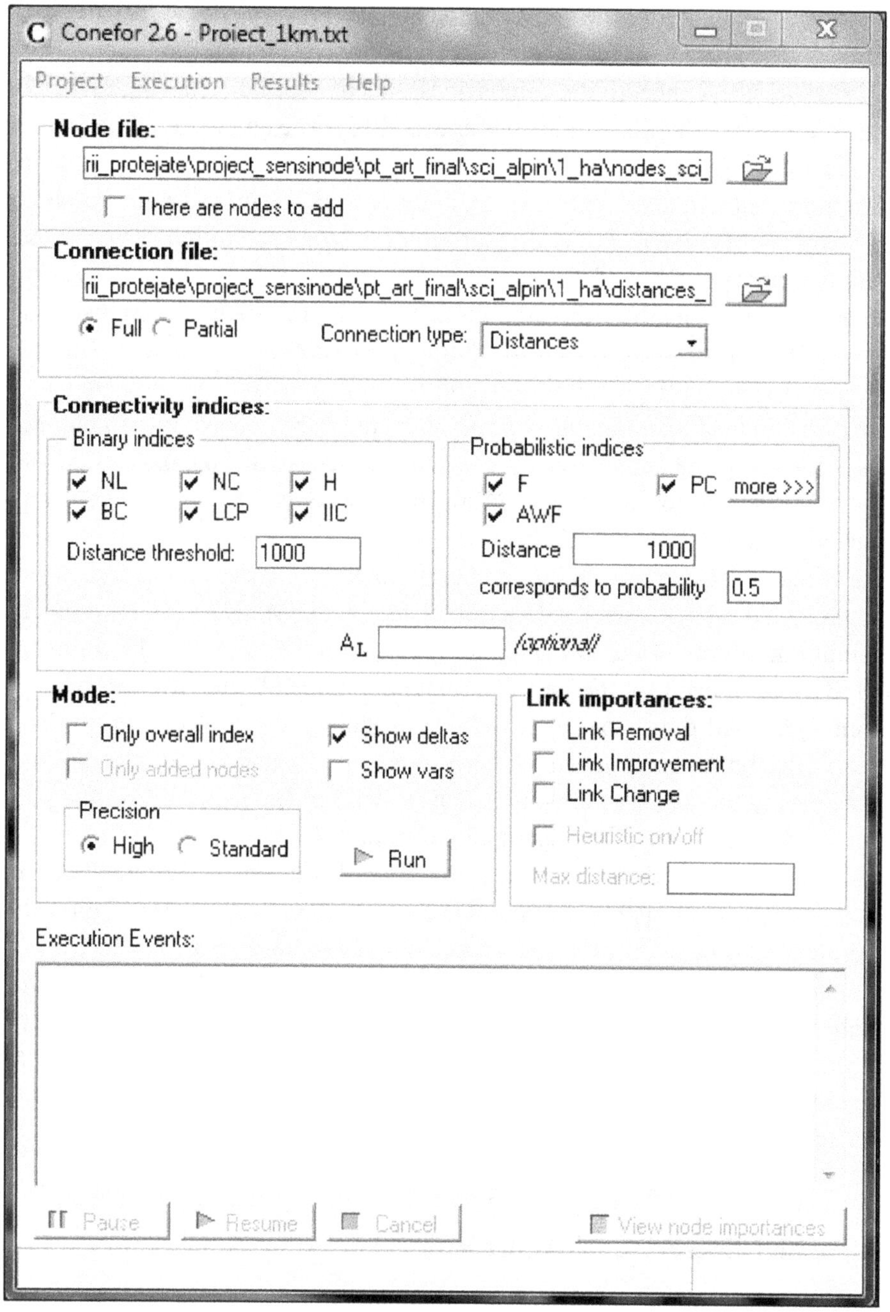

Habitat fragmentation ranks among the highest threats to global biodiversity and is the process by which a large and continuous habitat formation is divided into two or more patches (Boentje & Blinnikov, 2007). The main drivers of habitat fragmentation are the transportation infrastructure, zoning regulations and the land-ownership and use as the fragmentation process is positively related to population density and proximity to urban areas (Sims, 2013). Habitat losses appear in the existing habitats especially in the construction phase of the transportation infrastructure.

Among the resulting effects are: limiting dispersion and colonization, restricting species access to food sources or mating partners, dividing populations. In addition, a series of negative phenomena appear at the edges of new patches: changing climate parameters, appearance of invasive or exotic species, increase fire incidence, exposure to diseases.

Road mortality is influenced by factors such as the contact between habitats and roads, climate conditions or transportation characteristics (considering traffic volume the greatest contributor to road-kill). Amphibians and reptiles are the most affected groups of organisms, especially during their small scale migration related to breeding (for amphibians) or the temperature oscillation and using of roads as a thermoregulatory surface (for reptiles) (Garriga et al., 2012) while mammals and birds use roads as dispersal corridors or food sources.

In order to achieve a good conservation planning, decision makers should therefore consider both the total amounts of habitat conserved and the spatial configuration of that habitat in direct relation to the transportation infrastructure. In our case study we have chosen the Defileul Jiului Natura 2000 site (figure 2), fragmented by the national road no. 66, which follows mainly the center of the protected area

Using as inputs the dataset realized by the Ministry of Environment and Climate Change, containing the limits of natural and national parks from Romania, the limits of Natura 2000 sites (SPA and SCI), actualized for the year 2011, and the limits of scientific reserves, nature reserves and nature monuments

Figure 2. Defileului Jiului – Natura 2000 Site of Community Importance fragmented by a National Road

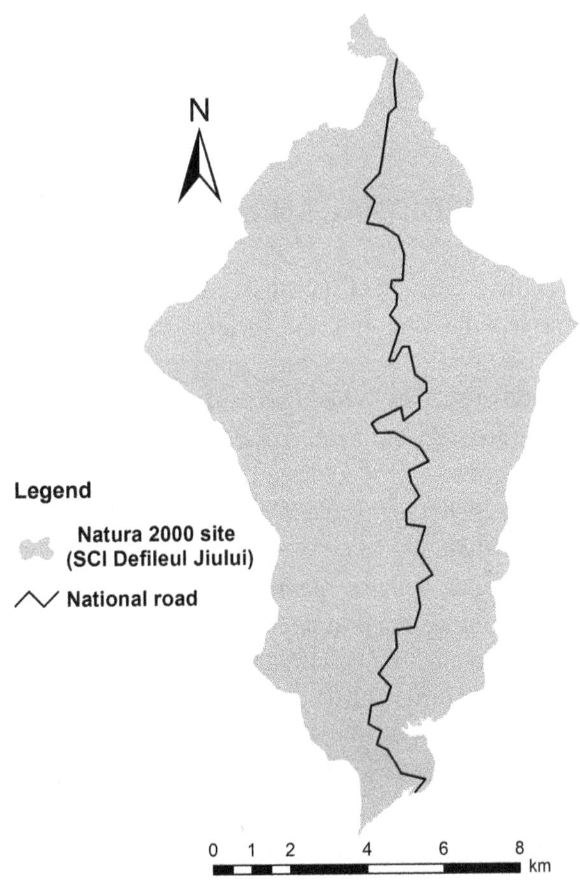

and datasets with the road network, represented only by national and European categories, the software was able to provide a clear image of the transport infrastructure in relation to the protected area it's overlapping. Local, regional and national institutions can use the map to plan better their decisions.

For example, the National Public Roads Commission can better correlate their road building policy with the existence of protected areas. When roads cannot be built to avoid large inclusive protected areas entirely, at least consider where the essential scientific reserves are. Since the Romanian and European Union's legislation is quite strict in regard to the protected areas safeguard, awareness of the protected areas presence in relation to existing and projected road is crucial and cost saving. Avoiding not only crossing of the protected areas by the transport infrastructure, but also negative impacts determined by transportation routes close to protected areas it is also embedded into the legislation.

Several issues emerge from the situation. Firstly, the two resulted fragments have a large edge length that the initial one, and thus a greater exposure to edge effects. This can only be reduced by choosing for the transportation infrastructure the shortest path (reducing its length and shifting to a straight form), but that's not the case in our study. Secondly, animal species have now have a barrier in the middle of their habitat that reduces their possibility of movement for food sources or mating, and increases road mortality, the solution being the installment of eco-ducts and eco-bridges allowing their safe passage over the transportation infrastructure. Traffic noise, especially disturbing for birds' species, can be reduced through the establishment of sound barriers along the infrastructure, while pollution can be controlled through protective barriers of resistant species.

All these represent simple solutions, but on one hand they are hard to enforce especially in a developing country where there is a substantial pressure on natural resources and legal provisions are both incomplete of poorly enforced, and on the other increase the general costs for the construction and maintenance of transportation infrastructure.

4. SPATIAL PLANNING AT REGIONAL AND NATIONAL LEVEL

At regional and national level, the existence of multiple or even conflicting goals between the ecological ones of the protected areas and the economic and social of the transportation network makes spatial planning even harder. While conservationist deal with protected areas design by involving strategies to achieve maximum representation of species in them given sets of constraints, the transportation network aims at connecting people and goods. As the main purpose of transportation infrastructures is to create a network establishing a form of connectivity between human settlements and their interests, one of the main objectives of protected areas is to create an ecological network providing ecological connectivity to species (Marcucci & Jordan, 2013).

Connectivity represents the main challenge in planning at regional and national scale. The connectivity of the protected areas network is fundamental in its long-term functioning and viability. Recent trends in transportation planning seek for their environmental linkages as a manner of developing an integrated transportation planning (Marcucci & Jordan, 2013) and novel spatially explicit approaches that analyze landscape connectivity have been developed that allow identifying those landscape elements that are of particular importance for habitat connectivity (Piquer-Rodríguez, Kuemmerle, Alcaraz-Segura, Zurita-Milla, & Cabello, 2012).

Criteria for assessing connectivity include rules to achieve contiguous reserve systems, contiguous areas of distribution for the focal species, or approaches where spatial criteria are incorporated in the objective function to be optimized (Alagador et al., 2012). Often the lack of data available makes it difficult to enforce or monitor the success of specific conservation measures related to the transportation infrastructure (Underwood, Wilson, Ludgerus, & Evans, 2013).

In Romania, the establishment and development of the Natura 2000 network beginning with 2006 determined numerous changes regarding the national policy of biodiversity conservation (Iojă et al., 2010). Maintaining and increasing the connectivity of protected areas network at national level represents a main requirement for achieving the conservation targets (Ioja et al., 2011). The efficiency of the protected areas network can be assessed using the connectivity as an efficiency index.

For the case study, we examined the connectivity of the Natura 2000 protected areas network for terrestrial mammals in the Alpine Biogeographic Region from Romania by evaluating the spatial distribution of Sites of Community Importance in the Region and assessing the functional connectivity of the protected areas network on two scenarios: with and without the impact induced by the roads network. Our analysis includes a number of 125 sites and the distribution of 17 species of terrestrial mammals, protected under the Habitats Directive, each having different dispersion distances and different home-range sizes (figure 3). We divided the 17 species in three distinct groups, according to their average dispersion distance and the average home-range (Minor & Lookingbill, 2010): small mammals (1 km, 1

Figure 3. The Natura 2000 network (SCI) for the small mammals]

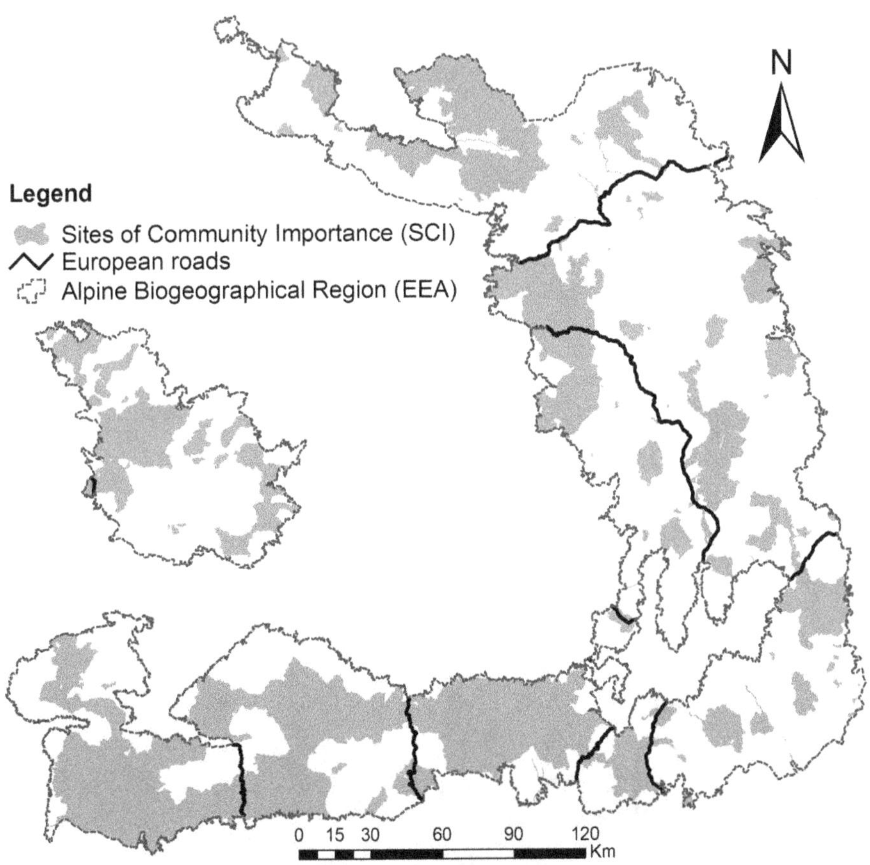

ha), intermediate mammals (10 km, 100 ha) and large mammals (100 km, 1000 ha). The protected areas crossed by the transportation infrastructure are considered to be disconnected (Gurrutxaga, Rubio, & Saura, 2011; Minor & Lookingbill, 2010) and eliminated from the analysis (figure 4).

Using a graph theory approach for measuring the connectivity of the protected areas network in the two cases mentioned above we used the Conefor 2.6 software (Saura & Torne, 2009) for identifying the connected protected areas and the ArcGis software for the spatial analysis. We computed the connectivity of the network by dividing the cumulated surface of protected areas from the largest component to the total surface of protected areas from the Alpine Biogeographic Region. Therefore the connectivity can take values between 0 – disconnected and 1 – connected (Minor & Lookingbill, 2010).

Regarding the spatial analysis of the Sites of Community Importance, in the Alpine Biogeographic Region the network occupies approximately 35% of the total surface, divided into 125 sites. There are 124 sites with a surface greater than 1 ha, 106 with a surface over 100 ha and only 71 with a surface over 1000 ha (the average surface of an individual site is 14034 ha). Our evaluation of the functional connectivity of the protected areas network in the Alpine Biogeographic Region revealed variations in the two cases, according to the category of mammals (table 1).

Figure 4. The distribution of the Natura 2000 network (SCI) for the small mammals after removing those disconnected reserves including European roads]

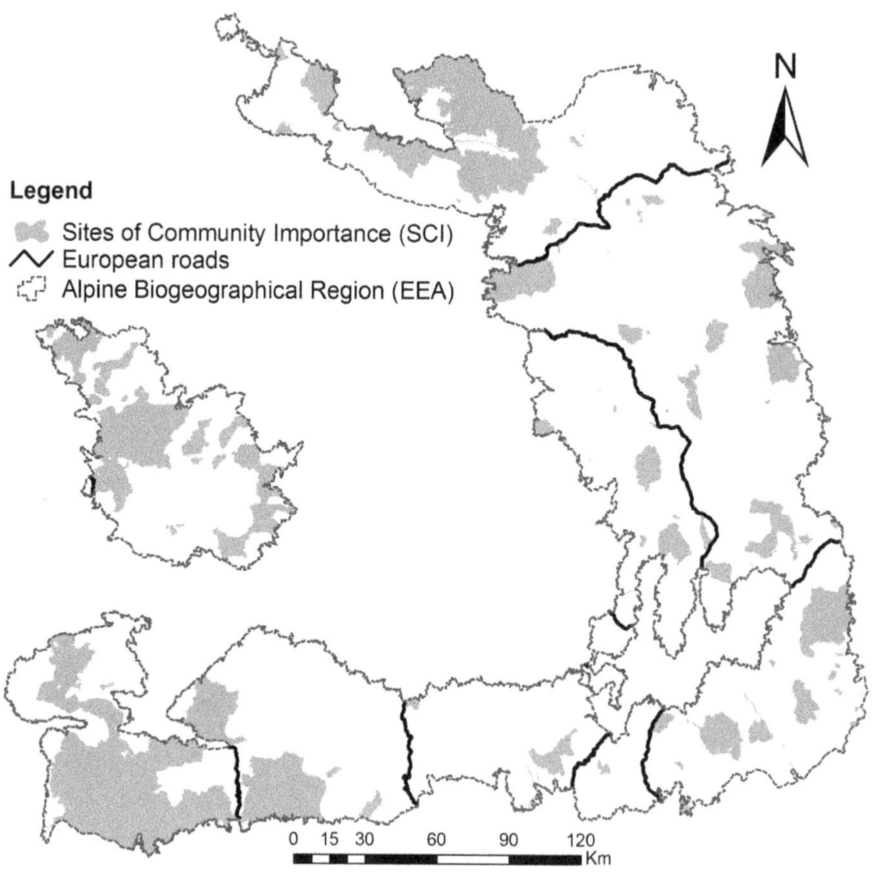

In this case, the transportation network represented by European roads significantly affects the connectivity score for small and intermediate mammals, reducing them by 38% and 27%, while for the large mammals the score remains the same. Furthermore, we can observe that the protected areas network is well connected functionally only for large mammals (score of 1), and has a reduced connectivity for small (0.47) and intermediate (0.50) mammals. The network therefore fails to achieve the connectivity objective. Further development and planning of the transportation infrastructure must be correctly assessed for increasing the connectivity of the Natura 2000 network in Romania.

DISCUSSION

While protected areas are trying to achieve a balanced cover of all essential natural habitats at the European Union scale and maintain vital wide life connections between these areas, the transport infrastructure is having a detrimental effect, working instead to support connection and accessibility for the human population and the corresponding immediate needs. Considering longer terms, the conservation of natural habitats and wildlife is paramount to ensure ecosystem outputs that the human population depends on. The conundrum is how to mitigate the two aspects and have the best possible in both of them: connectivity for natural habitats and wildlife and accessibility for the human population, offered by the transport infrastructure.

The case studies show the increase in fragmentation major transport infrastructure has on the protected areas. Habitat fragmentation greatly affects biodiversity and in this way the very purpose the protected areas were created for. While sometimes it's not possible to avoid this negative aspect, mitigation measures should be considered and put into place, such as safe crossing opportunities for wildlife. We have not looked into wildlife–vehicle collision cost, environmental costs in the building of transport infrastructure phase, social and economic costs determined by the barrier effect, all of which are higher in protected areas and important arguments when arguing for including protected areas management requirements into the transport infrastructure planning. Protected areas and transport infrastructures decision makers alike should be well aware of these aspects.

The software packages we presented are mainly integrated within GIS techniques and follow the elements of connectivity and corridors between ecological core areas. Their basic principles regard the identification of linkage areas between cores (based on the energetic costs of the difficulty of travelling between them) or the principle of centrality (a classification of priority surfaces according to their favorability and permeability). However, most of these planning instrument account only for the structure of infrastructure elements and only partially integrate their functionality (e.g. Linkage Mapper, Conefor).

Table 1. Functional connectivity of protected areas

Alpine Biogeographic Region	Small mammals		Intermediate mammals		Large mammals
	A*	B**	A	B	A and B
Connectivity score	0.47	0.29	0.50	0.36	1.00
Large components (ha)	825695	318791	868449	394851	1738474 and 1071798

A*- Initial distribution of protected areas (SCI)
B*- After removing all disconnected protected areas (SCI)

We used Conefor, a software package integrated with ArcGis. In our case study for the protected areas in the alpine biogeographic region and their relation with the transportation infrastructure the results are expressing the importance of habitats for maintaining and improving connectivity. These results can be integrated as a support element in the decision-making stage of the planning process. The obtained maps can be easily understand by stakeholders without specific knowledge in the field and can represent a strong instrument of support for planners in integrating conservation objectives with those of the transportation infrastructure.

The results will provide real support for administrative policy makers to sustain the importance of integrating these objectives in front of the increasing pressure from real estate market. Moreover, the decision makers can review the Master plans with the purpose of improving the quality of transportation infrastructure and reducing the maintenance costs. In using these types of instruments as DSS for the planning process difficulties can emerge from the with public administration institutions that must collaborate to provide information needed for each analysis. In addition, and inventory of administrative institutions and legislative tools existing would increase knowledge on the specific character of the planning process (hierarchies, streams of data, levels of decision-making etc.) thus allowing the placement of these types of analysis in the right place and to the right decision-makers.

Spatial planning in Romania has to address specific national, regional and local issues, while also considering the country's integration into the European Union. Some of the issues are the territorial disparities, with high standards of living in major cities, while the rest of the country is experiencing economic and social difficulties, unemployment and low economic development in some areas, urban sprawl around major cities, poorly developed public utilities systems in some regions, the lack of major road infrastructure to connect the main regions, cities and industrial production sites, important presence of protected areas. As a member state of the European Union, Romania has to connect it's infrastructure to that of its neighbor partner states of Hungary and Bulgaria. Convergence with the European Union's average standards on economic development and standards of living is also an important objective.

The most important planning authority is the Ministry of Regional Development and Public Administration, which is supposed to coordinate all sectorial policies. The Ministry of Agriculture and Rural Development, the Ministry of Economy, the Ministry of Environment and Climatic Change and the Ministry of Transport have all important attributes related to specific spatial planning domains. There are also some other national agencies with planning and implementation capabilities, such as the National Road Administration or the National Water Administration. The regional level of planning is less developed in Romania with a still going debate about how it should be implemented. Locally, planning is done at county level, by elected Councils, while the townships govern each their own territory, using mandatory master plans.

There is a National Planning Policy Framework Planning with strategic guidelines, principles and objectives. The National Planning Policy Framework is taken into account in the preparation of regional, local and neighborhood plans. It also reflects and promotes relevant EU obligations and statutory requirements. Most of the major infrastructure networks and projects are proposed and coordinated by a top down approach, by one of the aforementioned ministries or agencies. The projects must be approved by all public institutions that may be concerned and sometimes be subjected to public debate, leading to a generally slow approval process.

CONCLUSION

The main conclusion of our chapter is that although transportation planning has increasingly included environmental impact assessments, regarding the direct, indirect and cumulative impacts of project alternatives, conservation planning should shift from a site-based protection to an integrated approach considering the landscape as an undivided whole. The new tendency of replacing traditional infrastructure with green ones designed and constructed to have both connectivity and multifunctionality represents a further step in that direction.

The existing linkages between transportation infrastructure and protected areas are present both in the environmental and economic fields, but there is a consensus that infrastructures have more negative effects on the protected areas than positive influences. An intelligent spatial planning at local, regional and national level can help integrate the conservation objectives with the transportation infrastructure. The use of decision support systems, in the form of software packages and GIS techniques help planners cope with the multitude of conflicting objectives between them. The major scientific challenge is to discover the best ways of integrating the effective functionality of the protected areas network with the social, economic, politic and administrative objectives of the transportation infrastructure.

REFERENCES

Acevedo, M., Bairdcallicott, J., Monticino, M., Lyons, D., Palomino, J., & Rosales, J. et al. (2008). Models of natural and human dynamics in forest landscapes: Cross-site and cross-cultural synthesis. *Geoforum*, *39*(2), 846–866. doi:10.1016/j.geoforum.2006.10.008

Alagador, D., Triviño, M., Cerdeira, J. O., Brás, R., Cabeza, M., & Araújo, M. B. (2012). Linking like with like: Optimising connectivity between environmentally-similar habitats. *Landscape Ecology*, *27*(2), 291–301. doi:10.1007/s10980-012-9704-9

Antrop, M. (2005). Why landscapes of the past are important for the future. *Landscape and Urban Planning*, *70*(1-2), 21–34. doi:10.1016/j.landurbplan.2003.10.002

Audenaert, A., De Cleyn, S. H., & Vankerckhove, B. (2008). Economic analysis of passive houses and low-energy houses compared with standard houses. *Energy Policy*, *36*(1), 47–55. doi:10.1016/j.enpol.2007.09.022

Barber, J. R., Burdett, C. L., Reed, S. E., Warner, K. A., Formichella, C., & Crooks, K. R. et al. (2011). Anthropogenic noise exposure in protected natural areas: Estimating the scale of ecological consequences. *Landscape Ecology*, *26*(9), 1281–1295. doi:10.1007/s10980-011-9646-7

Beinat, E., & van Drunen, M. (1998). Spatial conflicts in transport policies: an exploration of the perspectives of regional and local actors. In E. Beinat & P. Nijkamp (Eds.), Multicriteria Analysis for Land-Use Management. Dordrecht: Kluwer Academic Publishers. Retrieved from http://link.springer.com/chapter/10.1007%2F978-94-015-9058-7_9 doi:10.1007/978-94-015-9058-7_9

Boentje, J. P., & Blinnikov, M. S. (2007). Post-Soviet forest fragmentation and loss in the Green Belt around Moscow, Russia (1991–2001): A remote sensing perspective. *Landscape and Urban Planning*, *82*(4), 208–221. doi:10.1016/j.landurbplan.2007.02.009

Calthrop, E., & Proost, S. (1998). Road transport externalities. *Environmental and Resource Economics, 11*(3/4), 335–348. doi:10.1023/A:1008267917001

Chape, S., Blyth, S., Fish, L., Fox, P., & Spalding, M. (2003). *United Nations List of Protected Areas.* International Union for Conservation of Nature and Natural Resources. Cambridge, UK: Thanet Press Ltd., UK. Retrieved from http://www.iucn.org/about/work/programmes/gpap_home/gpap_wcpa/gpap_wcpapub/?2256/2003-United-Nations-list-of-protected-areas

Ciarleglio, M., Sarkar, S., & Barnes, J. W. (2010). ConsNet Manual. Ver 2.0. Biodiversity and Biocultural Conservation Laboratory (BBCL). Austin, Texas: University of Texas. Retrieved from http://uts.cc.utexas.edu/~consbio/Cons/ConsNetManual_v200.pdf

Cui, Q., & Li, Y. (2014). The evaluation of transportation energy efficiency: An application of three-stage virtual frontier DEA. *Transportation Research Part D, Transport and Environment, 29,* 1–11. doi:10.1016/j.trd.2014.03.007

de Ceuster, G., Voge, T., Chen, M., de Kievit, M., Laird, J., Koh, A., et al. (2010). *Trans-European transport network planning methodology.* DG MOVE. Retrieved from http://www.tmleuven.com/project/tent/finalreport-tent.pdf

Dudley, N. (2008). *Guidelines for Applying Protected Area Management Categories.* Gland, Switzerland: IUCN. Retrieved from https://cmsdata.iucn.org/downloads/guidelines_for_applying_protected_area_management_categories.pdf

Ecosystems and Human Well-Being: Scenarios. (2005). Washington D.C.: Island Press. Retrieved from http://islandpress.org/ecosystems-and-human-well-being-scenarios

Game, E. T., & Grantham, H. S. (2008). *Marxan User Manual: For Marxan version 1.8.10.* University of Queensland, St. Lucia, Queensland, Australia, and Pacific Marine Analysis and Research Association. Retrieved from http://www.uq.edu.au/marxan/docs/Marxan_User_Manual_2008.pdf

Garriga, N., Santos, X., Montori, A., Richter-Boix, A., Franch, M., & Llorente, G. A. (2012). Are protected areas truly protected? The impact of road traffic on vertebrate fauna. *Biodiversity and Conservation, 21*(11), 2761–2774. doi:10.1007/s10531-012-0332-0

Gurrutxaga, M., Rubio, L., & Saura, S. (2011). Key connectors in protected forest area networks and the impact of highways: A transnational case study from the Cantabrian Range to the Western Alps (SW Europe). *Landscape and Urban Planning, 101*(4), 310–320. doi:10.1016/j.landurbplan.2011.02.036

Handy, S. L., & Mokhtarian, P. L. (1995). Planning for Telecommuting Measurement and Policy Issues. *Journal of the American Planning Association, 61*(1), 99–111. doi:10.1080/01944369508975623

Hensher, D. A., & Button, K. J. (2003). *Handbook of Transport and the Environment.* Amsterdam: Elsevier.

Ioja, C., Rozylowicz, L., Patroescu, M., Nita, M., & Onose, D. (2011). Agriculture and Conservation in the Natura 2000 Network: A Sustainable Development Approach of the European Union. *Agricultural and Environmental Informatics, Governance and Management: Emerging Research Applications,* 339-358. doi: 10.4018/978-1-60960-621-3.Ch018

Iojă, C. I., Pătroescu, M., Rozylowicz, L., Popescu, V. D., Vergheleţ, M., Zotta, M. I., & Felciuc, M. (2010). The efficacy of Romania's protected areas network in conserving biodiversity. *Biological Conservation, 143*(11), 2468–2476. doi:10.1016/j.biocon.2010.06.013

Junier, S., & Mostert, E. (2014). A decision support system for the implementation of the Water Framework Directive in the Netherlands: Process, validity and useful information. *Environmental Science & Policy, 40*, 49–56. doi:10.1016/j.envsci.2014.04.004

Lichtenberg, E., Tra, C., & Hardie, I. (2007). Land use regulation and the provision of open space in suburban residential subdivisions. *Journal of Environmental Economics and Management, 54*(2), 199–213. doi:10.1016/j.jeem.2007.02.001

Marcucci, D. J., & Jordan, L. M. (2013). Benefits and challenges of linking green infrastructure and highway planning in the United States. *Environmental Management, 51*(1), 182–197. doi:10.1007/s00267-012-9966-7 PMID:23135391

Matsatsinis, N. F., Ipsilantis, P., & Samaras, G. (2008). Special issue on Multicriteria Decision Support Systems. *Operations Research, 8*(1), 3–4. doi:10.1007/s12351-008-0002-x

Minor, E. S., & Lookingbill, T. R. (2010). A Multiscale Network Analysis of Protected-Area Connectivity for Mammals in the United States. *Conservation Biology, 24*(6), 1549–1558. doi:10.1111/j.1523-1739.2010.01558.x PMID:20666801

Nacu, A. (2013). National roads of Romania (vectorial data set) [Data file]. Retrieved from http://www.geo-spatial.org/download/re-eaua-rutiera-na-ionala-a-romaniei-set-de-date-vectorial

Niă, M. R. (2008). Analyze model for the ecological footprint of new residential spaces in the Bucharest Metropolitan Area. *Forum Geografic,* (7), 8, 200-207.

Niă, M. R. (2011). *Dinamica rezidenialului în zona metropolitană a municipiului Bucureti i proiecia ei în starea mediului.* (PH.D. Thesis). University of Bucharest.

Nocera, S., Maino, F., & Cavallaro, F. (2012). A heuristic method for determining CO2 efficiency in transportation planning. *European Transport Research Review, 4*(2), 91–106. doi:10.1007/s12544-012-0073-x

Ontario Protected Areas Planning Manual. (2009). Ontario Ministry of Natural Resources. Peterborough: Queen's Printer for Ontario. Retrieved from http://www.ontarioparks.com/english/planning_pdf/papm/papm_approved.pdf

Pascual-Hortal, L., & Saura, S. (2006). Comparison and development of new graph-based landscape connectivity indices: Towards the priorization of habitat patches and corridors for conservation. *Landscape Ecology, 21*(7), 959–967. doi:10.1007/s10980-006-0013-z

Pătroescu, M. Niă, M. R., Iojă, I. C., & Vânău, G. O. (2009). The ecological footprint-indicator for analysing the environmental impact of residential surfaces in metropolitan areas [Case study: Bucharest Metropolitan Area]. *Strategies, Concepts and Technologies for Planning the Urban Future Conference, CITIES 3.0*

Petersen, M. S., Sessa, C., Enei, R., Ulied, A., Larrea, E., Obisco, O., et al. (2009). *Report on Transport Scenarios with a 20 and 40 year Horizon.* DG TREN. Retrieved from http://ec.europa.eu/transport/themes/strategies/studies/doc/future_of_transport/ 2009_02_transvisions_report.pdf

Piquer-Rodríguez, M., Kuemmerle, T., Alcaraz-Segura, D., Zurita-Milla, R., & Cabello, J. (2012). Future land use effects on the connectivity of protected area networks in southeastern Spain. *Journal for Nature Conservation*, *20*(6), 326–336. doi:10.1016/j.jnc.2012.07.001

Possingham, H., Ball, I., & Andelman, S. (2000). Mathematical methods for identifying representative reserve netorks. In S. Ferson & M. Burgman (Eds.), *Quantitative methods for conservation biology* (pp. 291–305). New York: Springer-Verlag. doi:10.1007/0-387-22648-6_17

Pressey, R. L., Watts, M., Ridges, M., & Barrett, T. (2005). *C-Plan conservation planning software. User Manual.* NSW Department of Environment and Conservation Retrieved from http://www.edg.org.au/software/cplan/C-Plan_Manual_2005.pdf

Primack, R., Pătroescu, M., Rozylowicz, L., & Iojă, I. C. (2008). *Fundamentele conservării diversităţii biologice.* Bucureşti: Editura AGIR.

Saura, S., & Rubio, L. (2010). A common currency for the different ways in which patches and links can contribute to habitat availability and connectivity in the landscape. *Ecography*, *33*(3), 523–537. doi:10.1111/j.1600-0587.2009.05760.x

Saura, S., & Torne, J. (2009). Conefor Sensinode 2.2: A software package for quantifying the importance of habitat patches for landscape connectivity. *Environmental Modelling & Software*, *24*(1), 135–139. doi:10.1016/j.envsoft.2008.05.005

Saz-Salazar, S., & Rausell-Köster, P. (2008). A Double-Hurdle model of urban green areas valuation: Dealing with zero responses. *Landscape and Urban Planning*, *84*(3-4), 241–251. doi:10.1016/j.landurbplan.2007.08.008

Sevenant, M., & Antrop, M. (2010). Transdisciplinary landscape planning: Does the public have aspirations? Experiences from a case study in Ghent (Flanders, Belgium). *Land Use Policy*, *27*(2), 373–386. doi:10.1016/j.landusepol.2009.05.005

Sims, K. R. E. (2013). Do Protected Areas Reduce Forest Fragmentation? A Microlandscapes Approach. *Environmental and Resource Economics*, *58*(2), 303–333. doi:10.1007/s10640-013-9707-2

Taleai, M., Sharifi, A., Sliuzas, R., & Mesgari, M. (2007). Evaluating the compatibility of multi-functional and intensive urban land uses. *International Journal of Applied Earth Observation and Geoinformation*, *9*(4), 375–391. doi:10.1016/j.jag.2006.12.002

Underwood, J. N., Wilson, S. K., Ludgerus, L., & Evans, R. D. (2013). Integrating connectivity science and spatial conservation management of coral reefs in north-west Australia. *Journal for Nature Conservation*, *21*(3), 163–172. doi:10.1016/j.jnc.2012.12.001

Urban, D., & Keitt, T. (2001). Landscape connectivity: A graph-theoretic perspective. *Ecology*, *82*(5), 1205–1218. Retrieved from http://amadeus.ecs.umass.edu/mie373/landscapes_msts.pdf doi:10.1890/0012-9658(2001)082[1205:LCAGTP]2.0.CO;2

Watts, M., Stewart, R., Dan, S., & Kircher, L. (2011). Using the Zonae Cogito Decision Support System. The Ecology Centre, University Of Queensland. Retrieved from http://www.uq.edu.au/marxan/docs/ZonaeCogitoManual_8March2011.pdf

Watts, M. E., Ball, I. R., Stewart, R. S., Klein, C. J., Wilson, K., & Steinback, C. et al. (2009). Marxan with Zones: Software for optimal conservation based land- and sea-use zoning. *Environmental Modelling & Software*, *24*(12), 1513–1521. doi:10.1016/j.envsoft.2009.06.005

Zhu, X., & Dale, A. P. (2000). Identifying Opportunities for Decision Support Systems in Support of Regional Resource Use Planning: An Approach through Soft Systems Methodology. *Environmental Management*, *26*(4), 371–384. doi:10.1007/s002670010094 PMID:10954800

KEY TERMS AND DEFINITIONS

Connectivity: Characteristic of habitats and protected areas networks giving species the possibility of moving in their biogeographical area.

Decision Support Systems (DSS): Computerized information systems used to support decision making of different stakeholders, and often used by public institutions.

GIS (Geographical Information System): System designed to capture, store, manipulate, analyze, manage and present spatial and geographical data.

Habitat Fragmentation: Process by which a continuous habitat or groups of habitats is split into at least two smaller parts mainly due to the development of linear infrastructures or other types of human activities.

Natura 2000: Network of natural protected areas in the European Union aimed at nature conservation and sustainable development of human activities.

Protected Area: Defined geographical spaces established through legal or other means with the objective of achieving long term nature conservation and protect ecosystem services and cultural values.

Spatial Planning: Methods and techniques used with the purpose of balancing the distribution of people and activities in space—s of various scales.

Transport Infrastructure: Human built framework supporting the transportation needs of people and goods, including roads, railways, ports, airports and their associated areas.

Chapter 13
Multi-Objective Optimizer for Multimodal Distribution Networks:
Operating Cost, Carbon Footprint and Delivery Time

Mauro Gamberi
University of Padova, Italy

Francesco Pilati
University of Padova, Italy

Marco Bortolini
University of Bologna, Italy

Alberto Regattieri
University of Bologna, Italy

ABSTRACT

A multi-objective optimizer Decision Support System (DSS) to minimize the operating cost, the carbon footprint and the delivery time in the design of multi-modal Distribution Networks (DNs) is presented to overcome the widely adopted methodologies focused on the cost minimization, only. The proposed approach simultaneously assesses three independent objective functions, evaluating the network costs, the Carbon Footprint (CO_2 emissions) and the shipping time from the producers to the final retailers. The DSS manages multimodal four-level (three-stage) DNs, best connecting the producers to the final retailers, through a set of Distribution Centres (DCs). It allows multiple transport modes and inter-modality options looking to the most effective DN configuration from the introduced multi-objective perspective. The three optimization criteria can be considered independently or solved simultaneously, through the so-called Pareto frontier approach. Finally, the proposed DSS is validated against a case study about the delivery of Italian fresh food to several European retailers.

INTRODUCTION

Distribution Networks (DNs) are the basic link between supply and demand. They include the physical systems and the intelligence tools able to ship products and services from the producers to the final consumers. DN theory belongs to the general Supply Chain Management (SCM) introduced by Lambert et al. (1998) as "the integration of key business processes from end-user through original suppliers, that provides product, service and information that add value for customers and other stakeholders".

DOI: 10.4018/978-1-4666-8648-9.ch013

Such a definition points out the key elements and of all DNs:

- End-Users: they represent the target of the SCM activities, to whom the value is directed, appreciated and experienced. They are typically a constraint out of the direct control of the decision makers and express expectations, demand of products, information and auxiliary services;
- Suppliers: they represent the active actors of the DNs, feeding it with products, information and services to deliver high value to the end-users. Typical suppliers are the primary resource producers, the process and manufacturing companies, the logistic providers and freight brokers, the last-mile distributors, etc. Such actors are responsible of the DN service level and rising costs and their effective integration and optimization is the primary goal of SCM;
- Supplied products: they represent the shipped goods in traditional DNs where the end-users look for physical items. Nevertheless, in this context, products are to be considered from an extensive perspective to include, packages, auxiliary tools, services and complementary products fully satisfying the end-user needs and representing a value for them for which they are ready to pay a monetary offset to have their ownership. Looking at the DN flows, the supplied products move forward from the primary producers to the end-users, i.e. direct logistic flow;
- Information: it represents the intangible element of the DNs directly associated to the knowledge moving forward from the suppliers to the end-users and backward from the end-users to the suppliers. The accurate information flow is essential to coordinate the DN stakeholders matching their interests and expectations. The information flow management is an essential part of the value transfer in SCM;
- Stakeholders: they represent the group of actors, directly and indirectly operating on the DN, having economic and non-economic interests on it and that are relevant in the decision making process. Stakeholders are from both the private and the public sectors; they include the control institutions and the general public to represent the community interests. SCM interacts with the DN stakeholders, balancing their needs and requirements with the purpose of increasing the overall system value;
- Business processes: they represent, by means of a general concept, the bundle of methods, models, practices and rules to be adopted by the stakeholders, at all levels, to reach their purposes, optimizing the DN, and adding value to the supplied products to meet the end-user expectations. Both qualitative and quantitative processes are developed through coordinated and integrated supporting decision tools and actions. Both the scientific community and the practitioners look for the most effective business processes to optimize the existent and the new DNs.

The effective design of DNs deals with and integrated set of actions looking for the whole system optimum. Manzini et al. (2007) highlight the trend of the literature about the importance of adopting an integrated approach not focused on a single element of the DN and based on coordination and information sharing. The authors further propose a reference quantitative approach for the integrated supply chain design.

Among the issues to be considered in the DN design, a basic one is the perspective to be adopted and the targets to look for (Ramaa et al., 2009). Three main perspectives can be followed.

- Single-objective, single-criterion perspective;
- Multi-objective, single-criterion perspective;

- Multi-objective, multi-criteria perspective.

In single-objective, single-criterion DN design perspective a sole driver is considered and, consequently, a unique function is managed and optimized. Standard approaches based on such a perspective look at the minimization of the DN length or shipping time, the unitary equivalent cost reduction, etc. No trade-off among different criteria is done. On the contrary, multi-objective approaches are based on the inclusion of more than one driver. Typically, drivers with divergent trends are considered, e.g. shipping time vs. shipping cost, DN cost vs. DN environmental impact, etc., looking for intermediate trade-off configurations. From such a perspective, there are basically two ways in which two or more objectives can be dealt with, simultaneously. The former involves transforming the multi-objective problem into a single-objective problem that aggregates all the objectives through a weighted sum procedure in which every objective is multiplied by a weighting factor and the objective function is calculated as the sum of the weighted objectives, i.e. multi-objective, single-criterion perspective. This requires the *a-priori* knowledge of the relative importance of the different criteria. An alternative method is to accept several objectives simultaneously and determine a non-dominated set of alternatives. This set is a collection of alternative solutions that represent potential trade-offs among objectives, i.e. multi-objective, multi-criteria perspective (Moncayo-Martínez & Zhang, 2011).

Both the literature and the standard practice convey to identify the following criteria driving the DN design, management and optimization:

- Distance: the purpose is to minimize the total travelled distance between the suppliers and the end-users, avoiding long routes, un-useful trips, partial load shipments, etc. Grouping strategies to supply more than one customer within the same trip are implemented. Typical objective functions are expressed in travelled kilometers/miles per year;
- Time: the purpose is to minimize the total or average shipping time to supply the end-users. Respect to the distance criterion, the inclusion of the vehicle average commercial speed, the load/unload times, the route congestions, etc., is done. Typical objective functions are expressed in hours per supplied costumer;
- Cost: the purpose is to minimize the annual equivalent cost for the DN activities. The boundary limits to define the cost drivers to be included generally vary among the approaches and models and they are generally classified between fix and variable costs, level and stage costs, operational and investment costs. Typical objective functions are expressed in euro/dollars per year or shipped product;
- Service level: differently from cost based DN design, in the case of the service level the target is the minimization of the gap between the end-user expectations and the experienced logistic service, trying to match them. Typical objective functions are expressed in percentage of satisfaction respect to predefined assessment metrics;
- Environmental impact: a different and emerging perspective is about the direct and indirect impact on the environment of the DNs. The environmental burdens of production, shipment and storage processes for products and services are minimized. Behind such a criterion the concept of Green Supply Chain (GSC) and Sustainable Supply Chain (SSC) are spread (Seuring, 2013). Typical objective functions refer to the emitted pollutant quantities, e.g. CO_2, particulate, green-house-gasses (GHGs), etc., or to standard environmental assessment metrics, e.g. CO_2 equivalent, mid-point and end-point indicators from the standardized life cycle assessment (LCA) methodology, etc.

Finally, the structure and features of the DNs heavily affect the design process and the achievable performances. A wide set of DN configurations exist and are reviewed by the literature (Melo et al.,2009). From the DN design and SCM view-point the most relevant features to consider deal with:

- **Flow Direction:** Typical DNs are also called forward logistic networks and consider the supply of products from the sources to the end-users. Nevertheless, from an extended view-point, the flow direction may be from the end-users to the sources in the case of reverse logistic networks. This is the case of waste collection and end-of-life product recovery and post-treatment. Finally, in the recent past the concept of closed-loop DN emerged to jointly consider both the direct and the reverse flows of goods and services facing an integrated design and DN optimization (Faccio et al., 2014);
- **DN Levels and Stages:** In SCM theory, each level is the origin and/or the destination of the supplied products. The levels typically refer to the primary source suppliers, the process and manufacturing plants, the distribution facilities, the buyers, the clients and the final-users. The physical and informative connections among two subsequent levels define a stage of the DN. Single and multi-stage DNs exist and define the so-called direct shipment short distribution channel and the long indirect distribution channel. Finally, the intermediate level by-pass is, sometime, allowed;
- **Shipping Modes:** In the case of physical goods, three shipment modes are usable. Mono-modal shipping, in which a single carrier is used (typically trucks), only, intermodal shipping, involving the transportation of freight through intermodal containers and vehicles, using multiple modes of transportation (rail, ship, air and truck) without any handling of the freight itself when changing modes, and multimodal shipping, defined as the carriage of goods by at least two different modes of transport on the basis of a multimodal transport contract from a place to another designated for delivery;
- **Supplied Products:** DNs for single and multi-product supply exist. The former considers a single good or service, the latter optimizes the distribution of a mix of goods and services optimizing the freight composition and mixes together with the vehicle routes;
- **Static vs. Dynamic DN:** in static DNs the temporal evolution of both the demand and the shipments is neglected considering a unique delivery time window. In the case of dynamic DNs the temporal variable is properly included and managed to define the distribution schedule (Ambrosino & Scutellà, 2005).

According to the introduced topic and classification criteria this chapter faces the optimal design of forward DNs. A set of fixed end-users with their own demands for a mix of physical products is considered best-building the static, multi-stage and multi-modal network to supply them. The intermediate level by-pass is, further, allowed making the DN flexible. The DN design follows a multi-objective, multi-criteria perspective considering three drivers. The network operating unitary cost function looks for an efficient DN, not to heavily affect the delivered product full cost and market price, the carbon footprint function looks for a SSC reducing the emitted pollutants. Finally, the delivery time function looks for a quick and safe supply of the end-user demand. Such three drivers are jointly considered through the Pareto frontier multi-criteria approach to identify effective trade-off DN configurations. The whole approach is based on a linear programming model and it is developed through a multi-objective optimizer Decision Support System (DSS) assisting the logistic designers and managers.

In the following sections the background about both the topic and the methods is provided before introducing the model formulation and the DSS features and interface. A consistent case study taken from the fresh food European Union (EU) supply chain is, then, discussed to present a full application of the whole approach and tool. Further research directions and final remarks conclude this chapter.

BACKGROUND

The background behind the proposed approach and DSS includes parallel topics faced, separately, in the following sub-sections. At first, an essential review of the existent DN design methodologies is provided stressing the difference between optimal and heuristic approaches. In the following, the multi-objective network design trends are introduced revising the existent criteria, methods and outcomes. The final sub-section focuses on the Pareto frontier multi-criteria approach providing the key concepts to support the model and the approach description.

DN Design Methodologies

The aim of all the DN design methodologies is to study the most effective ways to ship products from the supply centers to the end-users facing the related decisions (Manzini et al., 2007), e.g. the allocation of the market demand to the producers considering their capacity and peculiarities (Apaiah & Hendrix, 2005), the shipment mode choice (Eskigun et al., 2005), the distribution center (DC) inclusion (Cheng & Tsai, 2009; Ioannou, 2005), etc. A deep review of the DN and SCM design decisions is in Klibi et al. (2010) pointing out a set of strategic questions to be addressed: "Which markets should we target? What delivery time should we provide in different product markets and at what price? How many production and DCs should be implemented? Where should they be located? Which activities should be externalized? Which partners should we select? What production, storage and handling technologies should we adopt and how much capacity should we have? Which products should be produced/stocked in each location? Which factory/DC/demand zones should be supplied by each supplier/factory/DC? What means of transportation should be used?" Answering to all questions with a unique approach and methodology is almost impossible in practice especially if more than one decisional criterion is adopted (cost, time, environmental impact, etc.). The current literature agrees to split the problem into multiple parts proposing dedicated methodologies for each of them. Useful approaches are in Manzini et al. (2014), presenting strategic, tactical and operational DN planning horizons to focus on the long-term, mid-term and short-term network design separately, and in Mehrjerdi (2009) classifying the DN components and related separate management actions. Temporal and functional classifications of the problem are introduced.

Beamon (1998) reviews the existing models and methods for the supply chain design and analysis proposing a reference classification against the following categories:

- Deterministic analytical models;
- Stochastic analytical models;
- Economic analytical models;
- Simulation models.

Such categories cross with two other basic methods to tackle the problem:

- Optimal models;
- Heuristics.

Deterministic analytical models are based on fix input data. No data variability is modeled differently from the stochastic analytical models in which some of the variables are stochastically distributed, e.g. the expected end-user future demand levels. Economic analytical models are considered separately due to their importance and high adoption. The economic perspective is frequently the unique considered decisional criterion. Finally, simulation models refer to a different approach respect to analytical models. The former builds models reproducing the DN behavior to infer *ex-post* general trends, while the latter builds conceptual mathematical models to predict *ex-ante* the best values for the decisional variables. Finally, optimal models look for the optimization of the considered objective functions and are, generally, based on linear programming models, while heuristics propose good solutions and practical rules-of-thumb to face the considered DN design issue. Almost all the existing methodologies for the DN best design may be grouped according to the previous drivers. The extensive review proposed by Soni & Kodali (2012), considering 619 empirical studies on supply chain management, follows such categories. Similarly, Appelqvist et al. (2004) follow similar criteria in their 83 article review with the, further, inclusion of the industrial sector the studies refer to. Such contribution highlights the range of applicability of the design methodologies as another relevant issue to consider. Both general and sector dependent models are proposed.

According to the introduced categories the model behind the proposed DSS is a deterministic, optimal, linear programming model facing the DN strategic planning from a general and non sector dependent view-point. It, further, includes a multi-objective perspective, as discussed in the following.

Multi-Objective Network Optimization Trends

Multi-objective optimization follows the basic idea of matching multiple criteria in the DN design (Deb, 2011). The recent literature on SCM stresses the importance of adopting multiple criteria discussing about its superiority toward single-objective approaches (Harris et al., 2009). Nevertheless, the most of the existing design methods still considers a single objective function, typically related to the DN total annual cost. Melo et al. (2009) estimate that, approximately, 75% of the articles published on this topic in the last decade focuses on cost minimization, 16% on profit maximization and 9%, only, adopts a multi-objective perspective. A gap between the current research state of the art and the expectations exists and needs to be filled.

Aslam & Ng (2010) review the literature about multi-objective optimization in SCM and point out a standard procedure to adopt such method in the DN design. They highlight the following five steps:

1. Formulation of the multi-objective optimization problem;
2. Multi-objective optimization;
3. Trade-off solutions;
4. High-level user information;
5. Selection of the final solution.

Starting from the model formulation and the choice of the optimization criteria two sequential and separate phases occur. The former deals with the multi-objective model optimization, i.e. step 2 and 3. Such phase does not univocally identify an unique solution, i.e. DN structure, but it is able to calculate a set of optimal trade-off solutions among the considered design criteria. The choice of the final solution is done in the latter phase by the panel of the decision makers, i.e. step 4 and 5. Despite the trade-off solution finding is objective and based on operation research algorithms, the final choice among such solutions is not univocal and depends on the personal preferences of the decision makers.

The recent literature facing the DN design through multi-objective optimization follows such a procedure. The basic difference among contributions is due to the considered decisional criteria to optimize. A first set of models focuses on logistic cost and delivery performance bi-objective optimization. Pokharel (2008) presents a two-objective model for decision making in supply chain based on cost minimization and delivery reliability maximization, while Xu et al. (2008) focus on fix and variable logistic costs and delivery time. Furthermore, Rajabalipour et al. (2011) stress that time and cost are familiar criteria for every logistic provider and propose a three-level DN design bi-objective model based on the joint optimization of response time to consumers, transportation and facility costs. Similarly, Olivares-Benitez et al. (2013) focus on the same criteria presenting a mathematical model for DN design and a metaheuristic algorithm for its solution. Concerning the service level experienced by the end-users, Franca et al. (2010) propose a bi-objective model for the DN profit and quality level maximization, while Pishvaee et al. (2010) look for the total cost minimization and the maximization of the responsiveness of the network. The difficulties in expressing through a quantitative function the end-user experienced service level make such models almost qualitative or focused on partial aspects of this issue, e.g. the incidence of *faulties* on the total delivered products.

The increasing interest on environmental sustainability and low carbon patterns influences the SCM and DN design introducing the aforementioned concepts of GSC and SSC. The recent literature investigated such aspects through reviews and reference frameworks (Seuring & Müller, 2008; Ageron et al., 2012; Dekker et al., 2012; Hassini et al., 2012; Rizet et al., 2012; Dasaklis & Pappis, 2013; Seuring, 2013). Following such trend, multi-objective DN design models including the environmental criterion, recently appear on the literature. Frota Neto et al. (2008) develop a bi-objective model for paper industry DN design including production and logistic costs together with the correspondent environmental impacts. Ramudhin et al. (2009) look for the best trade-off between logistic costs and GHG emissions, while Chaabane et al. (2011) present a bi-objective model for SSC design considering costs and emissions during the production, storage and shipment phases. Similarly, Bouzembrak et al. (2011) present a conceptual model for the GSC design with the aim of capturing good compromises between the total cost and the DN environmental influence.

The proposed analysis of the recent literature about multi-objective network design points out the following research trends:

- The DN economic perspective, i.e. cost minimization or profit maximization, is essential and it must not be neglected;
- Promising optimization criteria to look for are about the delivery time and the network environmental impact;
- The service level is difficult to quantify and it is generally included among the model constraints;

- The most of the multi-objective models include two functions, i.e. bi-objective optimization, even if among the encouraged further developments several authors state about the inclusion of a third criterion.

Pioneering tri-objective contributions are outlined by Altiparmak et al. (2006) considering cost, customer service and vehicle capacity utilization joint optimization, Harris et al. (2009), considering cost, environmental impact and uncovered demand functions and, recently, Xifeng et al. (2013), considering cost, service reliability and carbon dioxide emissions. All authors recommend further studies in such a direction.

The proposed DSS follows the reviewed trends and suggestions considering a tri-objective DN optimization model dealing with the operating cost, the carbon footprint and the delivery time minimization. The Pareto frontier approach is adopted to calculate the trade-off efficient solution (previous step 2 and 3 of the Aslam & Ng (2010) procedure). Such a solving approach is known as effective and it is frequently adopted by the literature (Frota Neto et al., 2008; Bouzembrak et al., 2011). A description of its key concepts is in the last part of this section.

Pareto Frontier Multi-Criteria Solving Approach

The generic multi-objective minimization problem states as follows:

$$\min_{x \in S} \left\{ f_1(x), f_2(x), \dots, f_n(x) \right\} \tag{1}$$

$$S = \left\{ x \in R^m : h(x) = 0, g(x) \leq 0 \right\} \tag{2}$$

n objective functions are to be minimized (with $n > 1$) subject to a set of constraints of equality and inequality defining the $x^* \in S \subseteq R^m$ m-dimension admissible region, S (also called objective space). The image of the objective space is the n-dimension region, C, univocally defined through the so-called attained set. Formally:

$$C = \left\{ y \in R^n : y = f(x), x \in S) \right\} \tag{3}$$

For the single-objective scalar space, the concept of optimality is univocal. In the case of minimization problems, $x^* \in S \subseteq R^m$ is the minimum of the scalar objective function $y = f(x) \in C \subseteq R$ if $\forall x \in S : f(x) \geq f(x^*)$. Such a concept does not apply directly to the multi-objective setting. The notion of Pareto optimality is necessary. Basically, a vector $x^* \in S$ is a Pareto optimal for the multi-objective problem (1)-(2) if all other vectors, $x \in S$, have a higher value for at least one of the objective functions f_i in (1), with $i = 1, \dots\dots, n$ or they have the same value for all the objective functions. More properly, a vector $x^* \in S$ is said to be a *weak* Pareto optimum or a *weak* efficient solution of the multi-objective problem if there is no other $x \in S$ such that $f_i(x) < f_i(x^*)$ for all $i = 1,.......,n$. Furthermore, a vector $x^* \in S$ is said to be a *strict* Pareto optimum or a *strict* efficient solution of the multi-objective problem if there is no other $x \in S$ such that $f_i(x) \leq f_i(x^*)$ for all, ith at least one strict inequality. Finally, local Pareto-optimal points are defined as above, except for the attention restriction to a feasible neighborhood of x^*.

The image of the efficient set, i.e. the image of all the efficient solutions, is called Pareto frontier, curve or surface. Its shape indicates the nature of the best trade-off among the different objective functions. The points of the Pareto frontier are called non-dominated points or, shortly, Pareto points. Figure 1 exemplifies such concept for a bi-objective minimization problem.

A subset of the Pareto points is made of the so-called anchor points. Given a multi-objective problem with n objective functions, n anchor points exist. The i-th anchor point comes from the solution of the following single-objective model.

$$\min_{x \in S} f_i(x) \tag{4}$$

$$S = \{x \in R^m : h(x) = 0, g(x) \leq 0\} \tag{5}$$

Figure 1. Pareto frontier for a bi-objective minimization problem, example

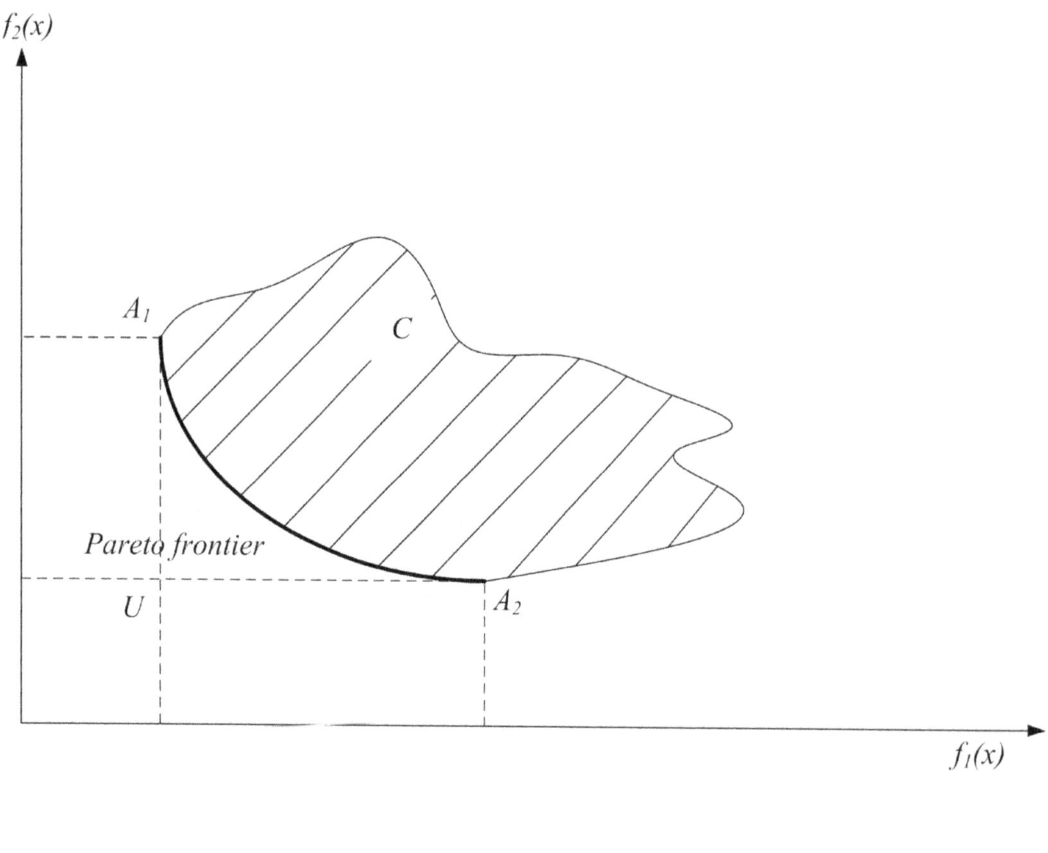

Given the optimal solution $\tilde{x}_i$ of (4)-(5) the related anchor point coordinates in the C space are $A_i\left(f_1\left(\tilde{x}_i\right), f_2\left(\tilde{x}_i\right), \ldots\ldots, f_n\left(\tilde{x}_i\right)\right)$. $\tilde{x}_i$ minimizes $f_i\left(x\right)$ so that all the other admissible solutions for the multi-objective problem are with $f_i\left(x\right) \geq f_i\left(\tilde{x}_i\right)$. As a consequence, each anchor point A_i is the best solution looking at the correspondent objective function, $f_i\left(x\right)$. The point $U\left(f_1\left(\tilde{x}_1\right), f_2\left(\tilde{x}_2\right), \ldots\ldots, f_n\left(\tilde{x}_n\right)\right)$ is the so called utopia point, representing the *virtual* desirable solution for the multi-objective problem. Generally, $U \notin C$ so that $\nexists x \in S : f\left(x\right) = \left(f_1\left(\tilde{x}_1\right), f_2\left(\tilde{x}_2\right), \ldots\ldots, f_n\left(\tilde{x}_n\right)\right)$. On the contrary, the Pareto frontier is the *locus of points* admissible for the multi-objective problem and most desirable given the n objective functions. In Figure 1, $A_1\left(f_1\left(\tilde{x}_1\right), f_2\left(\tilde{x}_2\right)\right)$ and $A_2\left(f_1\left(\tilde{x}_1\right), f_2\left(\tilde{x}_2\right)\right)$ are the two anchor points and $U\left(f_1\left(\tilde{x}_1\right), f_2\left(\tilde{x}_2\right)\right)$ is the utopia point.

Solving a multi-objective optimization model means finding its Pareto frontier. Such curve cannot be computed efficiently in almost all the practical cases. Even if it is theoretically possible to find all these points exactly, they are often of exponential size. Theory demonstrates that they are NP-hard to compute. Approximation techniques to face such goal are used. Generally, starting from the anchor points, a defined number of Pareto points is found, approximating the Pareto frontier. The detailed discussion of the existing methods and techniques to face the Pareto frontier definition is out the aim of the present chapter. In the following, the normalized normal constraint method introduced by Messac, Ismail-Yahaya, & Mattson (2003) is shortly presented, only, as the method implemented in the proposed DSS.

Normalized Normal Constraint Method for Generating the Pareto Frontier

Starting from the multi-objective minimization problem in (1)-(2), the anchor points, A_i, and the utopia point, U, the normalized normal constraint method firstly normalizes the R^n dimensions through the following metrics. Each dimension $f_i\left(x\right)$ is scaled as follows:

$$\varphi_i\left(x\right) = \frac{f_i\left(x\right) - f_i\left(\tilde{x}_i\right)}{\max_j\left\{f_i(\tilde{x}_j)\right\} - f_i\left(\tilde{x}_i\right)} \tag{6}$$

$\varphi_i\left(x\right)$ ranges in $\left[0,1\right]$.

In the normalized R^n hyper-space, the so-called utopia hyper-plane is defined (called utopia line in bi-objective optimization and utopia plane in tri-objective optimization). All the anchor points belong to this hyper-plane and its directions along the axes are univocally identified by n - 1 vectors. For the k-th direction ($k = 1,\ldots\ldots n$ - 1) such vector is:

$$N_k = \varphi\left(\tilde{x}_n\right) - \varphi\left(\tilde{x}_k\right) \tag{7}$$

Furthermore, the Pareto frontier is approximated through a set of Pareto points. To get each of them, the introduced utopia hyper-plane is evenly sampled along each of its directions moving stepwise with a normalized constant increment δ_k so that m_k samples are obtained per each direction. Follows that:

$$\delta_k = \frac{1}{m_k - 1} \quad k = 1,......n - 1 \tag{8}$$

Each sample corresponds to a point of the utopia hyper-plane, called utopia point, X_j, univocally identified by its n coordinates respecting the following equation.

$$X_j = \sum_{k=1}^{n} \alpha_{kj} \bullet \varphi\left(\tilde{x}_k\right) \tag{9}$$

where $0 \le \alpha_{kj} \le 1$ and $\sum_{k=1}^{n} \alpha_{kj} = 1$.

For each utopia point the correspondent Pareto point in the normalized hyper-space comes by solving the following auxiliary single-objective model.

$$\min_{x \in S} \varphi_n\left(x\right) \tag{10}$$

$$S = \left\{x \in R^m : h\left(x\right) = 0, g\left(x\right) \le 0\right\} \tag{11}$$

$$N_k\left(\varphi\left(x\right) - X_j\right)^T \le 0, \; k = 1, \ldots, n - 1 \tag{12}$$

$$\varphi\left(x\right) = \left\{\varphi_1\left(x\right), \varphi_2\left(x\right), \ldots, \varphi_n\left(x\right)\right\} \tag{13}$$

The model minimizes the n-th normalized objective function, $\varphi_n\left(x\right)$, (10), considering the objective space, S, (11). The $n - 1$ vectorial constraints in (12) limit the image of the objective space, C, through a same number of hyper-planes orthogonal to the utopia hyper-plane and including the utopia point, X_j. Figure 2 graphically exemplifies such concepts for the bi-objective and tri-objective cases.

Given the j-th solution, x_j^*, of the model in (10)-(13) and the optimal value $\varphi_n\left(x_j^*\right)$ of the n-th objective function in the normalized hyper-space the correspondent Pareto point is $\left(\varphi_1\left(x_j^*\right), \varphi_2\left(x_j^*\right), \ldots, \varphi_n\left(x_j^*\right)\right)$ in the normalized hyper-space and $\left(f_1\left(x_j^*\right), f_2\left(x_j^*\right), \ldots, f_n\left(x_j^*\right)\right)$ in the R^n original hyper-space. The metric in (6) allows the switch between the two coordinate systems.

Figure 2. Bi and tri-objective configurations for the normalized normal constraint method.

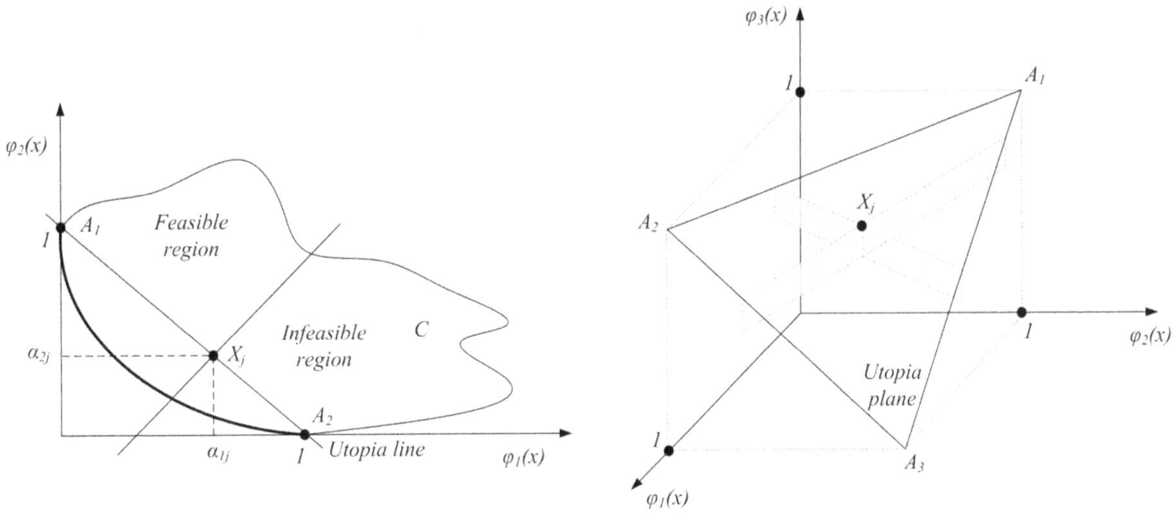

Finally, the introduced normalized normal constraint method for generating the Pareto frontier guarantees that the obtained Pareto points are local Pareto-optimal points. All the anchor points belong to this set. The final check of each local Pareto-optimal point against the previous notion of Pareto global optimality allows to discharge the so-called dominated points, i.e. non Pareto global optimal points, fixing the Pareto frontier. Such a last checking step is, also, called, Pareto filtering procedure.

MULTI-OBJECTIVE MODEL FORMULATION AND DECISION SUPPORT SYSTEM

Starting from the problem statement to define the overall DN structure and allowed product flows, the present section firstly introduces the multi-objective model formulation and, then, presents the DSS features and interfaces. The model formulation discussion is divided between the feasibility constraints, making the so-called *feasibility problem* and the three objective functions about operative cost, carbon footprint and delivery time minimization. Two different formulations of the multi-objective optimization problem are discussed. The former comes from the literature but presents a lack of applicability, the latter overcomes such a weakness. The DSS follows the model formulations and the required input parameters supporting the logistic planner in consolidating the input database, solving the model in accordance with the Pareto optimality concepts and, finally, saving and managing the output to draw useful conclusions about the most effective DN structure and management.

Problem Statement and Network Configuration

The reference network structure consists of a multi-modal four-level (three-stage) forward DN (see Figure 3).

The DN levels include:

Figure 3. Distribution network reference structure

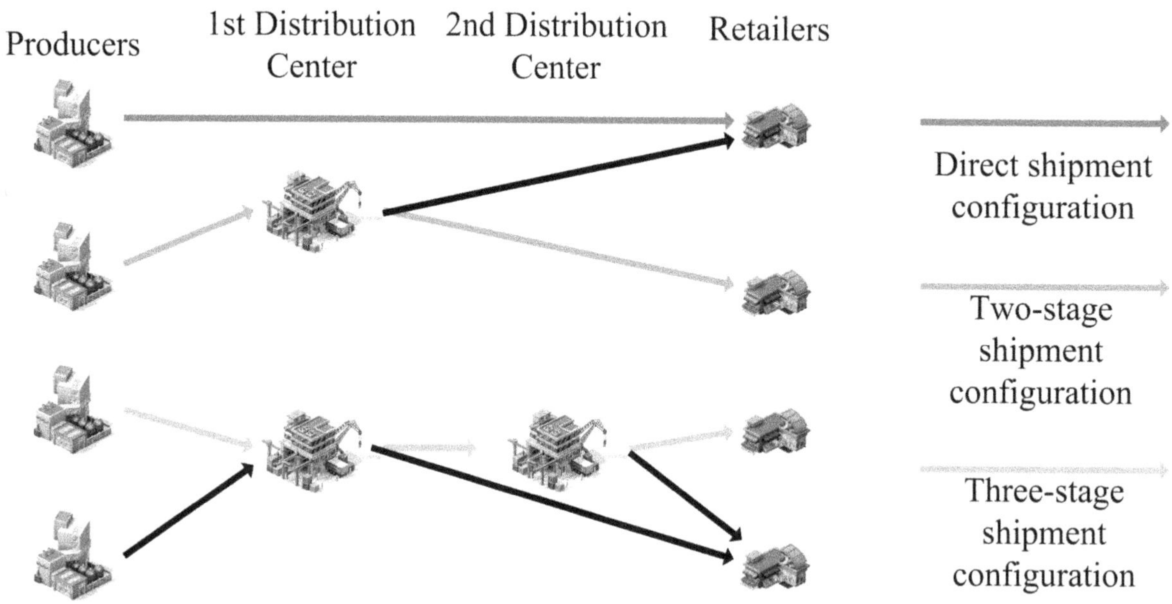

- **Producers:** This level deals with the product, i.e. goods with possible auxiliary services, manufacturing and assembly making it ready for the market. All the production processes are supposed to be concentrated at this level and performed by a production company leading the production process. Outsourcing and part supply are not included in the model focusing on the distribution phase;
- **Central Distributors:** This level, called 1st distribution center in Figure 3, is used in the case of indirect distribution channels to collect the producer output, sorting and delivery the loads to save shipping time and costs and eventually changing the shipping mode (i.e. intermodality concept). In dynamic models the central distributors may be, also, temporary storage centers. Central distributors are generally large logistic platforms, close to the production districts and with high level facilities and available services, e.g. railway, air and/or naval terminals, custom, high handling capacity, etc.;
- **Regional Distributors:** This level, called 2nd distribution center in Figure 3, is used in the case of long channels to further handle the products before the final consumption level. Regional distributors are closer to the end-users and represent the terminal for the so-called *last-mile shipment.* In the case of multi-modal shipments, the frequently allow the further shipping mode change;
- **Retailers:** In this context, the retailers are the DN end-users, having their demand profile to be supplied. The retailers may be large consumers or local vendors selling products to the single end-users and families. They represent the destination of the logistic fluxes.

In this network, there are three possible distribution configurations. Direct shipment (red arrow in Figure 3) links the retailers to the producers, two-stage shipment (blue arrow in Figure 3) considers a short indirect distribution channel with central distributors between producers and retailers. Finally, the three-stage shipment further includes regional distributors. Such distribution configurations define the correspondent network stages and introduce the distributor by-pass option making the DN more flexible.

Furthermore, the problem allows considering a generic number of shipping modes and delivered products. The intermodality concept is included through the possibility of changing the shipping mode in the indirect channels thanks to the available facilities at the distributor sites. As example a load can be shipped by truck from the producer to the central distributors, by train from the central to the regional distributor and, finally, by truck between the regional distributor and the retailer. All combinations are admissible limited by the infrastructures available at each level. Finally, for the sake of simplicity and to face the problem from a long-term strategic view-point, the problem and DSS adopt a static perspective. Expected extensions of the problem statement need to include the suppliers as a further level before the producers and the dynamic DN modeling.

Feasibility Model Analytic Formulation

Aim of the feasibility model is to define the aforementioned objective space, S, in which the admissible solutions lie. In this model, the objective functions are not included and they are detailed, separately, in the following to build the full model formulation, i.e. objective functions and feasibility model. Two analytic formulations of the feasibility model are given starting from the literature and overcoming one of its major lack. Some common notations are, preliminarily, introduced.

Indices

c product index, $c = 1,....,C$;
e central distributor index, $e = 1,....,E$;
h regional distributor index, $h = 1,....,H$;
u shipping mode index, $u = 1,....,U$;
p producer index, $p = 1,....,P$;
r retailer/end-user index, $r = 1,....,R$.

Parameters

hcc_{ec} handling capacity of central distributor e for product c [units/year];
hcr_{hc} handling capacity of regional distributor h for product c [units/year];
pcp_{pc} production capacity of producer p for product c [units/year];
dem_{rc} market demand of retailer/end-user r for product c [units/year].

Feasibility Model Formulation #1

The feasibility model formulation #1 is based on the so-called location allocation problem (LAP) widely discussed by the literature (Manzini, 2012; Manzini et al., 2014). The goals are to define the logistic nodes to activate, the origin, path and destination of the product flows through the DN. The following decisional variables are necessary.

x_{pecu} product c flow from producer p to central distributor e with mode u [units/year];
x_{phcu} product c flow from producer p to regional distributor h with mode u [units/year];

x_{prcu} product c flow from producer p to retailer/end-user r with mode u [units/year];

x_{ehcu} product c flow from central distributor e to regional distributor h with mode u [units/year];

x_{ercu} product c flow from central distributor e to retailer/end-user r with mode u [units/year];

x_{hrcu} product c flow from regional distributor h to retailer/end-user r with mode u [units/year];

The feasibility model formulation, making the DN consistent, is in the following.

$$\sum_{p=1}^{P}\sum_{u=1}^{U} x_{prcu} + \sum_{e=1}^{E}\sum_{u=1}^{U} x_{ercu} + \sum_{h=1}^{H}\sum_{u=1}^{U} x_{hrcu} = dem_{rc} \quad \forall r,c \tag{14}$$

$$\sum_{e=1}^{E}\sum_{u=1}^{U} x_{pecu} + \sum_{h=1}^{H}\sum_{u=1}^{U} x_{phcu} + \sum_{r=1}^{R}\sum_{u=1}^{U} x_{prcu} \le pcp_{pc} \quad \forall p,c \tag{15}$$

$$\sum_{e=1}^{E}\sum_{u=1}^{U} x_{pecu} + \sum_{h=1}^{H}\sum_{u=1}^{U} x_{phcu} + \sum_{r=1}^{R}\sum_{u=1}^{U} x_{prcu} \le pcp_{pc} \quad \forall e,c \tag{16}$$

$$\sum_{r=1}^{R}\sum_{u=1}^{U} x_{hrcu} \le hcr_{hc} \quad \forall h,c \tag{17}$$

$$\sum_{p=1}^{P}\sum_{u=1}^{U} x_{pecu} \ge \sum_{h=1}^{H}\sum_{u=1}^{U} x_{ehcu} + \sum_{r=1}^{R}\sum_{u=1}^{U} x_{ercu} \quad \forall e,c \tag{18}$$

$$\sum_{p=1}^{P}\sum_{u=1}^{U} x_{phcu} + \sum_{e=1}^{E}\sum_{u=1}^{U} x_{ehcu} \ge \sum_{r=1}^{R}\sum_{u=1}^{U} x_{hrcu} \quad \forall h,c \tag{19}$$

$$x_{pecu}, x_{phcu}, x_{prcu}, x_{ehcu}, x_{ercu}, x_{hrcu} \ge 0 \quad \forall p,e,h,r,c,u \tag{20}$$

Equation (14) guarantees the retailer/end-user demand is fully satisfied. In (15) the producer capacity per each product is not exceeded, while the central and regional distributor handling capacity constraints are in (16)-(17). Equations (18)-(19) balance the entry and exit flows at the distribution levels so that all the shipped products are, also, received. Finally, equation (20) give consistence to the decisional variables. The feasibility model includes $C \cdot \left(R + P + 2E + 2H\right)$ constraints (excluding the variable consistence definition), while the decisional variables are

$C \cdot U \cdot \left(PE + PH + PR + EH + ER + HR\right)$.

The proposed formulation of the feasibility model allows knowing the aggregate quantities of product flowing through the DN from the producers to the retailers/end-users. The nodes to activate are, consequently, defined. The key lack of the proposed model formulation is the impossibility to track the flow of each single product unit, especially in the case of indirect distribution channels. This is due to the presence of the distributors, acting as decoupling entities, so that, given the values of the flow variables, it is not possible to retrace the *journey* of each product unit, e.g. unit-load, container, etc. In the recent past, traceability rises as a key factor to consider in the DN design (Bevilacqua et al., 2009; Cheng & Tsai, 2009; Mehrjerdi, 2009). Such an issue is particularly relevant for high value products, with safety and quality properties rapidly decreasing within the lifetime. A brightly example is from the food sector. Nowadays, food traceability is mandatory to get high quality standards and to abide by the regulations actually in force. Both hardware solutions, protocols and reference frameworks to face the food traceability topic are introduced by the recent literature (Regattieri et al., 2007; Grunow & Piramuthu, 2013; Manzini & Accorsi, 2013; Storøy et al., 2013). The following alternative feasibility model formulation allows product traceability and it is suitable for applications in high value and critic sectors such as the food DNs.

Feasibility Model Formulation #2

The model goal is the same as before. This formulation is based on the concept of *path*. A path univocally connects a producer to a retailer/end-user through a shipment configuration. It is identified by its origin and destination, the intermediate distributors (if an indirect channel is used) and the shipping mode used between each couple of consecutive logistic nodes. Each unit of product flows the DN through one and only one path. Data about each path are known in advance so that traceability is ensured. According to the path concept, the model decisional variable become the following:

w_{zc} product c flow through path z [units/year];

where $z = 1,\ldots..,Z$ is the index for the paths. Given a DN, a possible way to define the set of the available paths is to enumerate them starting from the permutations of all the DN entities and discarding those which are infeasible e.g. a combination of shipping mode and logistic node do not match with the available facilities.

The feasibility model formulation, making the DN consistent, is in the following. The operator $a \rhd z$ is used to indicate if the entity a belongs to the path z.

$$\sum_{z : r \rhd z} w_{zc} = dem_{rc} \quad \forall r, c \tag{21}$$

$$\sum_{z:p \triangleright z} w_{zc} \leq pcp_{pc} \ \forall p,c \tag{22}$$

$$\sum_{z:e \triangleright z} w_{zc} \leq hcc_{ec} \ \forall e,c \tag{23}$$

$$\sum_{z:h \triangleright z} w_{zc} \leq hcr_{hc} \ \forall h,c \tag{24}$$

$$w_{zc} \geq 0 \ \forall z,c \tag{25}$$

Equation (21) fully satisfies the demand, while the node capacity constraints are in (22)-(24). Finally, the variables are defined in (25).

The feasibility model includes $C \cdot (R + P + E + H)$ constraints (excluding the variable consistence definition), while the decisional variables are $C \cdot Z$.

Similarities between the two feasibility model formulations are evident. In the latter formulation the constraints about flow balances at the nodes (see (18)-(19)) are unnecessary and already integrated within the path concept.

Tri-Objective Function Formulation

The proposed multi-objective model is based on three functions to minimize. The operating cost function looks for the most effective DN configuration, leading to the lowest global cost of the network. The carbon footprint function aims at minimizing the CO_2 equivalent emissions from the shipments and inbound activities. Finally, the delivery time function forces the DN to be fast with short transit times. The analytic formulation for such objective functions is provided in the following sub-sections.

Operating Cost Objective Function

The following drivers of cost are introduced.

hoc_{ec} inbound handling cost of central distributor *e* for product *c* [€/unit];
hor_{hc} inbound handling cost of regional distributor *h* for product *c* [€/unit];
pop_{pc} inbound production cost of producer *p* for product *c* [€/unit];
so_{pecu} shipping cost from producer *p* to central distributor *e* for product *c* and mode *u* [€/unit]; so_{phcu} shipping cost from producer *p* to regional distributor *h* for product *c* and mode *u* [€/unit];
so_{prcu} shipping cost from producer *p* to retailer/end-user *r* for product *c* and mode *u* [€/unit];

so_{ehcu} shipping cost from central distributor *e* to regional distributor *h* for product *c* and mode *u* [€/unit];
so_{ercu} shipping cost from central distributor *e* to retailer/end-user *r* for product *c* and mode *u* [€/unit];
so_{hrcu} shipping cost from regional distributor *h* to retailer/end-user *r* for product *c* and mode *u* [€/unit].

According to the previous formulation #1 of the feasibility model the operating cost objective function is expressed as:

$$OC^{I} = \sum_{p=1}^{P}\sum_{c=1}^{C} pop_{pc} \cdot \left(\sum_{e=1}^{E}\sum_{u=1}^{U} x_{pecu} + \sum_{h=1}^{H}\sum_{u=1}^{U} x_{phcu} + \sum_{r=1}^{R}\sum_{u=1}^{U} x_{prcu} \right) + \tag{26}$$

$$\sum_{e=1}^{E}\sum_{c=1}^{C} hoc_{ec} \cdot \left(\sum_{h=1}^{H}\sum_{u=1}^{U} x_{ehcu} + \sum_{r=1}^{R}\sum_{u=1}^{U} x_{ercu} \right) + \tag{27}$$

$$\sum_{h=1}^{H}\sum_{c=1}^{C} hor_{hc} \cdot \sum_{r=1}^{R}\sum_{u=1}^{U} x_{hrcu} + \tag{28}$$

$$\sum_{c=1}^{C}\sum_{u=1}^{U}\sum_{p=1}^{P} \left(\sum_{e=1}^{E} so_{pecu} x_{pecu} + \sum_{h=1}^{H} so_{phcu} x_{phcu} + \sum_{r=1}^{R} so_{prcu} x_{prcu} \right) + \tag{29}$$

$$\sum_{c=1}^{C}\sum_{u=1}^{U}\sum_{e=1}^{E} \left(\sum_{h=1}^{H} so_{ehcu} x_{ehcu} + \sum_{r=1}^{R} so_{ercu} x_{ercu} \right) + \tag{30}$$

$$\sum_{c=1}^{C}\sum_{u=1}^{U}\sum_{h=1}^{H}\sum_{r=1}^{R} so_{hrcu} x_{hrcu} \tag{31}$$

The inbound production cost is in (26) for the producers, while the handling one is in (27) for the central distributors and in (28) for the regional distributors. For each entity the correspondent processed units are considered. Finally, the shipment cost for the stages starting from a producer is in (29), for the stages starting from a central distributor is in (30) and for the stages starting from a regional distributor is in (31).

Considering the previous formulation #2 of the feasibility model the variable unitary operating cost associated to each path and product, vo_{zc}, is the following, in [€/unit]:

$$vo_{zc} = \sum_{p:p \rhd z} pop_{pc} + \sum_{e:e \rhd z} hoc_{ec} + \sum_{h:h \rhd z} hor_{hc} + \tag{32}$$

$$\sum_{p:p \rhd z} \sum_{e:e \rhd z} \sum_{u:u \rhd z^{pe}} so_{pecu} + \sum_{p:p \rhd z} \sum_{h:h \rhd z} \sum_{u:u \rhd z^{ph}} so_{phcu} + \sum_{p:p \rhd z} \sum_{r:r \rhd z} \sum_{u:u \rhd z^{pr}} so_{prcu} + \tag{33}$$

$$\sum_{e:e \rhd z} \sum_{h:h \rhd z} \sum_{u:u \rhd z^{eh}} so_{ehcu} + \sum_{e:e \rhd z} \sum_{r:r \rhd z} \sum_{u:u \rhd z^{er}} so_{ercu} + \tag{34}$$

$$\sum_{h:h \rhd z} \sum_{r:r \rhd z} \sum_{u:u \rhd z^{hr}} so_{hrcu} \tag{35}$$

where the notation z^{ab} indicates the part of the path z from node a to node b.

In (32) the production and handling costs for the nodes visited by the path z are included, while (33)-(35) consider the shipping cost for the visited stages. In the case a stage is skipped the correspondent terms are null.

The operating cost objective function becomes the following:

$$OC^{II} = \sum_{z=1}^{Z} \sum_{c=1}^{C} vo_{zc} w_{zc} \tag{36}$$

The variable cost is grouped by path and product in (36).

Carbon Footprint Objective Function

The carbon footprint objective function formulation follows the structure of the operating cost function. The sole differences are the inclusion of the emissions instead of the costs. The unitary emissions are the following.

e_{ec} handling emissions of central distributor e for product c [kg CO_2/unit];

e_{hc} handling emissions of regional distributor h for product c [kg CO_2/unit];

e_{pc} production emissions of producer p for product c [kg CO_2/unit];

e_{pecu} emissions from producer p to central distributor e for product c and mode u [kg CO_2/unit];

e_{phcu} emissions from producer p to regional distributor h for product c and mode u [kg CO_2/unit];
e_{prcu} emissions from producer p to retailer/end-user r for product c and mode u [CO_2/unit];
e_{ehcu} emissions from central distributor e to regional distributor h for product c and mode u [kg CO_2/unit];
e_{ercu} emissions from central distributor e to retailer/end-user r for product c and mode u [kg CO_2/unit];
e_{hrcu} emissions from regional distributor h to retailer/end-user r for product c and mode u [kg CO_2/unit].

According to the previous formulation #1 of the feasibility model the carbon footprint objective function is expressed as:

$$CF^{I} = \sum_{p=1}^{P}\sum_{c=1}^{C} e_{pc} \cdot \left(\sum_{e=1}^{E}\sum_{u=1}^{U} x_{pecu} + \sum_{h=1}^{H}\sum_{u=1}^{U} x_{phcu} + \sum_{r=1}^{R}\sum_{u=1}^{U} x_{prcu} \right) + \tag{37}$$

$$\sum_{e=1}^{E}\sum_{c=1}^{C} e_{ec} \cdot \left(\sum_{h=1}^{H}\sum_{u=1}^{U} x_{ehcu} + \sum_{r=1}^{R}\sum_{u=1}^{U} x_{ercu} \right) + \tag{38}$$

$$\sum_{h=1}^{H}\sum_{c=1}^{C} e_{hc} \cdot \sum_{r=1}^{R}\sum_{u=1}^{U} x_{hrcu} + \tag{39}$$

$$\sum_{c=1}^{C}\sum_{u=1}^{U}\sum_{p=1}^{P} \left(\sum_{e=1}^{E} e_{pecu} x_{pecu} + \sum_{h=1}^{H} e_{phcu} x_{phcu} + \sum_{r=1}^{R} e_{prcu} x_{prcu} \right) + \tag{40}$$

$$\sum_{c=1}^{C}\sum_{u=1}^{U}\sum_{e=1}^{E} \left(\sum_{h=1}^{H} e_{ehcu} x_{ehcu} + \sum_{r=1}^{R} e_{ercu} x_{ercu} \right) + \tag{41}$$

$$\sum_{c=1}^{C}\sum_{u=1}^{U}\sum_{h=1}^{H}\sum_{r=1}^{R} e_{hrcu} x_{hrcu} \tag{42}$$

The emissions at the three levels of the DN are in (37)-(39), while the emissions due to shipments through the stages are in (40)-(42).

Considering the previous formulation #2 of the feasibility model the unitary emissions associated to each path and product, e_{zc}, are the following, in [kg CO_2/unit]:

$$e_{zc} = \sum_{p:p\rhd z} e_{pc} + \sum_{e:e\rhd z} e_{ec} + \sum_{h:h\rhd z} e_{hc} + \quad (43)$$

$$\sum_{p:p\rhd z}\sum_{e:e\rhd z}\sum_{u:u\rhd z^{pe}} e_{pecu} + \sum_{p:p\rhd z}\sum_{h:h\rhd z}\sum_{u:u\rhd z^{ph}} e_{phcu} + \sum_{p:p\rhd z}\sum_{r:r\rhd z}\sum_{u:u\rhd z^{pr}} e_{prcu} + \quad (44)$$

$$\sum_{e:e\rhd z}\sum_{h:h\rhd z}\sum_{u:u\rhd z^{eh}} e_{ehcu} + \sum_{e:e\rhd z}\sum_{r:r\rhd z}\sum_{u:u\rhd z^{er}} e_{ercu} + \quad (45)$$

$$\sum_{h:h\rhd z}\sum_{r:r\rhd z}\sum_{u:u\rhd z^{hr}} e_{hrcu} \quad (46)$$

The unitary emissions for the levels visited by path z for product c are in (43). In (44)-(46) the unitary emissions for the visited stages, path and product are introduced.

The carbon footprint objective function becomes the following:

$$CF^{II} = \sum_{z=1}^{Z}\sum_{c=1}^{C} e_{zc} w_{zc} \quad (47)$$

Delivery Time Objective Function

The delivery time objective function aims to reduce the supply time lapse. The following delivery time, experienced through the DN, are introduced:

t_{ec} storage fix time of central distributor e for product c [h];

t_{hc} storage fix time of regional distributor h for product c [h];

t_{pecu} shipping time from producer p to central distributor e for product c and mode u [h];

t_{phcu} shipping time from producer p to regional distributor h for product c and mode u [h];

t_{prcu} shipping time from producer p to retailer/end-user r for product c and mode u [h];

t_{ehcu} shipping time from central distributor e to regional distributor h for product c and mode u [h];

t_{ercu} shipping time from central distributor e to retailer/end-user r for product c and mode u [h];

t_{hrcu} shipping time from regional distributor h to retailer/end-user r for product c and mode u

[h].

According to the previous formulation #1 of the feasibility model the delivery time objective function is expressed as:

$$DT^{I} = \sum_{e=1}^{E}\sum_{c=1}^{C} t_{ec}\left(\sum_{h=1}^{H}\sum_{u=1}^{U} x_{ehcu} + \sum_{r=1}^{R}\sum_{u=1}^{U} x_{ercu}\right) + \tag{48}$$

$$\sum_{h=1}^{H}\sum_{c=1}^{C} t_{hc}\sum_{r=1}^{R}\sum_{u=1}^{U} x_{hrcu} + \tag{49}$$

$$\sum_{c=1}^{C}\sum_{u=1}^{U}\sum_{p=1}^{P}\left(\sum_{e=1}^{E} t_{pecu}x_{pecu} + \sum_{h=1}^{H} t_{phcu}x_{phcu} + \sum_{r=1}^{R} t_{prcu}x_{prcu}\right) + \tag{50}$$

$$\sum_{c=1}^{C}\sum_{u=1}^{U}\sum_{e=1}^{E}\left(\sum_{h=1}^{H} t_{ehcu}x_{ehcu} + \sum_{r=1}^{R} t_{ercu}x_{ercu}\right) + \tag{51}$$

$$\sum_{c=1}^{C}\sum_{u=1}^{U}\sum_{h=1}^{H}\sum_{r=1}^{R} t_{hrcu}x_{hrcu} \tag{52}$$

DT is expressed in [h·unit] and it is the weighed sum of the storage and shipping times for all levels and stages of the DN. The storage times are in (48)-(49), while the shipping times are in (50)-(52). The weights are the amount of product units flowing through the considered level and stage. Low storage time nodes and short connections are, consequently, preferred.

Considering the previous formulation #2 of the feasibility model the delivery time for each path and product, t_{zc}, is the following, in [h]:

$$t_{zc} = \sum_{e:e\triangleright z} t_{ec} + \sum_{h:h\triangleright z} t_{hc} + \tag{53}$$

$$\sum_{p:p\triangleright z} \sum_{e:e\triangleright z} \sum_{u:u\triangleright z^{pe}} t_{pecu} + \sum_{p:p\triangleright z} \sum_{h:h\triangleright z} \sum_{u:u\triangleright z^{ph}} t_{phcu} + \sum_{p:p\triangleright z} \sum_{r:r\triangleright z} \sum_{u:u\triangleright z^{pr}} t_{prcu} + \tag{54}$$

$$\sum_{e:e\triangleright z} \sum_{h:h\triangleright z} \sum_{u:u\triangleright z^{eh}} t_{ehcu} + \sum_{e:e\triangleright z} \sum_{r:r\triangleright z} \sum_{u:u\triangleright z^{er}} t_{ercu} + \tag{55}$$

$$\sum_{h:h\triangleright z} \sum_{r:r\triangleright z} \sum_{u:u\triangleright z^{hr}} t_{hrcu} \tag{56}$$

The storage fix time for the visited nodes is in (53) and the shipping time for the stages is in (54)-(56). The delivery time objective function becomes the following:

$$DT^{II} = \sum_{z=1}^{Z} \sum_{c=1}^{C} t_{zc} w_{zc} \tag{57}$$

Similarities between the carbon footprint and the delivery time analytic formulations are evident.

Given the model solution, the *ex-post* calculation of the average operating cost, carbon footprint and delivery time per unit of product is possible, for the two model formulations, as follows:

$$AOC^{I} = \frac{OC^{I}}{2\left(\sum_{c=1}^{C} \sum_{u=1}^{U} \left(\sum_{p=1}^{P} \left(\sum_{e=1}^{E} x_{pecu} + \sum_{h=1}^{H} x_{phcu} + \sum_{r=1}^{R} x_{prcu}\right) + \sum_{e=1}^{E} \left(\sum_{h=1}^{H} x_{ehcu} + \sum_{r=1}^{R} x_{ercu}\right) + \sum_{h=1}^{H} \sum_{r=1}^{R} x_{hrcu}\right)\right)} \tag{58}$$

$$AOC^{II} = \frac{OC^{II}}{\sum_{z=1}^{Z} \sum_{c=1}^{C} w_{zc}} = \frac{OC^{II}}{\sum_{r=1}^{R} \sum_{c=1}^{C} dem_{rc}} \tag{59}$$

$$ACF^{I}=\frac{CF^{I}}{2\left(\sum_{c=1}^{C}\sum_{u=1}^{U}\left(\sum_{p=1}^{P}\left(\sum_{e=1}^{E}x_{pecu}+\sum_{h=1}^{H}x_{phcu}+\sum_{r=1}^{R}x_{prcu}\right)+\sum_{e=1}^{E}\left(\sum_{h=1}^{H}x_{ehcu}+\sum_{r=1}^{R}x_{ercu}\right)+\sum_{h=1}^{H}\sum_{r=1}^{R}x_{hrcu}\right)\right)} \tag{60}$$

$$ACF^{II}=\frac{CF^{II}}{\sum_{z=1}^{Z}\sum_{c=1}^{C}w_{zc}}=\frac{CF^{II}}{\sum_{r=1}^{R}\sum_{c=1}^{C}dem_{rc}} \tag{61}$$

$$ADT^{I}=\frac{DT^{I}}{2\left(\sum_{c=1}^{C}\sum_{u=1}^{U}\left(\sum_{p=1}^{P}\left(\sum_{e=1}^{E}x_{pecu}+\sum_{h=1}^{H}x_{phcu}+\sum_{r=1}^{R}x_{prcu}\right)+\sum_{e=1}^{E}\left(\sum_{h=1}^{H}x_{ehcu}+\sum_{r=1}^{R}x_{ercu}\right)+\sum_{h=1}^{H}\sum_{r=1}^{R}x_{hrcu}\right)\right)} \tag{62}$$

$$ADT^{II}=\frac{DT^{II}}{\sum_{z=1}^{Z}\sum_{c=1}^{C}w_{zc}}=\frac{DT^{II}}{\sum_{r=1}^{R}\sum_{c=1}^{C}dem_{rc}} \tag{63}$$

From the delivery time perspective, the advantages of adopting the formulation #2 of the optimization model are double.

- The model formulation allows tracking the product shipments. Each unit of product is associated to exactly one path so that the whole chain is monitored and the delivery time is exactly known, i.e. the value of t_{zc} for the correspondent path. As discussed above, in formulation #1 the presence of the distributors, acting as decoupling entities, causes the loss of the product traceability for the indirect channels so that only the average delivery time is computable;
- the model formulation is more compact than in formulation #1.

The overall tri-objective model formulation includes the three objective functions, i.e. $\{OC^{I}, CF^{I}, DT^{I}\}$ and $\{OC^{II}, CF^{II}, DT^{II}\}$, respectively, and the correspondent feasibility model, i.e. (14)-(20) and (21)-(25). Such a model is behind the DSS described in the next sub-section supporting the decision makers in the design process, from the data loading, through the Pareto frontier generation, to the final DN configuration choice.

Decision Support System for the Distribution Network Design

To support the decision makers in the DN design process a DSS is developed and proposed in this sub-section. Aim of the DSS is to assist the logistic planner through all the DN design phases, from the input database consolidation to the DN final configuration definition. Figure 4 represents the DSS structure.

The first DSS phase deals with the input database consolidation. The information required concerns the network nodes and the delivered products. The formers are distinguished by their geographical location, by the transport modes and available logistic facilities and by the handling-production capacity or market demand. The latter relevant features are the production cost and emissions. The second DSS phase is the model formulation selection. If product traceability is a required feature of the considered DN, the decision maker has to adopt the model formulation #2. On the contrary, the model formulation #1 is usable. Model formulation #2 requires an additional DSS phase, i.e. the path feature evaluation. First of all, the feasible paths are enumerated. As introduced, a feasible path is distinguished by a combination of transport modes and logistic facilities available at the network nodes that belong to the path. For each couple of nodes a, b the travelled distance d_{ab}^{u} is evaluated. Finally, the path cost (vo_{zc}), emissions (e_{zc}) and delivery time (t_{zc}) are calculated. The DSS integrates transport cost $\left(\alpha^{uc}\left(d_{ab}^{u}\right)\right)$, emission $\left(\gamma^{uc}\left(d_{ab}^{u}\right)\right)$ and time $\left(\beta^{uc}\left(d_{ab}^{u}\right)\right)$ functions to support the users in the case of lack of field data. Figure 5 represents the DSS structure and the travelled distance evaluation dashboard

Furthermore, the third DSS phase purpose is the DN optimization problem definition. The input data, collected in the first phase, enable to calculate the model parameters. Each constraint and the objective functions are formulated according to these values. The DSS phases from 1 to 3 are necessary to point out the DN optimal configurations, representing the aforementioned Pareto points. Each point is generated by the DSS phase 4. Both the anchor points and the Pareto frontier come from the normalized normal constraint method presented in the Background section. Given the Pareto frontier, the DN final configuration has to be selected. The selection criteria is subjective. The DSS fifth phase enables the decision maker to select the preferred Pareto point. Finally, DSS last phase analyzes the DN final configuration in terms of number of DN level and transport modes providing aggregate performance indicators.

The DSS phases are coded in a customized tool developed using MS Visual Studio™ integrated development environment. The tool provides a user-friendly interface (Figure 5) and it is connected to several auxiliary softwares. The input database is uploaded using MS Access™ and this application collects the DN final configuration features. The path travelled distance as well the geographical representation of the DN final configuration is obtained using MS MapPoint™. The normalized normal constraint method is implemented using AMPL™ modeling language, whereas the Pareto points are determined by GUROBI™ optimization solver.

The next section applies the proposed multi-objective model and DSS to design a case study DN for the fresh food industry. Particularly, to guarantee product high quality standards as well to abide the regulations actually in force, traceability is a required feature. Thus, the adopted model formulation is the formulation #2.

CASE STUDY

A case study about fresh food DN design is the focus of the present section. The first sub-section investigates perishability, known as the product most relevant feature. An approach is proposed to quantify it and the multi-objective optimization model is updated to include such a characteristic. The second sub-section presents the case study input data, whereas the results, obtained adopting the optimization model and the DSS, are, finally, proposed.

Figure 4. DSS structure for DN design

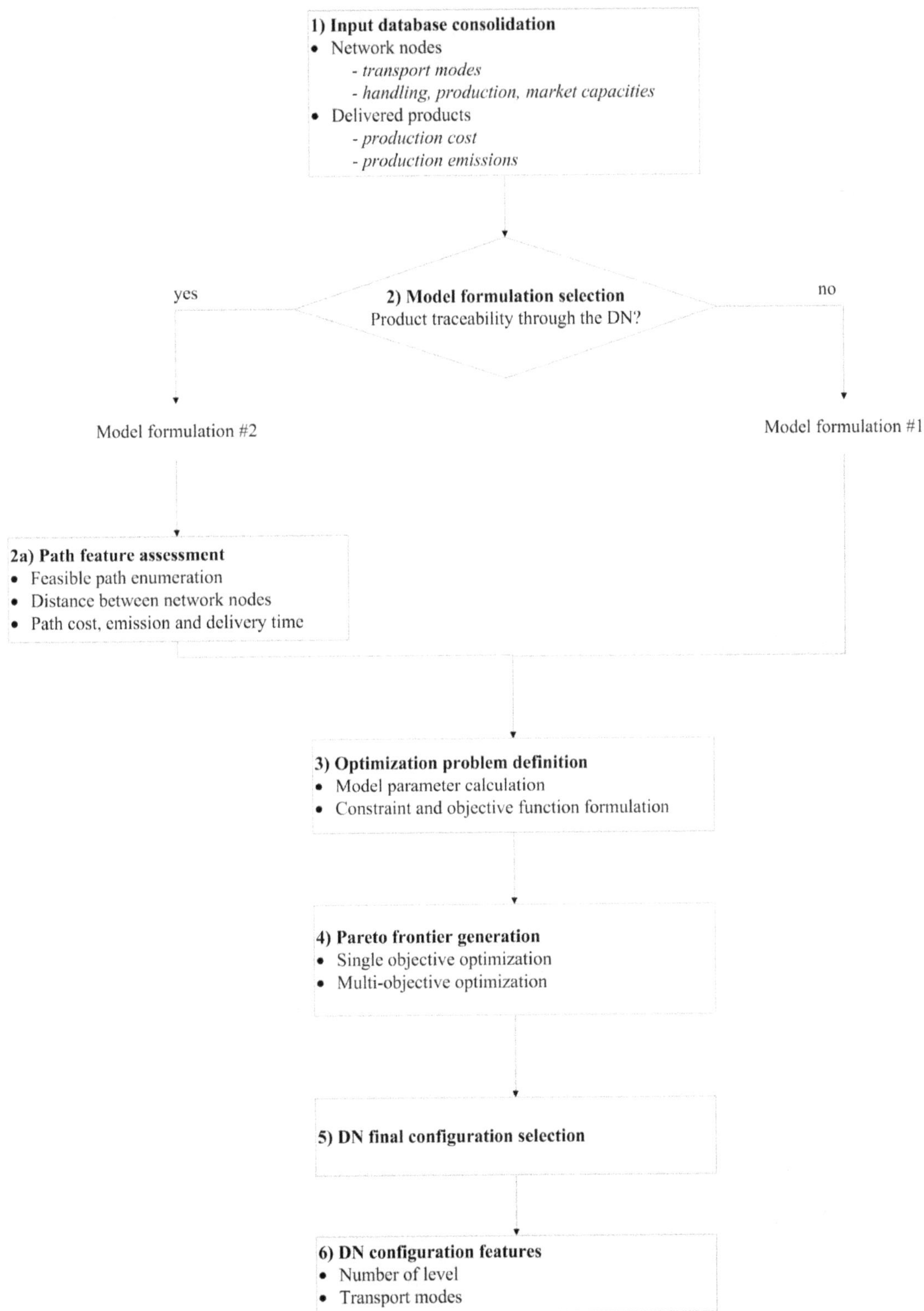

Figure 5. DSS structure and the travelled distance evaluation dashboard

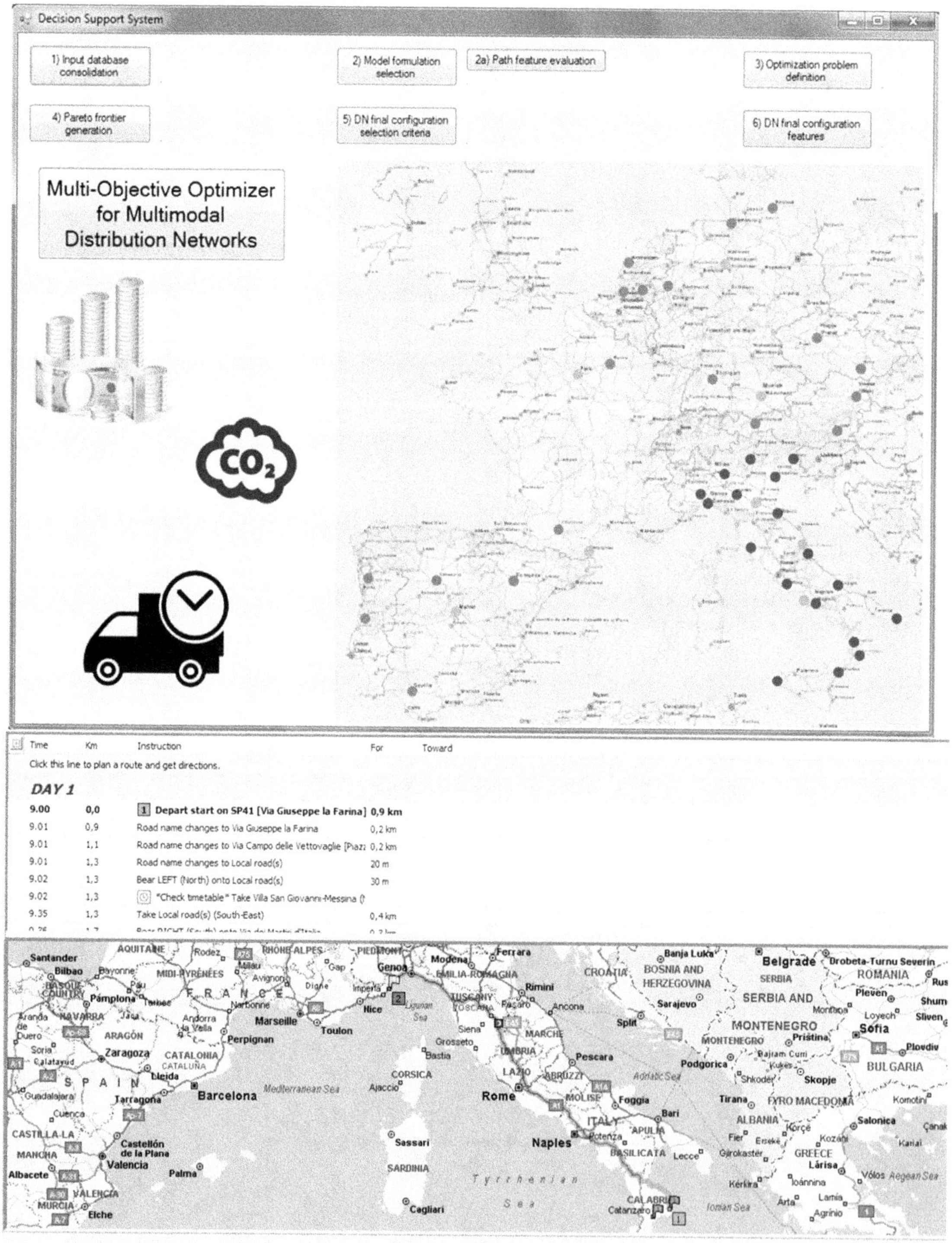

Fresh Food Distribution Network

The fresh food DNs differ from traditional DNs because of the peculiarity of the food produces toward manufacture goods. The fresh food quality is not constant over the produce lifetime. It rapidly decreases reaching a value of zero after the so-called shelf life (Osvald & Stirn, 2008). Several models express the correlation between quality and time. According to Osvald & Stirn (2008), this chapter adopts a linear relation, called quality loss function. Such a function is defined in (64) and shown in Figure 6.

$$\xi_{zc} = \min\left\{\frac{1}{QRP_c}\left(\frac{sl_c - t_{zc}}{sl_c}\right), 1\right\} \tag{64}$$

ξ_{zc} is a function of the shelf life, sl_c, the delivery time, t_{zc}, and the so-called Quality Reduction Point, QRP_c, representing the produce shelf life limit not affecting the produce quality. The greater the transport time, the lower the quality of delivered food is (Widodo, Nagasawa, Morizawa, & Ota, 2006).

Considering the quality loss function, the multi-objective optimization model upgrades are in the following. The equation numbers refer to the previous correspondent equation formulations.

$$\sum_{z\,:r \rhd z} \xi_{zc} w_{zc} = dem_{rc} \qquad \forall r,c \,(22')$$

Figure 6. Quality loss function concept

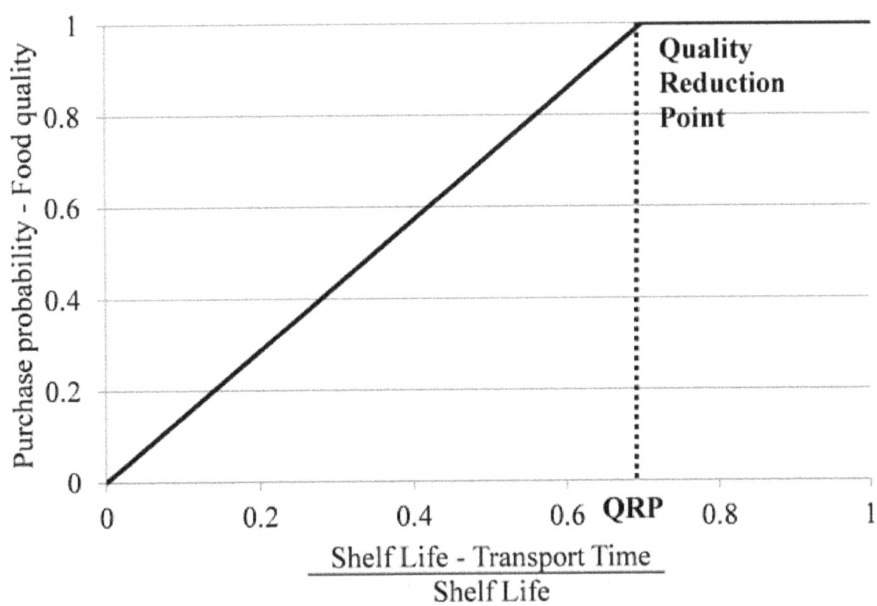

$$vo_{zc} = \sum_{p:p\rhd z} \left(1 - \xi_{zc}\right) pop_{pc} + \sum_{e:e\rhd z} hoc_{ec} + \sum_{h:h\rhd z} hor_{hc} + (35')$$

$$e_{zc} = \sum_{p:p\rhd z} \left(1 - \xi_{zc}\right) e_{pc} + \sum_{e:e\rhd z} e_{ec} + \sum_{h:h\rhd z} e_{hc} + (47')$$

The full market demand supply constraint, considering the produce quality decrease over the delivery time, is in (22'). Only ξ_{zc} of the delivered produces are purchased by the retailers. Thus, $(1 - \xi_{zc})$ of these amounts increase the path operating cost (35') and the carbon footprint (47') with no contribution to the market demand satisfaction. Consequently, the production cost, pop_{pc}, and emissions, e_{pc}, of the produces not affected by the quality reduction are not included in the objective functions, because of they are constant and do not depend on the decisional variables.

Input Data

The proposed case study focuses on the DN to distribute a variety of six fruit and vegetable produces from a set of Italian producers to a set of European retailers. Both short shelf life produces, i.e. Brussels sprouts and tomatoes, and long shelf life produces, i.e. apples, oranges, pears and potatoes are included in the mix. The aforementioned Figure 5 shows the network geography. 18 producers, i.e. production areas, 18 retailers, i.e. consumption areas, and 8 DCs are considered. The producers are medium farmers joining a national consortium, while the retailers are, generally, middle-size city companies located in Austria, the Czech Republic, France, Germany, the Netherlands, Portugal and Spain. Finally, three alternative shipment modes are possible: truck, train and plane. The detail of the DN nodes and features are in Table 1.

In Table 1, for each of the production areas the available shipment facilities are shown (truck is available for all facilities) together with the production capacities related to the crop yield of a reference common month. An interesting further development deals with the analysis of the DN changes among time periods due to the produce seasonality effects. Similarly, this table shows the retailer available shipment facilities (truck is available for all facilities) together with the aggregate market demand for each produce. The table data refers to the same reference month.

Eight are the DCs: Bologna, Hannover, Madrid, Munich, Napoli, Paris, Perpignan, and Teramo. These are for both two-stage and three-stage shipment strategies, i.e. all the DCs can be a central or regional hub. In addition, for all the DCs the three shipment modes are supposed to be available. For the sake of simplicity it is, further, supposed that all the DCs are able to handle all the quantities of produces it is convenient to be shipped through such nodes.

Concerning the produces, Table 2 summarizes their features, cost and emissions per ton.

Particularly, the produce shelf life is from (Caccioni, 2005), the QRPs are from a customized market analysis and the production costs are from a recent agri-food market analysis (ISMEA, 2012). Finally, the production emissions are from (Pimentel, 2006). Furthermore, the production cost and emissions are considered the same for all producers. The cost, time and emission parameters used to define the objective functions are in Table 3.

Time and cost functions are from a market survey, summarizing information provided by several international carriers. The emission data are available from the Ecoinvent (2007). All the transport func-

Table 1. Producers and retailers production capacity, market demand and transport facilities.

	Network node	Airport	Railway station	Potatoes	Apples	Pears	Brussels sprouts	Oranges	Tomatoes
				[ton/ month]	[ton/ month]	[ton/ month]	[ton/ month]	[ton/ month]	[ton/ month]
Producers	Alessandria	×	×	892	32	94	26	-	52
	Brescia	×	×	95	20	-	-	-	12
	Catanzaro	×	×	507	71	66	63	7350	150
	Cesena	×	×	181	178	235	-	-	124
	Cosenza	×	×	785	25	101	151	9761	405
	Foggia	×	×	296	14	-	836	669	273
	Genoa	✓	✓	434	-	-	24	-	178
	Grosseto	×	×	43	15	98	-	-	15
	Latina	✓	✓	102	64	32	54	635	861
	Lecce	×	×	689	-	40	36	649	143
	Messina	×	×	409	27	169	48	3269	124
	Parma	×	×	28	-	-	-	-	-
	Pescara	×	×	173	66	32	37	-	149
	Salerno	×	×	1147	98	240	391	1226	1364
	Trapani	×	×	122	-	-	67	1231	179
	Trento	×	×	209	7212	-	-	-	-
	Udine	×	×	330	282	108	-	-	-
	Venice	✓	✓	52	90	469	21	-	173
Retailers	Amsterdam	✓	✓	-	2182	-	-	6986	-
	Braga	×	×	740	102	-	-	122	-
	Breda	×	×	-	383	-	-	1226	-
	Brno	×	×	-	134	154	190	559	326
	Coimbra	×	×	412	154	-	-	112	-
	Eindhoven	×	×	-	596	-	-	1909	-
	Essen	×	×	-	391	190	130	313	377
	Graz	×	×	-	-	-	108	127	171
	Innsbruck	×	×	-	-	-	104	158	133
	Prague	✓	✓	-	334	134	225	1398	815
	Reims	×	×	-	-	135	-	677	-
	Rostock	×	×	-	138	132	111	110	133
	Seville	×	×	1319	271	-	-	-	-
	Stuttgart	×	×	-	394	191	130	316	380
	Toulouse	×	×	-	-	185	-	1637	-
	Valladolid	×	×	591	122	-	-	-	-
	Wien	✓	✓	-	-	-	152	831	468
	Zaragoza	×	×	1263	260	-	-	-	-

Table 2. Produce shelf life, QRP, production cost and emissions

Produces		Potatoes	Apples	Pears	Brussels sprouts	Oranges	Tomatoes
Shelf Life [h]	sl_c	5760	2880	2880	120	1440	168
QRP [%]	QRP_c	12.5	25.0	50.0	80.0	50.0	85.7
Production cost [€/ton]	pop_{pc}	190	540	580	370	260	550
Production emissions [kg $CO_{2\ EQ.}$/ton]	e_{pc}	34.68	50.38	54.11	34.04	21.17	90.24

Table 3. Transport cost, time and emission functions

		Truck	Train	Airplane
Transport cost [€/ton]	$\alpha^{uc}\left(d^u_{ab}\right)$	$0.2872 \cdot d^{u\ -0.183}_{ab}$	$0.8705 \cdot d^{u\ -0.373}_{ab}$	$869.32 \cdot d^{u\ -0.861}_{ab}$
Transport time [h]	$\beta^{uc}\left(d^u_{ab}\right)$	$0.0067 \cdot d^{u\ 1.1465}_{ab}+2.823$	$10^{-10} \cdot d^{u\ 3}_{ab}-10^{-5} \cdot d^{u\ 2}_{ab}+0.0829 \cdot d^u_{ab}+22.914$	$0.0023 \cdot d^u_{ab}+19.254$
Transport emissions [kg $CO_{2\ EQ.}$/ton]	$\gamma^{uc}\left(d^u_{ab}\right)$	$d^u_{ab} \cdot 0.484$	$d^u_{ab} \cdot 1.670$	$d^u_{ab} \cdot 0.03920$

tions are considered equal for all produces. Finally, the inbound handling cost (hoc_{ec}, hor_{hc}), emissions (e_{ec}, e_{hc}) and the storage fix time (t_{ec}, t_{hc}) are assumed equal for all the produces and DCs. According to the data provided by several DCs, hoc_{ec}, hor_{hc} = 2.641 €/ton, e_{ec}, e_{hc} = 0.6826 kg $CO_{2\ EQ.}$/ton, t_{ec}, t_{hc} = 12 h.

The introduced input parameters lead to, approximately, 101,700 feasible paths, 610,308 decisional variables and 610,524 constraints. The following sub-section presents and discusses the case study key results and the DN most effective configuration.

Results and Discussion

According to the network design strategy outlined in the DSS, the first step of the analysis deals with the independent optimization of the three functions leading to the best solutions in terms of time, cost and environmental impact, i.e. anchor points. Such solutions are called *Time solution, Cost solution* and *Environmental solution*, respectively. The values of the three objective functions $\left(AOC^{II}, ACF^{II}, ADT^{II}\right)$ are computed and compared. Particularly, the *Time solution* is (620.6,1116.0,33.4), the *Cost solution* is (102.6,111.2,323.7) and the *Environmental solution* is (118.9,136.1, 274.5). For each optimization function, Figure 7 compares the three anchor points.

Results highlight two opposite trends. On one side by achieving the goal of the delivery time minimization a very high increase of both the network cost and carbon footprint is experienced (the blue bars in Figure 7). On the other side, the cost and environmental solutions lead to comparable network

Figure 7. Comparison among time, cost and impact DN solutions

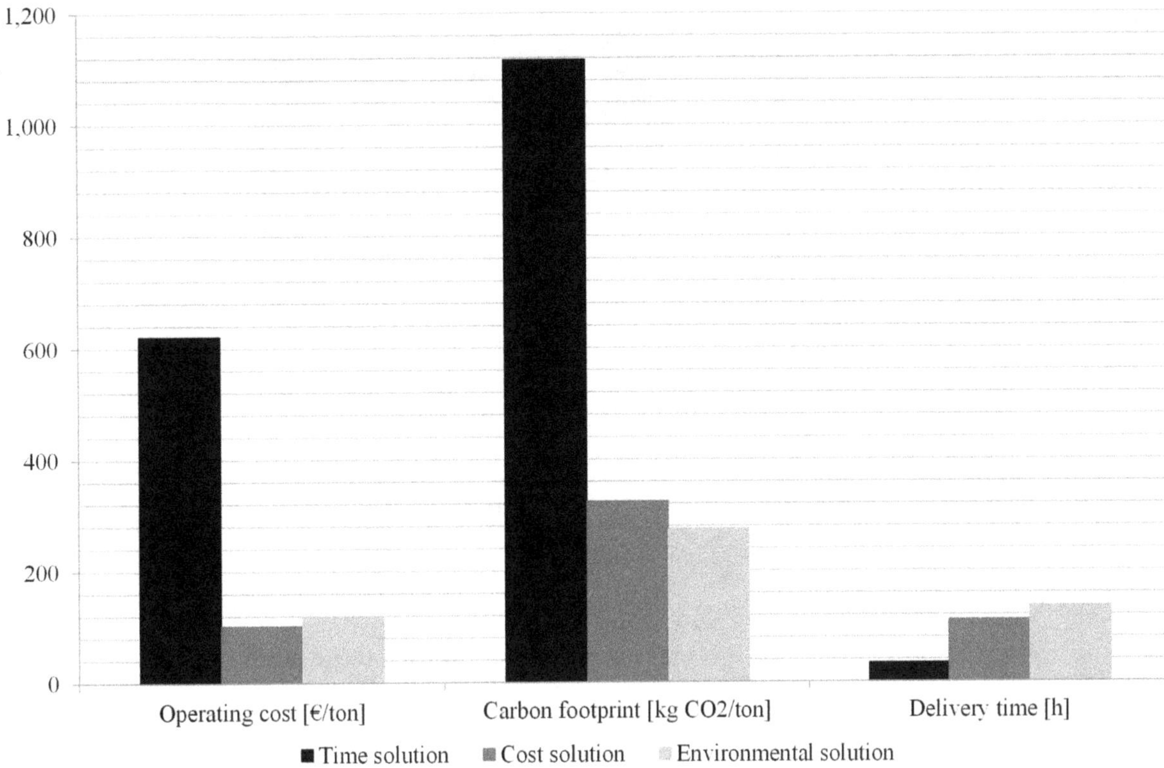

configurations, so that no relevant performance decrease is experienced on one objective function by minimizing the other one (the red and green bars in Figure 7). On the contrary, in such solutions the delivery time is significantly higher respect to the *Time solution*. Such outcomes are in accordance with the chosen shipment strategies, as shown in the following Figure 8, distinguishing between short and long shelf life produces.

The *Time solution* and the *Cost solution* prefer truck direct shipments for short shelf life produces, while the two-stage and three-stage shipments are adopted for long shelf life produces, except for the *Time solution*, where approximately the 70% of the long shelf life produces is shipped through a truck direct shipment and the 27% is shipped through truck and plane, to save time.

Given the three anchor points and applying the normalized normal constrain method, the Pareto frontiers, both aggregately and split by produce are determined by the DSS. Figure 9 depicts such frontier for tomatoes in the X-carbon footprint, Y-delivery time, Z-operating cost diagram.

Each Pareto point represents a feasible DN configuration distinguished by a tuple of operating cost, carbon footprint and delivery time values. The DN final configuration has to be selected among these points.

Despite the Pareto point selection criteria is subjective and based on the logistic manager preferences, to identify an effective trade-off DN configuration the DSS can implements a weighted sum approach. Considering one produce at time, the function $G(pp)$ is calculated for each Pareto point (

Figure 8. Shipment strategies for Time, Cost and Environmental solution

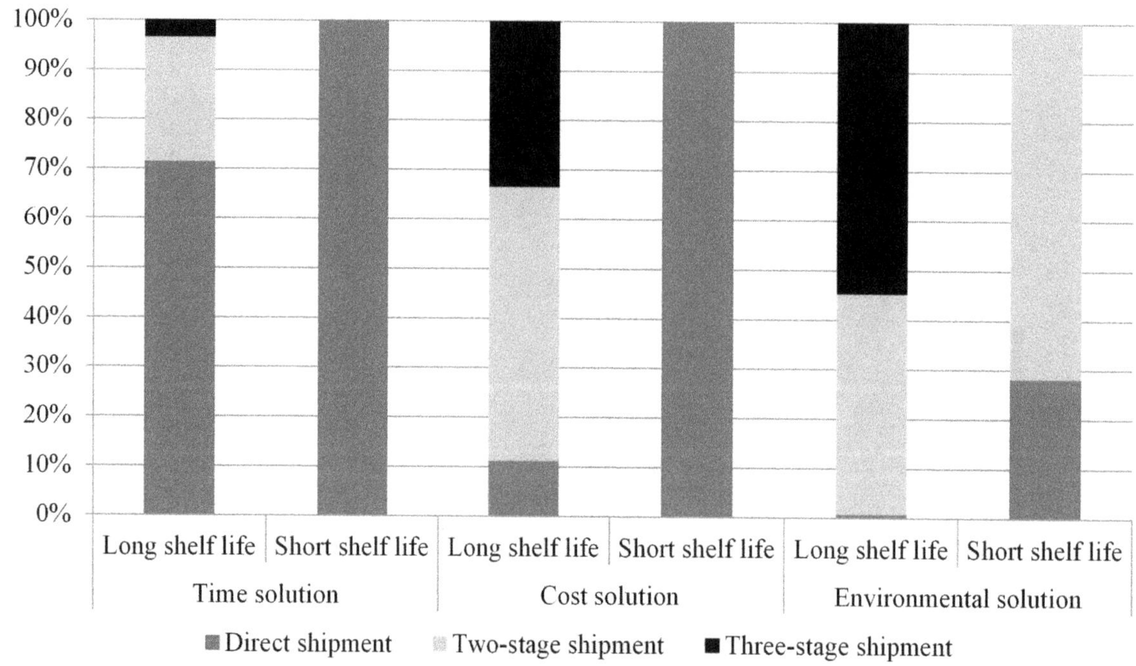

pp) as presented in (65). The weighting factors w , w_{CO_2} and w_t are tuned according to the relevance of the cost, environmental and time objective functions for the planner of the DN. The proposed DN final configuration minimizes $G(pp)$ value.

$$G(pp) = w \frac{OC^{II}(pp)}{\min OC^{II}(pp)} + w_{CO_2} \frac{CF^{II}(pp)}{\min CF^{II}(pp)} + w_t \frac{DT^{II}(pp)}{\min DT^{II}(pp)} \quad (65)$$

According to the defined approach, seven scenarios are tested varying the weighting coefficient set $\left(w , w_{CO_2}, w_t\right)$. to stress the relevance of each objective function. Table 4 provides the tested sets.

In Set 1 the three functions have the same relevance, while in Sets 2 to 4 one weighting coefficient is doubled respect to the others. Finally, in Sets 5 to 7 two weighting coefficients are doubled respect to the remaining one. For each set, produce and Pareto point, the function $G(pp)$ is computed. Figure 10 represents, for each set, the DN objective function values determined by of the selected Pareto points.

The DN objective functions are differently affected by the weighting sets. The operating cost is almost equal for all the tested set. On the contrary, the carbon footprint and the delivery time are significantly affected by the set choice. As example, Set 5 limits the environmental emissions but it increases the shipment time. On the contrary the sets with lowest delivery time, e.g. Sets 4 and 6, lead to a high emission DN.

As discussed, the DN final configuration is selected according to the relevance of each objective function from the logistic planner view-point. In this context, the Set 5 scenario appears a good compromise.

Figure 9. Multi-objective optimization: Pareto frontier for tomatoes

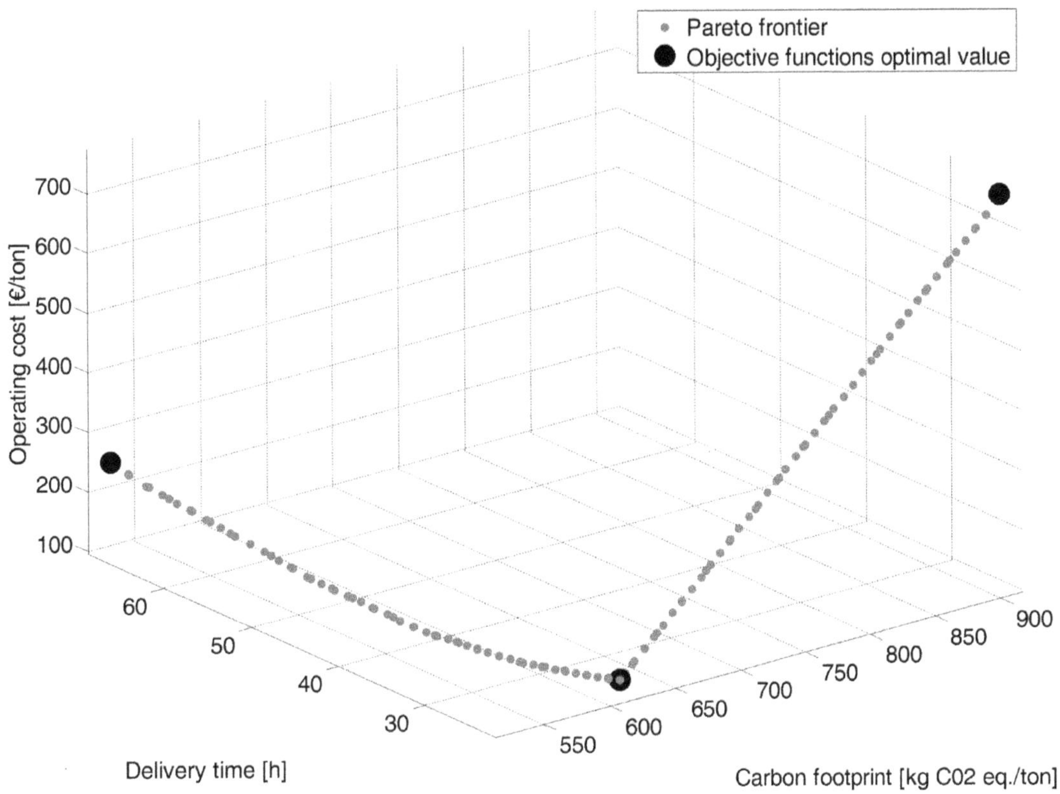

Table 4. Weighting coefficient sets

Set	1	2	3	4	5	6	7
w	0.33	0.50	0.25	0.25	0.40	0.40	0.20
w_{CO_2}	0.33	0.25	0.50	0.25	0.40	0.20	0.40
w_t	0.33	0.25	0.25	0.50	0.20	0.40	0.40

The three objective function values are (102.9, 326.1, 110.1). In such a DN configuration, the carbon footprint reduction balances the delivery time decrease. Furthermore, the shipment time objective function increase does not affect the produce quality level. Finally, no produce is wasted during shipment. Table 5 presents the detail of the DN final configuration split by shipped produce.

The truck direct shipment is the delivery strategy suggested for the short shelf life produces, while the truck and rail mode, with two-stage or three-stage shipment strategy, is preferred for the most of the long shelf life produces.

Figure 10. Operating cost, carbon footprint and delivery time values for the tested sets

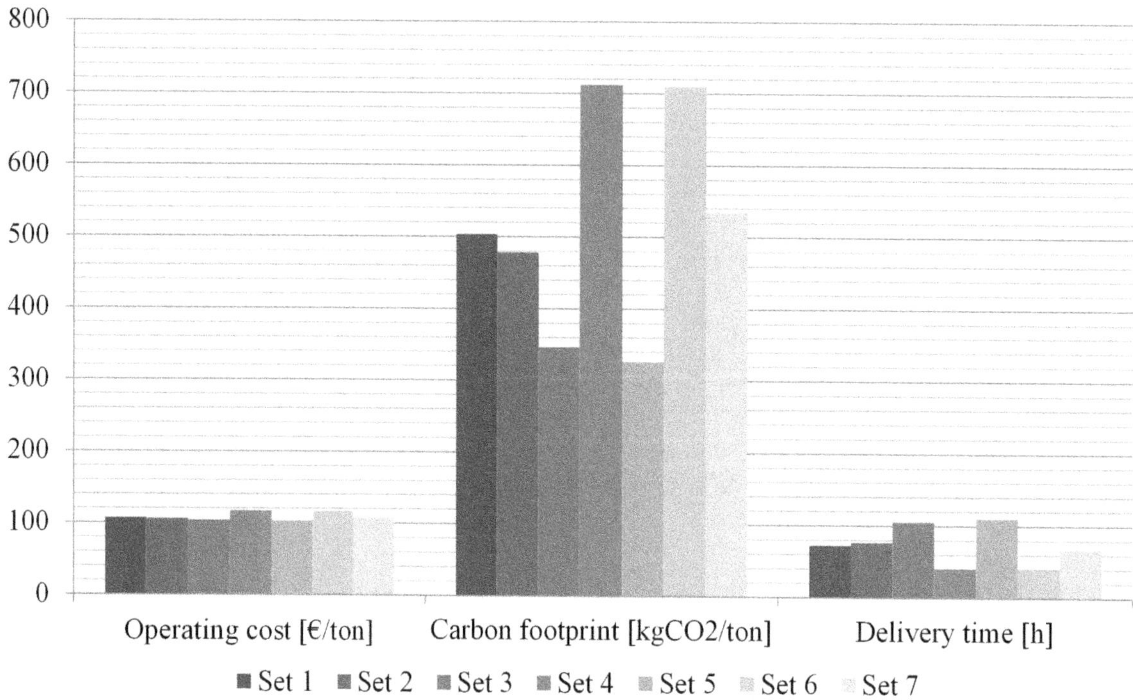

Table 5. DN final configuration

	Tomatoes	Brussels sprouts	Oranges	Apples	Potatoes	Pears
Operating cost [€/ton]	105.7	120.8	106.7	81.4	116.8	71.9
Carbon footprint [kg $CO_{2\,EQ}$/ton]	625.7	687.2	256.1	307.9	344.4	254.1
Delivery time [h]	24.6	27.3	141.3	68.1	134.1	68.1
Truck shipment [%]	100%	100%	0%	28%	0%	31%
Truck and rail shipment [%]	0%	0%	95%	52%	79%	56%

FUTURE RESEARCH DIRECTIONS

The DN design and SCM receive a growing attention by the literature due to their high impact on the producer, distributor and retailer performances. The proposed tri-objective model and DSS support the strategic design of a DN from the operating cost, environmental impact and delivery time view-points. Starting from this contribution and the reviewed state of the art, several research opportunities emerge.

Following the classification criteria proposed in the Introduction and Background sections of the present chapter some future research guidelines are outlined in the following.

- Concerning the DN structure, the inclusion of a further level and stage to include the producer suppliers is possible. In such a way the focus is extended from the final products to include the raw materials and the supplied components. The concept of extended supply chain is, consequently, modeled;
- The inclusion of the reverse loop flows, from the end-users to the collectors and recyclers, is possible to convey to the frequently discussed close-loop supply chain network. This DN structure is proposed by the literature on green SCM, following a *cradle-to-cradle* paradigm, while lower attention is paid on optimizing such a network from a multi-objective perspective;
- The proposed model and DSS do not include a service level objective function. The necessity to overcome the approach based on the full demand supply by introducing penalties and stochastic distributions of the unsuccessful deliveries is to be addressed;
- The inclusion in both the model and the DSS of the temporal dynamics, typical of the DN tactical and operational planning, is a further possible extension of the present chapter.

The proposed future research directions, together with the others emerging in the next future, are to be addressed following a structured quantitative methodology. The background analysis, to better fix the problem, is preliminary to the analytic model update and the subsequent inclusion of new DSS modules. Finally, the model and DSS validation against multiple consistent case studies, taken from several industrial sectors with their own features and peculiarities, is to be done. Feedbacks and upgrades complete the research methodology.

CONCLUSION

This chapter addresses the effective design of multimodal distribution networks (DNs) from a multi-objective perspective. Three of the most relevant optimization drivers are jointly considered looking for the multi-objective optimal network configuration. The operating cost function stresses the importance of effective DNs to save money making the retailer/end-user full price convenient, the carbon footprint looks for environmentally sustainable DNs able to mitigate the impact on the climate change and to reduce emissions due to production, storage and shipment activities. Finally, the delivery time function forces to speed the distribution process to promptly supply the market demand.

Starting from the analysis of the current scenario on the DN design methodologies and the review of the most recent state of the art on the multi-objective network optimization trends, the Pareto frontier multi-criteria solving approach is described and fully adopted to support the DN configuration though a multi-objective linear programming model. Two model formulations are discussed stressing the key differences in terms of product traceability. Both the feasibility model and the objective function analytic formulations are fully described and commented. The tri-objective model is behind an optimizer decision support system (DSS) supporting the planner in the steps to design an effective supply chain network. Both the DSS logic of working and the supporting user-friendly customized tool, implementing the process, are extensively discussed and their application is exemplified through a case study taken from the fresh food industry. Particularly, the peculiarities of food produces are considered to update the general multi-objective model, e.g. produce perishability. The data are from Italian producers distributing products to multiple European countries. Short and long shelf life products are studied to highlight differences in the most effective shipping strategies. The key outcomes stress opposite trends between operating cost and environmental impact on one side and the time objective function on the other. Fast DNs are, generally, expensive and with high impact on the carbon emissions. Effective trade-offs are found to speed the distribution process for the sole high perishable fresh produces.

Concluding remarks highlight opportunities to enlarge the research boundaries to include further DN stages and the so called reverse logistics, whose aim is to collect end-of-life products to be recovered and regenerated before being distributed to the market with relevant cost and environmental savings.

REFERENCES

Ageron, B., Gunasekaran, A., & Spalanzani, A. (2012). Sustainable supply management: An empirical study. *International Journal of Production Economics*, *140*(1), 168–182. doi:10.1016/j.ijpe.2011.04.007

Altiparmak, F., Gen, M., Lin, L., & Paksoy, T. (2006). A genetic algorithm approach for multi-objective optimization of supply chain networks. *Computers & Industrial Engineering*, *51*(1), 196–215. doi:10.1016/j.cie.2006.07.011

Ambrosino, D., & Grazia Scutellà, M. (2005). Distribution network design: New problems and related models. *European Journal of Operational Research*, *165*(3), 610–624. doi:10.1016/j.ejor.2003.04.009

Apaiah, R. K., & Hendrix, E. M. T. (2005). Design of a supply chain network for pea-based novel protein foods. *Journal of Food Engineering*, *70*(3), 383–391. doi:10.1016/j.jfoodeng.2004.02.043

Appelqvist, P., Lehtonen, J.-M., & Kokkonen, J. (2004). Modelling in product and supply chain design: Literature survey and case study. *Journal of Manufacturing Technology Management*, *15*(7), 675–686. doi:10.1108/17410380410555916

Aslam, T., & Ng, A. H. C. (2010). Multi-Objective Optimization for Supply Chain Management : A Literature Review and New Development. *Proceedings of the 8th International Conference on Supply Chain Management and Information Systems* (pp. 1–8). IEEE.

Beamon, B. M. (1998). Supply chain design and analysis : Models and methods. *International Journal of Production Economics*, *55*(3), 281–294. doi:10.1016/S0925-5273(98)00079-6

Bevilacqua, M., Ciarapica, F. E., & Giacchetta, G. (2009). Business process reengineering of a supply chain and a traceability system: A case study. *Journal of Food Engineering*, *93*(1), 13–22. doi:10.1016/j.jfoodeng.2008.12.020

Bouzembrak, Y., Allaoui, H., Goncalves, G., & Bouchriha, H. (2011). A multi-objective green supply chain network design. *Proceedings of the 2011 4th International Conference on Logistics* (pp. 357–361). IEEE. doi:10.1109/LOGISTIQUA.2011.5939315

Caccioni, D. R. (2005). Ortofrutta & marketing. Rome: Agra.

Chaabane, A., Ramudhin, A., & Paquet, M. (2011). Designing supply chains with sustainability considerations. *Production Planning and Control*, *22*(8), 727–741. doi:10.1080/09537287.2010.543554

Cheng, Y.-H., & Tsai, Y.-L. (2009). Factors influencing shippers to use multiple country consolidation services in international distribution centers. *International Journal of Production Economics*, *122*(1), 78–88. doi:10.1016/j.ijpe.2008.12.022

Dasaklis, T. K., & Pappis, C. P. (2013). Supply chain management in view of climate change : An overview of possible impacts and the road ahead. *Journal of Industrial Engineering and Management*, *6*(4), 1124–1138. doi:10.3926/jiem.883

Deb, K. (2011). Multi-Objective Optimization Using Evolutionary Algorithms : An Introduction. In L. Wang, A.H.C. Ng, K. Deb (Ed.), Multi-objective Evolutionary Optimisation for Product Design and Manufacturing (pp. 1–24). London: Springer-Verlag.

Dekker, R., Bloemhof, J., & Mallidis, I. (2012). Operations Research for green logistics – An overview of aspects, issues, contributions and challenges. *European Journal of Operational Research*, *219*(3), 671–679. doi:10.1016/j.ejor.2011.11.010

Ecoinvent data v.2.0: ecoinvent reports no. 1-25. (2007). Ecoinvent. Dübendorf: Swiss center for life cycle inventory.

Eskigun, E., Uzsoy, R., Preckel, P. V., Beaujon, G., Krishnan, S., & Tew, J. D. (2005). Outbound supply chain network design with mode selection, lead times and capacitated vehicle distribution centers. *European Journal of Operational Research*, *165*(1), 182–206. doi:10.1016/j.ejor.2003.11.029

Faccio, M., Persona, A., Sgarbossa, F., & Zanin, G. (2014). Sustainable SC through the complete reprocessing of end-of-life products by manufacturers: A traditional versus social responsibility company perspective. *European Journal of Operational Research*, *233*(2), 359–373. doi:10.1016/j.ejor.2013.03.027

Franca, R. B., Jones, E. C., Richards, C. N., & Carlson, J. P. (2010). Multi-objective stochastic supply chain modeling to evaluate tradeoffs between profit and quality. *International Journal of Production Economics*, *127*(2), 292–299. doi:10.1016/j.ijpe.2009.09.005

Frota Neto, J. Q., Bloemhof-Ruwaard, J. M., van Nunen, J. E. E., & van Heck, E. (2008). Designing and evaluating sustainable logistics networks. *International Journal of Production Economics*, *111*(2), 195–208. doi:10.1016/j.ijpe.2006.10.014

Grunow, M., & Piramuthu, S. (2013). RFID in highly perishable food supply chains – Remaining shelf life to supplant expiry date? *International Journal of Production Economics*, *146*(2), 717–727. doi:10.1016/j.ijpe.2013.08.028

Harris, I., Mumford, C., & Naim, M. (2009). The Multi-Objective Uncapacitated Facility Location Problem for Green Logistics. *Proceedings of the IEEE Congress on Evolutionary Computation* (pp. 2732–2739). Trondheim: IEEE. doi:10.1109/CEC.2009.4983285

Hassini, E., Surti, C., & Searcy, C. (2012). A literature review and a case study of sustainable supply chains with a focus on metrics. *International Journal of Production Economics*, *140*(1), 69–82. doi:10.1016/j.ijpe.2012.01.042

Ioannou, G. (2005). Streamlining the supply chain of the Hellenic sugar industry. *Journal of Food Engineering*, *70*(3), 323–332. doi:10.1016/j.jfoodeng.2004.01.041

ISMEA. (2012). *Prezzi medi mensili per prodotto - Frutta e ortaggi - Mercato all'origine.* Retrieved from http://www.ismea.it/flex/cm/pages/ServeBLOB.php/L/IT/IDPagina/133

Klibi, W., Martel, A., & Guitouni, A. (2010). The design of robust value-creating supply chain networks: A critical review. *European Journal of Operational Research, 203*(2), 283–293. doi:10.1016/j.ejor.2009.06.011

Lambert, D., Cooper, M., & Pagh, J. (1998). Supply chain management: Implementation issues and research opportunities. *International Journal of Logistic Management, 9*(2), 1–20. doi:10.1108/09574099810805807

Manzini, R. (2012). A top-down approach and a decision support system for the design and management of logistic networks. *Transportation Research Part E, Logistics and Transportation Review, 48*(6), 1185–1204. doi:10.1016/j.tre.2012.06.004

Manzini, R., & Accorsi, R. (2013). The new conceptual framework for food supply chain assessment. *Journal of Food Engineering, 115*(2), 251–263. doi:10.1016/j.jfoodeng.2012.10.026

Manzini, R., Accorsi, R., & Bortolini, M. (2014). Operational planning models for distribution networks. *International Journal of Production Research, 52*(1), 89–116. doi:10.1080/00207543.2013.828168

Manzini, R., Gamberi, M., Gebennini, E., & Regattieri, A. (2008). An integrated approach to the design and management of a supply chain system. *International Journal of Advanced Manufacturing Technology, 37*(5-6), 625–640. doi:10.1007/s00170-007-0997-9

ZareMehrjerdi, Y. Z. (2009). Excellent supply chain management. *Assembly Automation, 29*(1), 52–60. doi:10.1108/01445150910929866

Melo, M. T., Nickel, S., & Saldanha-da-Gama, F. (2009). Facility location and supply chain management – A review. *European Journal of Operational Research, 196*(2), 401–412. doi:10.1016/j.ejor.2008.05.007

Messac, A., Ismail-Yahaya, A., & Mattson, C. A. (2003). The normalized normal constraint method for generating the Pareto frontier. *Structural and Multidisciplinary Optimization, 25*(2), 86–98. doi:10.1007/s00158-002-0276-1

Moncayo-Martínez, L. A., & Zhang, D. Z. (2011). Multi-objective ant colony optimisation: A metaheuristic approach to supply chain design. *International Journal of Production Economics, 131*(1), 407–420. doi:10.1016/j.ijpe.2010.11.026

Olivares-Benitez, E., Ríos-Mercado, R. Z., & González-Velarde, J. L. (2013). A metaheuristic algorithm to solve the selection of transportation channels in supply chain design. *International Journal of Production Economics, 145*(1), 161–172. doi:10.1016/j.ijpe.2013.01.017

Osvald, A., & Stirn, L. Z. (2008). A vehicle routing algorithm for the distribution of fresh vegetables and similar perishable food. *Journal of Food Engineering, 85*(2), 285–295. doi:10.1016/j.jfoodeng.2007.07.008

Pimentel, D. (2006). *Impacts of organic farming on the efficiency of energy use in agriculture.* Itaca: University, Organic Center - Cornell.

Pishvaee, M. S., Farahani, R. Z., & Dullaert, W. (2010). A memetic algorithm for bi-objective integrated forward/reverse logistics network design. *Computers & Operations Research, 37*(6), 1100–1112. doi:10.1016/j.cor.2009.09.018

Pokharel, S. (2008). A two objective model for decision making in a supply chain. *International Journal of Production Economics, 111*(2), 378–388. doi:10.1016/j.ijpe.2007.01.006

Rajabalipour Cheshmehgaz, H., Desa, M. I., & Wibowo, A. (2013). A flexible three-level logistic network design considering cost and time criteria with a multi-objective evolutionary algorithm. *Journal of Intelligent Manufacturing, 24*(2), 277–293. doi:10.1007/s10845-011-0584-7

Ramaa, A., Rangaswamy, T. M., & Subramanya, K. N. (2009). A Review of literature on performance measurement of supply chain network. *Proceedings of the 2009 Second International Conference on Emerging Trends in Engineering & Technology* (pp. 802–807). IEEE. doi:10.1109/ICETET.2009.18

Ramudhin, A., Chaabane, A., & Paquet, M. (2009). On the design of sustainable green supply chains. In *Proceedings of the 2009 International Conference on Computers & Industrial Engineering* (pp. 979–984). Troyes: IEEE. doi:10.1109/ICCIE.2009.5223686

Regattieri, A., Gamberi, M., & Manzini, R. (2007). Traceability of food products: General framework and experimental evidence. *Journal of Food Engineering, 81*(2), 347–356. doi:10.1016/j.jfoodeng.2006.10.032

Rizet, C., Browne, M., Cornelis, E., & Leonardi, J. (2012). Assessing carbon footprint and energy efficiency in competing supply chains: Review – Case studies and benchmarking. *Transportation Research Part D, Transport and Environment, 17*(4), 293–300. doi:10.1016/j.trd.2012.01.002

Seuring, S. (2013). A review of modeling approaches for sustainable supply chain management. *Decision Support Systems, 54*(4), 1513–1520. doi:10.1016/j.dss.2012.05.053

Seuring, S., & Müller, M. (2008). From a literature review to a conceptual framework for sustainable supply chain management. *Journal of Cleaner Production, 16*(15), 1699–1710. doi:10.1016/j.jclepro.2008.04.020

Soni, G., & Kodali, R. (2012). A critical review of empirical research methodology in supply chain management. *Journal of Manufacturing Technology Management, 23*(6), 753–779. doi:10.1108/17410381211253326

Storøy, J., Thakur, M., & Olsen, P. (2013). The TraceFood Framework – Principles and guidelines for implementing traceability in food value chains. *Journal of Food Engineering, 115*(1), 41–48. doi:10.1016/j.jfoodeng.2012.09.018

Widodo, K. H., Nagasawa, H., Morizawa, K., & Ota, M. (2006). A periodical flowering–harvesting model for delivering agricultural fresh products. *European Journal of Operational Research, 170*(1), 24–43. doi:10.1016/j.ejor.2004.05.024

Xifeng, T., Ji, Z., & Peng, X. (2013). A multi-objective optimization model for sustainable logistics facility location. *Transportation Research Part D, Transport and Environment, 22*, 45–48. doi:10.1016/j.trd.2013.03.003

Xu, J., Liu, Q., & Wang, R. (2008). A class of multi-objective supply chain networks optimal model under random fuzzy environment and its application to the industry of Chinese liquor. *Information Sciences*, *178*(8), 2022–2043. doi:10.1016/j.ins.2007.11.025

KEY TERMS AND DEFINITIONS

Carbon Footprint: The set of the environmental emissions caused by a certain entity, e.g. organization, product, person, etc., over a defined time period or for a predefined activity.

Delivery Time: The time lapse to supply a product from the producer to the retailer/end-user. It includes shipping time, storage time, handling and processing time at the intermediate distributors.

Distribution Network: The physical structures and the intelligence tools necessary to ship products and services from the producers to the final consumers.

Decision Support System: The customized ICT platform, based on the integration of different languages and auxiliary tools, supporting the decision making process.

Operating Delivery Cost: The fix and variable costs to supply a product from the producer to the retailer/end-user. It includes, level costs, e.g. production, handling, storage, etc., and stage costs due to the product shipment.

Level: In distribution networks, a level is the origin and/or the destination of the supplied products. Each level includes nodes of the same typology, e.g. suppliers, producers, distributors, buyers, final-users, etc.

Multi-Objective Optimization: A multi-criteria decision making strategy, based on mathematical optimization problems, involving two or more objective functions to be optimized, simultaneously.

Pareto Frontier: In multi-objective optimization, it is the *locus of point*, in the objective space, of the non-dominated efficient solutions so that it is impossible to make any target better off without making at least another target worse off.

Stage: In distribution networks, a stage connects two subsequent levels defining the product flow direction.

Supply Chain Management: The integration of key business processes from end-user through original suppliers, that provides product, service and information that add value for customers and other stakeholders.

Chapter 14
A Literature Survey for Hazardous Materials Transportation

Serpil Erol
Gazi University, Turkey

Zafer Yilmaz
Turkish Military Academy, Turkey

ABSTRACT

Transportation has the greatest importance in logistics. The main focus for the carriers is the cost of transportation. Transportation of hazardous materials (hazmat) is a special kind of transportation due to freight transported. Causalities due to the accidents caused by vehicles that are carrying hazardous materials will be intolerable. For hazmat transportation, in addition to transport costs, risk of transporting hazmat also has to be considered. Many researchers studied on hazmat transportation problems in order to propose optimal solutions with respect to cost, risk, emergency response, facility location etc. In this study, a literature survey of articles about hazmat transportation was prepared. The articles published in refereed journal from 1973 to 2014 were taken into consideration. The articles were also classified according to their main focuses and hazmat type carried.

INTRODUCTION

Hazardous materials (also called as dangerous goods) are defined as any substance or material capable of causing harm to people, property and environment (Erkut et al., 2007). Dangerous goods are sorted into nine classes according to their properties by United Nations. These are explosive substances and articles, gases, flammable liquids, flammable solids, oxidizing substances, toxic substances, infectious substances, radioactive material, corrosive substances, miscellaneous dangerous substances and articles. The rulers in countries and the carriers must take a special care for transporting hazmats. If there is no accident then transportation of hazmats will not be different from transporting other goods. When an accident occurs then the harm to population and environment will be inevitable. Although it is vital to

DOI: 10.4018/978-1-4666-8648-9.ch014

keep information of hazmats only few countries have special departments for transporting hazmats. These departments keep a database for all hazmat incidents and their results. Figure 1 shows the distribution of accidents by hazmat class in 2013 in USA according to information of PHMSA - US Department of Transportation Pipeline and Hazardous Materials.

The Figure 1 illustrates that the majority of the accidents/incidents are flammable-combustible liquids incidents.

Hazmat can be transported by all kind of modes (air, highway, railway, water and pipeline), but they are mainly transported by road. Most of the incidents caused by hazmat carrying vehicles were resulted in fatalities, injuries and environmental damage. If an accident cause hazmat release from the vehicle, then these accidents are called as incidents. In 2013, there were 15,919 total incidents resulting 12 fatalities and $81,365,866 damages in USA (Hazmat Intelligence Portal, U.S. Department of Transportation).

The special importance of hazmat transportation forces the researchers to study on how to decrease risk or which route is the best for hazmat transportation. Risk reduction is the main subject for hazmat transportation but the cost of the transportation has to be considered equally. The researchers also searched the emergency response models and evacuation methodologies after hazmat accidents. Accident analysis was studied by the researchers to have an understanding of the results after the hazmat accidents. Not only the hazmat transportation but also hazmat waste transportation was studied by the researchers especially about the nuclear waste. From 1973 to 2014, there were 179 articles which were published in refereed journals about hazmat transportation.

Figure 1. Accidents by hazmat class in 2013 (PHMSA, 2014)

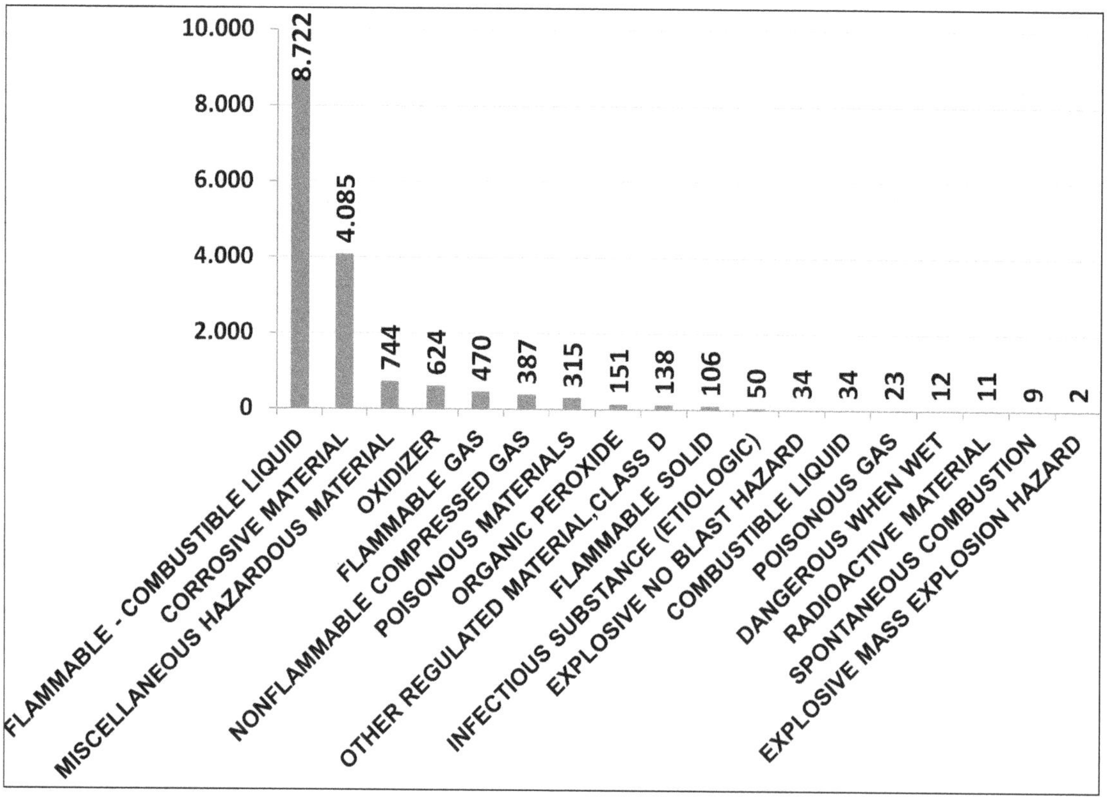

In this study, a literature survey of articles about hazmat transportation was prepared. The articles published in refereed journals were taken into consideration. First, the articles were classified into eight parts (risk, routing, routing and scheduling, network design, emergency response, accident analysis) and the general explanations of these subjects were given. Later, the articles were listed in detail according to these eight classifications. The figures about the classification according to publication years and hazmat type of the articles were also given.

HAZMAT TRANSPORTATION LITERATURE

Hazmat transportation was one of the main subjects of the researchers especially for the last 30 years. The special importance of hazmat transportation forced the researchers to study on how to decrease risk or which route is the best for hazmat transportation. From early 1970s hazmat transportation was one of the main concerns of researchers. Hazmat transportation literature from 1973 to 2014 can be seen in Figure 2. Figure 2 shows the number of hazmat-transportation related papers published in refereed journals.

When the articles were classified according to hazmat class, most of the researchers did not take into consideration the type of hazmat carried by vehicles. Figure 3. shows the classification of articles with respect to the type of hazmat carried. Mainly class-2 and class-3 hazmats are researched by the authors. It can be understood from the Figure 3 that none of the researchers have studied on the hazmat type class-8 and class-9.

The transport cost is one of the main concerns of logistic firms. But when it comes to hazmat transportation, risk of transporting hazardous materials has to be considered as important as transport cost. Causalities due to the accidents caused by trucks that are carrying hazardous materials will be intolerable. Majority of the researchers have studied on risk assessment of hazmat transportation to find optimal networks for reducing hazardous material (hazmat) transportation risks. Routing of hazmat vehicles considering cost and risk also studied by the researchers. Emergency response models, scheduling the hazmat transportation, facility location, hazmat waste transportation and network design problems for hazmat transportation were the other subjects studied by some researchers.

Figure 2. Hazmat articles with respect to publication years

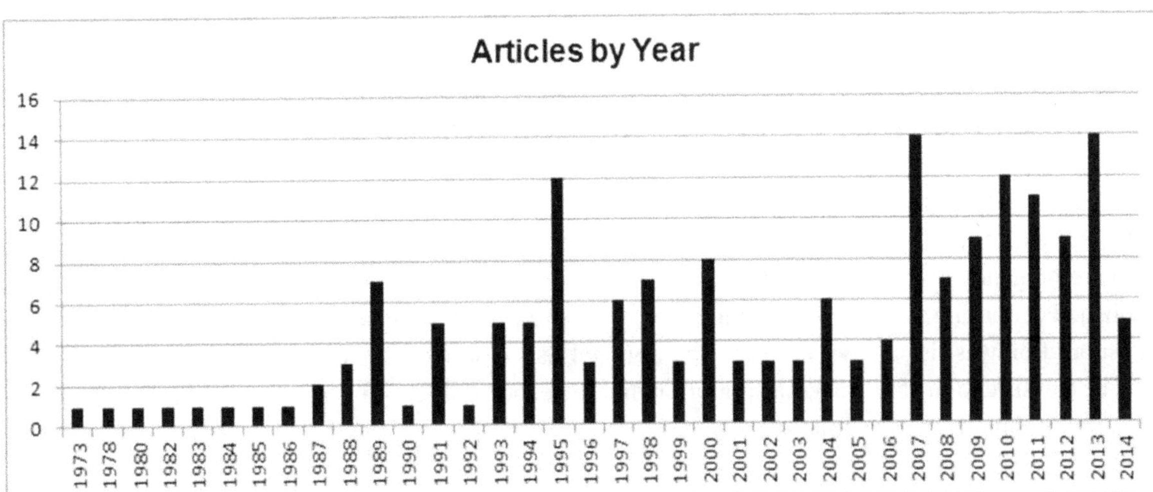

Figure 3. Number of articles with respect to hazmat type

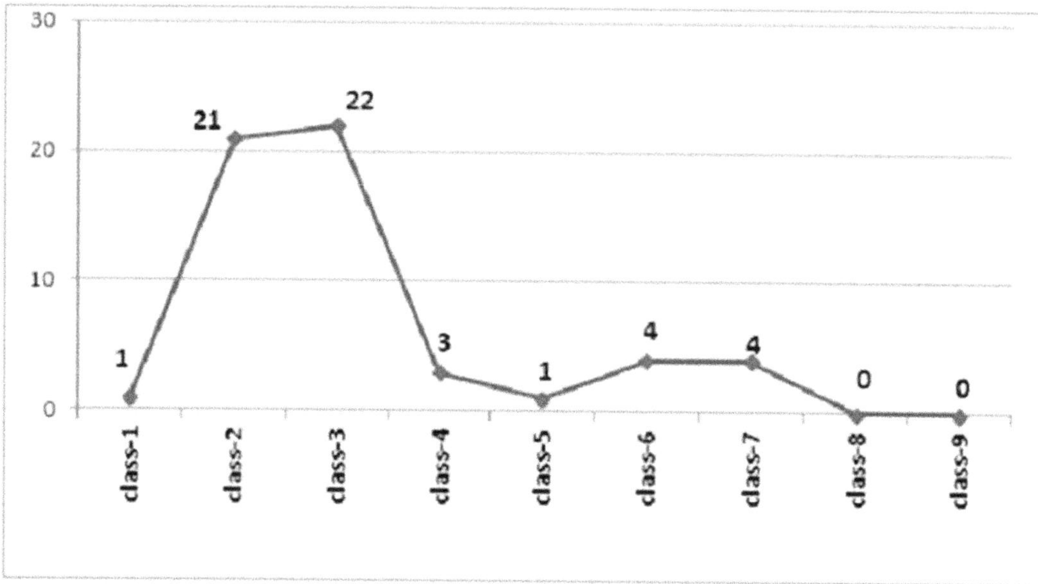

One of the classifications of hazmat transportation can be done according to the mode of transport. Hazmat can be carried by road, railway, air, water and pipeline. Road transportation has the greatest percentage in hazmat transportation. In the second place comes the rail transportation of hazmats. Hazmats can be transported via marine in huge amounts. Some shipments are intermodal. Rail-road transportation and marine-road modes are the most common way of intermodal transportation. The researchers studied mainly the road transportation of hazmats in their articles. In 29 articles out of 179 the rail transportation of hazmats are examined. The number of articles about marine transportation of hazmats is really too low. For all the transport modes, many efforts were being spent by researchers to find optimal routes for hazmat carrying vehicles with the objective of minimizing cost and transport risk. For these purposes, in the articles which are examined, the optimization models (vehicle routing, shortest path, network design, etc.) were proposed to have an optimal solution to hazmat transportation problems.

Hazmat transportation problems are usually bi-objective problems. While carriers are trying to find best routes for cost minimization, the local or global authorities have to consider minimizing the risk on the population living along the path of hazmat carrying vehicles. Therefore, in most of the articles, risk and routing were the main concern. Emergency response, scheduling, facility location, waste transportation, accident analysis, and network design were the second concern of researchers. In these articles, linear, non-linear and heuristic models were proposed to find optimal routes.

As a second classification, these articles can be classified as; risk (risk assessment, risk reduction, risk analysis etc.), routing, routing and scheduling, routing and facility location, hazmat waste transportation, emergency response, network design, accident analysis. Figure 4 shows the number of articles published according to this classification. According to Figure 4, most researched subject for hazmat transportation was risk analysis (80 articles were published about this subject).

The classification of hazmat transportation may differ according to authors. One of the classifications proposed in this study is as follows;

Figure 4. Number of articles published according to main focus of the articles

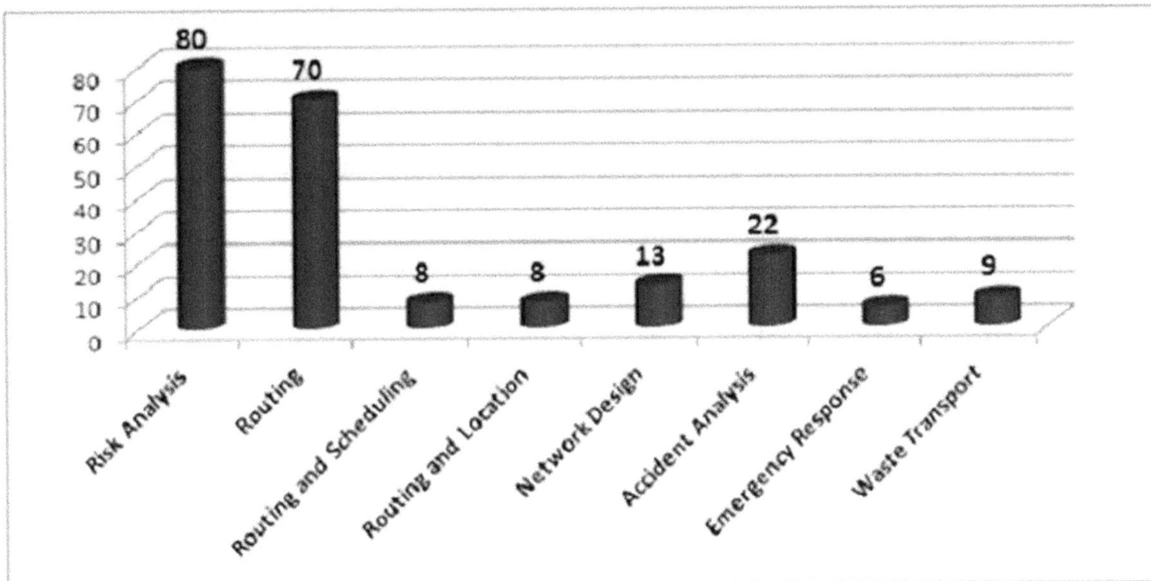

- Risk (risk assessment, risk reduction, risk analysis etc.)
- Routing
- Routing and Scheduling
- Routing and Facility Location
- Hazmat Waste Transportation
- Emergency Response
- Network Design
- Accident Analysis

It would be better to explain these 8 main focuses of articles briefly to make it clearer. Of course many explanations of these subjects can be found in details in many other sources; however the main idea to give brief explanations of 8 subjects is to make it easy for the readers to understand the articles published.

Risk (Risk Assessment, Risk Reduction, Risk Analysis Etc.)

Hazmat transportation differs from normal transportation within the frame of risk. Ang explains the risk in three parts (Zhang et al., 2000):

- Probability of accident involving release of hazmat,
- The level of potential exposure,
- The consequences (injuries, fatalities, property damage)

Erkut et al. (2007) focused on Quantative Risk Assessment (QRA) with three steps:

- Hazard and exposed receptor identification

- Frequency analysis
- Consequence modeling and risk calculation (Risk of a road was calculated by Erkut as edge risk and path risk).
    - Modeling the impact area,
    - Gaussian Plume Model,
    - Risk Cost (injuries and fatalities, cleanup costs, property damage, evacuation, product loses, traffic incident delay, environmental damage)

Risk of a road was calculated by Erkut et al. (2007) as edge risk and path risk. Risk has been main concern of articles about hazmat transportation.

Routing

Vehicles routing problems were being studied for more than 60 years. The focus to hazmat vehicle routing was initiated in 1978. From 1978 to 2014 the researchers have studied on routing the hazardous materials and they mainly focused on local-global routing, transportation mode, vehicle type, single-multi commodity, single-multi objective routing of hazmat transportation. Although the mode of transporting hazmat can differ; main idea for routing hazmat vehicles was to minimize transport cost and risk.

Traditionally, main consideration for hazmat transportation was cost and safety. (Huang, 2006) Therefore for routing hazmat vehicles while carriers are trying to decrease their transport cost, local or global authorities should care about the safety of the population.

Routing and Scheduling

Government and local authorities serve as supervisor to hazmat carriers and, they allocate road segments or available population zones within a time period. (Shen et al., 2013). The carriers should route and schedule their hazmat vehicles according to restrictions of the authorities to minimize transport risk.

The traffic conditions and other risk factors in hazmat transportation often vary with time and can at best be known a priori with uncertainty (Erkut et al., 2007). For example, for a hazmat truck, the travel time can be uncertain and depend on traffic congestion, weather conditions, and road conditions during the vehicle's trip across those links. Hence, the transport risk and arrival time at the destination can vary with the dispatch schedule from the origin. All these effects must be considered while routing and scheduling the hazmat vehicles for the arrival of vehicles to their destinations in scheduled time. So, time limitations of local or global authorities also could be met by correct scheduling.

Routing and Facility Location

Besides routing the hazmat carrying vehicles, locating the places where the hazmat should be stored or used has also great importance to reduce the risk of hazmat. It was one of the focuses of researchers to locate the facilities in order to decrease the risk of hazmat on population. Locating facilities should be considered with routes of vehicle which carry hazmat.

Hazmat shipments usually initiated from facilities which are also potentially harmful to public and environmental safety. These facilities can be petroleum refineries, nuclear power plants, ammunition storages, gun factories etc.

The location decisions pertaining to such facilities have a considerable effect on the routing of hazmat shipments. Therefore, integration of facility location and routing decisions can be an effective means to mitigate the total risk in a region where hazmats are processed and transported (Erkut et al., 2007).

Hazmat Waste Transportation

The first studies were focused on the routing of hazmat to the destinations. As the countries became more industrialized, the amount of hazmat wastes has reached to a remarkable level. Therefore, in recent years besides the hazmat transportation to demand points; the collection, treatment, recycling and disposal of hazmat wastes in an organized manner has gained importance.

Increasing developments in technology and industry have led to a significant hazardous waste management problem, demanding a more structured and scientific manner of managing hazmat (Samanlıoglu, 2013). For this reason, in recent years, the researchers have begun to focus on hazmat waste transportation also.

Emergency Response

To establish a hazmat emergency response system constitutes a building block toward a comprehensive public and environmental safety infrastructure (Berman et al., 2007). Emergency response departments are fire-fighting departments, first-aid teams and police stations (Xu et al., 2013). One of the main concerns of emergency response researches is to locate response departments to best places in which the risk of the link is highest. As the amount of hazmat shipments increased, the hazmat accidents are also expected to be increased. To minimize the effects of hazmat accidents, emergency response subject should be considered as important as the risk reduction and routing of hazmat vehicles. If the emergency response teams are located in right places, this will first decrease the number of hazmat accidents due to police controls and it will also help to decrease the effects of accidents by response teams' fast reaction.

Network Design

Designing a hazmat transportation network is first introduced by Kara & Verter (2004). They proposed a bi-level integer programming model and considered the problem with two sides: government and carriers. Their study proposed a model for minimizing risk while considering the carriers' concern about minimizing cost. Erkut & Alp (2007) proposed an "optimum communication spanning tree problem" to design a network for hazmat shipments. Bianco et al. (2009) introduced a bi-level network flow model.

Accident Analysis

Ronza et al. (2007), used transportation accident databases to investigate ignition and explosion probabilities of flammable spills. Ghazinoory & Kheirkhah (2008) focused on transportation risk as the main parameter and they developed strategies for reaching the optimal condition and they laid out the impacts of these strategies on the reduction of accidents are analyzed.

For all subjects mentioned above, the researchers have tried to find optimal solutions to hazmat transportation in their articles. They proposed linear or non-linear solutions to find best route for the

Table 1. Detailed list of articles published according to their main focuses

Risk (risk assessment, risk reduction, risk analysis, etc.)
Luckritz & Schneider, (1980); Philipson & Napadensky, (1982); Scanlon & Cantilli, (1985); Swoveland, (1987); Abkowitz & Cheng, (1988); Saccomanno et al., (1989); Vanaerde et al., (1989); Gopalan et al., (1990); Glickman, (1991); Glickman & Golding, (1991); List et al., (1991); Purdy, (1993); Saccomanno & Shortreed, (1993); Leeming & Saccomanno, (1994); Current & Ratick, (1995); Erkut, (1995); Erkut & Verter, (1995); Moore et al., (1995); Spadoni et al., (1995); Ashtakala & Eno,(1996); Dennis, (1996); Pine & Marx, (19979; Lovett et al., (1997); Bonvicini et al., (1998); Cassini, (1998); Mills & Neuhauser, (1998); Leonelli et al., (1999); Pet-Armacost et al., (1999); Leonelli et al., (2000); Zhang et al., (2000); Bubbico et al., (2000); Raj & Pritchard, (2000); Verter & Kara, (2001); Hwang et al, (2001); Milazzo et al,(2002); Fabiano et al., (2002); Kara et al, (2003); Martinez-Alegria et al., (2003); Nardini et al, (2003); Bubbico et al., (2004); Fabiano et al., (2005); Gheorghea et al., (2005); Boulmakoul, (2006); Bubbico et al.,(2006); Cozzani et al.,(2007); Glickman et al.,(2007); Verma & Verter,(2007); Godoy et al.,(2007); Zhang & Zhao, (2007); Verma, (2009); Paltrinieri et al.,(2009); Kheirkhah et al., (2009); Clark et al., (2009); Guo & Verma,(2010); Verma & Verter,(2010); Reniers et al., (2010); Milazzo et al., (2010); Dzemydienė & et al.,(2011); Verma, (2011); L´opez-Atamoros & Fernandez- Villagómez, (2011); Das et al.,(2012); Chakrabarti & Parikh, (2012); Wang et al.,(2012); Reniers & Dullaert,(2013); Tena-Chollet et al., (2013); Raemdonck et al.,(2013); Liu et al.,(2013); Toumazis & Kwon,(2013); Chakrabarti & Parikh, (2013a); ; Chakrabarti & Parikh, (2013b); Saat et al.,(2014); Cappanera & Nonato,(2014); Kang et al.,(2014)
Routing
Kalelkar & Brinks, (1978); Glickman, (1983); Coleman, (1984); Haas & Kichner, (1987); Abkowitz & Cheng, (1988); Chin & Cheng, (1989); Zografos & Davis, (1989); Zografos & Samara, (1989); List & Mirchandani, (1991); List et al., (1991); Miaou & Chin, (1991); Abkowitz et al.,(1992); Wijeratne et al., (1993); Lassarre et al., (1993); Sivakumar et al., (1993); Beroggi, (1994); Patel & Horowitz, (1994); Andersson, (1994); Jacobs & Warmerdam, (1994); Karkazis & Boffey, (1995); Boffey & Karkazis, (1995); Glickman & Sontag, (1995); Karkazis & Boffey, (1995); Moore et al., (1995); Ashtakala & Eno,(1996); Jin et al., (1996); Erkut & Glickman, (1997); Helander & Melachrinoudis, (1997); Nembhard & White, (1997); Sherali et al., (1997); Cassini, (1998); Erkut & Verter, (1998); Giannikos, (1998); Miller-Hooks & Mahmassani, (1998); Iakovou et al., (1999); Frank et al., (2000); Leonelli et al., (2000); Zografos et al., (2000); Erkut & Ingolfsson, (2000); Akgun et al., (2000); Iakovou, (2001); Zografos & Androutsopoulos, (2004); Huang & Cheu, (2004); Huang, (2006); Carotenuto et al., (2007); Akgun et al., (2007); Glickman et al., (2007); Alumur & Kara, (2007); Bell,(2007); Dadkar et al., (2008); Zografos & Androutsopoulos, (2008); Bonvicini & Spadoni, (2008); Verma & Verter, (2010); Dadkar et al.,(2010); Caramia, (2010); Chakrabarti & Parikh, (2011a); Chakrabarti & Parikh, (2011b); Chakrabarti & Parikh, (2011c); Li & Leung, (2011); Xie et al.,(2012); Verma et al., (2012); Das et al., (2012); Reilly et al., (2012); Chakrabarti & Parikh, (2013a); Samanlioglu, (2013); Desai & Lim, (2013); Toumazis & Kwon, (2013); Desai & Lim, (2013); Chakrabarti & Parikh, (2013b); Mahmoudabadi & Seyedhosseini, (2014); Kang et al., (2014)
Routing and Scheduling
Cox & Turnquist,(1986); Sulijoadikusumo & Nozick,(1998); Carotenuto et al.,(2007); Erkut & Alp, (2007); Pradhanang et al., (2010); Androutsopoulos & Zografos, (2010); Szeto, (2013); Pradhanang et al., (2014)
Routing and Facility Location
Zografos & Samara (1989); List et al., (1991); Current & Ratick, (1995); Helander & Melachrinoudis, (1997); Alumur & Kara, (2007); Xie et al., (2012); Samanlioglu, (2013); Cappanera & Nonato, (2014)
Emergency Response
Ogle et al., (2005); Berman et al., (2007); Zografos & Androutsopoulos, (2008); Planas et al., (2008); Ren et al., (2012); Chakrabarti & Parikh, (2013)
Network Design
List & Mirchandani, (1991); Wijeratne et al., (1993); Sivakumar et al., (1993); Miller-Hooks & Mahmassani, (1998); Zhang et al., (2000); Kara & Verter, (2004); Berman et al., (2007); Erkut & Alp, (2007); Erkut & Gzara, (2008); Verter & Kara, (2008); Bianco et al., (2009); Kazantzi et al., (2011); Gzara, (2013)
Accident Analysis
Aerde et al.,(1988); Shappert et al., (1973); Glickman, (1988); Harwood et al., (1989); Vilchez et al., (1995); Chakraborty & Armstrong, (1995); Romer et al., (1995); Shorten et al., (2002); Anderson & Barkan, (2004); Schweitzer, (2006); Ronza et al.,(2007); Ghazinoory & Kheirkhah, (2008); Trépaniera et al.,(2009); Qiaoa et al., (2009); Samuel et al.,(2009); Welles et al.,(2009); Lozanoa et al.,(2010); Yang et al.,(2010); Ellis,(2011); Pesic,(2011); Zhao et al.,(2012); Shen et al.,(2013)
Hazmat Waste Transportation
Zografos & Samara, (1989); List & Mirchandani, (1991); Jacobs & Warmerdam, (1994); Lovett et al., (1997); Giannikos, (1998); Alumur & Kara, (2007); Das et al.,(2012b); Samanlioglu, (2013)

hazmat carrying vehicles. The main contribution of this work is to present the articles about hazmat transportation in details, so that the researchers who will work on hazmat transportation in their future works will have detailed information about hazmat transportation.

Table 1 provides the classification of articles which are published in refereed journals between 1973 and 2014 using the above eight main subjects.

80 of articles over 179 articles deal mainly with risk of hazmat transportation. 70 of the articles take into consideration the routing of hazmat vehicles mainly. So, it can be understood from Tabe1 that main focus for hazmat transportation is fixed to risk analysis and routing of transporting hazardous materials. The researches about hazmat transportation mainly focused on finding optimal routes for the vehicles in which cost and risk were minimized. In 6 articles emergency response, in 13 articles network design, in 8 articles routing and facility location, in 8 articles routing and scheduling, in 9 articles hazmat waste transportation and in 22 articles accident analysis were studied. Since in some articles two subjects were studied at the same time, total number of the articles in the Table 1 is more than 179.

CONCLUSION

Hazmat transportation is a crucial subject since the incidents caused by hazmat carrying vehicles resulted in huge amount of fatalities and environmental damage. Although its importance few countries take serious steps towards hazmat transportation risk. Only this few countries has special departments about hazmat transportation. Hazmat transportation has attracted the attention of researchers since the early 1970s and it did not lose its popularity since then. The researchers focused mainly risk reduction and routing of hazardous materials.

There are so many articles, books, websites, seminar reports, congress reports about hazardous materials transportation. In this study only the articles which are published in refereed journals were examined.

It is also important for the researchers to publish their articles in refereed journals. This brings the question where the hazmat transportation articles were being published permanently. The future researcher can refer to Table 2 where to publish their articles about hazmat transportation.

Table 2. Popular journals in which hazmat transportation articles published

Order	Name of Journal in which The Analyzed Articles Published	Number of Articles Published
1	Journal of Hazardous Materials	21
2	Journal of Loss Prevention in the Process Industries	18
3	Transportation Research Part B,C,D,E	16
4	Transportation Research Record	15
5	Operations Research	15
6	European Journal of Operational Research	13
7	Transportation Science	9
8	Computers & Operations Research	8
9	Accident Analysis and Prevention	4
10	Risk Analysis	4
11	Others	56

In introduction part of this study, it was figured out the importance of hazmat transportation with respect to fatalities. In second part, a literature survey was prepared about the hazmat transportation articles since 1973. 179 articles were analyzed in details, and the articles were classified according to publication year, hazmat class and the main concern of the article. Short briefs about risk, routing, routing and scheduling, routing and facility location, hazmat waste transportation, network design, emergency response and accident analysis were given. At the end of this study, all the analyzed articles were given in a Table 1 according to 8 classification subject.

Transportation models require minimizing the cost or time to find optimal routes for ordinary freights. It becomes bi-objective when the vehicles carry hazmat. In future articles, both cost and risk have to be considered. Usually, before studying on a subject, the researchers need to learn what have been done before about that subject. This study is supposed to assist the authors who are going to work on hazmat transportation.

REFERENCES

Abkowitz, M., & Cheng, P. D. M. (1988). Developing a risk cost framework for routing truck movements of hazardous materials. *Accident; Analysis and Prevention*, *20*(1), 39–51. doi:10.1016/0001-4575(88)90013-9 PMID:3337764

Abkowitz, M., & Cheng, P. D. M. (1989). Hazardous materials transport risk estimation under conditions of limited data availability. *Transportation Research Record*, *1245*, 14–22.

Abkowitz, M., Lepofsky, M., & Cheng, P. (1992). Selecting criteria for designating hazardous materials highway routes. *Transportation Research Record*, *1333*, 30–35.

Akgun, V., Erkut, E., & Batta, R. (2000). On finding dissimilar paths. *European Journal of Operational Research*, *121*(2), 232–246. doi:10.1016/S0377-2217(99)00214-3

Akgün, V., Parekh, A., Batta, R., & Rump, C. M. (2007). Routing of a hazmat truck in the presence of weather systems. *Computers & Operations Research*, *34*(5), 1351–1373. doi:10.1016/j.cor.2005.06.005

Alumur, S., & Kara, B. Y. (2007). A new model for the hazardous waste location-routing problem. *Computers & Operations Research*, *34*(5), 1406–1423. doi:10.1016/j.cor.2005.06.012

Anderson, R. T., & Barkan, C. P. L. (2004). Railroad accident rates for use in transportation risk analysis. *Transportation Research Record*, *1863*, 88–98. doi:10.3141/1863-12

Andersson, S. E. (1994). Safe transport of dangerous goods – road, rail or sea – a screening of technical and administrative factors. *European Journal of Operational Research*, *75*(3), 499–507. doi:10.1016/0377-2217(94)90292-5

Androutsopoulos, K. N., & Zografos, K. G. (2010). Solving the bicriterion routing and scheduling problem for hazardous materials distribution. *Transportation Research Part C, Emerging Technologies*, *18*(5), 713–726. doi:10.1016/j.trc.2009.12.002

Ashtakala, B., & Eno, L. A. (1996). Minimum risk route model for hazardous materials. *Journal of Transportation Engineering*, *122*(5), 350–357. doi:10.1061/(ASCE)0733-947X(1996)122:5(350)

Berman, O., Verter, V., & Kara, B. Y. (2007). Designing Emergency Response Networks for Hazardous Materials Transportation. *Computers & Operations Research*, *34*(5), 1374–1388. doi:10.1016/j.cor.2005.06.006

Beroggi, G. E. G. (1994). A real-time routing model for hazardous materials. *European Journal of Operational Research*, *75*(3), 508–520. doi:10.1016/0377-2217(94)90293-3

Bianco, L., Caramia, M., & Giordani, S. (2009). A bilevel flow model for hazmat transportation network design. *Transportation Research Part C, Emerging Technologies*, *17*(2), 175–196. doi:10.1016/j.trc.2008.10.001

Boffey, T. B., & Karkazis, J. (1995). Linear versus nonlinear models for hazardous materials routing. *INFOR*, *33*, 114–117.

Bonvicini, S., Leonelli, P., & Spadoni, G. (1998). Risk analysis of hazardous materials transportation: Evaluating uncertainty by means of fuzzy logic. *Journal of Hazardous Materials*, *62*(1), 59–74. doi:10.1016/S0304-3894(98)00158-7

Bonvicini, S., & Spadoni, G. (2008). A hazmat multi-commodity routing model satisfying risk criteria: A case study. *Journal of Loss Prevention in the Process Industries*, *21*(4), 345–358. doi:10.1016/j.jlp.2007.11.009

Boulmakoul, A. (2006). Fuzzy graphs modelling for HazMat telegeomonitoring. *European Journal of Operational Research*, *175*(3), 1514–1525. doi:10.1016/j.ejor.2005.02.025

Bubbico, R., Di Cave, S., & Mazzarotta, B. (2004). Risk analysis for road and rail transport of hazardous materials: A GIS approach. *Journal of Loss Prevention in the Process Industries*, *17*(6), 483–488. doi:10.1016/j.jlp.2004.08.011

Bubbico, R., Ferrari, C., & Mazzarotta, B. (2000). Risk analysis of LPG transport by road and rail. *Journal of Loss Prevention in the Process Industries*, *13*(1), 27–31. doi:10.1016/S0950-4230(99)00057-1

Bubbico, R., Maschio, G., Mazzarotta, B., Milazzo, M. F., & Parisi, E. (2006). Risk management of road and rail transport of hazardous materials in Sicily. *Journal of Loss Prevention in the Process Industries*, *19*(1), 32–38. doi:10.1016/j.jlp.2005.05.011

Cappanera, P., & Nonato, M. (2014). The gateway location problem: A cost oriented analysis of a new risk mitigation strategy in hazmat transportation. *Procedia: Social and Behavioral Sciences*, *111*, 918–926. doi:10.1016/j.sbspro.2014.01.126

Caramia, M., Giordani, S., & Iovanella, A. (2010). On the selection of k routes in multiobjective hazmat route planning. *IMA Journal of Management Mathematics*, *21*(3), 239–251. doi:10.1093/imaman/dpp017

Carotenuto, P., Giordani, S., & Ricciardelli, S. (2007). Finding minimum and equitable risk routes for hazmat shipments. *Computers & Operations Research*, *34*(5), 1304–1327. doi:10.1016/j.cor.2005.06.003

Carotenuto, P., Giordani, S., Ricciardelli, S., & Rismondo, S. (2007). A tabu search approach for scheduling hazmat shipments. *Computers & Operations Research*, *34*(5), 1328–1350. doi:10.1016/j.cor.2005.06.004

Cassini, P. (1998). Road transportation of dangerous goods: Quantitative risk assessment and route comparison. *Journal of Hazardous Materials*, *61*(1–3), 133–138. doi:10.1016/S0304-3894(98)00117-4

Chakrabarti, U. K., & Parikh, J. K. (2011a). Class-2 hazmat transportation consequence assessment on surrounding population. *Journal of Loss Prevention in the Process Industries*, *24*(6), 758–766. doi:10.1016/j.jlp.2011.04.011

Chakrabarti, U. K., & Parikh, J. K. (2011b). Route evaluation for hazmat transportation based on total risk - A case of Indian State Highways. *Journal of Loss Prevention in the Process Industries*, *24*(5), 524–530. doi:10.1016/j.jlp.2011.03.002

Chakrabarti, U. K., & Parikh, J. K. (2011c). Route risk evaluation on class-2 hazmat transportation. *Process Safety and Environmental Protection*, *89*(4), 248–260. doi:10.1016/j.psep.2011.04.003

Chakrabarti, U. K., & Parikh, J. K. (2012). Applying HAZAN methodology to hazmat transportation risk assessment. *Process Safety and Environmental Protection*, *90*(5), 368–375. doi:10.1016/j.psep.2012.05.011

Chakrabarti, U. K., & Parikh, J. K. (2013). A societal risk study for transportation of class-3 hazmats – A case of Indian state highways. *Process Safety and Environmental Protection*, *91*(4), 275–284. doi:10.1016/j.psep.2012.06.009

Chakrabarti, U. K., & Parikh, J. K. (2013). Risk-based route evaluation against country-specific criteria of risk tolerability for hazmat transportation through Indian State Highways. *Journal of Loss Prevention in the Process Industries*, *26*, 723–736.

Chakrabarti, U. K., & Parikh, J. K. (2013). Using consequence e based hazard zone assessment for effective evacuation planning of vulnerable settlements along hazmat transport corridors through industrial city of Surat in western India. *Journal of Loss Prevention in the Process Industries*, *26*(5), 941–947. doi:10.1016/j.jlp.2013.04.003

Chakraborty, J., & Armstrong, M. P. (1995). Using geographic plume analysis to assess community vulnerability to hazardous accidents. *Computers, Environment and Urban Systems*, *19*(5-6), 341–356. doi:10.1016/0198-9715(95)00018-6

Chin, S. M., & Cheng, P. D. M. (1989). Bicriterion routing scheme for nuclear spent fuel transportation. *Transportation Research Record*, *1245*, 60–64.

Clark, R. M., & Besterfield-Sacre, M. E. (2009). A new approach to hazardous materials transportation risk analysis: Decision modeling to ıdentify critical variables. *Risk Analysis*, *29*(3), 344–354. doi:10.1111/j.1539-6924.2008.01163.x PMID:19087232

Coleman, J. A. (1984). Railroad–highway crossings and route selection for transporting hazardous materials. *Public Roads*, *48*(2), 63–71.

Cox, R., & Turnquist, M. (1986). Scheduling truck shipments of hazardous materials in the presence of curfews. *Transportation Research Record*, *1063*, 21–26.

Cozzani, V., Bonvicini, S., Spadoni, G., & Zanelli, S. (2007). Hazmat transport: A methodological framework for the risk analysis of marshalling yards. *Journal of Hazardous Materials*, *147*(1-2), 412–423. doi:10.1016/j.jhazmat.2007.01.065 PMID:17418942

Current, J. R., & Ratick, S. (1995). A model to assess risk, equity, and efficiency in facility location and transportation of hazardous materials. *Location Science*, *3*(3), 187–201. doi:10.1016/0966-8349(95)00013-5

Dadkar, Y., Jones, D., & Nozick, L. (2008). Identifying geographically diverse routes for the transportation of hazardous materials. *Transportation Research Part E, Logistics and Transportation Review*, *44*(3), 333–349. doi:10.1016/j.tre.2006.10.010

Dadkar, Y., Nozick, L., & Jones, D. (2010). Optimizing facility use restrictions for the movement of hazardous materials. *Transportation Research Part B: Methodological*, *44*(2), 267–281. doi:10.1016/j.trb.2009.07.006

Das, A., Gupta, A. K., Mazumder, T. N. (2012a). A comprehensive risk assessment framework for offsite transportation of inflammable hazardous waste. *Journal of Hazardous Materials, 227– 228*, 88– 96.

Das, A., Mazumder, T. N., & Gupta, A. K. (2012b). Pareto frontier analyses based decision making tool for transportation of hazardous waste. *Journal of Hazardous Materials, 227– 228*, 341– 352.

Dennis, S. M. (1996). Estimating risk costs per unit of exposure for hazardous materials transported by rail. *Logistics and Transportation Review*, *32*(4), 351–375.

Desai, S., & Lim, G. (2013). Solution time reduction techniques of a stochastic dynamic programming approach for hazardous material route selection problem. *Computers & Industrial Engineering*, *65*(4), 634–645. doi:10.1016/j.cie.2013.05.002

Desai, S., & Lim, G. J. (2013). Solution time reduction techniques of a stochastic dynamic programming approach for hazardous material route selection problem. *Computers & Industrial Engineering*, *65*(4), 634–645. doi:10.1016/j.cie.2013.05.002

Dzemydienė, D., & Dzindzalieta, R. (2010). Development of architecture of embedded decision support systems for risk evaluation of transportation of dangerous goods. *Technological and Economic Development of Economy*, *16*(4), 654–671. doi:10.3846/tede.2010.40

Ellis, J. (2011). Analysis of accidents and incidents occurring during transport of packaged dangerous goods by sea. *Safety Science*, *49*(8-9), 1231–1237. doi:10.1016/j.ssci.2011.04.004

Erkut, E. (1995). On the credibility of the conditional risk model for routing hazardous materials. *Operations Research Letters*, *18*(1), 49–52. doi:10.1016/0167-6377(95)00030-N

Erkut, E., & Alp, O. (2007). Integrated routing and scheduling of hazmat trucks with stops en route. *Transportation Science*, *41*(1), 107–122. doi:10.1287/trsc.1060.0176

Erkut, E., & Alp, O. (2007). Designing a Road Network for Hazardous Materials Shipments. *Computers & Operations Research*, *34*(5), 1389–1405. doi:10.1016/j.cor.2005.06.007

Erkut, E., & Glickman, T. (1997). Minimax population exposure in routing highway shipments of hazardous materials. *Transportation Research Record*, *1602*, 93–100. doi:10.3141/1602-14

Erkut, E., & Gzara, F. (2008). Solving the hazmat transport network design problem. *Computers & Operations Research*, *35*(7), 2234–2247. doi:10.1016/j.cor.2006.10.022

Erkut, E., & Ingolfsson, A. (2000). Catastrophe avoidance models for hazardous materials route planning. *Transportation Science, 34*(2), 165–179. doi:10.1287/trsc.34.2.165.12303

Erkut, E., Tjandra, A. S., & Verter, V. (2007). Hazardous Materials Transportation. In C. Barnhart & G. Laporte (Eds.), *Handbook in OR & MS* (Vol. 14, pp. 539–621). UK: Elsevier.

Erkut, E., & Verter, V. (1995b). A framework for hazardous materials transport risk assessment. *Risk Analysis, 15*(5), 589–601. doi:10.1111/j.1539-6924.1995.tb00755.x

Erkut, E., & Verter, V. (1998). Modeling of transport risk for hazardous materials. *Operations Research, 46*(5), 625–642. doi:10.1287/opre.46.5.625

Fabiano, B., Curro, F., Palazzi, E., & Pastorino, R. (2002). A framework for risk assessment and decision making strategies in dangerous good transportation. *Journal of Hazardous Materials, 93*(1), 1–15. doi:10.1016/S0304-3894(02)00034-1 PMID:12062950

Fabiano, B., Curro`, F., Reverberi, A. P., & Pastorino, R. (2005). Dangerous good transportation by road: From risk analysis. *Journal of Loss Prevention in the Process Industries, 18*(4-6), 403–413. doi:10.1016/j.jlp.2005.06.031

Fabiano, B., & Palazzi, E. (2010). Hazmat transportation by heavy vehicles and road tunnels: A simplified modeling procedure to risk assessment and mitigation applied to an Italian case study. *International Journal of Heavy Vehicle Systems, 17*(3-4), 216–236. doi:10.1504/IJHVS.2010.035988

Frank, W. C., Thill, J. C., & Batta, R. (2000). Spatial decision support system for hazardous material truck routing. *Transportation Research Part C, Emerging Technologies, 8*(1–6), 337–359. doi:10.1016/S0968-090X(00)00007-3

Ghazinoory, S., & Kheirkhah, A. S. (2008). Transportation of hazardous materials in Iran: A strategic approach for decreasing accidents. *Transport, 23*(2), 104–111. doi:10.3846/1648-4142.2008.23.104-111

Gheorghea, A. V., Birchmeier, J., & Vamanu, D. (2005). Comprehensive risk assessment for rail transportation of dangerous goods: A validated platform for decision support. *Reliability Engineering & System Safety, 88*(3), 247–272. doi:10.1016/j.ress.2004.07.017

Giannikos, I. (1998). A multiobjective programming model for locating treatment sites and routing hazardous wastes. *European Journal of Operational Research, 104*(2), 333–342. doi:10.1016/S0377-2217(97)00188-4

Glickman, T. S. (1983). Rerouting railroad shipments of hazardous materials to avoid populated areas. *Accident; Analysis and Prevention, 15*(5), 329–335. doi:10.1016/0001-4575(83)90012-X

Glickman, T. S. (1988). Benchmark estimates of release accident rates in hazardous materials transportation of rail and truck. *Transportation Research Record, 1193*, 22–28.

Glickman, T. S. (1991). An expeditious risk assessment of the highway transportation of flammable liquids in bulk. *Transportation Science, 25*(2), 115–123. doi:10.1287/trsc.25.2.115

Glickman, T. S., Erkut, E., & Zschocke, M. S. (2007). The cost and risk impacts of rerouting railroad shipments of hazardous materials. *Accident; Analysis and Prevention*, *39*(5), 1015–1025. doi:10.1016/j.aap.2007.01.006 PMID:17349610

Glickman, T. S., & Golding, D. (1991). For a few dollars more: Public trust and the case for transporting nuclear waste in dedicated trains. *Policy Studies Review*, *10*(4), 127–138. doi:10.1111/j.1541-1338.1991.tb00285.x

Glickman, T. S., & Sontag, M. A. (1995). The tradeoffs associated with rerouting highway shipments of hazardous materials to minimize risk. *Risk Analysis*, *15*(1), 61–67. doi:10.1111/j.1539-6924.1995.tb00093.x

Godoy, S. M., Santa Cruz, A. S. M., & Scenna, N. J. (2007). STRRAP system—A software for hazardous materials risk assessment. *Reliability Engineering & System Safety*, *92*(7), 847–857. doi:10.1016/j.ress.2006.02.012

Gopalan, R., Kolluri, K. S., Batta, R., & Karwan, M. H. (1990a). Modeling equity of risk in the transportation of hazardous materials. *Operations Research*, *38*(6), 961–973. doi:10.1287/opre.38.6.961

Guo, X., & Verma, M. (2010). Choosing vehicle capacity to minimize risk for transporting flammable materials. *Journal of Loss Prevention in the Process Industries*, *23*(2), 220–225. doi:10.1016/j.jlp.2009.07.007

Gzara, F. (2013). A cutting plane approach for bilevel hazardous material transport network design. *Operations Research Letters*, *41*(1), 40–46. doi:10.1016/j.orl.2012.10.007

Haas, T. J., & Kichner, J. J. (1987). Hazardous materials in marine transportation – a practical course. *Journal of Chemical Education*, *64*(1), 34–35. doi:10.1021/ed064p34

Harwood, D. W., Russell, E. R., & Viner, J. G. (1989). Characteristics of accidents and incidents in highway transportation of hazardous materials. *Transportation Research Record*, *1245*, 23–33.

Helander, M. E., & Melachrinoudis, E. (1997). Facility location and reliable route planning in hazardous material transportation. *Transportation Science*, *31*(3), 216–226. doi:10.1287/trsc.31.3.216

Huang, B. (2006). GIS-Based Route Planning for Hazardous Material Transportation. *Journal of Environmental Informatics*, *8*(1), 49–57. doi:10.3808/jei.200600076

Huang, B., Cheu, R. L., & Liew, Y. S. (2004). GIS and genetic algorithms for HAZMAT route planning with security considerations. *International Journal of Geographical Information Science*, *18*(8), 769–787. doi:10.1080/13658810410001705307

Hwang, S. T., Brown, D. F., O'Steen, J. K., Policastro, A. J., & Dunn, W. E. (2001). Risk assessment for national transportation of selected hazardous materials. *Transportation Research Record*, *1763*, 114–124. doi:10.3141/1763-17

Iakovou, E., Douligeris, C., Li, H., Ip, C., & Yudhbir, L. (1999). A maritime global route planning model for hazardous materials transportation. *Transportation Science*, *33*(1), 34–48. doi:10.1287/trsc.33.1.34

Iakovou, E. T. (2001). An interactive multiobjective model for the strategic maritime transportation of petroleum products: Risk analysis and routing. *Safety Science*, *39*(1–2), 19–29. doi:10.1016/S0925-7535(01)00022-4

Jacobs, T. L., & Warmerdam, J. M. (1994). Simultaneous routing and siting for hazardous-waste operations. *Journal of Urban Planning and Development*, *120*(3), 115–131. doi:10.1061/(ASCE)0733-9488(1994)120:3(115)

Jia, H., Zhang, L., Lou, X., & Cao, H. (2011). A fuzzy-stochastic constraint programming model for hazmat road transportation considering terrorism attacking. *Systems Engineering Procedia*, *1*, 130–136. doi:10.1016/j.sepro.2011.08.022

Jin, H. H., Batta, R. J., & Karwan, M. H. (1996). On the analysis of two new models for transporting hazardous materials. *Operations Research*, *44*(5), 710–723. doi:10.1287/opre.44.5.710

Juiping, X., Gang, J., & Lei, X. (2013). Hazmat transportation network design model with emergency response under complex fuzzy environment. Hindawi Publishing Cooperation Mathematical Problems in Engineering [Article ID 517372], 2013.

Kalelkar, A. S., & Brooks, R. E. (1978). Use of multidimensional utility functions in hazardous shipment decisions. *Accident; Analysis and Prevention*, *10*(3), 251–265. doi:10.1016/0001-4575(78)90016-7

Kang, Y., Batta, R., & Kwon, C. (2014). Generalized route planning model for hazardous material transportation with VaRand equity considerations. *Computers & Operations Research*, *43*, 237–247. doi:10.1016/j.cor.2013.09.015

Kara, B. Y., Erkut, E., & Verter, V. (2003). Accurate calculation of hazardous materials transport risks. *Operations Research Letters*, *31*(4), 285–292. doi:10.1016/S0167-6377(02)00238-9

Kara, B. Y., & Verter, V. (2004). Designing a road network for hazardous materials transportation. *Transportation Science*, *38*(2), 188–196. doi:10.1287/trsc.1030.0065

Karkazis, J., & Boffey, T. B. (1995). Optimal location of routes for vehicles transporting hazardous materials. *European Journal of Operational Research*, *86*(2), 201–215. doi:10.1016/0377-2217(94)00099-X

Karkazis, J., & Boffey, T. B. (1995). Optimal location of routes for vehicles transporting hazardous materials. *European Journal of Operational Research*, *86*(2), 201–215. doi:10.1016/0377-2217(94)00099-X

Kazantzi, V., Kazantzis, N., & Gerogiannis, V. C. (2011). Risk informed optimization of a hazardous material multi-periodic transportation model. *Journal of Loss Prevention in the Process Industries*, *24*(6), 767–773. doi:10.1016/j.jlp.2011.05.006

Kheirkhah, A. S., Esmailzadeh, A., & Ghazinoory, S. (2009). Developing strategies to reduce the risk of hazardous materials transportation in Iran using the method of fuzzy swot analysis. *Transport*, *24*(4), 325–332. doi:10.3846/1648-4142.2009.24.325-332

Lassarre, S., Fedra, K., & Weigkricht, E. (1993). Computer-assisted routing of dangerous goods for Haute-Normandie. *Journal of Transportation Engineering*, *119*(2), 200–210. doi:10.1061/(ASCE)0733-947X(1993)119:2(200)

Leeming, D. G., & Saccomanno, F. F. (1994). Use of quantified risk assessment in evaluating the risks of transporting chlorine by road and rail. *Transportation Research Record, 1430*, 27–35.

Leonelli, P., Bonvicini, S., & Spadoni, G. (1999). New detailed numerical procedures for calculating risk measures in hazardous materials transportation. *Journal of Loss Prevention in the Process Industries, 12*(6), 507–515. doi:10.1016/S0950-4230(99)00023-6

Leonelli, P., Bonvicini, S., & Spadoni, G. (2000). Hazardous materials transportation: A risk-analysis-based routing methodology. *Journal of Hazardous Materials, 71*(1-3), 283–300. doi:10.1016/S0304-3894(99)00084-9 PMID:10677666

Li, R., & Leung, Y. (2011). Multi-objective route planning for dangerous goods using compromise programming. *Journal of Geographical Systems, 13*(3), 249–271. doi:10.1007/s10109-010-0124-6

List, G., & Mirchandani, P. (1991). An integrated network planar multiobjective model for routing and siting for hazardous materials and wastes. *Transportation Science, 25*(2), 146–156. doi:10.1287/trsc.25.2.146

List, G. F., Mirchandani, P. B., Turnquist, M. A., & Zografos, K. G. (1991). Modeling and analysis for hazardous materials transportation-risk analysis, routing scheduling and facility location. *Transportation Science, 25*(2), 100–114. doi:10.1287/trsc.25.2.100

Liu, X., Saat, M. R., & Barkan, C. P. L. (2013). Integrated risk reduction framework to improve railway hazardous materials transportation safety. *Journal of Hazardous Materials, 260*, 131–140. PMID:23747471

López-Atamoros, L. G., Fernandez-Villagómez, G. F., Cruz-Gómez, M. J., & Durán-de-Bazúa, C. (2011). Development of a relative ranking risk ındex for risk assessment in the transportation of liquefied petroleum gas in Mexico City's metropolitan area. *Human and Ecological Risk Assessment, 17*(6), 1193–1209. doi:10.1080/10807039.2011.618383

Lovett, A. A., Parfitt, J. P., & Brainard, J. S. (1997). Using GIS in risk analysis: A case study of hazardous waste transport. *Risk Analysis, 17*(5), 625–633. doi:10.1111/j.1539-6924.1997.tb00903.x PMID:9404052

Lozano, A., Muñoz, A., Macías, L., & Antún, J. P. (2011). Hazardous materials transportation in Mexico City: Chlorine and gasoline cases. *Transportation Research Part C, Emerging Technologies, 19*(5), 779–789. doi:10.1016/j.trc.2010.09.001

Lozanoa, A., Muñoza, A., Antúna, J. P., Granadosa, F., & Guarnerosa, L. (2010). Analysis of hazmat transportation accidents in congested urban areas, based on actual accidents in Mexico. *Procedia: Social and Behavioral Sciences, 2*(3), 6053–6064. doi:10.1016/j.sbspro.2010.04.018

Luckritz, R. T., & Schneider, A. L. (1980). Decision making in hazardous materials transportation. *Journal of Hazardous Materials, 4*(2), 129–143. doi:10.1016/0304-3894(80)80029-X

Mahmoudabadi, A., & Seyedhosseini, S. M. (2014). Developing a chaotic pattern of dynamic hazmat routing problem. *IATSS Research, 37*, 110–118.

Martinez-Alegria, R., Ordonez, C., & Taboada, J. (2003). A conceptual model for analyzing the risks involved in the transportation of hazardous goods: Implementation in a geographic information system. *Human and Ecological Risk Assessment, 9*(3), 857–873. doi:10.1080/713609970

Miaou, S. P., & Chin, S. M. (1991). Computing k-shortest path for nuclear spent fuel highway transportation. *European Journal of Operational Research, 53*(1), 64–80. doi:10.1016/0377-2217(91)90093-B

Michael, G., & Bell, H. (2007). Mixed routing strategies for hazardous materials: Decision-making under complete uncertainty. *International Journal of Sustainable Transportation, 1*(2), 133–142. doi:10.1080/15568310601092013

Milazzo, M. F., Lisi, R., Maschio, G., Antonioni, G., Bonvicini, S., & Spadoni, G. (2002). Hazmat transport through, Messina town: From risk analysis suggestions for improving territorial safety. *Journal of Loss Prevention in the Process Industries, 15*(5), 347–356. doi:10.1016/S0950-4230(02)00028-1

Milazzo, M. F., Lisi, R., Maschio, G., Antonioni, G., & Spadoni, G. (2010). A study of land transport of dangerous substances in Eastern Sicily. *Journal of Loss Prevention in the Process Industries, 23*(3), 393–403. doi:10.1016/j.jlp.2010.01.007

Miller-Hooks, E. D., & Mahmassani, H. S. (1998). Optimal routing of hazardous materials in stochastic, time-varying transportation networks. *Transportation Research Record, 1645*, 143–151. doi:10.3141/1645-18

Mills, G. S., & Neuhauser, K. S. (1998). Urban risks of truck transport of radioactive material. *Risk Analysis, 18*(6), 781–785. doi:10.1111/j.1539-6924.1998.tb01120.x

Moore, J. E., Sandquist, G. M., & Slaughter, D. M. (1995). A route-specific system for risk assessment of radioactive materials transportation accidents. *Nuclear Technology, 112*(1), 63–78.

Nardini, L., Aparicio, L., Bandoni, A., & Tonelli, S. M. (2003). Regional risk associated with the transport of hazardous materials. *Latin American Applied Research, 33*(3), 213–218.

Nembhard, D. A., & White, C. C. III. (1997). Applications of non-order-preserving path selection to hazmat routing. *Transportation Science, 31*(3), 262–271. doi:10.1287/trsc.31.3.262

Ogle, R. A., Morrison, D. R., & Viz, M. J. (2005). Emergency response to a noncollision hazmat release from a railcar. *Process Safety Progress, 24*(2), 81–85. doi:10.1002/prs.10065

Paltrinieri, N., Landucci, G., Molag, M., Bonvicini, S., & Spadonia, G., & Cozzani, V. (2009). Risk reduction in road and rail LPG transportation by passive fire protection. *Journal of Hazardous Materials, 167*(1-3), 332–344. doi:10.1016/j.jhazmat.2008.12.122 PMID:19188020

Patel, M. H., & Horowitz, A. J. (1994). Optimal routing of hazardous materials considering risk of spill. *Transportation Research Part A, Policy and Practice, 28*(2), 119–132. doi:10.1016/0965-8564(94)90033-7

Pesic, D. J., Blagojevic, M. D., & Glisovic, S. M. (2011). The model of air pollution generated by fire chemical accident in an urban street canyon. *Transportation Research Part D, Transport and Environment, 16*(4), 321–326. doi:10.1016/j.trd.2011.01.012

Pet-Armacost, J. J., Sepulveda, J., & Sakude, M. (1999). Monte Carlo sensitivity analysis of unknown parameters in hazardous materials transportation risk assessment. *Risk Analysis, 19*(6), 1173–1184. doi:10.1111/j.1539-6924.1999.tb01136.x PMID:10765455

Philipson, L. L., & Napadensky, H. S. (1982). The methodologies of hazardous materials transportation risk assessment. *Journal of Hazardous Materials*, *6*(4), 361–382. doi:10.1016/0304-3894(82)85003-6

Pine, J. C., & Marx, B. D. (1997). Utilizing state hazardous materials transportation data in hazardous analysis. *Journal of Hazardous Materials*, *54*(1–2), 113–122. doi:10.1016/S0304-3894(96)01871-7

Planas, E., Pastor, E., Presutto, F., & Tixier, J. (2008). Results of the MITRA project: Monitoring and intervention for the transportation of dangerous goods. *Journal of Hazardous Materials*, *152*(2), 516–526. doi:10.1016/j.jhazmat.2007.07.032 PMID:17884284

Pradhananga, R., Taniguchi, E., & Yamada, T. (2010). Ant colony system based routing and scheduling for hazardous material transportation. *Procedia: Social and Behavioral Sciences*, *2*(3), 6097–6108. doi:10.1016/j.sbspro.2010.04.022

Pradhananga, R., Taniguchi, E., Yamada, T., & Qureshi, A. G. (2014). Bi-objective decision support system for routing and scheduling of hazardous materials. *Socio-Economic Planning Sciences*, *48*(2), 1–14. doi:10.1016/j.seps.2014.02.003

Purdy, G. (1993). Risk analysis of the transportation of dangerous goods by road and rail. *Journal of Hazardous Materials*, *33*(2), 229–259. doi:10.1016/0304-3894(93)85056-K

Qiao, Y., Keren, N., & Mannan, M. S. (2009). Utilization of accident databases and fuzzy sets to estimate frequency of hazmat transport accidents. *Journal of Hazardous Materials*, *167*(1-3), 374–382. doi:10.1016/j.jhazmat.2009.01.097 PMID:19250750

Raj, P. K., & Pritchard, E. W. (2000). Hazardous materials transportation on US railroads: Application of risk analysis methods to decision making in development of regulations. *Transportation Research Record*, *1707*, 22–26. doi:10.3141/1707-03

Reilly, A., Nozick, L., Xu, N., & Jones, D. (2012). Game theory-based identification of facility use restrictions for the movement of hazardous materials under terrorist threat. *Transportation Research Part E, Logistics and Transportation Review*, *48*(1), 115–131. doi:10.1016/j.tre.2011.06.002

Ren, C., Yuan, X., Wang, J., Zhang, X., & Li, J. (2012). Study on emergency response rank mode of flammable and explosive hazardous materials road transportation. *Procedia Engineering*, *45*, 830–835. doi:10.1016/j.proeng.2012.08.246

Reniers, G. L. L., & Dullaert, W. (2013). A method to assess multi-modal hazmat transport security vulnerabilities: Hazmat transport SVA. *Transport Policy*, *28*, 103–113. doi:10.1016/j.tranpol.2012.05.002

Reniers, G. L. L., Jongh, K.D., Gorrens, B., Lauwers, D., Leest, M.V., & Witlox, F. (2010). Transportation Risk Analysis tool for hazardous Substances (TRANS) – A user-friendly, semi-quantitative multi-mode hazmat transport route safety risk estimation methodology for Flanders. *Transportation Research Part D, Transport and Environment*, *15*(8), 489–496. doi:10.1016/j.trd.2010.07.001

Romer, H., & Haastrup, P., & StyhrPetersen, H. J. (1995). Accidents duringmarine transport of dangerous goods – distribution of fatalities. *Journal of Loss Prevention in the Process Industries*, *8*(1), 29–34. doi:10.1016/0950-4230(95)90059-X

Ronza, A., Vilchez, J. A., & Casal, J. (2007). Using transportation accident databases to investigate ignition and explosion probabilities of flammable spills. *Journal of Hazardous Materials*, *146*(1-2), 106–123. doi:10.1016/j.jhazmat.2006.11.057 PMID:17240052

Saat, M. R., Werth, C. J., Schaeffer, D., Yoon, H., & Barkan, C. P. L. (2014). Environmental risk analysis of hazardous material rail transportation. *Journal of Hazardous Materials*, *264*, 560–569. doi:10.1016/j.jhazmat.2013.10.051 PMID:24239259

Saccomanno, F. F., & Shortreed, J. H. (1993). Hazmat transport risks – societal and individual perspectives. *Journal of Transportation Engineering*, *119*(2), 177–188. doi:10.1061/(ASCE)0733-947X(1993)119:2(177)

Saccomanno, F. F., Shortreed, J. H., Van Aerde, M., & Higgs, J. (1989). Comparison of risk measures for the transport of dangerous commodities by truck and rail. *Transportation Research Record*, *1245*, 1–13.

Samanlioglu, F. (2013). A multi-objective mathematical model for the industrial hazardous waste location-routing problem. *European Journal of Operational Research*, *226*(2), 332–340. doi:10.1016/j.ejor.2012.11.019

Samuel, C., Keren, N., Shelley, M. C., & Freeman, S. A. (2009). Frequency analysis of hazardous material transportation incidents as a function of distance from origin to incident location. *Journal of Loss Prevention in the Process Industries*, *22*(6), 783–790. doi:10.1016/j.jlp.2009.08.013

Scanlon, R. D., & Cantilli, E. J. (1985). Assessing the risk and safety in the transportation of hazardous materials. *Transportation Research Record*, (1020).

Schweitzer, L. (2006). Environmental justice and hazmat transport: A spatial analysis in southern California. *Transportation Research Part D, Transport and Environment*, *11*(6), 408–421. doi:10.1016/j.trd.2006.08.003

Shappert, L. B., Brobst, W. A., Langhaar, J. W., & Sisler, J. A. (1973). Probability and consequences of transportation accidents involving radioactive-material shipments in the nuclear fuel cycle. *Nuclear Safety*, *14*(6), 597–604.

Shen, X., Xie, C., Liu, H., & Qui, Z. (2013). Model and algorithm for routing and scheduling problem in hazardous materials transportation network. *Journal of Networks*, *8*(5), 1027–1034. doi:10.4304/jnw.8.5.1027-1034

Shen, X., Yan, Y., Li, X., Xie, C., & Wang, L. (2014). Analysis on tank truck accidents involved in road hazardous materials transportation in China. *Traffic Injury Prevention*, 15(7), 762–768. PMID:24380669

Sherali, H. D., Brizendine, L. D., Glickman, T. S., & Subramanian, S. (1997). Low-probability high consequence considerations in routing hazardous materials shipments. *Transportation Science*, *31*(3), 237–251. doi:10.1287/trsc.31.3.237

Shorten, C. V., Galloway, J., Krebs, J. G., & Fleming, R. S. (2002). A 12-year history of hazardous materials incidents in Chester County, Pennsylvania. *Journal of Hazardous Materials*, *89*(1), 29–40. doi:10.1016/S0304-3894(01)00302-8 PMID:11734344

Sivakumar, R. A., Batta, R., & Karwan, M. H. (1993). A network-based model for transporting extremely hazardous materials. *Operations Research Letters*, *13*(2), 85–93. doi:10.1016/0167-6377(93)90034-E

Spadoni, G., Leonelli, P., Verlicchi, P., & Fiore, R. (1995). A numerical procedure for assessing risks from road transport of dangerous substances. *Journal of Loss Prevention in the Process Industries*, *8*(4), 245–252. doi:10.1016/0950-4230(95)00029-Z

Sulijoadikusumo, G. S., & Nozick, L. K. (1998). Multiobjective routing and scheduling of hazardous materials shipments. *Transportation Research Record*, *1613*, 96–104. doi:10.3141/1613-13

Swoveland, C. (1987). Risk analysis of regulatory options for the transport of dangerous commodities by rail. *Interfaces*, *17*(4), 90–107. doi:10.1287/inte.17.4.90

Szeto, W. Y. (2013). Routing and scheduling hazardous material shipments: Nash game approach. *Transportmetrica B. Transport Dynamics*, *1*(3), 237–260. doi:10.1080/21680566.2013.861330

Tena-Chollet, F., Tixier, J., Dusserre, G., & Mangin, J.-F. (2013). Development of a spatial risk assessment tool for the transportation of hydrocarbons: Methodology and implementation in a geographical information system. *Environmental Modelling & Software*, *46*, 61–74. doi:10.1016/j.envsoft.2013.02.010

Toumazis, I., & Kwon, C. (2013). Routing hazardous materials on time-dependent networks using conditional value-at-risk. *Transportation Research Part C, Emerging Technologies*, *37*, 73–92. doi:10.1016/j.trc.2013.09.006

Trépanier, M., & Leroux, M., & MarcellisWarin, N. (2009). Cross-analysis of hazmat road accidents using multiple databases. *Accident; Analysis and Prevention*, *41*(6), 1192–1198. doi:10.1016/j.aap.2008.05.010 PMID:19819367

Van Aerde, M., Stewart, A., & Saccomanno, F. (1988). Estimating the impacts of L.P.G. spills during transportation accidents. *Journal of Hazardous Materials*, *20*, 375–392. doi:10.1016/0304-3894(88)87023-7

Van Raemdonck, K., Macharis, C., & Mairesse, O. (2013). Risk analysis system for the transport of hazardous materials. *Journal of Safety Research*, *45*, 55–63. doi:10.1016/j.jsr.2013.01.002 PMID:23708476

Vanaerde, M., Shortreed, J., Stewart, A. M., & Matthews, M. (1989). Assessing the risks associated with the transport of dangerous goods by truck and rail using the riskmod model. *Canadian Journal of Civil Engineering*, *16*(3), 326–334. doi:10.1139/l89-058

Verma, M. (2009). A cost and expected consequence approach to planning and managing railroad transportation of hazardous materials. *Transportation Research Part D, Transport and Environment*, *14*(5), 300–308. doi:10.1016/j.trd.2009.03.002

Verma, M. (2011). Railroad transportation of dangerous goods: A conditional exposure approach to minimize transport risk. *Transportation Research Part C, Emerging Technologies*, *19*(5), 790–802. doi:10.1016/j.trc.2010.07.003

Verma, M., & Verter, V. (2007). Railroad transportation of dangerous goods: Population exposure to airborne toxins. *Computers & Operations Research*, *34*(5), 1287–1303. doi:10.1016/j.cor.2005.06.013

Verma, M., & Verter, V. (2010). A lead-time based approach for planning rail–truck intermodal transportation of dangerous goods. *European Journal of Operational Research*, *202*(3), 696–706. doi:10.1016/j.ejor.2009.06.005

Verma, M., Verter, V., & Zufferey, N. (2012). A bi-objective model for planning and managing rail-truck intermodal transportation of hazardous materials. *Transportation Research Part E, Logistics and Transportation Review*, *48*(1), 132–149. doi:10.1016/j.tre.2011.06.001

Verter, V., & Kara, B. Y. (2001). A GIS-based framework for hazardous materials transport risk assessment. *Risk Analysis*, *21*(6), 1109–1120. doi:10.1111/0272-4332.216179 PMID:11824686

Verter, V., & Kara, B. Y. (2008). A path-based approach for Hazmat transport network design. *Management Science*, *54*(1), 29–40. doi:10.1287/mnsc.1070.0763

Vilchez, J. A., Sevilla, S., Montiel, H., & Casal, J. (1995). Historical analysis of accidents inchemical plants and in the transportation of hazardous materials. *Journal of Loss Prevention in the Process Industries*, *8*(2), 87–96. doi:10.1016/0950-4230(95)00006-M

Wang, J., Kang, Y., Kwon, C., & Batta, R. (2012). Dual toll pricing for hazardous materials transport with linear delay. *Networks and Spatial Economics*, *12*(1), 147–165. doi:10.1007/s11067-011-9156-9

Welles, W. L., Wilburn, R. E., Ehrlich, J. K., & Kamara, J. M. (2009). New York hazardous substances emergency events surveillance (HSEES) data support emergency response, promote safety and protect public health. *Journal of Loss Prevention in the Process Industries*, *22*(6), 728–734. doi:10.1016/j.jlp.2008.07.013

Wijeratne, A. B., Turnquist, M. A., & Mirchandani, P. B. (1993). Multiobjective routing of hazardous materials in stochastic networks. *European Journal of Operational Research*, *65*(1), 33–43. doi:10.1016/0377-2217(93)90142-A

Xie, Y., Lu, W., Wang, W., & Quadrifoglio, L. (2012). A multimodal location and routing model for hazardous materials transportation. *Journal of Hazardous Materials, 227– 228*, 135– 141.

Yang, J., Li, F., Zhou, L., Zhang, L., Huang, L., & Bi, J. (2010). A survey on hazardous materials accidents during road transport in China from 2000 to 2008. *Journal of Hazardous Materials*, *184*(1-3), 647–653. doi:10.1016/j.jhazmat.2010.08.085 PMID:20863620

Zhang J., & Zhao L. (2007). Risk analysis of dangerous chemicals transportation. *Systems Engineering - Theory & Practice, 27*(12), 117-122.

Zhang, J. J., Hodgson, J., & Erkut, E. (2000). Using GIS to assess the risks of hazardous materials transport in networks. *European Journal of Operational Research*, *121*(2), 316–329. doi:10.1016/S0377-2217(99)00220-9

Zhao, L., Wang, X., & Qian, Y. (2012). Analysis of factors that influence hazardous material transportation accidents based on Bayesian networks: A case study in China. *Safety Science*, *50*(4), 1049–1055. doi:10.1016/j.ssci.2011.12.003

Zografos, K. G., & Androutsopoulos, K. N. (2002). Heuristic algorithms for solving hazardous materials logistical problems. *Transportation Research Record*, *1783*, 158–166. doi:10.3141/1783-20

Zografos, K. G., & Androutsopoulos, K. N. (2008). A decision support system for integrated hazardous materials routing and emergency response decisions. *Transportation Research Part C, Emerging Technologies*, *16*(6), 684–703. doi:10.1016/j.trc.2008.01.004

Zografos, K. G., & Davis, C. (1989). Multi-objective programming approach for routing hazardous materials. *Journal of Transportation Engineering*, *115*(6), 661–673. doi:10.1061/(ASCE)0733-947X(1989)115:6(661)

Zografos, K. G., & Samara, S. (1989). Combined location–routing model for hazardous waste transportation and disposal. *Transportation Research Record*, *1245*, 52–59.

Zografos, K. G., Vasilakis, G. M., & Giannouli, I. M. (2000). A methodological framework for developing a DSS for hazardous material emergency response operations. *Journal of Hazardous Materials*, *71*, 503–521. doi:10.1016/S0304-3894(99)00096-5 PMID:10677678

KEY TERMS AND DEFINITIONS

Hazmat: An abbreviation for Hazardous Materials.

Hazmat Class: Shows each of the classification of Hazardous materials out of 9 different hazmat type.

Hazmat Waste: All kind of remaining from hazardous materials.

Hazmat Transportation Accidents: All kind of accidents that happen during transporting hazardous materials by all kind of vehicles.

Intermodal Transportation: Transporting the goods by using at least two different transport modes without handling the freight.

Routing: Choosing a path out of alternative paths for a vehicle from a start point to its destination.

Transport Mode: Shows the way in which freight is transported. The transport modes are road, railway, marine, air and pipeline.

Chapter 15
A Tool for GIS Based Risk Analysis for Transportation of Dangerous Goods on Road (the RAGISADR):
A Case Study for Fuel Products

Serhan Karabulut
Turkish Land Forces, Turkey

Ebru V. Ocalir-Akunal
Gazi University, Turkey

ABSTRACT

In this study, an environmental risk analysis model is developed as a decision support system (the RAGISADR) in order to assess environmental risks for transportation of dangerous goods. Transport of dangerous goods needs to be regulated to prevent accidents. Moving dangerous goods by road is governed by international regulations. GIS are used to quantify the factors on each link in the network that contribute to each of the evaluation criteria for a possible route. The Analytic Hierarchy Process (AHP) is used to assign weights to factors. The priority weights of each environmental criteria is calculated by using Analytical Hierarchy Process (AHP) before risk models are implemented in the road network of Izmir-Manisa-Aydın-Mugla-Denizli (Aegean Region of Turkey). The most convenient route according to the criteria set is determined. The results give decision maker the possibility to choose the best alternative among possible routes according to certain criteria.

INTRODUCTION

In most of the cases dangerous goods have to be transported from a point of origin to one or more destinations points. The dangerous goods are then transported from a production facility to storage, distribution, or another facility where the dangerous good is required. For example, many oil products are refined at the refinery and then shipped to storage tanks at different locations within a country. Dangerous goods

DOI: 10.4018/978-1-4666-8648-9.ch015

transport risk is anything that product the probability of undesirable event during dangerous goods transportation on road network and the consequence of that event. Due to the nature of dangerous goods every production, storage, and transportation activity related to the use of these materials inherits many risks for both society and environment.

In this case, the problem is the definition of the desired criteria between two points regarding the most appropriate route selection process. Dangerous goods transportation generally involves multiple stakeholders such as shippers, carriers, and governments; each having its own priority and perspective, and each playing a role in the safe movement of the dangerous goods from origins to destinations over a transportation network. Therefore, dangerous goods transportation is a typical multistakeholder and multiobjective problem which is generally complicated to solve. Given the multiobjective nature of the dangerous goods transportation, it is usually impossible to identify a single optimal route that can optimize every single objective. In most cases, risk and safety interests conflict with economic interests, rendering the decision making process a complex task. Decision support systems such as routing and risk analysis models are used in management of dangerous goods transport operetions, which allow decision makers determine the best route for the transport of dangerous goods.

The transportation of hazmats can be classified according to the mode of transport, namely: road, rail, water, air, and pipeline. Approximately 450 million tons of hazardous materials are transported annually in the world. Historical evidence has shown that over 63 percent of dangerous goods shipments are moved by road (Alisan, 2009). In the United States, according to a report published in 2010 by the carrying amount of dangerous goods transportation mode portion of the 53% is transported by road (USDOT & USDOC, 2010). Also, annually 750,000 tonnes of hazardous materials are transported by approximately 15,000 on-road vehicles in Turkey (Alisan, 2009).

When dangerous goods are transported by road, an accident could cause considerable harm to people and the environment. Spillage of the goods could lead to fires, explosions and chemical poisoning or burning. In order to minimize any risk related to the use of dangerous goods, safety standards and codes are often implemented to regulate the proper use of dangerous goods. The purpose of these regulations and codes is to minimize the risk attached to the use of dangerous goods. There is an international agreement, published by United Nations Economic Commission for Europe, called ADR-European Agreement, concerning the International Carriage of Dangerous Goods by Road. The purpose of ADR is to ensure that dangerous goods being carried by road can cross international borders freely, as long as goods, vehicles and drivers comply with its provisions (UN, 2013).

ADR sets out the requirements for classifying, packaging, labelling and certifying dangerous goods. These requirements are set out in Annex A to ADR. Vehicles carrying dangerous goods must comply with the provisions of Annex B to ADR, which includes vehicle and tank specifications and other operational requirements.

Classification of dangerous goods is divided into nine classes based on their predominant hazard. The classes of dangerous goods according to ADR are the following (UN, 2013):

Class 1, Explosives are solid or liquid substances (or mixtures of substances) capable by chemical reaction of producing gases at such a temperature and pressure and at such a speed as to cause damage to the surroundings.

Class 2, Gases covers pure gases, mixtures of gases, mixtures of one or more gases with one or more other substances and articles containing such substances.

Class 3, Flammable liquids covers substances and articles containing have at 50 °C a vapour pressure of not more than 300 kPa (3 bar) and are not completely gaseous at 20 °C and at standard pressure of 101.3 kPa; and have a flash-point of not more than 60 °C.

Class 4, Flammable solids, substances liable to spontaneous combustion, substances that in contact with water emit flammable gases. Flammable solids are readily combustible solids and solids which may cause fire through friction. Also included are solid substances that ignite on contact with air, and substances which react with water to emit flammable gases liable to form explosive mixtures with air.

Class 5, Oxidizing substances; organic peroxides. Oxidizing substances cover substances which may, generally by yielding oxygen, cause or contribute to the combustion of other material. Organic peroxides are organic substances which may be considered derivatives of hydrogen peroxide, where one or both of the hydrogen atoms have been replaced by organic radicals.

Class 6, Toxic substances; infectious substances. Toxic substances are able by a single action or by action of short duration to cause damage to human health, or death, by inhalation, by cutaneous absorption or by ingestion. Infectious substances are substances which are known or are reasonably expected to contain pathogens.

Class 7, Radioactive material. Radioactive material means any material containing radionuclides where both the activity concentration and the total activity in the consignment exceed the defined values.

Class 8, Corrosives are substances which by chemical action attack epithelial tissue with which they are in contact, or which in the event of leakage are capable of damaging or destroying other goods, or means of transport.

Class 9, Miscellaneous dangerous substances and articles covers substances and articles which, during carriage, present a danger not covered by the heading of other classes.

Among hazardous materials types, petroleum products and LPG constitute approximately 80% of total hazardous materials consumed in Turkey (Umit, 2003). Automotive fuel consumption is continuously increasing every year in Turkey. According to PETDER (http://www.petder.org.tr/petder-sektor-raporlari, 2012), total consumption of the white products (Gasoline, diesel fuels and LPG auto-gas) increased by 3.9% and reached to 18.2 million tons in 2012. The magnitude of petroleum and LPG products in Turkey is significant enough to incorporate in a study. Therefore, in this paper, fuel and LPG products transportation are focused. LPG products belong to the class of gases (Class 2) and fuel products belong to the class of flammable-combustible liquids (Class 3).

In order to be able to choose economic and safe routes for the transport of dangerous goods it is necessary to use tools that allow planners and decision makers to evaluate all possible routes, and to assist in finding a solution to a particular problem. In this study, transport risk is analyzed when dangerous goods shipments contain different type of dangerous good and multiple origin-destination pair. A GIS based environmental risk analysis model (the RAGISADR) has been developed to evaluate the environmental risks of transportation of dangerous goods on road. The model is integrated into GIS in order to enable large-scale applications. The developed model can be used as a tool for route decision making of fuel products so far. A risk based decision support model is presented for transportation of dangerous goods on road. The relevance of the model as a decision support tool is demonstrated through a case application to a road network in a region in western Turkey. The case study has demonstrated some useful decision support features of the proposed model concerning issues.

Geographic Information Systems (GIS) are utilized when using the existing models as well as during the development of the environmental risk analysis model. The MapInfo Professional 10.5 software is used as the primary GIS in order to visualize and analyse the data by mapping and graphically demonstrating the different methods. The Izmir-Manisa-Aydın-Mugla-Denizli road network is selected for the implementation of various risk analysis models and afterwards for the assessment of the results. The priority weights of each environmental criterion entered into the model are calculated by using Analytical Hierarchy Process (AHP). The most convenient route according to the criteria set is determined after assessing the outcomes of the risk analyses models.

The risk value for each shipment are computed by taking into consideration five objectives that minimize (1) the travel distance, (2) the population exposure, (3) the incident probability, (4) societal risk (accident probability multiplied by the number of people in the danger zone), and (5) the pollution of environmental elements was evaluated.

Literature Survey

Hazardous materials transportation problem has become significantly important for decades due to the consequences of possible incidents of transportation. Thus, the aspects of the hazardous materials transportation attract an increasing attention from researchers in years. Risk is the major component that separates hazmat transportation problems from other transportation problems in the literature.

Decision making process related to the route selection for the transport of dangerous goods are complex due to the fact that controversial issues (economic and risk related) can be taken into account; and when controversial issues are dealt with in relation to a given phenomenon, making decisions must be sustained in formal analysis (Bohnenblust & Slovic, 1998). The risk analysis model for dangerous goods transport therefore can serve as a tool to ease the decision making process and may help to reach consensus among the parties involved. The risk analysis model can give an overview of the consequences when using certain routes for the transport of dangerous goods. Even though the risk analysis model does not generate the solution to the problem related to the economic and safe transport of dangerous goods; it can provide valuable information related to the routing problem, thus contributing to the decision making process.

Risk is a measure of the probability and severity of harm to an exposed receptor due to potential undesired events involving dangerous goods (Alp, 1995). Most commonly, risk is defined as the product of the probability of an undesirable event and the consequence of that event (Erkut & Verter, 1998). In the context of dangerous goods transportation, an undesirable event is an accident followed by the release of a hazardous substance. The consequences of a hazmat r(P)elease can include economic or environmental losses as well as damage to human population. According to the traditional risk definition, the risk of transporting dangerous goods (R) is defined as accident probability multiplied by consequences of the accident (C)(Barilla et al., 2009). Therefore, (R) is given by:

$$R = P.C \tag{1}$$

Transportation of dangerous goods is a continuous activity. Note that in the context of hazmat routing there is linear risk (risk along and edge and route). Path risk can be expressed as follows (Erkut & Verter, 1998):

$$\sum_{s=1}^{r} C_{s,m} \tag{2}$$

This model is often referred to as the traditional risk model. Although the traditional risk model has been the most popular one, many other hazmat transport risk models have been proposed in the literature. Table 1 summarizes models and cites studies that have used each model. Most of the models use population exposure as the consequence measure.

This study presents a risk based decision support model that can be used to select economic and safe routes for the transport of dangerous goods on a regional road network. The developed model consists mainly of quantitative risk assessment (QRA) model that estimates the risk for the entire network and at its specific nodes. In this study, applied five models were applied on the data that were collected for the

Table 1. Risk models for hazmat transportation

Model	Mathematical Expression	Criteria	References
Population exposure	$\sum_{s=1}^{n} d_s\left(2l_s\lambda_m + \pi\lambda_m^2\right)$	The number of people living in exposure zone	Batta & Chiu (1988), Revelle et al. (1991), Erkut & Ingolfsson (2000), Verter & Kara (2001), Jassbi et al.(2010)
Traditional risk	$\left(\sum_{s=1}^{n} l_s p_s d_s\right)\pi\lambda_m^2$	The number of people in the danger zone. The number of having an incident on link	Pijawka et al. (1985), Alp (1995), Zhang et al. (2000), Erkut & Verter (1995), Verter & Kara (2001), Jassbi et al.(2010)
Incident probability	$\sum_{s=1}^{n} l_s p_s$	The number of having an incident on link	Saccomanno & Chan (1985), Abkowitz et al. (1992), Verter& Kara (2001), Jassbi et al.(2010)
Perceived risk	$\sum_{s=1}^{n} p_s c_s^{\alpha}, \alpha > 0$	The number of people that will be affected in respect to low probability–high consequence event	Abkowitz et al. (1992), Erkut & Verter (1995)
Conditional risk	$\sum_{s=1}^{n} p_s c_s / \sum_{s=1}^{n} p_s$	The expected risk of the first accident	Sivakumar et al. (1993), Sherali et al. (1997)
Maximum population exposure	$\max_{e_s \in P^{C_s}}$	The maximum consequence per unit population	Erkut & Ingolfsson (2000)
Expected disutility	$\sum_{s=1}^{n} p_s\left(\exp\left(\alpha c_s\right) - \right), \alpha > 0$	The number of people affected by a catastrophic incidents	Erkut & Ingolfsson (2000)
Mean –variance	$\sum_{s=1}^{n}\left(p_s c_s + \beta p_s c_s^2\right)4, \beta > 0$	The expected value and variance of the number of people affected by an incident	Sivakumar & Batta (1994)
Demand satisfaction	$\sum_{s=1}^{n}\left(1-\exp\left(-p_s\right)\right)c_s \prod_{j=1}^{n}\exp\left(p_j\right)$	The expected total consequence from all trips required to fulfill demand	Erkut & Ingolfsson (2005)
Time-based risk	$\left(L_s V_s\right) c_{sm}$	Duration of risk exposure	Ak & Bozkaya (2008)

continued on following page

Table 1. Continued

Model	Mathematical Expression	Criteria	References
Cumulative weights and scores	$\sum_{s=1}^{n_s}\left(w_s \sum_{cf=1}^{n_{sf}} w_{cf} s_{cf}\right)$	Population exposure, socio-economic impact, risks of hijack, traffic conditions, emergency response	Huang (2004), Huang et al. (2006), Huang et al. (2004)
Environmental risk	$\frac{k_s}{K}\left(\sum_{s=1}^{n_s}\left(w_s \sum_{sf=1}^{n_{sf}} w_{sf} A_{s,m}\right)\right)$	Environmental factors in exposure zone	

m hazmat type
s road segment
λ the impact radius
l_s l_s length of link s
d_s population density on link s
p_s probability of accident on link s
c_s cycle of the routes on link s
L_s the length of the route segment l
V_s the truck speed on link l
c_{sm} the total population within a danger circle.
c criteria
w_f weight of criteria c
c_f factor under criteria c
s_{cf} score of factor f under criteria c
k_s number of accidents on link s
K number of total accidents
$A_{s,m}$ area of environmental factors on road segment s of hazmat m

study area. In the present risk assessment methodology for dangerous goods the following most common risk models are utilized: societal risk, population exposure, and incident probability. In addition to the risk models mentioned above, shortest travel distance is taken into account for choosing a dangerous goods shipment route. Under the Shortest Path model, it is assumed that dangeros goods carriers choose to use the shortest path between an origin and a destination. Also, the developed environmental risk analysis model (the RAGISADR) has been applied to evaluate the environmental risks of transportation of dangerous goods on road.

Population Exposure Model

Consider a hazmat truck moving on a link. The circular area around the truck is where the population is exposed to risk. An impact zone of a single dangerous goods incident is marked out with a circle of radius centered at the accident location. Consider the union of all such circles centered at all points on link *(i, j)*. The impact area is the "semicircular exposure zone". Figure 1 shows the semicircular impact zone of radius (λ) for link *(i, j)* (Kara et al., 2003). A danger zone of a single hazmat incident is defined with a circle of (λ) radius, which is centred at the position where the accident takes place. (λ) is the impact radius of the hazmat that is being carried, which may change according to the hazmat type. The exposure zone of a link is considered as "risk corridor" on both sides of a link. Finally, the number of people within an exposure zone is the population exposure of the corresponding link (Erkut & Verter, 1995).

Population exposure is one of the risk definitions in the hazmat literature. According to this risk definition, the area within (λ) neighborhood of a point c on link *(i,j)* is under potential risk, where (λ) varies for different hazmat types (Batta & Chiu, 1988).

Note that it is possible to preprocess the population density data and compute $C_{s,m}$ for all links in the transport network. This is rather tempting since it can be accomplished easily using a GIS. Let d_s indicate the population density on link s. Let $C_{s,m}$ indicate the number of people in the risk corridor of dangerous goods type m around link s, and λ_m indicate the impact radius of dangerous goods type m (Verter & Kara, 2001).

$$C_{s,m} = \pi \lambda_m^2 d_s \tag{3}$$

When the links are not straight line, then it is needed to compute the area within the danger zone by integration ($\sum_{s=1}^{r} C_{s,m}$), which may be time-consuming. There is however an easier way of calculating population exposure when GIS tools are utilized.

Incident Probability Model

The incident probability of a path is computed by adding the incident probabilities along each link of that path. The probability of incident on a given link, however, depends on the incident probabilities of f all links leading up to that link. Hence, the incident probabilities on links are path-dependent (Kara et al., 2003). The probability of the dangerous goods transport incident depends on the type of dangerous good.

Figure 1. Population exposure of a single truck on link (i,j)

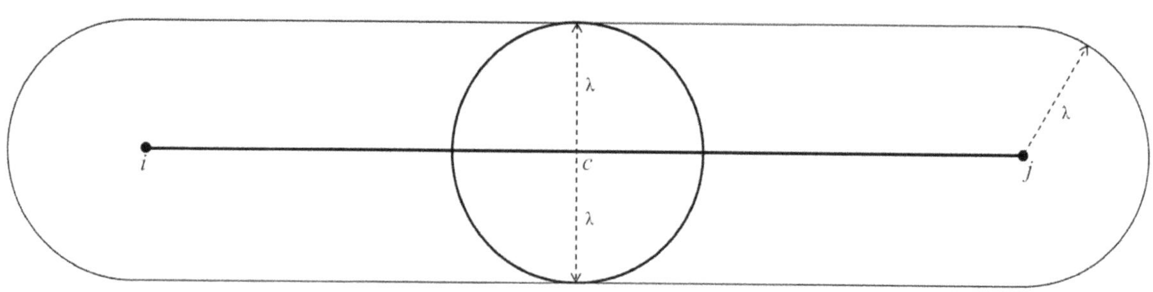

The Incident Probability for each road segments are typically computed as (Saccomanno & Chan, 1985):

$$R_{ip} = \sum_{s=1}^{r} l_s p_s \tag{4}$$

Where p_s indicates the probability of accident, m is hazmat type, l_s express the length of each arc. The algorithm adjusts the link impedances (incident probabilities) at each iteration by multiplying them with the probability of safely arriving at the starting node of the arc.

Societal Risk Model

Traditional definition of risk arises as Societal Risk (accident probability multiplied by the number of people in the danger zone) in the hazmat literature. Several authors preferred to use this risk definition as it combines both incident probability and number of people exposed, which in most cases become more effective and justifiable.

According to the definitions described above, the societal risk is (Erkut & Verter, 1998):

$$R_{sr} = \sum_{s=1}^{r} R_{ip} C_{s,m} = \left(\sum_{s=1}^{r} l_s p_s d_s \right) \pi \lambda_m^2 \tag{5}$$

In practice, societal risk is calculated for each segment of uniform accident environment and uniform population density, and the segment societal risks are summed to arrive at the total network societal risk.

In recent years, large numbers of studies on risk assessment of the dangerous goods transportion have been reported in literature (Saccomanno & Chan, 1985; Pijawka et al., 1985; Batta & Chiu, 1988; ReVelle et al., 1991; Abkowitz et al., 1992; Sivakumar, et al., 1993; Sivakumar & Batta, 1994; Erkut & Verter, 1998; Bonvicini et al., 1998; Erkut & Ingolfsson, 2000; Barilla et al., 2009; Jassbi & Makvandi, 2010.). Also, the use of GIS to aid dangerous goods route planning is not new. In many studies GIS is applied in the risk analysis of dangerous goods transportation (Lovett et al.,1997; Zhang et al., 2000; Verter & Kara, 2001, Umit & Kara, 2003; Huang, 2004, 2006; Huang et al., 2006; Monprapussorn et al., 2009).

The vast collection of literature involving dangerous goods transportation risk analysis reveals that the main considerations are economic issues and population exposure. Nevertheless, the environmental risks in transporting dangerous goods have been studied in literature (Wessberg et al., 2008; Topuz et al., 2011; Li et al., 2013).

The literature dealing with the risk analysis of dangerous goods is rich and numerous models have been proposed in recent years. However, it still lacks of the environment factors being taken into consideration in the proposed framework. The need to include environmental considerations has only recently gained greater attention. In this connection, this study seeks to identify a set of evaluation criteria that can be used to route the petroleum product and LPG trucks, incorporating emphasis on the environment aspect in addition to costs and exposure. The identified criteria will then be tested using a GIS coupled with the Analytic Hierarchy Process (AHP) to assess their suitability. The major aim of this study is to consider and to integrate environment factors into the dangerous goods transport framework with the simplified approach.

Proposed Environmental Risk Analysis Model (the RAGISADR)

Environmental risk (ER) can be defined as a risk whose results affect human health, life conditions, environment, soil, surface water, ground water, air, climate, flora/fauna, biodiversity, structure of community, buildings, view of city, cultural heritage and the relationships between those components (Wessberg et al., 2008). Risk assessment forms the core of the contaminated land investigation and assessment process. The purpose is to obtain information on sources, pathways and receptors present on or adjacent to a site. The presence of a linkage between these may then constitute a risk, the significance of which must be assessed and is dependent upon a number of factors.

In the context of environment contamination, there are three essential elements to any risk: source (contaminant), pathway, and receptor. The environmental risk model developed in this study can be expressed as the following steps. The main steps are the analysis of the risk sources, pathway analysis and receptor analysis.

Risk source analysis refers to identifying and analyzing the possible interference which have a negative effect on the environment or its components in the region (Ruifang, 2010). Contaminant is a substance which has the potential to cause harm to environmental elements. In this study, liquid fuel (gasoline and diesel) and LPG are considered to be the source of environmental risk.

Pathway analysis refers to analyzing the route by which a source of contamination can migrate; an identified receptor can be exposed to, or be affected by an identified source.

Receptor analysis means the risk which bears the risk. It refers to the component parts of disadvantage effects of the risk sources on the environment in risk assessment. In literature, the receptors in the environmental risk analysis of dangerous goods during transportation consist of three aspects, i.e. water environment, soil environment and ecosystem (Ruifang, 2010; Gunasekera et al., 2003; Andersson et al., 2007; Rygnestad, 2000; Liu et al., 2009; Ashtakala et al., 1996).

The study proposes a procedure and a model to analyze the environmental consequences of accidents on the basis of the environmental elements, such as water element, soil element, and featured areas. During assessing the environmental risk, the factors sensitive to the risk, which have an important status in environment must be chosen by judgement and analysis. Scope of the study covered all of the environmental criteria were identified as two levels (Figure 2).

According to the definition of risk, the appropriate environmental risk (R_{er}) can be expressed as the unexpected degree of harm on the environment. It is defined as the product of the accident probability P (degree of risk) and environmental consequences of the accident, with the characterization of R (Ruifang, 2010).

$$R_{er} = P_s C \tag{6}$$

In developed environmental risk analysis model, environmental consequences have been adapted from the cumulative weights and scores model (Huang, 2004, 2006; Huang, et al., 2004). The cumulative weights and scores represent a cost function for each route, which is given by:

$$Cost\ of\ route(R) = \sum_{c=1}^{n_c} \left(w_c \sum_{cf=1}^{n_{cf}} w_{cf} s_{cf} \right) \tag{7}$$

Figure 2. Hierarchy of environmental main and sub-criteria

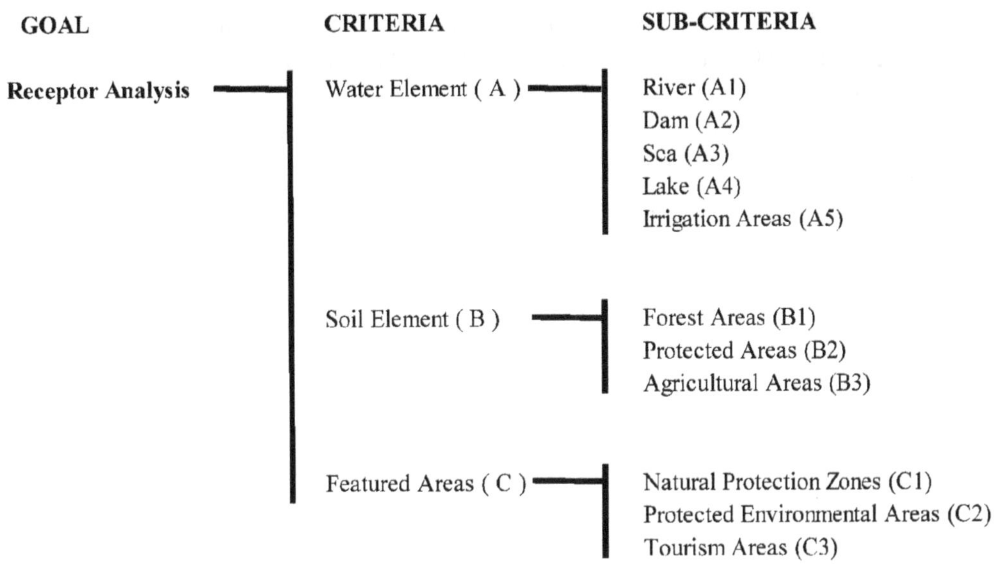

where c = criteria; nc = number of criteria; wc = weight of criteria c; cf = factor under criteria c; ncf = number of factors under criteria c; wcf = weight of factor f under criteria c; and scf = score of factor f under criteria c.

The key element in the proposed environmental risk analysis model is environmental consequences of the accident. The environmental consequence in the environmental risk analysis model is described similar to the cumulative weights and scores model. But instead of scoring function calculation factors, a spatial calculation is used. Hence, the environmental risk on road segment s of hazmat m can be expressed as

$$R_{er} = \frac{k_s}{K}\left(\sum_{k=1}^{n_k}\left(w_k \sum_{kf=1}^{n_f} w_{kf} A_{s,m}\right)\right) \tag{8}$$

k_s number of accidents on link s
K number of total accidents
k criteria
n_k number of criteria
w_k weight of criteria
k_f factor under criteria k
n_f number of factors under criteria k
w_{kf} weight of factor f under criteria k
$A_{s,m}$ area of environmental factors on road segment s of hazmat m

Environmental risk is estimated by multiplying accident probability for each link by the consequences of a dangerous goods accident. The impacts of dangerous goods during transportation on environment mainly embody on environmental elements, i.e. water, soil and featured areas. Environmental risk means the assessment of the damaging contents and the occurring probability of the risk event. Consequences of accidents for environmental elements can be characterized by the total damage of water element, soil element, and featured areas.

The minimum environmental risk route between an o-d pair is obtained by using the risk unit on each link as impedance. The optimal route according to the criteria set is determined after assessing the outcomes of the environmental risk analyses model.

Analysis Method

The methodology is integrated with GIS for efficient implementation. The role of GIS in the risk analysis procedure is to provide the road network and its surrounding information. It is also used to facilitate the quantification of routing objectives by creating a buffer area for the identification of the impact zone in the event of a dangerous goods incident, providing data query for corresponding attributes, and formulating a procedure for the calculation of relevant attribute values. Moreover, GIS facilitates the implementation of the algorithm to derive and visualize optimal routes.

To perform all the calculations and analysis in this paper, a GIS software package named MapInfo Professional 10.5 was used. This software, along with other applications included in the product suite, allows us to create, visualize, analyze and in general manage all geographic data. Also, the required functions needed for hazmat risk calculations are added on some abilities to software.

There are many undesirable consequences of a hazmat transportation accident, such as economic losses, injuries, environmental pollution, damage to wildlife, and fatalities. These consequences are a function of the impact area and population, property, and environmental assets within the impact area.

An impact zone of a single dangerous goods incident is marked out with a circle of radius centered at the accident location. The radius of a danger circle could be diverse depending on the type of the dangerous good. In this study, population centres are represented as polygons in the risk analysis (Erkut & Verter, 1998, Kara et al., 2003).

A impact area of a single hazmat incident is defined with a circle of λ radius, which is centred at the position where the accident takes place. λ, is the impact radius of the hazmat that is being carried, which may change according to the hazmat type. The assumption in this model is that each individual within the danger zone will be impacted equally and no one outside of this area will be impacted.

GIS is used to create a buffer area for the identification of the impact zone in the event of a dangerous goods incident. The potential impact area for liquid fuel (gasoline and diesel) is commonly taken at 800 m, for LPG is 1600 m in all directions (USDOT, 2008). When a hazmat truck uses a road link, all the people within the associated distance from that highway segment are exposed to the risk of being evacuated. Using the GIS-based model, the exposure zones around each link for evacuation distances of 800 meters and 1600 meters were determined. Accordingly, a buffer of 800 meters, and 1600 meters widths are generated using GIS for each road segment.

There are two main decision-makers in the dangerous goods literature: the regulators and the dangerous goods carriers. In many multiobjective dangerous goods transportation problems, it is necessary to provide a set of alternative paths. This is due to the case of a possible infeasibility of the "best" route(s) that is selected by the carrier.

In this study, transport risk when dangerous goods shipments contain different type of dangerous goods and multiple origin-destination pair was analyzed.

In this study, a series of shortest path problems with multiple objectives for various o-d pairs existed. The transport risk when dangerous goods shipments contain different type of dangerous goods and multiple origin-destination pairs was analyzed. Under the Shortest Path model, it is assumed that dangerous goods carriers choose to use the shortest path between an origin and a destination. The shortest path might not be a good choice for transporting dangerous goods. However, it was used as a benchmark to compare the optimal solutions found by the other four criteria. The methodology developed in this study aims to minimize of population exposure risk, incident probability, societal risk, and environmental risk.

Assessment of Criteria Weights

The Analytical Hierarchy Process (AHP) is used to determine the weights of evaluation criteria. The AHP, introduced by Saaty (1980), is known as a pairwise comparison method and a popular method in evaluation problems (Saaty, 2001, 2006). The complex structure of environmental risk assessment process, that is affected by multiple factors, leads to using the AHP in order to define and assess factors in a systematic manner. Priority weights of risk factors were determined by using pairwise comparisons and so, a more realistic calculation has been inferred.

Application of AHP to a weighting criteria involves four steps.

Step 1 is structuring of the decision problem into a hierarchical model. It includes decomposition of the decision problem into elements according to their common characteristics and the formation of a hierarchical model having different levels. Each level in the hierarchy corresponds to the common characteristic of the elements in that level. The topmost level is the 'focus' of the problem. In this study, focus of goal is "to weight the environmental risk factors". The goal is characterised by several main criteria and sub-criteria. The second level lists the main environmental elements; water, soil, and featured areas. The third level is sub-criteria of environmental elements.

Step 2 is making pair-wise comparisons and obtaining the judgmental matrix. In this step, the elements of a particular level are compared pairwise, with respect to a specific element in the immediate upper level. The priorities were derived according to pairwise assessments based on judgement. The relative importances were judged based on a scale of 1 to 9. From the pair-wise comparisons, a judgmental matrix was formed for each environmental element.

Step 3 is undertaking a consistency test. The consistency test is one of the essential features of the AHP method which aims to eliminate the possible inconsistency revealed in the criteria weights through the computation of consistency level of each matrix. The consistency ratio (CR) was used to determine and justify the inconsistency in the pair-wise comparison made by the respondents. Saaty has set the acceptable CR values for different matrix's sizes. If the CR value is ower than the acceptable value, the weight results are valid and consistent. In contrast, if the CR value is larger than the acceptable value, the matrix results are inconsistent and were exempted for the further analysis.

Step 4 is estimating relative weights of the components of each level. Once the local priorities of elements of different levels are available as outlined in the previous step, they are aggregated to obtain final priorities of the alternatives.

AHP works by developing priorities, which are derived for the criteria in terms of their importance to achieve the goal. The priorities were derived according to pairwise assessments based on judgement. AHP has the flexibility to combine quantitative and qualitative factors, to handle different groups of actors, to combine the opinions expressed by many experts. The multiattribute evaluation of environmental factors was performed by the experts from different decision making groups, such as academic staff, and environmental experts who is working in decision-making positions in various organizations. They assessed the relative importance for each of the criteria. The consensus of the groups was calculated using the geometric mean of individual judgements. Table 2 presents the average values of weights obtained in the weighting the environmental risk factors for the criteria and sub-criteria.

Findings relating to relative importance of weighting criteria and sub-criteria are summarized below:

- It was observed that 'Water Element' was continuously perceived as the most important main criterion.
- The sub-criterion, Agricultural Areas (under main criteria 'Soil Element'), was perceived as the most important weighting sub-criteria.
- As shown in the table, dam (0.367) was justified as the most important subcriteria for Water Element. Agricultural areas (0.731) was the highest value for Soil Element. Nature protection zone (0.439) was justified as the most important subcriteria for Featured Areas.

CASE STUDY FOR FUEL PRODUCTS TRANSPORTATION ON ROAD

To test the feasibility of the proposed evaluation method and criteria used, an area of westhern Turkey, Aegean Region, is chosen. Inner-city streets were not included in study area road network. The routes calculated were illustrated according to routing criteria that are based on various risk analysis models, including the one which was proposed.

Study Area

The westhern part of Turkey, Aegean Region, is chosen as the study area. Supply point is identified as point of origin and one of the major cities is identified as point of destination on the transportation network. The dangerous goods belonging to two different classes liquid fuel (gasoline and diesel), and LPG

Table 2. Weighting values for main and sub-criteria

Main-criteria / Relative Weights	Sub-criteria	Relative Weights
Water Element (A) 0.393 *CR = 0.019 < 0,1*	River (A1)	0.162
	Dam (A2)	0.367
	Sea (A3)	0.108
	Lake (A4)	0.151
	Irrigation Areas (A5)	0.212
Soil Element (B) 0.223 *CR = 0.033 < 0.1*	Forest Areas (B1)	0.156
	Protected Areas (B2)	0.113
	Agricultural Areas (B3)	0.731
Featured Areas (C) 0.384 *CR = 0.005 < 0.1*	Natural Protection Zones (C1)	0.439
	Protected Environmental Areas (C2)	0.376
	Tourism Areas (C3)	0.184

through the transportation network of Izmir-Manisa-Aydın-Mugla-Denizli were analyzed. An overall view of the study area, with the transportation network and the environmental elements, is depicted in Figure 3.

Risk analysis of dangerous goods transportation requires several data sources such as population density, length of road segments, impact radius by hazmat type, accident rates, environmental characteristics and origin-destination locations. In order to make the analysis as realistic as possible, all kinds of relevant and realistic data were collected as much as possible. Information concerning accident rates, population, and environmental characteristics was not directly available and data have to be collected from different sources and referred to the proper route segments using specific procedures.

We use the road map provided in MapInfo Professional 10.5 as a basis. Study area is examined in terms of road network; the total length of the transport network is approximately 5,357 km. The road network consists of 502 links. 16 highway links, 57 state road links, and 110 provincial road links are taken into consideration.

In the study area 4 LPG plants and 2 fuel oil plants are in use. The (o-d) shipments are between origins and 815 destinations, which are located on their exact geographical positions.

The population density values relevant to each road segment were evaluated based on census data, respectively. A digitized population center census division map of Turkey (5 cities) is used, which is obtained from Address Based Population Registration System (https://adres.nvi.gov.tr/) for 2011. In order to calculate population in buffer zone, two different distance ranges were assumed, respectively, equal to 800 meters and 1,600 meters, extending on both sides of the route. The population densities relevant to these buffer zones were separately calculated.

For accident probabilities, the accident data for 2010 used in this study was acquired from Turkish National Police Traffic Education and Research Department. Accident rate data needed for risk analysis have to be expressed as number of accidents occurred at each road segment. Available data include the number of accidents occurred in one year at each kilometre. First, the starting and ending points of each

Figure 3. Study area

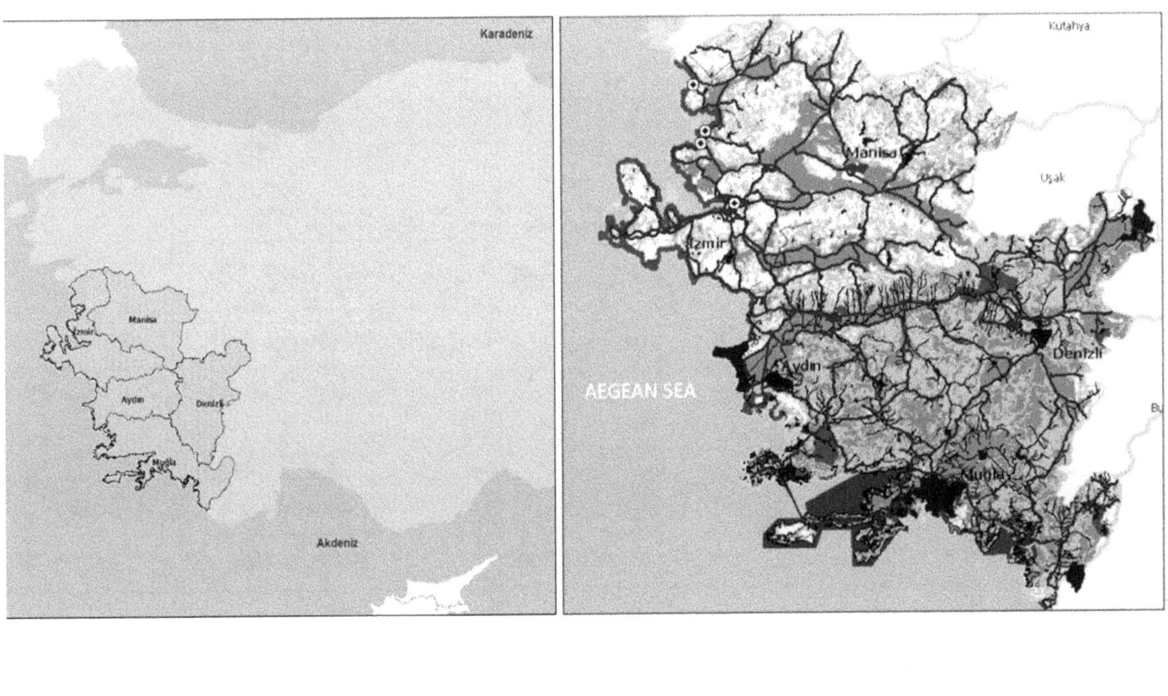

road have to be assessed, to refer the accident data to the correct route portion. Then, each portion of the road under exam has to be assigned to one or more kilometre numbers, associating the corresponding annual number of accidents. Finally, the accident rate of each route portion is calculated.

Data related to environmental elements were provided from environmental plans generated by Ministry of Environment and Urbanization in the form of digital data base.

The RAGISADR Results

A case study has been tested with the five different objectives; shortest path, min. population exposure, min. incident probability, min. societal risk, and min. environmental risk. Two illustrative examples of path selections ae provided using the objectives described earlier. Aliaga to Denizli is the o-d pair chosen as an example to illustrate the risk analysis for determining the minimum risk route by the optimization technique.

Two different hazardous materials transportation has been analyzed between the same o-d pair. Fuel and LPG are chosen as an example for the case study. In the case study, the risk analysis models are employed to search optimal routes for transporting fuel and LPG from Aliaga refinery to the designated fuel and LPG stations in Denizli. The optimal routes indicate the differences in terms of risk involved in transporting fuel and LPG. Based on fuel and LPG shipments, the routes generated using various models for transport between Aliaga and Denizli are compared in terms of the five objectives described earlier. Figure 4 shows the fuel shipments between Aliaga and Denizli with respect to each routing objective.

Figure 4. Alternative routes according to model results for fuel shipment between Aliaga – Denizli

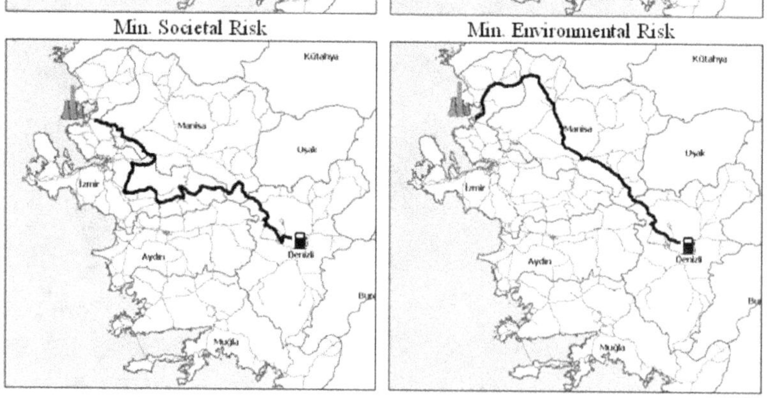

Objective	Distance (km)
Shortest Path	270
Min. Population Exposure	297
Min. Incident Probability	398
Min. Societal Risk	431
Min. Environmental Risk	332

In Figure 4, observe that the Shortest Path is 270 km. Minimum population exposure path has a length of 297 km, minimum incident probability path 398 km, minimum societal risk 431 km, and minimum environmental risk 332 km respectively. Table 3 summarizes the attributes of the routes obtained by risk analysis models for Aliaga-Denizli fuel shipment. The bold value in each column represents the minimum value of each objective.

The table reveals that the minimization of single objective function gives rise to a significant achievement of the chosen criterion, while heavily compromizing the others. In order to compare risk values easily, all values are divided by the corresponding figure. For example, first risk value has the shortest travel path. This leads to the highest societal risk and the highest incident probability. Similarly, Min. Societal Risk minimizes the expected societal risk in the event of an accident. This leads to the longest travel distance.

- According to the results from the table, it is clear that different routes are likely to be generated for hazmat delivery based on different criteria.
- To compare routes generated using various models, one can optimize the route under one risk model, while collecting the statistics under the remaining models. This information for the five routes is given in Figure 4. This information allows the decision maker to see how a route optimized by one criterion is performing against the other criteria. In our case, for instance, the route that has minimum environmental risk (10.77) has a population exposure risk measure of 95,909, and this value is the largest value among all five routes.
- The minimum societal risk path is 60% longer than the shortest path, which is very undesirable from shipper's perspective. Also, shortest path's societal risk value is approximately 12 times greater than minimum societal risk.

Table 3. Risk values of the routes obtained by risk analysis models for Aliaga-Denizli fuel shipment

Risk Values					
	Distance (km)	**Population Exposure**	**Incident Probability**	**Societal Risk**	**Environmental Risk**
Shortest Path	**270**	62,957	5.47	652,177	16.90
Min. Population Exp.	297	**27,616**	5.38	569,430	10.83
Min. Incident Prob.	398	44,252	**1.12**	65,028	12.24
Min. Societal Risk	431	50,735	1.30	**55,703**	11.97
Min. Environ. Risk	332	95,909	3.57	383,938	**10.77**
Normalized Risk Values					
	Distance (km)	**Population Exposure**	**Incident Probability**	**Societal Risk**	**Environmental Risk**
Shortest Path	1.00	2.27	4.88	11.70	1.56
Min. Population Exp.	1.10	1.00	4.80	10.22	1.01
Min. Incident Prob.	1.47	1.60	1.00	1.16	1.13
Min. Societal Risk	1.59	1.83	1.16	1.00	1.11
Min. Environ. Risk	1.22	3.47	3.18	6.89	1.00

According to model results for LPG shipment between Aliaga – Denizli, Minimum population exposure path has a length of 486 km., minimum incident probability path 471 km, minimum societal risk 463 km, and minimum environmental risk 332 km.

Table 4 provides the normalized risk values for the five objectives found in implementation. In order to compare risk values easily, all values are divided by the corresponding figure.

Figure 5 illustrates two different hazardous materials transportation analysis between the same o-d pair. The result of risk analysis show that the minimum risk routes for transporting different dangerous goods between a given o-d pair are different. Therefore it is not possible to designate a single optimal route for transportation of all types of dangerous goods. These routes are unique to the given type of dangerous goods.

The case study presented here demonstrates the efficacy of the proposed model in dangerous goods routing. Our computational experiments evince the behavior of the risk analysis as applied to the multiobjective dangerous goods routing problem in terms of optimality of solutions and computational complexity.

CONCLUSION

Dangerous goods transportation problem is an important problem that should be handled with extreme care since it is different from regular freight transportation as dangerous goods may harm to people, property or environment when an incident occurs. In this study, a GIS-based risk assessment framework for dangerous goods transportation is used. Certainly the decision making process related to the transport of dangerous goods on road can greatly benefit from the use of the risk analysis model.

Table 4. Risk values of the routes obtained by risk analysis models for Aliaga-Denizli LPG shipment

Risk Values					
	Distance (km)	Population Exposure	Incident Probability	Societal Risk	Environmental Risk
Shortest Path	**270**	222,474	5.54	1,503,405	34.45
Min. Population Exp.	486	**67,609**	2.93	298,148	32.65
Min. Incident Prob.	471	145,815	**1.38**	703,113	31.44
Min. Societal Risk	463	88,534	2.44	**164,687**	28.04
Min. Environ. Risk	332	196,043	3.32	876,008	**26.52**
Normalized Risk Values					
	Distance (km)	Population Exposure	Incident Probability	Societal Risk	Environmental Risk
Shortest Path	1.00	3.29	4.01	9.12	1.29
Min. Population Exp.	1.80	1.00	2.12	1.81	1.23
Min. Incident Prob.	1.74	2.15	1.00	4.26	1.18
Min. Societal Risk	1.71	1.30	1.77	1.00	1.05
Min. Environ. Risk	1.22	2.89	2.40	5.31	1.00

Figure 5. Comparison of routes generated for fuel and LPG shipment between Aliaga – Denizli

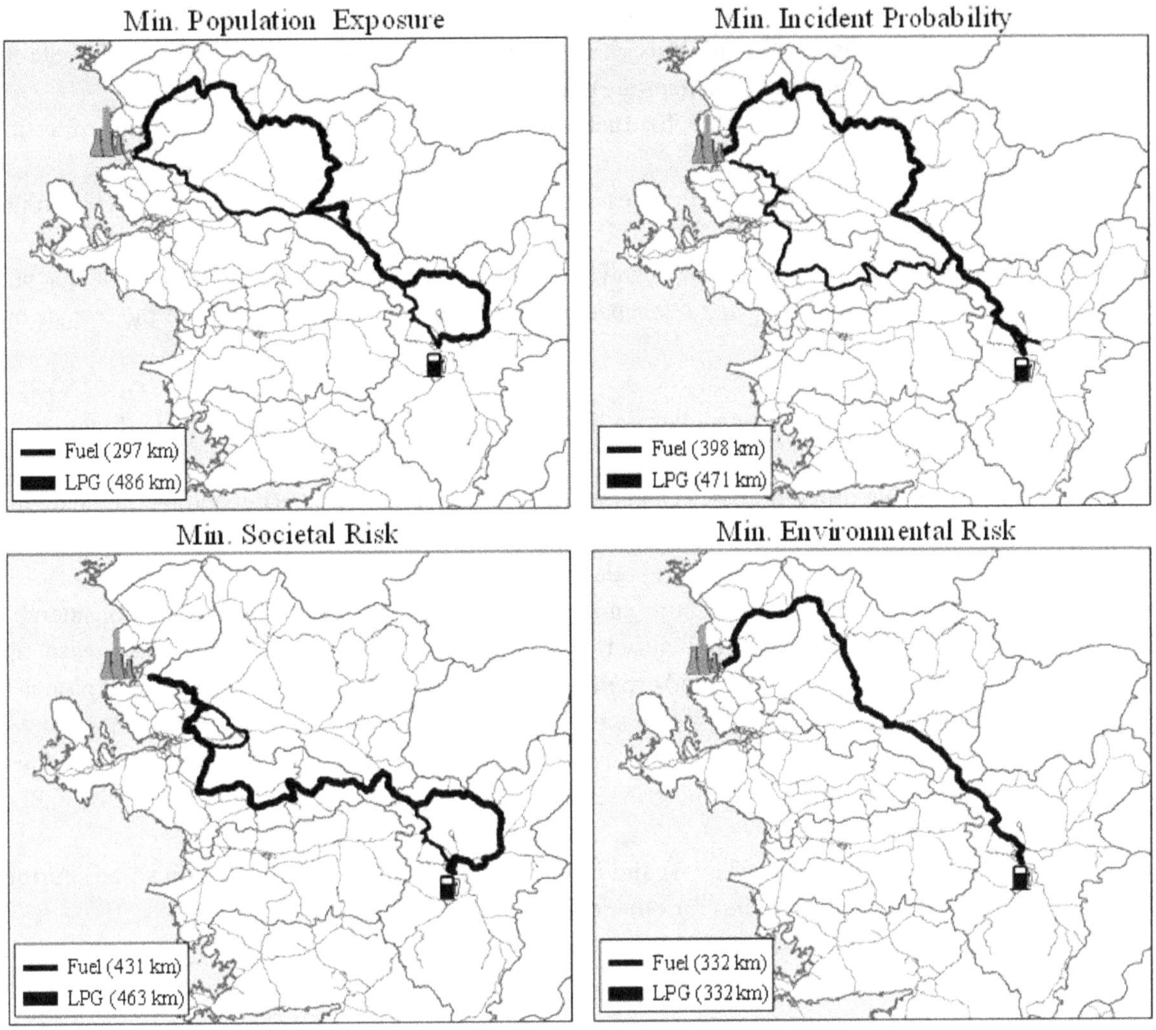

The results reveal the fact that different models usually lead to different optimal routes for a dangerous goods shipment between the designated origin and destination. Moreover, the optimal path obtained by one model might perform very poorly under another.

The following conclusions were obtained based on the results of large-scale application:

- Different models result in different routes, and the models do not tolerate one another very well. Unfortunately, it is impossible to suggest one risk model for all instances of dangerous goods transport.
- In order to be able to evaluate environmental consequences of accidents an environmental risks analysis model, the RAGISADR is proposed to determine the environmental risk factors.

- The minimum risk routes for transporting different dangerous goods between a given o-d pair are different. This is because the consequences of an accident on the network segment are different for different dangerous goods. Therefore, it is not possible to designate a single optimal route for the transportation of all types of dangerous goods.
- The application of the RAGISADR for fuel products transportation helped to create a decision support system.
- The created decision support system gives authourites the chance to define the route with minimum risks.
- The proposed model can be customized and applied for other case studies and not just the one presented in this document is of great relevance when considering the application of the framework in other regions.

Future work will focus on refining and extending the proposed model in the following directions:

- It is well known that different routes can emerge as minimum risk routes depending on the definition of risk used. Hence, it is crucial to use multiple measures and provide decision-makers with a set of efficient solutions instead of a single "risk minimizing" route.
- Traffic conditions and other risk factors in dangerous goods transportation involve considerable uncertainty, which amplifies the difficulty involved in routing decisions. To make the present approach more flexible, dangerous goods route planning under uncertainty needs to be explored in further research.
- In addition, the methodology used can be extended to cover a much wider extent of the road network in Turkey. This will require more extensive data collection to ensure that the GIS database is accurate and up to date.
- The model is applied to fuel (Class 3) and LPG products (Class 2) transportation so far. Further research can be devoted to studies for other classes of dangerous goods.
- Finally, there is still an effort needed to turn the proposed solution methodology into a full decision support system in which users can utilize the methodologies presented in this paper in a real application setting.

REFERENCES

Abkowitz, M., Lepofsky, M., & Cheng, P. (1992). Selecting criteria for designating hazardous materials highway routes. *Transportation Research Record, 1333*, 30–35.

Ak, R., & Bozkaya, B. (2008, June). A proposed risk model and a GIS framework for hazardous materials transportation. *Engineering Management Conference, 2008. IEMC Europe 2008. IEEE International* (pp. 1-5). IEEE. doi:10.1109/IEMCE.2008.4617959

Alisan, D. (2009). *Tehlikeli kimyasal maddelerin taşınması ve depolanması* (in Turkish) (pp. 257–266). Ankara: Tehlikeli Kimyasalların Yönetimi Sempozyumu.

Alp, E. (1995). Risk-based transportation planning practice overall methodology and a case example. *INFOR, 33*(1), 4–19.

Ashtakala, B., & Eno, L. A. (1996). Minimum risk route model for hazardous materials. *Journal of Transportation Engineering*, *122*(5), 350–357. doi:10.1061/(ASCE)0733-947X(1996)122:5(350)

Barilla, D., Leonardi, G., & Puglisi, A. (2009). Risk assessment for hazardous materials *Transportation Applied Mathematical Sciences, 3*(46), 2295-2309.

Batta, R., & Chiu, S. S. (1988). Optimal obnoxious paths on a network: Transportation of hazardous materials. *Operations Research*, *36*(1), 84–92. doi:10.1287/opre.36.1.84

Bohnenblust, H., & Slovic, P. (1998). Integrating technical analysis and public values in risk-based decision making. *Reliability Engineering & System Safety*, *59*(1), 151–159. doi:10.1016/S0951-8320(97)00136-1

Bonvicini, S., Leonelli, P., & Spadoni, G. (1998). Risk analysis of hazardous materials transportation: Evaluating uncertainty by means of fuzzy logic. *Journal of Hazardous Materials*, *62*(1), 59–74. doi:10.1016/S0304-3894(98)00158-7

Castillo, J. E. A. (2004). *Route optimization for hazardous materials transport* [Unpublished master's thesis]. International Institute for Geo-information Science and Earth Observation, Urban Planning and Land Administration, The Netherlands.

Erkut, E., & Ingolfsson, A. (2000). Catastrophe avoidance models for hazardous materials route planning. *Transportation Science*, *34*(2), 165–179. doi:10.1287/trsc.34.2.165.12303

Erkut, E., & Ingolfsson, A. (2005). Transport risk models for hazardous materials. *Operations Research Letters*, *33*(1), 81–89. doi:10.1016/j.orl.2004.02.006

Erkut, E., Tjandra, A. S., & Verter, V. (2007). Hazardous Materials Transportation. In C. Barnhart & G. Laporte (Eds.), *Handbook in OR & MS* (Vol. 14, pp. 539–621). UK: Elsevier.

Erkut, E., & Verter, V. (1998). Modeling of transport risk for hazardous materials. *Operations Research*, *46*(5), 625–642. doi:10.1287/opre.46.5.625

Gunasekera, M. Y., & Edwards, D. W. (2003). Estimating the environmental impact of catastrophic chemical releases to the atmosphere: An index method for ranking alternative chemical process routes. *Trans Institution of Chemical Engineers*, *81*, 463–474.

Huang, B. (2004). A GIS-AHP method for HAZMAT route planning with consideration of security. *Environmental Informatics Archives*, *2*, 818–830.

Huang, B. (2006). GIS-based route planning for hazardous material transportation. *Journal of Environmental Informatics*, *8*(1), 49–57. doi:10.3808/jei.200600076

Huang, B., Cheu, R. L., & Liew, Y. S. (2004). GIS and genetic algorithms for HAZMAT route planning with security considerations. *International Journal of Geographical Information Science*, *18*(8), 769–787. doi:10.1080/13658810410001705307

Jassbi, J., & Makvandi, P. (2010). Route selection based on soft MODM framework in transportation of hazardous materials. *Applied Mathematical Sciences*, *4*(63), 3121–3132.

Kara, B. Y., Erkut, E., & Verter, V. (2003). Accurate calculation of hazardous materials transport risks. *Operations Research Letters*, *31*(4), 285–292. doi:10.1016/S0167-6377(02)00238-9

Li, R., Leung, Y., Huang, B., & Lin, H. (2013). A genetic algorithm for multiobjective dangerous goods route planning. *International Journal of Geographical Information Science*, *27*(6), 1073–1089. doi:10.1080/13658816.2012.737921

Liu, K. F. R., Liang, H. H., Yeh, K., & Chen, C. W. (2009). A qualitative decision support for environmental impact assessment using fuzzy logic. *Journal of Environmental Informatics*, *13*(2), 93–103. doi:10.3808/jei.200900144

Lovett, A. A., Parfitt, J. P., & Brainard, J. S. (1997). Using GIS in risk analysis: A case study of hazardous waste transport. *Risk Analysis*, *17*(5), 625–633. doi:10.1111/j.1539-6924.1997.tb00903.x PMID:9404052

Monprapussorn, S., Watts, D. J., & Banomyong, R. (2009). Sustainable hazardous materials route planning with environmental consideration. *Asian Journal on Energy and Environment*, *10*(02), 122–132.

Pijawka, K. D., Foote, S., Soesilo, A., Scanlon, R. D., & Cantilli, E. J. (1985). *Improving transportation of hazardous materials through risk assessment and routing. National Research Council.* Washington, DC, USA: Transportation Research Board.

ReVelle, C., Cohon, J., & Shobrys, D.ReVelle. (1991). Simultaneous siting and routing in the disposal of hazardous wastes. *Transportation Science*, *25*(2), 138–145. doi:10.1287/trsc.25.2.138

Ruifang, M. O. U. (2010). Environmental risk assessment model on dangerous goods in logistics. *ICLEM*, *2010*, 2023–2029.

Rygnestad, H. (2000). *Integrating Environmental Economics and Policy Analyses in a Geographical Information System.* Statens Jordbrugs-og Fiskeriøkonomiske Institut.

Saaty, T. L. (2001). The seven pillars of the analytic hierarchy process. *Proceedings of the 15th Int'l Conf on Multiple Criteria Decision Making (MCDM)* (pp.1-15). New York: Springer.

Saaty, T. L. (2006). Rank from comparisons and from ratings in the analytic hierarchy /network processes. *European Journal of Operational Research*, *168*(2), 557. doi:10.1016/j.ejor.2004.04.032

Saccomanno, F., & Chan, A. (1985). Economic evaluation of routing strategies for hazardous road shipments. *Transportation Research Record*, *1020*, 12–18.

Scott-Andersson, A., Tysklind, M., & Fangmark, I. (2007). A method to relate chemical accident properties and expert judgements in order to derive useful information for the development of environment-accident index. *Journal of Hazardous Materials*, *147*(1-2), 524–533. doi:10.1016/j.jhazmat.2007.01.052 PMID:17320285

Sherali, H. D., Brizendine, L. D., Glickman, T. S., & Subramanian, S. (1997). Low-probability high-consequence considerations in routing hazardous materials shipments. *Transportation Science*, *31*(3), 237–251. doi:10.1287/trsc.31.3.237

Sivakumar, R. A., & Batta, R. (1994). The variance-constrained shortest path problem. *Transportation Science*, *28*(4), 309–316. doi:10.1287/trsc.28.4.309

Sivakumar, R. A., Batta, R., & Karwan, M. H. (1993). A Network-based model for transporting extremely hazardous materials. *Operations Research Letters*, *13*(2), 85–93. doi:10.1016/0167-6377(93)90034-E

Topuz, E., Talinli, I., & Aydin, E. (2011). Integration of environmental and human health risk assessment for industries using hazardous materials: A quantitative multi criteria approach for environmental decision makers. *Environment International*, *37*(2), 393–403. doi:10.1016/j.envint.2010.10.013 PMID:21111481

Umit, H., & Kara, B. Y. (2003). Büyük ölçekli bir tehlikeli madde taşımacılığının coğrafi bilgi sistemleri tabanlı analizi: Shell Türkiye örneği [in Turkish]. *Yöneylem Araştırma Dergisi*, *15*, 35–47.

United Nations. (2013). *European agreement concerning the international carriage of dangerous goods by road*. Geneva: Author.

U.S. Department of Transportation & U.S. Department of Commerce. (2010). *Hazardous materials transportation 2007 commodity flow survey*.

USDOT (U.S. Department of Transportation). (2008). *Emergency response guidebook*. Washington, DC, USA: PHMSA.

Verter, V., & Kara, B. Y. (2001). A GIS-based framework for hazardous materials transport risk assessment. *Risk Analysis*, *21*(6), 1109–1120. doi:10.1111/0272-4332.216179 PMID:11824686

Wessberg, N., Molarius, R., Seppala, J., Koskela, S., & Pennanen, J. (2008). Environmental risk analysis for accidental emissions. *Journal of Chemical Health and Safety*, *15*(1), 24–31. doi:10.1016/j.jchas.2007.07.001

Zhang, J. J., Hodgson, J., & Erkut, E. (2000). Using GIS to assess the risks of hazardous materials transport in networks. *European Journal of Operational Research*, *121*(2), 316–329. doi:10.1016/S0377-2217(99)00220-9

KEY TERMS AND DEFINITIONS

Arc: Composed of a pair of nodes called "starting node" and "ending node" respectively.

Incident: An occurrence or event that requires action by emergency service personnel to prevent or minimize loss of life or damage to property and/or natural resources.

Liquefied Petroleum Gas: Petroleum gases which can be liquefied under moderate pressures.

Node: A point in a network topology where lines intersect.

Optimal Route: The route that best fulfils the objectives set by the stakeholders related with the process of transporting dangerous goods.

Population Exposure: The number of people impacted in the event of an accident.

Priority Weights: A relative weight for measuring criterion relative importance.

Quantitative Risk Assessment: Incorporates numerical estimates of frequency and the consequence in a sophisticated but approximate manner to estimate quantitative measure of risks.

Compilation of References

Abbas, M. (2014). Challenges in implementation of TVM (Ticket Vending Machine) in developing countries for mass transport system: A study of human behavior while interacting with ticket vending machine-TVM. In *Design, User Experience, and Usability. User Experience Design for Everyday Life Applications and Services* (pp. 245–254). Switzerland: Springer International Publishing. doi:10.1007/978-3-319-07635-5_24

Abdinnour-Helm, S. (1998). A hybrid heuristic for the uncapacitated hub location problem. *European Journal of Operational Research, 106*(2-3), 489–499. doi:10.1016/S0377-2217(97)00286-5

Abdinnour-Helm, S., & Venkataramanan, M. A. (1998). Solution approaches to hub location problems. *Annals of Operations Research, 78*, 31–50. doi:10.1023/A:1018954217758

Abkowitz, M., & Cheng, P. D. M. (1988). Developing a risk cost framework for routing truck movements of hazardous materials. *Accident; Analysis and Prevention, 20*(1), 39–51. doi:10.1016/0001-4575(88)90013-9 PMID:3337764

Abkowitz, M., & Cheng, P. D. M. (1989). Hazardous materials transport risk estimation under conditions of limited data availability. *Transportation Research Record, 1245*, 14–22.

Abkowitz, M., Lepofsky, M., & Cheng, P. (1992). Selecting criteria for designating hazardous materials highway routes. *Transportation Research Record, 1333*, 30–35.

Acevedo, M., Bairdcallicott, J., Monticino, M., Lyons, D., Palomino, J., & Rosales, J. et al. (2008). Models of natural and human dynamics in forest landscapes: Cross-site and cross-cultural synthesis. *Geoforum, 39*(2), 846–866. doi:10.1016/j.geoforum.2006.10.008

Achour, N., Lakhdari, A., & Ferguene, F. (2013). Motion planning for car-like robots using hierarchical genetic algorithms. In S. I. Ao, L. Gelman, D. W. Hukins, A. Hunter, & A. M. Korsunsky (Eds.), *World Congress on Engineering - WCE 2013, Vol II, Lecture Notes in Engineering and Computer Science* (pp. 787-791). Hong Kong China. IAENG Press.

Adamski, A. (2014). DISCON: public transport dispatching robust control. *Procedia: Social and Behavioral Sciences, 111*, 1206–1216. doi:10.1016/j.sbspro.2014.01.155

Adler, N., & Hashai, N. (2005). Effect of open skies in the Middle East region. *Transportation Research, 39*, 878–894.

Ageron, B., Gunasekaran, A., & Spalanzani, A. (2012). Sustainable supply management: An empirical study. *International Journal of Production Economics, 140*(1), 168–182. doi:10.1016/j.ijpe.2011.04.007

Aggarwal, C. C. (2003). Towards systematic design of distance functions for data mining applications.*Proceedings of the ninth ACM SIGKDD international conference on Knowledge discovery and data mining*. ACM. doi:10.1145/956750.956756

Aggarwal, C. C., & Reddy, C. K. (Eds.). (2013). *Data Clustering: Algorithms and Applications*. CRC Press.

AhmadBeygi, S., Cohn, A., Guan, Y., & Belobaba, P. (2008). Analysis of the potential for delay propagation in passenger airline networks. *Journal of Air Transport Management*, *14*(5), 221–236. doi:10.1016/j.jairtraman.2008.04.010

Aho, A. V. (2004). Computer representations of graphs. In J. L. Gross, J. Yellen, & P. Zhang (Eds.), *Handbook of Graph Theory* (p. 62). USA: CRC Press.

Akgun, V., Erkut, E., & Batta, R. (2000). On finding dissimilar paths. *European Journal of Operational Research*, *121*(2), 232–246. doi:10.1016/S0377-2217(99)00214-3

Akgün, V., Parekh, A., Batta, R., & Rump, C. M. (2007). Routing of a hazmat truck in the presence of weather systems. *Computers & Operations Research*, *34*(5), 1351–1373. doi:10.1016/j.cor.2005.06.005

Akin, D. (2014). Is Istanbul's growth monocentric or multicentric? Analyses of intraurban travel data between 1996 and 2006. In M.G. Karlaftis, N.D. Lagaros, M. Papadrakakis (Eds.), Proceedings of *OPT-i, 1st International Conference on Engineering and Applied Sciences Optimization* (pp. 2773-2799).

Akın, D. (2011). Kentsel Yapının Değişiminin Kentsel Yolculukların Hiyerarşik Kümeleme Analizi ile İncelenmesiyle Ortaya Konması: İstanbul Metropoliten Alanı 1996-2006 Yılları (*Studying the Change of Urban Structure by Hierarchical Cluster Analysis: Istanbul Metropolitan Region between 1996 and 2006*), KBAM 2. Sempozyumu (In *2nd Symposium of Urban and Regional Studies*). Ankara, Turkey.

Ak, R., & Bozkaya, B. (2008, June). A proposed risk model and a GIS framework for hazardous materials transportation.*Engineering Management Conference, 2008. IEMC Europe 2008. IEEE International* (pp. 1-5). IEEE. doi:10.1109/IEMCE.2008.4617959

Aksakalli, V. (2007). The BAO* algorithm for stochastic shortest path problems with dynamic learning.*Proceedings of the 46th IEEE Conference on Decision and Control*(Vol 1-14, pp. 2460-2465). New York. IEEE. doi:10.1109/CDC.2007.4434409

Alagador, D., Triviño, M., Cerdeira, J. O., Brás, R., Cabeza, M., & Araújo, M. B. (2012). Linking like with like: Optimising connectivity between environmentally-similar habitats. *Landscape Ecology*, *27*(2), 291–301. doi:10.1007/s10980-012-9704-9

Al-Harbi, K. M. (2001). Application of AHP in project management. *International Journal of Project Management*, *19*(1), 19–27. doi:10.1016/S0263-7863(99)00038-1

Alisan, D. (2009). *Tehlikeli kimyasal maddelerin taşınması ve depolanması*(in Turkish) (pp. 257–266). Ankara: Tehlikeli Kimyasalların Yönetimi Sempozyumu.

Alp, E. (1995). Risk-based transportation planning practice overall methodology and a case example. *INFOR*, *33*(1), 4–19.

Alpkokin, P., Cheung, C., Black, J., & Hayashi, Y. (2008). Dynamics of clustered employment growth and its impacts on commuting patterns in rapidly developing cities. *Transportation Research Part A, Policy and Practice*, *42*(3), 427–444. doi:10.1016/j.tra.2007.11.002

Alpkokin, P., Hayashi, Y., Black, J., & Gercek, H. (2005). Polycentric employment growth and impacts on urban commuting patterns: Case study of Istanbul. *Journal of the Eastern Asia Society for Transportation Studies*, *6*(1), 3835–3851.

Altiparmak, F., Gen, M., Lin, L., & Paksoy, T. (2006). A genetic algorithm approach for multi-objective optimization of supply chain networks. *Computers & Industrial Engineering*, *51*(1), 196–215. doi:10.1016/j.cie.2006.07.011

Alumur, S., & Kara, B. Y. (2007). A new model for the hazardous waste location-routing problem. *Computers & Operations Research*, *34*(5), 1406–1423. doi:10.1016/j.cor.2005.06.012

Alumur, S., & Kara, B. Y. (2008). Network hub location problems: The state of the art. *European Journal of Operational Research, 190*(1), 1–21. doi:10.1016/j.ejor.2007.06.008

Alumur, S., Kara, B., & Karasan, O. (2009). The design of single allocation incomplete hub networks. *Transportation Research Part B: Methodological, 43*(10), 936–951. doi:10.1016/j.trb.2009.04.004

Alumur, S., Kara, B., & Karasan, O. (2012). Multimodal hub location and hub network design. *Omega, 40*(6), 927–939. doi:10.1016/j.omega.2012.02.005

Amancio, D. R., Aluisio, S. M., Oliviera, O. N., & Costa, L. D. (2012). Complex networks analysis of language complexity. *Europhysics Letters, 100*(5).

Amancio, D. R., Oliveira, O. N. Jr, & Costa, L. D. (2011). On the concepts of complex networks to quantify the difficulty in finding the way out of labyrinths. *Physica A-Statistical Mechanics and Its Applications, 390*(23-24), 4673–4683. doi:10.1016/j.physa.2011.06.079

Ambrosino, D., & Grazia Scutellà, M. (2005). Distribution network design: New problems and related models. *European Journal of Operational Research, 165*(3), 610–624. doi:10.1016/j.ejor.2003.04.009

Anderson, E. C., Williamson, E. G., & Thompson, E. A. (2000). Monte Carlo evaluation of the likelihood for N-ε from temporally spaced samples. *Genetics, 156*(4), 2109–2118. PMID:11102399

Anderson, R. T., & Barkan, C. P. L. (2004). Railroad accident rates for use in transportation risk analysis. *Transportation Research Record, 1863*, 88–98. doi:10.3141/1863-12

Andersson, E., Housos, E., Kohl, N., & Wedelin, D. (1998). *Crew pairing optimization Operations Research in the Airline Industry* (pp. 228–258). New York: Springer US. doi:10.1007/978-1-4615-5501-8_8

Andersson, S. E. (1994). Safe transport of dangerous goods – road, rail or sea – a screening of technical and administrative factors. *European Journal of Operational Research, 75*(3), 499–507. doi:10.1016/0377-2217(94)90292-5

Androutsopoulos, K. N., & Zografos, K. G. (2010). Solving the bicriterion routing and scheduling problem for hazardous materials distribution. *Transportation Research Part C, Emerging Technologies, 18*(5), 713–726. doi:10.1016/j.trc.2009.12.002

Ankara Metropolitan Municipality Catalogue. (2013). *Ankara Buyuksehir Bld.* Ankara: Bilgilendirme ve Faaliyetlerden Yararlanma Kilavuzu.

Antrop, M. (2005). Why landscapes of the past are important for the future. *Landscape and Urban Planning, 70*(1-2), 21–34. doi:10.1016/j.landurbplan.2003.10.002

Apaiah, R. K., & Hendrix, E. M. T. (2005). Design of a supply chain network for pea-based novel protein foods. *Journal of Food Engineering, 70*(3), 383–391. doi:10.1016/j.jfoodeng.2004.02.043

Appelqvist, P., Lehtonen, J.-M., & Kokkonen, J. (2004). Modelling in product and supply chain design: Literature survey and case study. *Journal of Manufacturing Technology Management, 15*(7), 675–686. doi:10.1108/17410380410555916

Arabameri, A. (2014). Application of Analytic Hierarchy Process (AHP) for locating fire stations: Case study Maku City. *Merit Research Journal of Art, Social Science and Humanities, 2*(1), 001-010.

Arnold, P., Peeters, D., & Thomas, I. (2004). Modelling a rail/road intermodal transportation system. *Transportation Research Part E, Logistics and Transportation Review, 40*(3), 255–270. doi:10.1016/j.tre.2003.08.005

Arriva Bus App. (2015). Google Play Store. Retrieved from https://play.google.com/store/apps/details?id=de.hafas.android.arriva

Ashtakala, B., & Eno, L. A. (1996). Minimum risk route model for hazardous materials. *Journal of Transportation Engineering, 122*(5), 350–357. doi:10.1061/(ASCE)0733-947X(1996)122:5(350)

Aslam, T., & Ng, A. H. C. (2010). Multi-Objective Optimization for Supply Chain Management : A Literature Review and New Development.*Proceedings of the 8th International Conference on Supply Chain Management and Information Systems* (pp. 1–8). IEEE.

Audenaert, A., De Cleyn, S. H., & Vankerckhove, B. (2008). Economic analysis of passive houses and low-energy houses compared with standard houses. *Energy Policy, 36*(1), 47–55. doi:10.1016/j.enpol.2007.09.022

Autolib' app. (2015). Google Play Store. Retrieved from https://play.google.com/store/apps/details?id=com.polyconseil.autolib

Aversa, R., Botter, R. C., Haralambides, H. E., & Yoshizaki, H. T. Y. (2005). A mixed integer programming model on the location of a hub port in the East Coast of South America. *Maritime Economics & Logistics, 7*(1), 1–18. doi:10.1057/palgrave.mel.9100121

Aydogdu, A. (2006). *Transportation and energy policies that Turkey should* European Commission. *develop in order to reduce the number of traffic accidents in the EU accession period* [Unpublished master's thesis]. Gazi University, Institute of Science, Ankara.

Aykin, T. (1994). Lagrangian relaxation based approaches to capacitated hub-and-spoke network design problem. *European Journal of Operational Research, 79*(3), 501–523. doi:10.1016/0377-2217(94)90062-0

Aykin, T. (1995). Networking policies for hub-and-spoke systems with application to the air transportation system. *Transportation Science, 29*(3), 201–221. doi:10.1287/trsc.29.3.201

Bagheri, N., Benwell, G. L., & Holt, A. (2006). *Primary health care accessibility for rural Otago: ''A spatial analysis''. Health Care & Informatics*. Review Online.

Baird, A. J. (2006). Optimizing the Container transshipment hub location in northern Europe. *Journal of Transport Geography, 14*(3), 195–214. doi:10.1016/j.jtrangeo.2004.12.004

Baker, B. D., & Richards, C. E. (1999). A comparison of conventional linear regression methods and neural networks for forecasting educational spending. *Economics of Education Review, 18*(4), 405–415. doi:10.1016/S0272-7757(99)00003-5

Balas, C. E., & Balas, L. (2002). Risk assessment of some revetments in Southwest Wales, United Kingdom. *Journal of Waterway, Port, Coastal, and Ocean Engineering, 128*(5), 216–223. doi:10.1061/(ASCE)0733-950X(2002)128:5(216)

Balas, C. E., & Ergin, A. (2002). Reliability-Based risk assessment in coastal projects: Case study in Turkey. *Journal of Waterway, Port, Coastal, and Ocean Engineering, 128*(2), 52–61. doi:10.1061/(ASCE)0733-950X(2002)128:2(52)

Balas, C. E., & Koc, L. (2002). Risk assessment of vertical breakwaters- A case study in Turkey. *China Ocean Engineering, 16*(1), 123–134.

Balas, C. E., Koc, L., & Balas, L. (2004). Predictions of missing wave data by recurrent neuronets. *Journal of Waterway, Port, Coastal, and Ocean Engineering, 130*(5), 256–265. doi:10.1061/(ASCE)0733-950X(2004)130:5(256)

Balas, C. E., Williams, A. T., Simmons, S. L., & Ergin, A. (2001). A statistical riverine litter propagation model. *Marine Pollution Bulletin, 42*(11), 1169–1176. doi:10.1016/S0025-326X(01)00133-3 PMID:11763231

Balli, S., & Korukoglu, S. (2009). Operating system selection using fuzzy AHP and TOPSIS methods. *Mathematical and Computational Applications, 14*(2), 119–130.

Bania, N., Bauer, P., & Zlatoper, T. U. (1998). Air passenger service: A taxonomy of route network, hub location, and competition. *Logistics and Transportation Review*, *34*(1), 53–74. doi:10.1016/S1366-5545(97)00037-9

Banister, D. (2008). The sustainable mobility paradigm. *Transport Policy*, *15*(2), 73–80. doi:10.1016/j.tranpol.2007.10.005

Barber, J. R., Burdett, C. L., Reed, S. E., Warner, K. A., Formichella, C., & Crooks, K. R. et al. (2011). Anthropogenic noise exposure in protected natural areas: Estimating the scale of ecological consequences. *Landscape Ecology*, *26*(9), 1281–1295. doi:10.1007/s10980-011-9646-7

Barilla, D., Leonardi, G., & Puglisi, A. (2009). Risk assessment for hazardous materials *Transportation Applied Mathematical Sciences, 3*(46), 2295-2309.

Barnhart, C., Cohn, A. M., Johnson, E. L., Klabjan, D., Nemhauser, G. L., & Vance, P. H. (2003). *Airline crew scheduling Handbook of Transportation Science* (pp. 517–560). New York: Springer US.

Barnhart, C., Farahat, A., & Lohatepanont, M. (2009). Airline fleet assignment with enhanced revenue modeling. *Operations Research*, *57*(1), 231–244. doi:10.1287/opre.1070.0503

Barnhart, C., Johnson, E. L., Nemhauser, G. L., Savelsbergh, M. W., & Vance, P. H. (1998). Branch-and-price: Column generation for solving huge integer programs. *Operations Research*, *46*(3), 316–329. doi:10.1287/opre.46.3.316

Barnhart, C., & Shenoi, R. G. (1998). An approximate model and solution approach for the long-haul crew pairing problem. *Transportation Science*, *32*(3), 221–231. doi:10.1287/trsc.32.3.221

Barth, M., & Boriboonsomsin, K. (2008). Real-world carbon dioxide impacts of traffic congestion. *Transportation Research Record*, *2058*, 163–171. doi:10.3141/2058-20

Bartlett, P. L. (1998). The sample complexity of pattern classification with neural networks: The size of the weights is more important than the size of the network. *IEEE Transactions on Information Theory*, *44*(2), 525–536. doi:10.1109/18.661502

Basheer, I. A., & Hajmeer, M. (2000). Artificial neural networks: Fundamentals, computing, design, and application. *Journal of Microbiological Methods*, *43*(1), 3–31. doi:10.1016/S0167-7012(00)00201-3 PMID:11084225

Batta, R., & Chiu, S. S. (1988). Optimal obnoxious paths on a network: Transportation of hazardous materials. *Operations Research*, *36*(1), 84–92. doi:10.1287/opre.36.1.84

Beamon, B. M. (1998). Supply chain design and analysis : Models and methods. *International Journal of Production Economics*, *55*(3), 281–294. doi:10.1016/S0925-5273(98)00079-6

Beasley, J. A. (2004). OR Notes - Data Envelopment Analysis. Retrieved from http://people.brunel.ac.uk/~mastjjb/jeb/or/dea.html

Bedrettin, M. (2004). *Traffic Law and Fundamental Traffic Knowledge* (in Turkish). Ankara: General Directorate of Security.

Beinat, E., & van Drunen, M. (1998). Spatial conflicts in transport policies: an exploration of the perspectives of regional and local actors. In E. Beinat & P. Nijkamp (Eds.), Multicriteria Analysis for Land-Use Management. Dordrecht: Kluwer Academic Publishers. Retrieved from http://link.springer.com/chapter/10.1007%2F978-94-015-9058-7_9 doi:10.1007/978-94-015-9058-7_9

Ben Amor, H., & Valério de Carvalho, J. (2005). Cutting Stock Problems. In G. Desaulniers, J. Desrosiers, & M. Solomon (Eds.), *Column Generation* (pp. 131–161). New York: Springer US. doi:10.1007/0-387-25486-2_5

Berger, A. (2002). *Embedded Systems Design: An introduction to processes, tools, and techniques*. Lawrence, KS: CMP Books.

Berköz, L. (2000). Location of financial, insurance, and real estate firms in Istanbul. *Journal of Urban Planning and Development, 126*(2), 75–88. doi:10.1061/(ASCE)0733-9488(2000)126:2(75)

Berkoz, L., & Turk, S. S. (2010). Intrametropolitan determinants of foreign investment firms in Istanbul. *Journal of Urban Planning and Development, 136*(3), 234–242. doi:10.1061/(ASCE)UP.1943-5444.0000025

Berman, O., Verter, V., & Kara, B. Y. (2007). Designing Emergency Response Networks for Hazardous Materials Transportation. *Computers & Operations Research, 34*(5), 1374–1388. doi:10.1016/j.cor.2005.06.006

Beroggi, G. E. G. (1994). A real-time routing model for hazardous materials. *European Journal of Operational Research, 75*(3), 508–520. doi:10.1016/0377-2217(94)90293-3

Berry, M., & Hamilton, M. (2010). Changing urban spaces: Mobile phones on trains. *Mobilities, 5*(1), 111–129. doi:10.1080/17450100903435078

Bertsimas, D., & Sim, M. (2004). The price of robustness. *Operations Research, 52*(1), 35–53. doi:10.1287/opre.1030.0065

Besussi, E., Chin, N., Batty, M., & Longley, P. (2010). The structure and form of urban settlements. In *Remote Sensing of Urban and Suburban Areas* (pp. 13–31). Netherlands: Springer. doi:10.1007/978-1-4020-4385-7_2

Beuthe, M., & Bouffioux, C. H. (2008). Analysing qualitative attributes of freight transport from stated orders of preference experiment. *Journal of Transport Economics and Policy, 42*, 105–128.

Bevilacqua, M., Ciarapica, F. E., & Giacchetta, G. (2009). Business process reengineering of a supply chain and a traceability system: A case study. *Journal of Food Engineering, 93*(1), 13–22. doi:10.1016/j.jfoodeng.2008.12.020

Bhandari, K., Alpkokin, P., Kachi, N., Jia, P., & Kato, H. (2007). *Dynamics of employment distribution and accessibility index: Case of Delhi.* JSCE.

Bhandari, K., Jia, P., Alpkokin, P., Mukhopadhyay, D., & Gangopadhyay, S. (2007). Policies, commuting patterns and accessibility in a non-monocentric city: Case study of Delhi.*Proceedings of the Eastern Asia Society for Transportation Studies* (Vol. 2007, pp. 192-192). Eastern Asia Society for Transportation Studies.

Bharati, P., & Chaudhury, A. (2004). An empirical investigation of decision-making satisfaction in web-based decision support systems. *Decision Support Systems, 37*(2), 187–197. doi:10.1016/S0167-9236(03)00006-X

Bhargava, H. K., & Power, D. J. (2004). Decision support systems and Web technologies: A status report.*Americas Conference on Information Systems (AMCIS).* Boston.

Bhargava, H. K., Power, D. J., & Sun, D. (2007). Progress in web-based decision support technologies. *Decision Support Systems, 43*(4), 1083–1095. doi:10.1016/j.dss.2005.07.002

Bianco, L., Caramia, M., & Giordani, S. (2009). A bilevel flow model for hazmat transportation network design. *Transportation Research Part C, Emerging Technologies, 17*(2), 175–196. doi:10.1016/j.trc.2008.10.001

BiciMAD. (n. d.). BiciMAD.com. Retrieved from http://www.bicimad.com

Bixby, L. R. (2004). Spatial access to health care in Costa Rica and its equity: A GIS-based study. *Social Science & Medicine, 58*(April), 1271–1284. doi:10.1016/S0277-9536(03)00322-8 PMID:14759675

Black, M., Ebener, S., Vidaurre, M., Aguilar, P. N., & El Morjani, Z. (2004). Using GIS to Measure Physical Accessibility to Health Care. *Proceedings of Health GIS Conference.* San Diego.

Boccardo, P., Arneodo, F., & Botta, D. (2014). Application of Geomatic techniques in Infomobility and Intelligent Transport Systems (ITS). *European Journal of Remote Sensing, 47*, 95–115. doi:10.5721/EuJRS20144707

Boentje, J. P., & Blinnikov, M. S. (2007). Post-Soviet forest fragmentation and loss in the Green Belt around Moscow, Russia (1991–2001): A remote sensing perspective. *Landscape and Urban Planning*, *82*(4), 208–221. doi:10.1016/j.landurbplan.2007.02.009

Boffey, T. B., & Karkazis, J. (1995). Linear versus nonlinear models for hazardous materials routing. *INFOR*, *33*, 114–117.

Bohnenblust, H., & Slovic, P. (1998). Integrating technical analysis and public values in risk-based decision making. *Reliability Engineering & System Safety*, *59*(1), 151–159. doi:10.1016/S0951-8320(97)00136-1

Bollapragada, R., Camm, J., Rao, U. S., & Wu, J. (2005). A two-phase greedy algorithm to locate and allocate hubs for fixed-wireless broad bound access. *Operations Research Letters*, *33*(2), 134–142. doi:10.1016/j.orl.2004.05.007

Bonczek, R. H., Holsapple, C. W., & Whinston, A. B. (1981). *Foundations of decision support systems*. Sandiego, CA: Academic Press.

Bonvicini, S., Leonelli, P., & Spadoni, G. (1998). Risk analysis of hazardous materials transportation: Evaluating uncertainty by means of fuzzy logic. *Journal of Hazardous Materials*, *62*(1), 59–74. doi:10.1016/S0304-3894(98)00158-7

Bonvicini, S., & Spadoni, G. (2008). A hazmat multi-commodity routing model satisfying risk criteria: A case study. *Journal of Loss Prevention in the Process Industries*, *21*(4), 345–358. doi:10.1016/j.jlp.2007.11.009

Bookbinder, J. H., & Fox, N. S. (1998). Intermodal routing of Canada–Mexico shipments under NAFTA. *Transportation Research Part E, Logistics and Transportation Review*, *34*(4), 289–303. doi:10.1016/S1366-5545(98)00017-9

Booth, T. E. (2000). Adaptive importance sampling with rapidly varying importance function. *Nuclear Science and Engineering*, *136*(3), 399–408.

Boulmakoul, A. (2006). Fuzzy graphs modelling for HazMat telegeomonitoring. *European Journal of Operational Research*, *175*(3), 1514–1525. doi:10.1016/j.ejor.2005.02.025

Boulos, M. N., Roudsari, A. V., & Carson, E. R. (2001). Health Geomatics: An Enabling Suite of Technologies in Health and Healthcare. *Journal of Biomedical Informatics*, *34*(3), 195–219. doi:10.1006/jbin.2001.1015 PMID:11723701

Bouzembrak, Y., Allaoui, H., Goncalves, G., & Bouchriha, H. (2011). A multi-objective green supply chain network design.*Proceedings of the 2011 4th International Conference on Logistics* (pp. 357–361). IEEE. doi:10.1109/LOGISTIQUA.2011.5939315

Brabyn, L. (2002). Modelling population access to New Zealand's general practitioners. Proceedings of *International ESRI User Conference. San Diego.*

Brabyn, L., & Skelly, C. (2002). Modeling population access to New Zealand public hospitals. *International Journal of Health Geography, 1*, 1, 3.

Bresfelean, V. P., & Ghisoiu, N. (2010). Higher education decision making and decision support systems. *WSEAS Transactions on Advances in Engineering Education*, 2(7), 43–52.

Broy, M., Fox, J., Hölzl, F., Koss, D., Kuhrmann, M., & Meisinger, M. et al. (2008). Service-oriented modeling of CoCoME with Focus and AutoFocus. In A. Rausch, R. Reussner, R. Mirandola, & F. Plasil (Eds.), Lecture Notes in Computer Science: Vol. 5153. *The Common Component Modeling Example* (pp. 177–206). Berlin Heidelberg, Germany: Springer. doi:10.1007/978-3-540-85289-6_8

Bubbico, R., Di Cave, S., & Mazzarotta, B. (2004). Risk analysis for road and rail transport of hazardous materials: A GIS approach. *Journal of Loss Prevention in the Process Industries*, *17*(6), 483–488. doi:10.1016/j.jlp.2004.08.011

Bubbico, R., Ferrari, C., & Mazzarotta, B. (2000). Risk analysis of LPG transport by road and rail. *Journal of Loss Prevention in the Process Industries, 13*(1), 27–31. doi:10.1016/S0950-4230(99)00057-1

Bubbico, R., Maschio, G., Mazzarotta, B., Milazzo, M. F., & Parisi, E. (2006). Risk management of road and rail transport of hazardous materials in Sicily. *Journal of Loss Prevention in the Process Industries, 19*(1), 32–38. doi:10.1016/j.jlp.2005.05.011

Caccioni, D. R. (2005). Ortofrutta & marketing. Rome: Agra.

Calişkan, C. (2012). A faster polynomial algorithm for the constrained maximum flow problem. *Computers & Operations Research, 39*(11), 2634–2641. doi:10.1016/j.cor.2012.01.010

Calthrop, E., & Proost, S. (1998). Road transport externalities. *Environmental and Resource Economics, 11*(3/4), 335–348. doi:10.1023/A:1008267917001

Calvo, R. F., De Luigi, F., Haastrup, P., & Maniezzo, V. (2004). A distributed geographic information system for the daily car pooling problem. *Computers & Operations Research, 31*(13), 2263–2278. doi:10.1016/S0305-0548(03)00186-2

Cameron, A. C., & Windmeijer, F. A. G. (1997). An R-squared measure of goodness of fit for some common nonlinear regression models. *Journal of Econometrics, 77*(2), 329–342. doi:10.1016/S0304-4076(96)01818-0

Campbell, J. F. (1991). *Hub location problems and the p-hub median problem* [Working Paper 91-06-21]. Center for Business and Industrial Studies. University of Missouri, St. Louis, MO.

Campbell, A. M., Lowe, T. J., & Zhang, L. (2007). The P-hub center allocation problem. *European Journal of Operational Research, 176*(2), 819–835. doi:10.1016/j.ejor.2005.09.024

Campbell, J. F. (1994). Integer programming formulations of discrete hub location problems. *European Journal of Operational Research, 72*(2), 387–405. doi:10.1016/0377-2217(94)90318-2

Campbell, J. F. (1996). Hub location and P-hub median problem. *Operations Research, 44*(6), 923–935. doi:10.1287/opre.44.6.923

Campbell, J. F., Erns, A., & Krishnamoorthy, M. (2002). Hub location problems. In Z. Drezner & H. Hammacher (Eds.), *Facility location: applications and theory*. Berlin: Springer. doi:10.1007/978-3-642-56082-8_12

Campbell, J. F., & O'Kelly, M. E. (2012). Twenty-five years of hub location research. *Transportation Science, 46*(2), 153–169. doi:10.1287/trsc.1120.0410

Canovas, L., Garcia, S., & Marin, A. (2007). Solving the uncapacitated multiple allocation hub location problem by means of a dual-ascent technique. *European Journal of Operational Research, 179*(3), 990–1007. doi:10.1016/j.ejor.2005.08.028

CanoViktorsson, C. (2013). Traffic Radio as a Precursor to Smart Travel Planning Systems: The Challenge of Organizing "Collective Intelligence". *Journal of Urban Technology, 20*(4), 43–55. doi:10.1080/10630732.2013.863439

Cantarella, G. E. (2013). Day-to-day dynamic models for intelligent transportation systems design and appraisal. *Transportation Research Part C, Emerging Technologies, 29*, 117–130. doi:10.1016/j.trc.2012.03.005

Cappanera, P., & Nonato, M. (2014). The gateway location problem: A cost oriented analysis of a new risk mitigation strategy in hazmat transportation. *Procedia: Social and Behavioral Sciences, 111*, 918–926. doi:10.1016/j.sbspro.2014.01.126

Caramia, M., Giordani, S., & Iovanella, A. (2010). On the selection of k routes in multiobjective hazmat route planning. *IMA Journal of Management Mathematics, 21*(3), 239–251. doi:10.1093/imaman/dpp017

Carotenuto, P., Giordani, S., & Ricciardelli, S. (2007). Finding minimum and equitable risk routes for hazmat shipments. *Computers & Operations Research*, *34*(5), 1304–1327. doi:10.1016/j.cor.2005.06.003

Carotenuto, P., Giordani, S., Ricciardelli, S., & Rismondo, S. (2007). A tabu search approach for scheduling hazmat shipments. *Computers & Operations Research*, *34*(5), 1328–1350. doi:10.1016/j.cor.2005.06.004

Carreno, M. (2012). 'The Key', travel smartcard scheme, UK. *Eltis.org*. [REMOVED HYPERLINK FIELD]Retrieved from http://www.eltis.org/discover/case-studies/key-travel-smartcard-scheme-uk

Carreno, M. (2012). Using social media to promote bus travel, Cardiff Bus, UK. *Eltis.org*. Retrieved from http://www.eltis.org/discover/case-studies/using-social-media-promote-bus-travel-cardiff-bus-uk

Casalegno, M., & Chiu, D. (2008). *White Paper on Sustainable Transportation*, MIT Mobile Experience Laboratory, Cisco Systems Inc., 1-18. Retrieved from http://www.connectedurbandevelopment.org/pdf/sust_trans/sustainable_transportation.pdf

Casey, R. F., Labell, L. N., Prensky, S. P., & Schweiger, C. L. (1991). *Advanced public transportation systems: the state of the art (No. UMTA-MA-06-0196-91-2)*. Umta.

Cassini, P. (1998). Road transportation of dangerous goods: Quantitative risk assessment and route comparison. *Journal of Hazardous Materials*, *61*(1–3), 133–138. doi:10.1016/S0304-3894(98)00117-4

Castellanos, J. C., & Fruett, F. (2014). Embedded system to evaluate the passenger comfort in public transportation based on dynamical vehicle behavior with user's feedback. *Measurement*, *47*, 442–451. doi:10.1016/j.measurement.2013.08.068

Castells, M. (2009). *Communication power*. Oxford: Oxford University Press.

Castells, M., Fernandez-Ardevol, M., Linchuan Qiu, J., & Sey, A. (2007). *Mobile communication and society: A global perspective*. London: The MIT Press.

Castillo, J. E. A. (2004). *Route optimization for hazardous materials transport* [Unpublished master's thesis]. International Institute for Geo-information Science and Earth Observation, Urban Planning and Land Administration, The Netherlands.

Catling, I. (1990). System convergence—The potential for an integrated road transport environment. *Transportation*, *17*(3), 285–299. doi:10.1007/BF02127040

Cavar, I., Baric, D., & Kavran, Z. (2008). Assessment of proposed business plans of on-trip and pre-trip information in the LRT system (public urban transit) of the city of Zagreb. In Urban Transport XIV: Urban Transport and the Environment in the 21st Century.

Caves, D. L., Christensen, L., & Diwert, W. E. (1982). The economic theory of index numbers and the measurement of input, output and productivity. *Econometrica*, *50*(6), 1393–1414. doi:10.2307/1913388

Chaabane, A., Ramudhin, A., & Paquet, M. (2011). Designing supply chains with sustainability considerations. *Production Planning and Control*, *22*(8), 727–741. doi:10.1080/09537287.2010.543554

Chakrabarti, U. K., & Parikh, J. K. (2011a). Class-2 hazmat transportation consequence assessment on surrounding population. *Journal of Loss Prevention in the Process Industries*, *24*(6), 758–766. doi:10.1016/j.jlp.2011.04.011

Chakrabarti, U. K., & Parikh, J. K. (2011b). Route evaluation for hazmat transportation based on total risk - A case of Indian State Highways. *Journal of Loss Prevention in the Process Industries*, *24*(5), 524–530. doi:10.1016/j.jlp.2011.03.002

Chakrabarti, U. K., & Parikh, J. K. (2011c). Route risk evaluation on class-2 hazmat transportation. *Process Safety and Environmental Protection*, *89*(4), 248–260. doi:10.1016/j.psep.2011.04.003

Chakrabarti, U. K., & Parikh, J. K. (2012). Applying HAZAN methodology to hazmat transportation risk assessment. *Process Safety and Environmental Protection*, *90*(5), 368–375. doi:10.1016/j.psep.2012.05.011

Chakrabarti, U. K., & Parikh, J. K. (2013). A societal risk study for transportation of class-3 hazmats – A case of Indian state highways. *Process Safety and Environmental Protection*, *91*(4), 275–284. doi:10.1016/j.psep.2012.06.009

Chakrabarti, U. K., & Parikh, J. K. (2013). Risk-based route evaluation against country-specific criteria of risk tolerability for hazmat transportation through Indian State Highways. *Journal of Loss Prevention in the Process Industries*, *26*, 723–736.

Chakrabarti, U. K., & Parikh, J. K. (2013). Using consequence e based hazard zone assessment for effective evacuation planning of vulnerable settlements along hazmat transport corridors through industrial city of Surat in western India. *Journal of Loss Prevention in the Process Industries*, *26*(5), 941–947. doi:10.1016/j.jlp.2013.04.003

Chakraborty, J., & Armstrong, M. P. (1995). Using geographic plume analysis to assess community vulnerability to hazardous accidents. *Computers, Environment and Urban Systems*, *19*(5-6), 341–356. doi:10.1016/0198-9715(95)00018-6

Chaloulakou, A., Saisana, M., & Spyrellis, N. (2003). Comparative assessment of neural networks and regression models for forecasting summertime ozone in Athens. *The Science of the Total Environment*, *313*(1-3), 1–13. doi:10.1016/S0048-9697(03)00335-8 PMID:12922056

Chang, C. L., & Kang, S. J. (2012). Using the reinforcement learning scheme to the priority-based routing and call admission control in WDM networks. *Journal of Internet Technology*, *13*(5), 793–801.

Chape, S., Blyth, S., Fish, L., Fox, P., & Spalding, M. (2003). *United Nations List of Protected Areas*. International Union for Conservation of Nature and Natural Resources. Cambridge, UK: Thanet Press Ltd., UK. Retrieved from http://www.iucn.org/about/work/programmes/gpap_home/gpap_wcpa/gpap_wcpapub/?2256/2003-United-Nations-list-of-protected-areas

Chapelet, P., & Lefebvre, B. (2005). Contextualizing the urban healthcare system. Methodology for developing a geodatabase of Delhi's healthcare system. *CSH Occasional Paper, N 11* (p.135). New Delhi: French Research Institutes in India, Rajdhani Art Press.

Charnes, A., Cooper, W. W., & Rhodes, E. (1978). Measuring the efficiency of decision making units. *European Journal of Operational Research*, *2*(6), 429–444. doi:10.1016/0377-2217(78)90138-8

Chatterjee, D., & Mukherjee, B. (2013). Potential hospital location selection using AHP: A study in Rural India. *International Journal of Computers and Applications*, *71*(17), 1–7. doi:10.5120/12447-9144

Chen, Q. (2000). *Measuring Accessibility in GIS*. Retrieved from http://mather.ar.utexas.edu/students/cadlab/chen/measuring_accessibility_in_gis.htm

Chen, Q., & Weng, W. (1999). *Urban Accessibility Evaluation in GIS*. Retrieved from http://mather.ar.utexas.edu/Courses/parmenter/gis/fall/index.html

Cheng, J., & Druzdzel, M. J. (2000). An adaptive importance-sampling algorithm for evidential reasoning in large Bayesian networks. *Journal of Artificial Intelligence Research*, *13*, 155–188.

Cheng, T. K., Leu, J. S., Chen, I. X., Wu, J. L. C., & Zhu, Z. Y. (2014). Practical context awareness information system for wireless vehicular networks based on a core telematics platform. *International Journal of Distributed Sensor Networks*, 2014.

Cheng, Y.-H., & Tsai, Y.-L. (2009). Factors influencing shippers to use multiple country consolidation services in international distribution centers. *International Journal of Production Economics*, *122*(1), 78–88. doi:10.1016/j.ijpe.2008.12.022

Chen, J. A., & Hu, J. H. (2014). Energy-efficient digital signal processing via voltage-overscaling-based residue number system. *IEEE Transactions on Very Large Scale Integration (VLSI). Systems, 21*(7), 1322–1332.

Chen, J. F. (2007). A hybrid heuristic for the uncapacitated single allocation hub location problem. *Omega, 35*(2), 211–220. doi:10.1016/j.omega.2005.05.004

Chen, P. C., Yu, M. M., Chang, C. C., & Hsu, S. H. (2008). Total factor productivity growth in China's agricultural sector. *China Economic Review, 19*(4), 580–583. doi:10.1016/j.chieco.2008.07.001

Chien, S. I., & Tang, Y. (2014). Scheduling highway work zones with genetic algorithm considering the impact of traffic diversion. *Journal of Advanced Transportation, 48*(4), 287–303. doi:10.1002/atr.213

Chien, S., & Schonfeld, P. (2001). Optimal work zone lengths for four-lane highways. *Journal of Transportation Engineering, 127*(2), 124–131. doi:10.1061/(ASCE)0733-947X(2001)127:2(124)

Chin, S. M., & Cheng, P. D. M. (1989). Bicriterion routing scheme for nuclear spent fuel transportation. *Transportation Research Record, 1245*, 60–64.

Chiraphadhanakul, V., & Eggenberg, N. (2009). *How to evaluate the robustness of airlines schedules* [Technical Report TRANSP–OR 091006]. École Polytechnique Fédérale de Lausanne, Switzerland.

Christiaens, J. (2012). Scotty, the multimodal journey planner and information platform (Belgium). *Eltis.org*. Retrieved from http://www.eltis.org/discover/case-studies/scotty-multimodal-journey-planner-and-information-platform-belgium

Ciarleglio, M., Sarkar, S., & Barnes, J. W. (2010). ConsNet Manual. Ver 2.0. Biodiversity and Biocultural Conservation Laboratory (BBCL). Austin, Texas: University of Texas. Retrieved from http://uts.cc.utexas.edu/~consbio/Cons/ConsNetManual_v200.pdf

CIVITAS forum, Nash, A. (2011) *Social media in transport*. Retrieved from http://www.andynash.com/nash-publications/2011-Nash-Social-Media-Trans13oct2011.pdf

Clark, R. M., & Besterfield-Sacre, M. E. (2009). A new approach to hazardous materials transportation risk analysis: Decision modeling to ıdentify critical variables. *Risk Analysis, 29*(3), 344–354. doi:10.1111/j.1539-6924.2008.01163.x PMID:19087232

Coelli, T. J. (1996). *A guide to DEAP version 2.1:1996 data envelopment analysis program* (CEPA working papers No:8/96). Retrieved from November 25, 2014, from http://www.une.edu.au/econometrics/cepa.htm

Cohen, G, Salomon, I. & Nijkamp, P. (2002). Information–communications technologies (ICT) and

Cohen, G., Salomon, I., & Nijkamp, P. (2002, February). Transport: Does knowledge underpin policy? *Telecommunications Policy, 26*(1-2), 31–52. doi:10.1016/S0308-5961(01)00052-0

Coleman, J. A. (1984). Railroad–highway crossings and route selection for transporting hazardous materials. *Public Roads, 48*(2), 63–71.

Conesa, J., Cavas-Martínez, F., & Fernández-Pacheco, D. G. (2013). An agent-based paradigm for detecting and acting on vehicles driving in the opposite direction on highways. *Expert Systems with Applications, 40*(13), 5113–5124. doi:10.1016/j.eswa.2013.03.026

Constantine, L. L., & Lockwood, L. A. (1999). *Software for use: A practical guide to the models and methods of usage-centered design*. New York, NY: Addison-Wesley.

Constantine, L. L., & Lockwood, L. A. (2002). Usage-centered engineering for Web applications. *IEEE Software, 19*(2), 42–50. doi:10.1109/52.991331

Contardo, C., Cordeau, J.-F., & Gendron, B. (2014). An exact algorithm based on cut-and-column generation for the capacitated location-routing problem. *INFORMS Journal on Computing*, *26*(1), 88–102. doi:10.1287/ijoc.2013.0549

Contreras, I., Fernández, E., & Marín, A. (2010). The tree of hubs location problem. *European Journal of Operational Research*, *202*(2), 390–400. doi:10.1016/j.ejor.2009.05.044

Cook, W. D., & Seiford, L. M. (2009). Data envelopment analysis (DEA) – Thirty years on. *European Journal of Operational Research*, *192*(1), 1–17. doi:10.1016/j.ejor.2008.01.032

Correia, I., Nickel, S., & Saldanha-da-Gama, F. (2010). Single-assignment hub location problems with multiple capacity levels. *Transportation Research Part B: Methodological*, *44*(8-9), 1047–1066. doi:10.1016/j.trb.2009.12.016

Correia, I., Nickel, S., & Saldanha-da-Gama, F. (2010). The capacitated single-allocation hub location problem revisited: A note on a classical formulation. *European Journal of Operational Research*, *207*(1), 92–96. doi:10.1016/j.ejor.2010.04.015

Costa, M. G., Captivo, M. E., & Climaco, J. (2008). Capacitated single allocation hub location problem-A bi-criteria approach. *Computers & Operations Research*, *35*(11), 3671–3695. doi:10.1016/j.cor.2007.04.005

Coulibaly, P., Anctil, F., & Bobee, B. (2000). Daily reservoir inflow forecasting using artificial neural networks with stopped training approach. *Journal of Hydrology (Amsterdam)*, *230*(3-4), 244–257. doi:10.1016/S0022-1694(00)00214-6

Council Directive 2001/29/EC on the framework for the deployment of Intelligent Transport Systems in the field of road transport and for interfaces with other modes of transport [2001] OJ L 207, 1-13.

Courtney, J. F. (2001). Decision making and knowledge management in inquiring organizations: Toward a new decision-making paradigm for DSS. *Decision Support Systems*, *31*(1), 17–38. doi:10.1016/S0167-9236(00)00117-2

Covan, P. (2014, June 11). Auditor General slams Victorian bus GPS tender. *ITnews*. Retrieved from http://www.itnews.com.au/News/387872,auditor-general-slams-victorian-bus-gps-tender.aspx

Cox, R., & Turnquist, M. (1986). Scheduling truck shipments of hazardous materials in the presence of curfews. *Transportation Research Record*, *1063*, 21–26.

Cozzani, V., Bonvicini, S., Spadoni, G., & Zanelli, S. (2007). Hazmat transport: A methodological framework for the risk analysis of marshalling yards. *Journal of Hazardous Materials*, *147*(1-2), 412–423. doi:10.1016/j.jhazmat.2007.01.065 PMID:17418942

Crainic, T. G., Kim, K. H., Barnhart, C., & Laporte, G. (2007). Intermodal Transportation, Chapter 8 in Transportation. In C. Barnhart and G. Laporte (Eds.), Handbooks in Operations Research and Management Science (pp. 467-537). North-Holland, Amsterdam.

Crainic, T. G., & Laporte, G. (1997). Planning models for freight transportation. *European Journal of Operational Research*, *97*(3), 409–438. doi:10.1016/S0377-2217(96)00298-6

Crossland, M. D., Wynne, B. E., & Perkins, W. C. (1995). Spatial decision support systems: An overview of technology and a test of efficacy. *Decision Support Systems*, *14*(3), 219–235. doi:10.1016/0167-9236(94)00018-N

Cubuk, K., & Cansiz, O. F. (2005). Energy status between Transportation Systems in Turkey. *Paper presented at theEnergy Efficiency Conference*. Ministry of Energy and Natural Resources, Ankara.

Cui, Q., & Li, Y. (2014). The evaluation of transportation energy efficiency: An application of three-stage virtual frontier DEA. *Transportation Research Part D, Transport and Environment*, *29*, 1–11. doi:10.1016/j.trd.2014.03.007

Cunha, C. B., & Silva, M. R. (2007). A genetic algorithm for the problem of configuring a hub-and-spoke network for a LTL trucking company in Brazil. *European Journal of Operational Research, 179*(3), 747–758. doi:10.1016/j.ejor.2005.03.057

Current, J. R., & Ratick, S. (1995). A model to assess risk, equity, and efficiency in facility location and transportation of hazardous materials. *Location Science, 3*(3), 187–201. doi:10.1016/0966-8349(95)00013-5

Curry, B., Morgan, P., & Silver, M. (2002). Neural networks and non-linear statistical methods: An application to the modelling of price quality relationships. *Computers & Operations Research, 29*(8), 951–969. doi:10.1016/S0305-0548(00)00096-4

Cyclopath. (n. d.). Cyclopath.org. Retrieved from http://www.cyclopath.org

Dadkar, Y., Jones, D., & Nozick, L. (2008). Identifying geographically diverse routes for the transportation of hazardous materials. *Transportation Research Part E, Logistics and Transportation Review, 44*(3), 333–349. doi:10.1016/j.tre.2006.10.010

Dadkar, Y., Nozick, L., & Jones, D. (2010). Optimizing facility use restrictions for the movement of hazardous materials. *Transportation Research Part B: Methodological, 44*(2), 267–281. doi:10.1016/j.trb.2009.07.006

Das, A., Gupta, A. K., Mazumder, T. N. (2012a). A comprehensive risk assessment framework for offsite transportation of inflammable hazardous waste. *Journal of Hazardous Materials, 227– 228*, 88– 96.

Das, A., Mazumder, T. N., & Gupta, A. K. (2012b). Pareto frontier analyses based decision making tool for transportation of hazardous waste. *Journal of Hazardous Materials, 227– 228*, 341– 352.

Dasaklis, T. K., & Pappis, C. P. (2013). Supply chain management in view of climate change : An overview of possible impacts and the road ahead. *Journal of Industrial Engineering and Management, 6*(4), 1124–1138. doi:10.3926/jiem.883

Daskin, M. S. (1995). *Network and discrete location: Models, algorithms, and applications*. New York: Wiley. doi:10.1002/9781118032343

Davari, S., Zarandi, M. H., & Turksen, I. B. (2013). The incomplete hub-covering location problem considering imprecise location of demands. *Scientia Iranica E, 20*(3), 983–991.

Davis, G. A., Hourdos, J., Xiong, H., & Chatterjee, I. (2011). Outline for a causal modal of traffic conflicts and crashes. *Accident; Analysis and Prevention, 43*(6), 1907–1919. doi:10.1016/j.aap.2011.05.001 PMID:21819818

de Ceuster, G., Voge, T., Chen, M., de Kievit, M., Laird, J., Koh, A., et al. (2010). *Trans-European transport network planning methodology*. DG MOVE. Retrieved from http://www.tmleuven.com/project/tent/finalreport-tent.pdf

Deb, K. (2011). Multi-Objective Optimization Using Evolutionary Algorithms : An Introduction. In L. Wang, A.H.C. Ng, K. Deb (Ed.), Multi-objective Evolutionary Optimisation for Product Design and Manufacturing (pp. 1–24). London: Springer-Verlag.

Debnath, A. K., Chin, H. C., Haque, M. M., & Yuen, B. (2014). A methodological framework for benchmarking smart transport cities. *Cities (London, England), 37*, 47–56. doi:10.1016/j.cities.2013.11.004

deCamargo, R. S., & Miranda, G. (2012). Single allocation hub location problem under congestion: Network owner and user perspectives. *Expert Systems with Applications, 39*(3), 3385–3391. doi:10.1016/j.eswa.2011.09.026

deCamargo, R. S., Miranda, J. G., & Luna, H. P. (2008). Benders decomposition for the uncapacitated multiple allocation hub location problem. *Computers & Operations Research, 35*(4), 1047–1064. doi:10.1016/j.cor.2006.07.002

Dekker, R., Bloemhof, J., & Mallidis, I. (2012). Operations Research for green logistics – An overview of aspects, issues, contributions and challenges. *European Journal of Operational Research*, *219*(3), 671–679. doi:10.1016/j.ejor.2011.11.010

Demissie, M. G., Correria, G. H. A., & Bento, C. (2013). Exploring cellular network handover information for urban mobility analysis. *Journal of Transport Geography*, *31*, 164–170. doi:10.1016/j.jtrangeo.2013.06.016

Dennis, S. M. (1996). Estimating risk costs per unit of exposure for hazardous materials transported by rail. *Logistics and Transportation Review*, *32*(4), 351–375.

Densham, P. J. (1991). Spatial decision support systems. In D. J. Maguire, M. F. Goodchild, & D. W. Rhind (Eds.), *Geographic information systems: Principles and applications* (pp. 403–412). London: Langman.

Deo, M. C., & Kumar, N. K. (2000). Interpolation of wave heights. *Ocean Engineering*, *27*(9), 907–919. doi:10.1016/S0029-8018(99)00023-2

Deo, M. C., & Naidu, C. S. (1999). Real time wave forecasting using neural networks. *Ocean Engineering*, *26*, 192–203.

Desai, S., & Lim, G. (2013). Solution time reduction techniques of a stochastic dynamic programming approach for hazardous material route selection problem. *Computers & Industrial Engineering*, *65*(4), 634–645. doi:10.1016/j.cie.2013.05.002

Desaulniers, G., Desrosiers, J., Dumas, Y., Marc, S., Rioux, B., Solomon, M. M., & Soumis, F. (1997). Crew pairing at air France. *European Journal of Operational Research*, *97*(2), 245–259. doi:10.1016/S0377-2217(96)00195-6

Dhillon, B. (2004). *Engineering usability: fundamentals, applications, human factors, and human error*. Stevenson Ranch, CA: American Scientific Publishers.

Díaz-Báñez, J. M. (2008). Optimization for continuous shortest paths in transportation. In F. Erdogdu (Ed.), *Optimization In Food Engineering* (pp. 497–519). USA: Taylor & Francis.

Dijkstra, E. (1959). A note on two problems in connexion with graphs. *Numerische Mathematik*, *1*(1), 269–271. doi:10.1007/BF01386390

Ding, S., Zhao, H., Zhang, Y., Xu, X., & Nie, R. (2013). *Extreme learning machine: algorithm, theory and applications*. Springer Science Business Media Dordrecht.

Dodge, M., & White, S. (1995). GIS and public services in Wales: Analyzing supply and demand for healthcare and education. *Mapping Awareness Journal*, *9*(7).

Dökmeci, V., & Berköz, L. (1994). Transformation of Istanbul from a monocentric to a polycentric city. *European Planning Studies*, *2*(2), 193–205. doi:10.1080/09654319408720259

Donath, U., Haufe, J., Blochwitz, T., & Neidhold, T. (2008). A new approach for modeling and verification of discrete control components within a Modelica Environment. *Proceedings of the 6th International Modelica Conference*. Retrieved from http://publica.fraunhofer.de/documents/N-70395.html

Dong, B., Bowman, J. L., & Walker, J. (1998). *Analysis of Activity-Based Accessibility*. Retrieved from http://web.mit.edu/jwalker/www/JW_ABA.pdf

Dong, X., Ben-Akiva, M. E., Bowman, J. L., & Walker, J. L.(2006). Moving from trip-based to activity-based measures of accessibility. *Transportation Research Part A, Policy and Practice*, *40*(2), 163–180. doi:10.1016/j.tra.2005.05.002

Don, T., Harit, S., English, J. R., & Whicker, G. (1995). Hub and spoke networks in truckload trucking: Configuration testing and operational concerns. *Logistics and Transportation*, *31*, 209–237.

Dourish, P., & Satchell, C. (2011). The moral economy of social media. In M. Foth et al. (Eds.), *From social butterfly to engaged citizen. Urban informatics, Social media, ubiquitous computing, and mobile technology to support citizen engagement* (pp. 21–37). London: The MIT Press.

Dragicevic, S. (2004). The potential of Web-based GIS. *Journal of Geographical Systems*, *6*(2), 79–81. doi:10.1007/s10109-004-0133-4

Dreyfus, S. E. (1967). *An appraisal of some shortest path algorithms*. USA: Memorandum. Rand Corporation.

Dreyfus, S. E. (1969). An appraisal of some shortest path algorithms. *Operations Research*, *17*(3), 395–412. doi:10.1287/opre.17.3.395

Drilo, B., Saric, D., & Filjar, R. (2009, May). The role of telecommunications in development of new-generation intelligent transport systems. Proceedings of *Wireless Communication, Vehicular Technology, Information Theory and Aerospace & Electronic Systems Technology, 1st International Conference* (pp. 125-127). IEEE. doi:10.1109/WIRELESSVITAE.2009.5172435

Dudley, N. (2008). *Guidelines for Applying Protected Area Management Categories*. Gland, Switzerland: IUCN. Retrieved from https://cmsdata.iucn.org/downloads/guidelines_for_applying_protected_area_management_categories.pdf

Duyguvar, A.S. & Erol, S. (2014). Evaluation of logistics hubs in Turkey and location model proposal for new sustainable logistics hubs. Proceedings of *3rd National Logistics & Supply Chain Congress* (pp. 261-271).

Dyer, R. F., & Forman, E. H. (1992). Group decision support with the analytic hierarchy process. *Decision Support Systems*, *8*(2), 99–124. doi:10.1016/0167-9236(92)90003-8

Dzemydienė, D., & Dzindzalieta, R. (2010). Development of architecture of embedded decision support systems for risk evaluation of transportation of dangerous goods. *Technological and Economic Development of Economy*, *16*(4), 654–671. doi:10.3846/tede.2010.40

Easterbrook, S. M. (2010). Climate change: A software grand challenge.*Proceedings of the FSE/SDP Workshop on the Future of Software Engineering Research*. New York, NY: ACM. doi:10.1145/1882362.1882383

Ebener, S., El Morjani, Z., Ray, N., & Black, M. (2005). Physical accessibility to health care: from isotropy to anisotropy. *GIS@ development, 9*(6).

Ebery, J. (2001). Solving large single allocation p-hub problems with two or three hubs. *European Journal of Operational Research*, *128*(2), 447–458. doi:10.1016/S0377-2217(99)00370-7

Ebery, J., Krishnamoorty, M., Ernst, A., & Boland, N. (2000). The capacitated multiple allocation hub location problem: Formulation and algorithms. *European Journal of Operational Research*, *120*(3), 614–631. doi:10.1016/S0377-2217(98)00395-6

Ecoinvent data v.2.0: ecoinvent reports no. 1-25. (2007). Ecoinvent. Dübendorf: Swiss center for life cycle inventory.

Ecosystems and Human Well-Being: Scenarios. (2005). Washington D.C.: Island Press. Retrieved from http://islandpress.org/ecosystems-and-human-well-being-scenarios

Ecotaxsi. (n. d.). Retrieved from http://www.ecotaxi.ch/

Effat, H. A., & Hassan, O. A. (2013). Designing and evaluation of three alternatives highway routes using the analytical hierarchy process and the least-cost path analysis, Application in Sinai, Peninsula, Egypt. *The Egyptian Journal of Remote Sensing and Space Sciences*, *16*(2), 141–151. doi:10.1016/j.ejrs.2013.08.001

EGO website, http://map.ego.gov.tr:8080/ego/index.aspx

Ehrgott, M., & Ryan, D. M. (2002). Constructing robust crew schedules with bicriteria optimization. *Journal of Multi-Criteria Decision Analysis, 11*(3), 139–150. doi:10.1002/mcda.321

Eiselt, H. A. (2007). Locating landfills – optimization vs. reality. *European Journal of Operational Research, 179*(3), 1040–1049. doi:10.1016/j.ejor.2005.11.039

Eiselt, H. A., & Marianov, V. (2009). A conditional p-hub location problem with attraction functions. *Computers & Operations Research, 36*(12), 3128–3135. doi:10.1016/j.cor.2008.11.014

El Moudani, W., Cosenza, C. A. N., de Coligny, M., & Mora-Camino, F. (2001). A bi-criterion approach for the airlines crew rostering problem. *Paper presented at the Evolutionary Multi-Criterion Optimization.* doi:10.1007/3-540-44719-9_34

Ellis, J. (2011). Analysis of accidents and incidents occurring during transport of packaged dangerous goods by sea. *Safety Science, 49*(8-9), 1231–1237. doi:10.1016/j.ssci.2011.04.004

Emrouznejad , A., Parker, B. R., & Tavares, G. (2008). Evaluation of research in efficiency and productivity: A survey and analysis of the first 30 years of scholarly literature in DEA. *Socio-Economic Planning Sciences, 42*(3), 151-157.

EPOMM. (2013) Managing mobility for a better future: Mobility management: The smart way to sustainable mobility in European countries, regions and cities. Retrieved May, 6, 2014 from http://epomm.eu/index.php?id=2790

Ercoskun, O. Y. (2010). Green Urban Planning and Design for Smarter Communities. In A. Hallin, & T. Karrbom Gustavsson (Eds.), *Organizational Communication and Sustainable Development* (pp. 41–59). Hershey, PA: Information Science Publishing. doi:10.4018/978-1-60566-822-2.ch003

Ercoskun, O. Y. (Ed.). (2011). *Green and ecological technologies for urban planning: Creating smart cities.* Hershey, PA: IGI Global; doi:10.4018/978-1-61350-453-6

Ergun, H., Rawn, B., Belmans, R., & Van Hertem, D. (2014). Technology and topology optimization for multizonal transmission systems. *IEEE Transactions on Power Systems, 29*(5), 2469–2477. doi:10.1109/TPWRS.2014.2305174

Erkut, E. (1995). On the credibility of the conditional risk model for routing hazardous materials. *Operations Research Letters, 18*(1), 49–52. doi:10.1016/0167-6377(95)00030-N

Erkut, E., & Alp, O. (2007). Designing a Road Network for Hazardous Materials Shipments. *Computers & Operations Research, 34*(5), 1389–1405. doi:10.1016/j.cor.2005.06.007

Erkut, E., & Alp, O. (2007). Integrated routing and scheduling of hazmat trucks with stops en route. *Transportation Science, 41*(1), 107–122. doi:10.1287/trsc.1060.0176

Erkut, E., & Glickman, T. (1997). Minimax population exposure in routing highway shipments of hazardous materials. *Transportation Research Record, 1602*, 93–100. doi:10.3141/1602-14

Erkut, E., & Gzara, F. (2008). Solving the hazmat transport network design problem. *Computers & Operations Research, 35*(7), 2234–2247. doi:10.1016/j.cor.2006.10.022

Erkut, E., & Ingolfsson, A. (2000). Catastrophe avoidance models for hazardous materials route planning. *Transportation Science, 34*(2), 165–179. doi:10.1287/trsc.34.2.165.12303

Erkut, E., & Ingolfsson, A. (2005). Transport risk models for hazardous materials. *Operations Research Letters, 33*(1), 81–89. doi:10.1016/j.orl.2004.02.006

Erkut, E., Tjandra, A. S., & Verter, V. (2007). Hazardous Materials Transportation. In C. Barnhart & G. Laporte (Eds.), *Handbook in OR & MS* (Vol. 14, pp. 539–621). UK: Elsevier.

Erkut, E., & Verter, V. (1995b). A framework for hazardous materials transport risk assessment. *Risk Analysis*, *15*(5), 589–601. doi:10.1111/j.1539-6924.1995.tb00755.x

Erkut, E., & Verter, V. (1998). Modeling of transport risk for hazardous materials. *Operations Research*, *46*(5), 625–642. doi:10.1287/opre.46.5.625

Ernst, A., & Krishnamoorthy, M. (1996). Efficient algorithms for the uncapacitated single allocation P-hub median problem. *Location Science*, *4*(3), 139–154. doi:10.1016/S0966-8349(96)00011-3

Ertugay, K., & Duzgun, S. (2011). GIS-based stochastic modeling of physical accessibility using GPS-based floating car data and Monte Carlo simulation. *International Journal of Geographical Information Science*, *25*(9), 1491–1506. doi:10.1080/13658816.2010.528419

Eskigun, E., Uzsoy, R., Preckel, P. V., Beaujon, G., Krishnan, S., & Tew, J. D. (2005). Outbound supply chain network design with mode selection, lead times and capacitated vehicle distribution centers. *European Journal of Operational Research*, *165*(1), 182–206. doi:10.1016/j.ejor.2003.11.029

EU Green Capital website. (n. d.). Retrieved from http://ec.europa.eu/environment/europeangreencapital/ljubljana-wins-european-green-capital-award-2016/index.html

EU transport in figures Statistical Pocket Book 2011. (2011). Belgium.

Fabiano, B., Curro`, F., Reverberi, A. P., & Pastorino, R. (2005). Dangerous good transportation by road: From risk analysis. *Journal of Loss Prevention in the Process Industries*, *18*(4-6), 403–413. doi:10.1016/j.jlp.2005.06.031

Fabiano, B., Curro, F., Palazzi, E., & Pastorino, R. (2002). A framework for risk assessment and decision making strategies in dangerous good transportation. *Journal of Hazardous Materials*, *93*(1), 1–15. doi:10.1016/S0304-3894(02)00034-1 PMID:12062950

Fabiano, B., & Palazzi, E. (2010). Hazmat transportation by heavy vehicles and road tunnels: A simplified modeling procedure to risk assessment and mitigation applied to an Italian case study. *International Journal of Heavy Vehicle Systems*, *17*(3-4), 216–236. doi:10.1504/IJHVS.2010.035988

Faccio, M., Persona, A., Sgarbossa, F., & Zanin, G. (2014). Sustainable SC through the complete reprocessing of end-of-life products by manufacturers: A traditional versus social responsibility company perspective. *European Journal of Operational Research*, *233*(2), 359–373. doi:10.1016/j.ejor.2013.03.027

Farahani, R., Hekmatfar, M., Arabani, A., & Nikbakhsh, E. (2013). Hub location problems: A review of models, classification, solution techniques, and applications. *Computers & Industrial Engineering*, *64*(4), 1096–1109. doi:10.1016/j.cie.2013.01.012

Fare, R., Grosskoppf, S., Norris, M., & Zhang, Z. (1994). Productivity growth, technical progress and efficiency change in industrialized countries. *The American Economic Review*, *84*(1), 66–83.

Fazlollahtabar, H., Eslami, H., & Salmani, H. (2010). Designing a fuzzy expert system to evaluate alternatives in fuzzy analytic hierarchy process. *J. Software Engineering & Applications*, *3*(04), 409–418. doi:10.4236/jsea.2010.34046

Feilkas, M., Hölzl, F., Pfaller, C., Rittmann, S., Schätz, B., Schwitzer, W., et al. (2011). A refined top-down methodology for the development of automotive software systems – the Keyless Entry System case study [Technical Report TUM-I1103]. *Technische Universität München*. Retrieved from http://mediatum.ub.tum.de/?id=1094230

Feng, G., Huang, G. B., Lin, Q., & Gay, R. (2009). Error minimized extreme learning machine with growth of hidden nodes and incremental learning.[PubMed]. *IEEE Transactions on Neural Networks*, *20*(8), 1352–1357. doi:10.1109/TNN.2009.2024147

Ferreira, J., Trigo, P., & Filipe, P. (2009). Collaborative Carpooling System. *World Academy of Science. Engineering and Technology*, *3*, 609–613.

Fick, G., & Sprague, R. H. (1980). *Decision support systems: issues and challenges*. Oxford, United Kingdom: Pergamon Press.

Figliozzi, M. A. (2011). The impacts of congestion on time-definitive urban freight distribution networks CO2 emission levels: Results from a case study in Portland, Oregon. *Transportation Research Part C, Emerging Technologies*, *19*(5), 766–778. doi:10.1016/j.trc.2010.11.002

Fihn, J., & Finndahl, J. (2011). *A framework for how to make use of an automatic passenger counting system* [Masters Thesis]. RMIT University. Retrieved from http://www.essays.se/essay/de25b1166d/

Fixmytransport. (n. d.). Retrieved from http://www.fixmytransport.com

Floyd, R. (1962). Algorithm 97: Shortest path. *Communications of the ACM*, *5*(6), 345. doi:10.1145/367766.368168

Følstad, A., Law, E., & Hornbæk, K. (2012). Analysis in practical usability evaluation: A survey study.*Proceedings of the SIGCHI Conference on Human Factors in Computing Systems*. New York, NY: ACM. doi:10.1145/2207676.2208365

Fortney, J., Rost, K., & Warren, J. (2000). Comparing Alternative Methods of Measuring Geographic Access to Health Services. *Journal Health Services and Outcomes Research Methodology*, *1*(2), 173–184. doi:10.1023/A:1012545106828

Franca, R. B., Jones, E. C., Richards, C. N., & Carlson, J. P. (2010). Multi-objective stochastic supply chain modeling to evaluate tradeoffs between profit and quality. *International Journal of Production Economics*, *127*(2), 292–299. doi:10.1016/j.ijpe.2009.09.005

Frank, W. C., Thill, J. C., & Batta, R. (2000). Spatial decision support system for hazardous material truck routing. *Transportation Research Part C, Emerging Technologies*, *8*(1–6), 337–359. doi:10.1016/S0968-090X(00)00007-3

Frey, H., Rouphail, N., & Zhai, H. (2008). Link-based emission factors for heavy-duty diesel trucks based on real-world data. *Transportation Research Record*, *2058*, 23–332. doi:10.3141/2058-04

Frota Neto, J. Q., Bloemhof-Ruwaard, J. M., van Nunen, J. E. E., & van Heck, E. (2008). Designing and evaluating sustainable logistics networks. *International Journal of Production Economics*, *111*(2), 195–208. doi:10.1016/j.ijpe.2006.10.014

Gade, D., & Kuecuekyavuz, S. (2013). Formulations for dynamic lot sizing with service levels. *Naval Research Logistics*, *60*(2), 87–101. doi:10.1002/nav.21519

Gal-Tzur, A., Grant-Muller, S. M., Kuflik, T., Minkov, E., Nocera, S., & Shoor, I. (2014). The potential of social media in delivering transport policy goals. *Transport Policy*, *32*, 115–123. doi:10.1016/j.tranpol.2014.01.007

Game, E. T., & Grantham, H. S. (2008). *Marxan User Manual: For Marxan version 1.8.10*. University of Queensland, St. Lucia, Queensland, Australia, and Pacific Marine Analysis and Research Association. Retrieved from http://www.uq.edu.au/marxan/docs/Marxan_User_Manual_2008.pdf

Gan, G., Ma, C., & Wu, J. (2007). *Data clustering: theory, algorithms, and applications* (Vol. 20). Siam. doi:10.1137/1.9780898718348

Gao, C., Johnson, E., & Smith, B. (2009). Integrated airline fleet and crew robust planning. *Transportation Science*, *43*(1), 2–16. doi:10.1287/trsc.1080.0257

Garey, M. R., & Johnson, D. S. (1979). *Computers and Intractability: A Guide to the Theory of NP-Completeness*. San Francisco, USA: W. H. Freeman and Co.

Garriga, N., Santos, X., Montori, A., Richter-Boix, A., Franch, M., & Llorente, G. A. (2012). Are protected areas truly protected? The impact of road traffic on vertebrate fauna. *Biodiversity and Conservation*, *21*(11), 2761–2774. doi:10.1007/s10531-012-0332-0

Gauvin, C., Desaulniers, G., & Gendreau, M. (2014). A branch-cut-and-price algorithm for the vehicle routing problem with stochastic demands. *Computers & Operations Research*, *50*, 141–153. doi:10.1016/j.cor.2014.03.028

Geissler, P. L., & Chandler, D. (2000). Importance sampling and theory of nonequilibrium solvation dynamics in water. *The Journal of Chemical Physics*, *113*(21), 9759–9765. doi:10.1063/1.1290136

Gelareh, S., & Nickel, S. (2011). Hub location problems in transportation networks. *Transportation Research Part E, Logistics and Transportation Review*, *47*(6), 1092–1111. doi:10.1016/j.tre.2011.04.009

Geoffrion, A. (1976). The purpose of mathematical programming is insight, not numbers. *Interfaces*, *7*(1), 81–92. doi:10.1287/inte.7.1.81

Ghazinoory, S., & Kheirkhah, A. S. (2008). Transportation of hazardous materials in Iran: A strategic approach for decreasing accidents. *Transport*, *23*(2), 104–111. doi:10.3846/1648-4142.2008.23.104-111

Gheorghea, A. V., Birchmeier, J., & Vamanu, D. (2005). Comprehensive risk assessment for rail transportation of dangerous goods: A validated platform for decision support. *Reliability Engineering & System Safety*, *88*(3), 247–272. doi:10.1016/j.ress.2004.07.017

Ghio, C., Mark, L., Abdulkadir, N., & Ahmed, A. (2007). *Health GIS Tools and applications informing decisions in Yemen*. Geospatial Application Papers, Retrieved from [REMOVED HYPERLINK FIELD]http://www.gisdevelopment.net/application/health/overview/me05_012a.htm

Giannikos, I. (1998). A multiobjective programming model for locating treatment sites and routing hazardous wastes. *European Journal of Operational Research*, *104*(2), 333–342. doi:10.1016/S0377-2217(97)00188-4

Gibin, M., Longley, P., & Atkinson, P. (2007). Kernel density estimation and percent volume contours in general practice catchment area analysis in urban areas. *Proceedings of the GIScience research UK conference (GISRUK)*. Ireland. Maynooth.

Girolami, M., & Rogers, S. (2006). Variational Bayesian multinomial probit regression with gaussian process priors. *Neural Computation*, *18*(8), 1790–1817. doi:10.1162/neco.2006.18.8.1790

Giuli, D., Paganelli, F., Cuomo, S., & Cianchi, P. (2013). Toward a Cooperative Approach for Continuous Innovation of Mobility Information Services. *IEEE Systems Journal*, *7*(4), 669–680. doi:10.1109/JSYST.2013.2255521

Glickman, T. S. (1983). Rerouting railroad shipments of hazardous materials to avoid populated areas. *Accident; Analysis and Prevention*, *15*(5), 329–335. doi:10.1016/0001-4575(83)90012-X

Glickman, T. S. (1988). Benchmark estimates of release accident rates in hazardous materials transportation of rail and truck. *Transportation Research Record*, *1193*, 22–28.

Glickman, T. S. (1991). An expeditious risk assessment of the highway transportation of flammable liquids in bulk. *Transportation Science*, *25*(2), 115–123. doi:10.1287/trsc.25.2.115

Glickman, T. S., Erkut, E., & Zschocke, M. S. (2007). The cost and risk impacts of rerouting railroad shipments of hazardous materials. *Accident; Analysis and Prevention*, *39*(5), 1015–1025. doi:10.1016/j.aap.2007.01.006 PMID:17349610

Glickman, T. S., & Golding, D. (1991). For a few dollars more: Public trust and the case for transporting nuclear waste in dedicated trains. *Policy Studies Review*, *10*(4), 127–138. doi:10.1111/j.1541-1338.1991.tb00285.x

Glickman, T. S., & Sontag, M. A. (1995). The tradeoffs associated with rerouting highway shipments of hazardous materials to minimize risk. *Risk Analysis*, *15*(1), 61–67. doi:10.1111/j.1539-6924.1995.tb00093.x

Godoy, S. M., Santa Cruz, A. S. M., & Scenna, N. J. (2007). STRRAP system—A software for hazardous materials risk assessment. *Reliability Engineering & System Safety*, *92*(7), 847–857. doi:10.1016/j.ress.2006.02.012

Goldberg, D. E. (1989). *Genetic algorithms in search, optimization and machine learning. MA*. Reading: Addison-Wesley.

Goncalves, W. N., & Bruno, O. M. (2013). Combining fractal and deterministic walkers for texture analysis and classification. *Pattern Recognition*, *46*(11), 2953–2968. doi:10.1016/j.patcog.2013.03.012

Gooley, T. B. (1997). One location, many options. *Logistics Management and Distribution Report*, *37*(11), 81–89.

Gopalakrishnan, B., & Johnson, E. L. (2005). Airline crew scheduling: State-of-the-art. *Annals of Operations Research*, *140*(1), 305–337. doi:10.1007/s10479-005-3975-3

Gopalan, R., Kolluri, K. S., Batta, R., & Karwan, M. H. (1990a). Modeling equity of risk in the transportation of hazardous materials. *Operations Research*, *38*(6), 961–973. doi:10.1287/opre.38.6.961

Goulias, K. G. (2007). *An Optimal Resource Allocation Tool for Urban Development Using GIS-based Accessibility Measures and Stochastic Frontier Analysis. California PATH Program.* Institute of Transportation Studies, University of California at Berkeley.

Grahn, M. (2009). Sunfleet Gothenburg - 37 carsharing pools using environmental friendly cars only (Sweden). *Eltis.org*. Retrieved from http://www.eltis.org/discover/case-studies/sunfleet-gothenburg-37-carsharing-pools-using-environmental-friendly-cars-only

Gramacy, R. B., & Lee, H. K. H. (2008). Bayesian treed Gaussian process models with an application to computer modelling. *Journal of the American Statistical Association*, *103*(483), 1119–1130. doi:10.1198/016214508000000689

Green City Streets – Information technology for better public transportation. (n. d.). Greenscitystreets.com. Retrieved from http://www.greencitystreets.com

GreenSoft. (2009). *Green Software Engineering*. Retrieved from http://www.green-software-engineering.de/en/project.html

Grünig, M., & Marcelino, D. (2009). A Missing Link to Sustainable Mobility: Connecting Public Transportation to Car and Bike Sharing Programs: Best Practice Examples from Europe and the U.S. Heinrich Böll Stiftung Washington, Retrieved August, 5, 2014 from http://www.ecologic.eu/3185

Grunow, M., & Piramuthu, S. (2013). RFID in highly perishable food supply chains – Remaining shelf life to supplant expiry date? *International Journal of Production Economics*, *146*(2), 717–727. doi:10.1016/j.ijpe.2013.08.028

Guagliardo, M. F. (2004). Spatial accessibility of primary care: Concepts, methods and Challenges. *International Journal of Health Geographics*, *3*(3). PMID:14987337

Guest, T. (2007). *A Matter of Time: Air Traffic Delay in Europe* (Vol. 2). Brussels, Belgium: Eurocontrol.

Gulenc, I. F., & Bilgin, G. (2010). A model proposal for investment decisions: AHP method. *Journal of Proposal - Marmara University, 34*(17), 97-107.

Gunasekera, M. Y., & Edwards, D. W. (2003). Estimating the environmental impact of catastrophic chemical releases to the atmosphere: An index method for ranking alternative chemical process routes. *Trans Institution of Chemical Engineers*, *81*, 463–474.

Guo, X., & Verma, M. (2010). Choosing vehicle capacity to minimize risk for transporting flammable materials. *Journal of Loss Prevention in the Process Industries*, *23*(2), 220–225. doi:10.1016/j.jlp.2009.07.007

Gupta, V., Mohapatra, D., Raghunathan, A., & Roy, K. (2013). Low-power digital signal processing using approximate adders. *IEEE Transactions on Computer-Aided Design of Integrated Circuits and Systems*, *32*(1), 124–137. doi:10.1109/TCAD.2012.2217962

Gurrutxaga, M., Rubio, L., & Saura, S. (2011). Key connectors in protected forest area networks and the impact of highways: A transnational case study from the Cantabrian Range to the Western Alps (SW Europe). *Landscape and Urban Planning*, *101*(4), 310–320. doi:10.1016/j.landurbplan.2011.02.036

Gutiérrez, V., Izaguirre, M., Pérez, J., Muñoz, L., López, D., & Sánchez, M. (2010, July). Ambient Intelligence in Intermodal Transport Services: A Practical Implementation in Road Logistics. Proceedings of *Sensor Technologies and Applications (SENSORCOMM) Fourth International Conference* (pp. 203-209). IEEE. doi:10.1109/SENSORCOMM.2010.107

Gzara, F. (2013). A cutting plane approach for bilevel hazardous material transport network design. *Operations Research Letters*, *41*(1), 40–46. doi:10.1016/j.orl.2012.10.007

Haas, T. J., & Kichner, J. J. (1987). Hazardous materials in marine transportation – a practical course. *Journal of Chemical Education*, *64*(1), 34–35. doi:10.1021/ed064p34

Hagerstrand, T. (1970). What about people in regional science? *Papers / Regional Science Association. Regional Science Association. Meeting*, *24*(1), 7–21. doi:10.1111/j.1435-5597.1970.tb01464.x

Hakimi, S. L. (1964). Optimum location of switching centers and the absolute centers and medians of a graph. *Operations Research*, *12*(3), 450–459. doi:10.1287/opre.12.3.450

Halden, D., Mcguigan, D., Nisbet, A., & Mckinnon, A. (2000). *Guidance on accessibility measuring techniques and their application*. Scottish Executive Central Research Unit.

Halkidi, M., Batistakis, Y., & Vazirgiannis, M. (2002). Cluster validity methods: Part I. *SIGMOD Record*, *31*(2), 40–45. doi:10.1145/565117.565124

Hamilton, M., & Berry, M. (2010). Transmesh: Creating a mobile technological urban space. *Proceedings of the Mobile Technology in the Urban Field Symposium*. Melbourne, Australia: RMIT University. Design Research Institute.

Hamilton, M., Salim, F., Cheng, E., & Choy, S. L. (2011). Transafe: A crowdsourced mobile platform for crime and safety perception management. *ACM SIGCAS Computers and Society*, *41*(2), 32–37. doi:10.1145/2095272.2095275

Handy, S. L., & Mokhtarian, P. L. (1995). Planning for Telecommuting Measurement and Policy Issues. *Journal of the American Planning Association*, *61*(1), 99–111. doi:10.1080/01944369508975623

Han, Y. (2004). Improved algorithm for all pairs shortest paths. *Information Processing Letters*, *91*(5), 245–250. doi:10.1016/j.ipl.2004.05.006

Haranczyk, M., & Sethian, J. A. (2009). Navigating molecular worms inside chemical labyrinths. *Proceedings of the National Academy of Sciences of the United States of America*, *106*(51), 21472–21477. doi:10.1073/pnas.0910016106 PMID:20018716

Härdle, W., & Simar, L. (2007). *Applied multivariate statistical analysis* (Vol. 22007). Berlin: Springer.

Harris, I., Mumford, C., & Naim, M. (2009). The Multi-Objective Uncapacitated Facility Location Problem for Green Logistics.*Proceedings of the IEEE Congress on Evolutionary Computation* (pp. 2732–2739). Trondheim: IEEE. doi:10.1109/CEC.2009.4983285

Harvey, D. (2008). The right to the city. *New Left Review*, *53*, 23–40.

Harwood, D. W., Russell, E. R., & Viner, J. G. (1989). Characteristics of accidents and incidents in highway transportation of hazardous materials. *Transportation Research Record*, *1245*, 23–33.

Hashimoto, A., & Haneda, S. (2008). Measuring the change in R&D efficiency of Japanese pharmaceutical industry. *Research Policy*, *37*(10), 1829–1836. doi:10.1016/j.respol.2008.08.004

Hassini, E., Surti, C., & Searcy, C. (2012). A literature review and a case study of sustainable supply chains with a focus on metrics. *International Journal of Production Economics*, *140*(1), 69–82. doi:10.1016/j.ijpe.2012.01.042

Hekmatfar, M., & Pishvaee, M. (2009). *Hub location problem in facility location: Concepts, models, algorithms and case studies* (pp. 243–270). Berlin: Physica-Verlag. doi:10.1007/978-3-7908-2151-2_11

Helander, M. E., & Melachrinoudis, E. (1997). Facility location and reliable route planning in hazardous material transportation. *Transportation Science*, *31*(3), 216–226. doi:10.1287/trsc.31.3.216

Helft, M. (2007). Google's Buses Help Its Workers Beat the Rush. *The New York Times*.

Hensher, D. A., & Button, K. J. (2003). *Handbook of Transport and the Environment*. Amsterdam: Elsevier.

Heumann, J. (2002). Use cases, usability requirements, and user interfaces [Tutorial Notes]. In *Proceedings of Object-Oriented Programming, Systems, Languages, and Applications (OOPSLA 2002)*. New York, NY: ACM.

Hilty, L. M., & Aebischer, B. (Eds.). (2015). *ICT innovations for sustainability. Advances in Intelligent Systems and Computing, 310*. Zurich, Switzerland: Springer; doi:10.1007/978-3-319-09228-7

Hoffman, K. L., & Padberg, M. (1993). Solving airline crew scheduling problems by branch-and-cut. *Management Science*, *39*(6), 657–682. doi:10.1287/mnsc.39.6.657

Höfner, P., & Möller, B. (2012). Dijkstra, Floyd and Warshall meet Kleene. *Formal Aspects of Computing*, *24*(4-6), 459–476. doi:10.1007/s00165-012-0245-4

Holland, J. H. (1975). *Adaptation in natural and artificial systems*. Michigan: University of Michigan Press.

Huang, B. (2004). A GIS-AHP method for HAZMAT route planning with consideration of security. *Environmental Informatics Archives*, *2*, 818–830.

Huang, B. (2006). GIS-Based Route Planning for Hazardous Material Transportation. *Journal of Environmental Informatics*, *8*(1), 49–57. doi:10.3808/jei.200600076

Huang, B., Cheu, R. L., & Liew, Y. S. (2004). GIS and genetic algorithms for HAZMAT route planning with security considerations. *International Journal of Geographical Information Science*, *18*(8), 769–787. doi:10.1080/136588104 10001705307

Huang, G. B. (2003). Learning capability and storage capacity of two hidden- layer Feedforward networks.[PubMed]. *IEEE Transactions on Neural Networks*, *14*(2), 274–281. doi:10.1109/TNN.2003.809401

Huang, G. B., & Babri, H. A. (1998). Upper bounds on the number of hidden neurons in feedforward networks with arbitrary bounded nonlinear activation functions.[PubMed]. *IEEE Transactions on Neural Networks*, *9*(1), 224–229. doi:10.1109/72.655045

Huang, G. B., & Chen, L. (2006). Convex incremental extreme learning machine. *Neurocomputing*, *70*(16-18), 3056–3062. doi:10.1016/j.neucom.2007.02.009

Huang, G. B., Li, M. B., Chen, L., & Siew, C. K. (2008). Incremental extreme learning machine with fully complex hidden nodes. *Neurocomputing*, *71*(4-6), 576–583. doi:10.1016/j.neucom.2007.07.025

Huang, G. B., Zhou, H., Ding, X., & Zhang, R. (2012). Extreme learning machine for regression and multiclass classification.[PubMed]. *IEEE Transactions on Systems, Man, and Cybernetics. Part B, Cybernetics*, *42*(2), 513–529. doi:10.1109/TSMCB.2011.2168604

Huang, G. B., Zhu, Q. Y., & Siew, C. K. (2006). Extreme learning machine: a new learning scheme of feed forward neural networks.*Proceedings of IEEE International Joint Conference on Neural Networks*, *2*, 985–990.

Huang, G. B., Zhu, Q. Y., & Siew, C. K. (2006). Extreme learning machine: Theory and applications. *Neurocomputing*, *70*(1-3), 489–501. doi:10.1016/j.neucom.2005.12.126

Hwang, S. T., Brown, D. F., O'Steen, J. K., Policastro, A. J., & Dunn, W. E. (2001). Risk assessment for national transportation of selected hazardous materials. *Transportation Research Record*, *1763*, 114–124. doi:10.3141/1763-17

Hwang, Y. H., & Lee, H. Y. (2012). Uncapacitated single allocation p-hub maximal covering problem. *Computers & Industrial Engineering*, *63*(2), 382–389. doi:10.1016/j.cie.2012.03.014

Iakovou, E. T. (2001). An interactive multiobjective model for the strategic maritime transportation of petroleum products: Risk analysis and routing. *Safety Science*, *39*(1–2), 19–29. doi:10.1016/S0925-7535(01)00022-4

Iakovou, E., Douligeris, C., Li, H., Ip, C., & Yudhbir, L. (1999). A maritime global route planning model for hazardous materials transportation. *Transportation Science*, *33*(1), 34–48. doi:10.1287/trsc.33.1.34

Ioannou, G. (2005). Streamlining the supply chain of the Hellenic sugar industry. *Journal of Food Engineering*, *70*(3), 323–332. doi:10.1016/j.jfoodeng.2004.01.041

Ioja, C., Rozylowicz, L., Patroescu, M., Nita, M., & Onose, D. (2011). Agriculture and Conservation in the Natura 2000 Network: A Sustainable Development Approach of the European Union. *Agricultural and Environmental Informatics, Governance and Management: Emerging Research Applications*, 339-358. doi: 10.4018/978-1-60960-621-3.Ch018

Iojă, C. I., Pătroescu, M., Rozylowicz, L., Popescu, V. D., Verghelеţ, M., Zotta, M. I., & Felciuc, M. (2010). The efficacy of Romania's protected areas network in conserving biodiversity. *Biological Conservation*, *143*(11), 2468–2476. doi:10.1016/j.biocon.2010.06.013

Ionescu, L., & Kliewer, N. (2011). Increasing flexibility of airline crew schedules. *Procedia: Social and Behavioral Sciences*, *20*, 1019–1028. doi:10.1016/j.sbspro.2011.08.111

Ishfaq, R., & Sox, C. (2010). Intermodal logistics: The interplay of financial, operational and service issues. *Transportation Research Part E, Logistics and Transportation Review*, *46*(6), 926–949. doi:10.1016/j.tre.2010.02.003

Islam, M. N., & Seneviratne, P. N. (1993). Work-zone traffic management with transportation planning software. *Canadian Journal of Civil Engineering*, *20*(3), 471–479. doi:10.1139/l93-061

ISMEA. (2012). *Prezzi medi mensili per prodotto - Frutta e ortaggi - Mercato all'origine*. Retrieved from http://www.ismea.it/flex/cm/pages/ServeBLOB.php/L/IT/IDPagina/133

Izmir Metropolitan Municipality website. (n. d.). Retrieved from http://asps.unibel.com.tr/anasayfa

Jacobs, T. L., & Warmerdam, J. M. (1994). Simultaneous routing and siting for hazardous-waste operations. *Journal of Urban Planning and Development*, *120*(3), 115–131. doi:10.1061/(ASCE)0733-9488(1994)120:3(115)

Jaillet, P., Song, G., & Yu, G. (1996). Airline network design and hub location problems. *Location Science*, *4*(3), 195–212. doi:10.1016/S0966-8349(96)00016-2

Jain, A. K., & Dubes, R. C. (1988). *Algorithms for clustering data*. Prentice-Hall, Inc.

Jain, A. K., Murty, M. N., & Flynn, P. J. (1999). Data clustering: A review.[CSUR]. *ACM Computing Surveys, 31*(3), 264–323. doi:10.1145/331499.331504

Jakubauskas, G. (2006). Improvement of urban passenger transport ticketing systems by deploying intelligent transport systems. *Transport, 21*(4), 252–259.

Janes, K. R., Yang, S. X., & Hacker, R. R. (2004). Single-component modelling of pig farm odour with statistical methods and neural networks. *Biosystems Engineering, 88*(3), 271–279. doi:10.1016/j.biosystemseng.2004.04.004

Jang, J. S. R., Sun, T. S., & Mizutani, E. (1997). *Neuro-fuzzy and soft computing - A computational approach to learn machine intelligence*. USA: Prentice-Hall.

Jarašūnienė, A., & Batarlienė, N. (2013). Lithuanian road safety solutions based on intelligent transport systems. *Transport, 28*(1), 97–107. doi:10.3846/16484142.2013.782895

Jassbi, J., & Makvandi, P. (2010). Route selection based on soft MODM framework in transportation of hazardous materials. *Applied Mathematical Sciences, 4*(63), 3121–3132.

JayadevaShah, S., Bhaya, A., Kothari, R., & Chandra, S. (2013). Ants find the shortest path: A mathematical proof. *Swarm Intelligence, 7*(1), 43–62. doi:10.1007/s11721-013-0076-9

Jia, H., Zhang, L., Lou, X., & Cao, H. (2011). A fuzzy-stochastic constraint programming model for hazmat road transportation considering terrorism attacking. *Systems Engineering Procedia, 1*, 130–136. doi:10.1016/j.sepro.2011.08.022

Jiang, N., & Xie, C. (2014). Computing and analyzing mixed equilibrium network flows with gasoline and electric vehicles. *Computer-Aided Civil and Infrastructure Engineering, 29*(8), 626–641.

Jin, H. H., Batta, R. J., & Karwan, M. H. (1996). On the analysis of two new models for transporting hazardous materials. *Operations Research, 44*(5), 710–723. doi:10.1287/opre.44.5.710

Johnson-Restrepo, B., Pacheco-Londono, L., & Olivero-Verbel, J. (2003). Molecular parameters responsible for the melting point of 1,2,3-diazaborine compounds. *Journal of Chemical Information and Computer Sciences, 43*(5), 1513–1519. doi:10.1021/ci020387k PMID:14502485

Joly, O. (1999). Recent Overview of Spatial Accessibility Indicators Applied. In *France: 1st Synthesis of the French Research Network Contributions. Anexe 5, Part II: Study Program on European Spatial Development (SPESD)*. France: SDEC.

Juiping, X., Gang, J., & Lei, X. (2013). Hazmat transportation network design model with emergency response under complex fuzzy environment. Hindawi Publishing Cooperation Mathematical Problems in Engineering [Article ID 517372], 2013.

Juliao, R. P. (1999). *Measuring Accessibility Using Gis*. Geo-Computation-99, Retrieved 03.09.2008, from http://www.Geovista.Psu.Edu/Sites/Geocomp99/Gc99/010/ Gc010.Ppt.

Junier, S., & Mostert, E. (2014). A decision support system for the implementation of the Water Framework Directive in the Netherlands: Process, validity and useful information. *Environmental Science & Policy, 40*, 49–56. doi:10.1016/j.envsci.2014.04.004

Kahraman, C., Cebeci, U., & Ruan, D. (2004). Multi-attribute comparison of catering service companies using fuzzy AHP: The case of Turkey. *International Journal of Production Economics, 87*(2), 171–184. doi:10.1016/S0925-5273(03)00099-9

Kajdanowicz, T., Indyk, W., Kazienko, P., & Kukul, J. (2012). Comparison of the Efficiency of MapReduce and Bulk Synchronous Parallel Approaches to Large Network Processing. In *12th IEEE International Conference on Data Mining Workshops*(pp. 218-225). Brussels, Belgium. doi:10.1109/ICDMW.2012.135

Kajdanowicz, T., Kazienko, P., & Indyk, W. (2014). Parallel processing of large graphs. *Future Generation Computer Systems-The International Journal of Grid Computing and Escience*, *32*, 324–337. doi:10.1016/j.future.2013.08.007

Kalelkar, A. S., & Brooks, R. E. (1978). Use of multidimensional utility functions in hazardous shipment decisions. *Accident; Analysis and Prevention*, *10*(3), 251–265. doi:10.1016/0001-4575(78)90016-7

Kang, J. E., Chow, J. Y. J., & Recker, W. W. (2014). On activity-based network design problems. *Transportation Research Part B: Methodological*, *57*, 398–418. doi:10.1016/j.trb.2013.05.009

Kang, Y., Batta, R., & Kwon, C. (2014). Generalized route planning model for hazardous material transportation with VaRand equity considerations. *Computers & Operations Research*, *43*, 237–247. doi:10.1016/j.cor.2013.09.015

Kara, B. Y., Erkut, E., & Verter, V. (2003). Accurate calculation of hazardous materials transport risks. *Operations Research Letters*, *31*(4), 285–292. doi:10.1016/S0167-6377(02)00238-9

Kara, B. Y., & Verter, V. (2004). Designing a road network for hazardous materials transportation. *Transportation Science*, *38*(2), 188–196. doi:10.1287/trsc.1030.0065

Karim, A., & Adeli, H. (2003). Radial Basis Function Neural Network for Work Zone Capacity and Queue Estimation. *Journal of Transportation Engineering*, *129*(5), 494–503. doi:10.1061/(ASCE)0733-947X(2003)129:5(494)

Karkazis, J., & Boffey, T. B. (1995). Optimal location of routes for vehicles transporting hazardous materials. *European Journal of Operational Research*, *86*(2), 201–215. doi:10.1016/0377-2217(94)00099-X

Kazantzi, V., Kazantzis, N., & Gerogiannis, V. C. (2011). Risk informed optimization of a hazardous material multi-periodic transportation model. *Journal of Loss Prevention in the Process Industries*, *24*(6), 767–773. doi:10.1016/j.jlp.2011.05.006

Kermode, R. H., & Myyra, W. A. (1970). Freeway Lane Closure. *Traffic Engineering, 40*(5).

Kheirkhah, A. S., Esmailzadeh, A., & Ghazinoory, S. (2009). Developing strategies to reduce the risk of hazardous materials transportation in Iran using the method of fuzzy swot analysis. *Transport*, *24*(4), 325–332. doi:10.3846/1648-4142.2009.24.325-332

Klibi, W., Martel, A., & Guitouni, A. (2010). The design of robust value-creating supply chain networks: A critical review. *European Journal of Operational Research*, *203*(2), 283–293. doi:10.1016/j.ejor.2009.06.011

Klincewicz, J. G. (1991). Heuristics for the P-hub location problem. *European Journal of Operational Research*, *53*(1), 25–37. doi:10.1016/0377-2217(91)90090-I

Klincewicz, J. G. (1992). Avoiding local optima in the P-hub location problem using tabu search and grasp. *Annals of Operations Research*, *40*(1), 283–302. doi:10.1007/BF02060483

Klincewicz, J. G. (1996). A dual algorithm for the uncapacitated hub location problem. *Location Science*, *4*(3), 173–184. doi:10.1016/S0966-8349(96)00010-1

Klincewicz, J. G. (1998). Hub location in backbone tributary network design: A review. *Location Science*, *6*(1-4), 307–335. doi:10.1016/S0966-8349(98)00042-4

Kohl, N., & Karisch, S. E. (2004). Airline crew rostering: Problem types, modeling, and optimization. *Annals of Operations Research*, *127*(1-4), 223–257. doi:10.1023/B:ANOR.0000019091.54417.ca

Ko, J., & Fox, D. (2009). GP-BayesFilters: Bayesian filtering using Gaussian process prediction and observation models. *Autonomous Robots*, *27*(1), 75–90. doi:10.1007/s10514-009-9119-x

Kostakos, V., Camacho, T., & Mantero, C. (2013). Towards proximity-based passenger sensing on public transport buses. *Personal and Ubiquitous Computing*, *17*(8), 1807–1816. doi:10.1007/s00779-013-0652-4

Krammes, R. A., Dudek, C. L., & Memmott, J. L. (1987). Computer model for evaluating and scheduling freeway work-zone lane closures. *Transportation Research Record*, *1148*, 18–24.

Kratica, J., Milanovi'c, M., Stanimirovi'c, Z., & Tosic, D. (2011). An evolutionary-based approach for solving a capacitated hub location problem. *Applied Soft Computing*, *11*(2), 1858–1866. doi:10.1016/j.asoc.2010.05.035

Kratica, J., Stanimirovic, Z., Tosic, D., & Filipovic, V. (2007). Two genetic algorithms for solving the uncapacitated single allocation p-hub median problem. *European Journal of Operational Research*, *182*(1), 15–28. doi:10.1016/j.ejor.2006.06.056

Krishnamoorthy, M., Mills, G., & Sier, D. (1994). *Strategic configuration of the mail processing network: location–allocation modeling-stage 1* [Technical Report DMS-C94/9]. Division of Mathematics and statistics CSIRO, Australia.

Kristianto, Y., Gunasekaran, A., & Jiao, J. (2014). Logical reconfiguration of reconfigurable manufacturing systems with stream of variations modelling: A stochastic two-stage programming and shortest path model. *International Journal of Production Research*, *52*(5), 1401–1418. doi:10.1080/00207543.2013.843798

Kuby, M. J., & Gray, R. G. (1993). Hub network design problem with stoppers and feeders: Case of federal express. *Transportation Research*, *27*, 1–12.

Kuntay, O. (1976a). Planlama kontrol araci olarak fiziksel erişilebilirlik ve çekim gücü (Planning tool as physical accessibility and traction control.). *Şehircilik Enstitüsü Dergisi* (Urban Development Institute Journal), 33-47.

Kuntay, O. (1976b). *Mekanda Fiziksel Erişilebilirlik ve Çekim Gücü*. (Physical accessibility and gravity in space.) (Unpublished dissertation). Black Sea Technical University, Trabzon, Turkey.

Kuntay, O. (1990). Erişilebilirlik Kesin Bir Öncelik. *Planlama Dergisi*, *90*(7), 1–2.

Kwan, M. P. (1998). Space-time and integral measures of individual accessibility: A comparative analysis using a point-based framework. *Geographical Analysis*, *30*(3), 191–216. doi:10.1111/j.1538-4632.1998.tb00396.x

Kwan, M. P. (2004). GIS Methods in Time-Geographic Research: Geocomputation and Geovisualization of Human Activity Patterns. *Geografiska Annaler. Series B, Human Geography*, *86*(4), 267–280. doi:10.1111/j.0435-3684.2004.00167.x

Labbe, M., Yaman, H., & Gourdin, E. (2005). A branch and cut algorithm for hub location problems with single assignment. *European Journal of Operational Research*, *102*, 371–405.

Lai, V. SWong, B. K., & Cheung, W. (2002). Group decision making in a multiple criteria environment: A case using the AHP in software selection. *European Journal of Operational Research*, *137*(1), 134–144. doi:10.1016/S0377-2217(01)00084-4

Lambert, D., Cooper, M., & Pagh, J. (1998). Supply chain management: Implementation issues and research opportunities. *International Journal of Logistic Management*, *9*(2), 1–20. doi:10.1108/09574099810805807

Lanckriet, G. R. G., Ghaoui, L. E., Bhattacharyya, C., & Jordan, M. I. (2002). A robust minimax approach to classification. *Journal of Machine Learning Research*, *3*, 555–582.

Lassarre, S., Fedra, K., & Weigkricht, E. (1993). Computer-assisted routing of dangerous goods for Haute- Normandie. *Journal of Transportation Engineering*, *119*(2), 200–210. doi:10.1061/(ASCE)0733-947X(1993)119:2(200)

Lathia, N., Capra, L., Magliocchetti, D., De Vigili, F., Conti, G., & De Amicis, R. et al. (2012). Personalizing mobile travel information services. *Procedia: Social and Behavioral Sciences*, *48*, 1195–1204. doi:10.1016/j.sbspro.2012.06.1095

Lauritzen, S. L. (1981). Time series analysis in 1880: A discussion of contributions. *T. N. Thiele. International Statistical Review*, *49*(3), 319–331. doi:10.2307/1402616

Leeming, D. G., & Saccomanno, F. F. (1994). Use of quantified risk assessment in evaluating the risks of transporting chlorine by road and rail. *Transportation Research Record*, *1430*, 27–35.

Lee, Y., Lim, B., & Park, J. (1996). A hub location problem in designing digital data service networks: Lagrangian relaxation approach. *Location Science*, *4*(3), 185–194. doi:10.1016/S0966-8349(96)00009-5

Lenntorp, B. (1976). *Paths in space-time environments: A time geographic study of movement possibilities of individuals*. Lund, Sweden: Gleerup.

Lenntorp, B. (1999). Time-geography-At the end of its beginning. *GeoJournal*, *48*(3), 155–158. doi:10.1023/A:1007067322523

Leonelli, P., Bonvicini, S., & Spadoni, G. (1999). New detailed numerical procedures for calculating risk measures in hazardous materials transportation. *Journal of Loss Prevention in the Process Industries*, *12*(6), 507–515. doi:10.1016/S0950-4230(99)00023-6

Leonelli, P., Bonvicini, S., & Spadoni, G. (2000). Hazardous materials transportation: A risk-analysis- based routing methodology. *Journal of Hazardous Materials*, *71*(1-3), 283–300. doi:10.1016/S0304-3894(99)00084-9 PMID:10677666

Letchford, A. N., Nasiri, S. D., & Oukil, A. (2014). Pricing routines for vehicle routing with time windows on road networks. *Computers & Operations Research*, *51*, 331–337. doi:10.1016/j.cor.2014.06.022

Leveson, N. G., & Turner, C. S. (1993). An investigation of the Therac-25 accidents. *IEEE Computer*, *26*(7), 18–41. doi:10.1109/MC.1993.274940

Liagouras, G. A., Sayegh, A. A., & Koutsakis, P. (2014). A new location-aware calendar-based application for dynamic minimum path trip planning. *Wireless Personal Communications*, *78*(1), 29–44. doi:10.1007/s11277-014-1732-0

Liang, Z., & Chaovalitwongse, W. A. (2009). *The Aircraft Maintenance Routing Problem Optimization and Logistics Challenges in the Enterprise* (pp. 327–348). New York: Springer. doi:10.1007/978-0-387-88617-6_12

Liang, Z., Chaovalitwongse, W. A., Cha, M., & Moon, S. B. (2010). Redundant multicast routing in multilayer networks with shared risk resource groups: Complexity, models and algorithms. *Computers & Operations Research*, *37*(10), 1731–1739. doi:10.1016/j.cor.2009.12.009

Liang, Z., Chaovalitwongse, W. A., Huang, H. C., & Johnson, E. L. (2011). On a new rotation tour network model for aircraft maintenance routing problem. *Transportation Science*, *45*(1), 109–120. doi:10.1287/trsc.1100.0338

Lichtenberg, E., Tra, C., & Hardie, I. (2007). Land use regulation and the provision of open space in suburban residential subdivisions. *Journal of Environmental Economics and Management*, *54*(2), 199–213. doi:10.1016/j.jeem.2007.02.001

Lin, C., Chiang, Y., & Lin, S. (2014). Efficient model and heuristic for the intermodal terminal location problem. *Computers & Operations Research*, *51*, 41–51. doi:10.1016/j.cor.2014.05.004

Li, R., & Leung, Y. (2011). Multi-objective route planning for dangerous goods using compromise programming. *Journal of Geographical Systems*, *13*(3), 249–271. doi:10.1007/s10109-010-0124-6

Li, R., Leung, Y., Huang, B., & Lin, H. (2013). A genetic algorithm for multiobjective dangerous goods route planning. *International Journal of Geographical Information Science*, *27*(6), 1073–1089. doi:10.1080/13658816.2012.737921

List, G. F., Mirchandani, P. B., Turnquist, M. A., & Zografos, K. G. (1991). Modeling and analysis for hazardous materials transportation-risk analysis, routing scheduling and facility location. *Transportation Science*, *25*(2), 100–114. doi:10.1287/trsc.25.2.100

List, G., & Mirchandani, P. (1991). An integrated network planar multiobjective model for routing and siting for hazardous materials and wastes. *Transportation Science*, *25*(2), 146–156. doi:10.1287/trsc.25.2.146

Litman, T. (2011). *Measuring Transportation; Traffic, Mobility and Accessibility*. Victoria Transport Policy Institute.

Litman, T. (2013). The New Transportation Planning Paradigm. *ITE Journal*, *83*(6), 20–28.

Liu, L., Hou, A., Biderman, A., Ratti, C., & Chen, J. (2009, October). Understanding individual and collective mobility patterns from smart card records: A case study in Shenzhen. Proceedings of *Intelligent Transportation Systems 12th International IEEE Conference* (pp. 1-6). IEEE. doi:10.1109/ITSC.2009.5309662

Liu, K. F. R., Liang, H. H., Yeh, K., & Chen, C. W. (2009). A qualitative decision support for environmental impact assessment using fuzzy logic. *Journal of Environmental Informatics*, *13*(2), 93–103. doi:10.3808/jei.200900144

Liu, S., Duffy, A. H. B., Whitfield, R. I., & Boyle, I. M. (2010). Integration of decision support systems to improve decision support performance. *Knowledge and Information Systems*, *22*(3), 261–286. doi:10.1007/s10115-009-0192-4

Liu, X., Saat, M. R., & Barkan, C. P. L. (2013). Integrated risk reduction framework to improve railway hazardous materials transportation safety. *Journal of Hazardous Materials*, *260*, 131–140. PMID:23747471

Li, Y., & Tan, N. (2013). Study on Fleet Assignment Problem Model and Algorithm. *Mathematical Problems in Engineering*, 2013.

Lohatepanont, M., & Barnhart, C. (2004). Airline schedule planning: Integrated models and algorithms for schedule design and fleet assignment. *Transportation Science*, *38*(1), 19–32. doi:10.1287/trsc.1030.0026

López-Atamoros, L. G., Fernandez-Villagómez, G. F., Cruz-Gómez, M. J., & Durán-de-Bazúa, C. (2011). Development of a relative ranking risk ındex for risk assessment in the transportation of liquefied petroleum gas in Mexico City's metropolitan area. *Human and Ecological Risk Assessment*, *17*(6), 1193–1209. doi:10.1080/10807039.2011.618383

Lotfi, S., & Koohsari, M. J. (2009). Measuring objective accessibility to neighborhood facilities in the city (A case study: Zone 6 in Tehran, Iran). *Cities (London, England)*, *26*(3), 133–140. doi:10.1016/j.cities.2009.02.006

Lovett, A. A., Parfitt, J. P., & Brainard, J. S. (1997). Using GIS in risk analysis: A case study of hazardous waste transport. *Risk Analysis*, *17*(5), 625–633. doi:10.1111/j.1539-6924.1997.tb00903.x PMID:9404052

Lozano, A., Muñoz, A., Macías, L., & Antún, J. P. (2011). Hazardous materials transportation in Mexico City: Chlorine and gasoline cases. *Transportation Research Part C, Emerging Technologies*, *19*(5), 779–789. doi:10.1016/j.trc.2010.09.001

Lozanoa, A., Muñoza, A., Antúna, J. P., Granadosa, F., & Guarnerosa, L. (2010). Analysis of hazmat transportation accidents in congested urban areas, based on actual accidents in Mexico. *Procedia: Social and Behavioral Sciences*, *2*(3), 6053–6064. doi:10.1016/j.sbspro.2010.04.018

Luckritz, R. T., & Schneider, A. L. (1980). Decision making in hazardous materials transportation. *Journal of Hazardous Materials*, *4*(2), 129–143. doi:10.1016/0304-3894(80)80029-X

Lüer-Villagra, A., & Marianov, V. (2013). A competitive hub location and pricing problem. *European Journal of Operational Research*, *231*(3), 734–744. doi:10.1016/j.ejor.2013.06.006

Lumsden, K, ., Dallari, F., & Ruggeri, R. (1999). Improving the efficiency of the hub and spoke system for the SKF European distribution network. *International Journal of Physical Distribution & Logistics Management*, *29*(1), 50–66. doi:10.1108/09600039910253878

Luo, W. (2004). Using a GIS-based floating catchment method to assess areas with shortage of physicians. *Health & Place*, *10*(1), 1–11. doi:10.1016/S1353-8292(02)00067-9 PMID:14637284

Luo, W., & Wang, F. (2003). Measures of spatial accessibility to health care in a GIS environment: Synthesis and a case study in the Chicago region. *Environment and Planning. B, Planning & Design*, *30*(6), 865–884. doi:10.1068/b29120

MacFarlane, R. (2005). A guide to GIS applications in integrated emergency management. *Emergency planning College.*

Machuca, E., Mandow, L., de la Cruz, J. L. P., & Ruiz-Sepulveda, A. (2012). A comparison of heuristic best-first algorithms for bicriterion shortest path problems. *European Journal of Operational Research*, *217*(1), 44–53. doi:10.1016/j.ejor.2011.08.030

MacKinnon, R. (2011). *Let's take back the Internet!* TEDglobal 2011, filmed July 2011. Retrieved from http:// www.ted.com/talks/rebecca_mackinnon_let_s_take_back_ the_internet.html

Mahendran, P., Mohorty, M. B. K., & Saravanan, S. (2014). A fuzzy AHP approach for selection of measuring instrument for engineering college selection. *Applied Mathematical Sciences*, *8*, 2149–2161.

Mahmoudabadi, A., & Seyedhosseini, S. M. (2014). Developing a chaotic pattern of dynamic hazmat routing problem. *IATSS Research*, *37*, 110–118.

Majhi, S. K., & Chaupattanayak, S. (2013). Some interesting and novel results from analysis of shortest path searching algorithms. *International Journal of Advances in Computer Science and Cloud Computing*, *1*(1), 57–62.

Makrí, M. B. (2001). *Accessibility Indices. A Tool for Comprehensive Land-Use Planning*. Retrieved from [REMOVED HYPERLINK FIELD]http://www.infra.kth.se/tlenet/meet5/papers/Makri.pdf

Makri, M. C., & Folkesson, C. (1999). *Accessibility Measures for Analyses of Land-Use and Travelling with Geographical Information Systems.* Retrieved from http://www.trafikdage.dk/td/papers/papers99/papers/paper/bpot/makri/ makri.pdf

Makrí, M. B. (2002). Accessibility indices and planning theory.*Proceedings of Eighth International Conference on Urban Transport and the Environment for the 21st Century, Urban Transport VIII.* Seville, UK. Witt Press.

Malmquist, S. (1953). Index numbers and indifference surfaces. *Trabajos de Estadistica*, *4*(2), 209–242. doi:10.1007/BF03006863

Manzini, R. (2012). A top-down approach and a decision support system for the design and management of logistic networks. *Transportation Research Part E, Logistics and Transportation Review*, *48*(6), 1185–1204. doi:10.1016/j.tre.2012.06.004

Manzini, R., & Accorsi, R. (2013). The new conceptual framework for food supply chain assessment. *Journal of Food Engineering*, *115*(2), 251–263. doi:10.1016/j.jfoodeng.2012.10.026

Manzini, R., Accorsi, R., & Bortolini, M. (2014). Operational planning models for distribution networks. *International Journal of Production Research*, *52*(1), 89–116. doi:10.1080/00207543.2013.828168

Manzini, R., Gamberi, M., Gebennini, E., & Regattieri, A. (2008). An integrated approach to the design and management of a supply chain system. *International Journal of Advanced Manufacturing Technology*, *37*(5-6), 625–640. doi:10.1007/s00170-007-0997-9

Marakas, G. M. (2003). *Decision Support Systems* (2nd ed.). Upper Saddle River, NJ: Prentice-Hall.

Marcucci, D. J., & Jordan, L. M. (2013). Benefits and challenges of linking green infrastructure and highway planning in the United States. *Environmental Management*, *51*(1), 182–197. doi:10.1007/s00267-012-9966-7 PMID:23135391

Marianov, V., & Serra, D. (2003). Location models for airline hubs behaving as M/D/c queues. *Computers & Operations Research*, *30*(7), 983–1003. doi:10.1016/S0305-0548(02)00052-7

Marín, A. (2005). Formulating and solving splittable capacitated multiple allocation hub location problems. *Computers & Operations Research*, *32*(12), 3093–3109. doi:10.1016/j.cor.2004.04.008

Marquand, A., Howard, M., Brammer, M., Chu, C., Coen, S., & Mourão Miranda, J. (2010). Quantitative prediction of subjective pain intensity from whole-brain fMRI data using Gaussian processes.[PubMed]. *NeuroImage*, *49*(3), 2178–2189. doi:10.1016/j.neuroimage.2009.10.072

Marques da Costa, E., Marques da Costa, N., Abrantes, P., & Louro, A. (2013). ICT *Tools to enhance the utility of a spatial decision support system-methodological essays*. Proceedings of AESOP-ACSP Joint Congress. Dublin.

Martinelli, R., Pecin, D., & Poggi, M. (2014). Efficient elementary and restricted non-elementary route pricing. *European Journal of Operational Research*, *239*(1), 102–111. doi:10.1016/j.ejor.2014.05.005

Martinez-Alegria, R., Ordonez, C., & Taboada, J. (2003). A conceptual model for analyzing the risks involved in the transportation of hazardous goods: Implementation in a geographic information system. *Human and Ecological Risk Assessment*, *9*(3), 857–873. doi:10.1080/713609970

Martín, J. C., & Roman, C. (2003). Hub location in the South-Atlantic airline market. *Transportation Research Part A, Policy and Practice*, *37*(10), 865–888. doi:10.1016/S0965-8564(03)00060-0

Mase, H., Sakamato, M., & Sakai, T. (1995). Neural network for stability analysis of rubble mound breakwaters. *Journal of Waterway, Port, Coastal, and Ocean Engineering*, *121*(6), 294–299. doi:10.1061/(ASCE)0733-950X(1995)121:6(294)

Masson, R., Ropke, S., Lehuede, F., & Peton, O. (2014). A branch-and-cut-and-price approach for the pickup and delivery problem with shuttle routes. *European Journal of Operational Research*, *236*(3), 849–862. doi:10.1016/j.ejor.2013.08.042

Matsatsinis, N. F., Ipsilantis, P., & Samaras, G. (2008). Special issue on Multicriteria Decision Support Systems. *Operations Research*, *8*(1), 3–4. doi:10.1007/s12351-008-0002-x

Matthew, R., McGrail, J., & Humphreys, S. (2009). Measuring spatial accessibility to primary care in rural areas: Improving the effectiveness of the two-step floating catchment area method. *Applied Geography (Sevenoaks, England)*, *29*(4), 533–541. doi:10.1016/j.apgeog.2008.12.003

Matzoros, A. (2002). Decision support systems for public transport management: The Athens public transport authority project. *Transportation Planning and Technology*, *25*(3), 215–237. doi:10.1080/0308106022000018973

Ma, Y., Wang, H. W., Xie, Y., & Guo, M. (2014). Path planning for multiple mobile robots under double-warehouse. *Information Sciences*, *278*, 357–379. doi:10.1016/j.ins.2014.03.058

Mayer, G., & Wagner, B. (2002). Hub Locator: An exact solution method for the multiple allocation hub location problem. *Computers & Operations Research*, *29*(6), 715–739. doi:10.1016/S0305-0548(01)00080-6

McBrierty, C. (2012). M-Ticketing, Arriva Buses, UK. *Eltis.org*. Retrieved from http://www.eltis.org/discover/case-studies/m-ticketing-arriva-buses-uk[REMOVED HYPERLINK FIELD]

Medlock, L. S. (2013). Mobile phones and society- how being constantly connected impacts our lives. *The Mobile Issue*, 29. Retrieved from http://source.southuniversity.edu/mobile-phones-and-society-how-being-constantly-connected-impacts-our-lives-137313.aspx

Mehrotra, K., Mohan, C. K., & Ranka, S. (1996). *Elements of artificial neural networks*. Cambridge: MIT Press.

Melchers, R. E. (1999). *Structural reliability analysis and prediction* (2nd ed., pp. 90–110). New York, USA: John Wiley& Sons.

Melo, M. T., Nickel, S., & Saldanha-da-Gama, F. (2009). Facility location and supply chain management – A review. *European Journal of Operational Research*, *196*(2), 401–412. doi:10.1016/j.ejor.2008.05.007

Memmott, J. L., & Dudek, C. L. (1984). Queue and user cost evaluation of work zones (QUEWZ). *Transportation Research Record*, *979*, 12–19.

Mercier, A., & Soumis, F. (2007). An integrated aircraft routing, crew scheduling and flight retiming model. *Computers & Operations Research*, *34*(8), 2251–2265. doi:10.1016/j.cor.2005.09.001

Messac, A., Ismail-Yahaya, A., & Mattson, C. A. (2003). The normalized normal constraint method for generating the Pareto frontier. *Structural and Multidisciplinary Optimization*, *25*(2), 86–98. doi:10.1007/s00158-002-0276-1

Messina, J. P., Shortridge, A. M., Groop, R. E., Varnakovida, P., & Finn, M. J. (2006). Evaluating Michigan's community hospital access: Spatial methods for decision support. *International Journal of Health Geographics*, *5*(1), 42. doi:10.1186/1476-072X-5-42 PMID:16995948

Messina, J., Shortridge, A. M., Groop, R. E., Varnakovida, P., & Finn, M. J. (2006). Evaluating michigan's community hospital access: Spatial methods for decision support. *International Journal of Health Geographics*, *5*(42), 1–18. PMID:16995948

Metting, H. J., van Zomeren, P. V., van der Ley, C. P., Coenegracht, P. M. J., & de Jong, G. J. (2000). Comparison of migration modeling in micellar electro kinetic chromatography by linear regression and by use of an artificial neural network. *Chromatographia*, *52*(9-10), 607–613. doi:10.1007/BF02789760

Miaou, S. P., & Chin, S. M. (1991). Computing k-shortest path for nuclear spent fuel highway transportation. *European Journal of Operational Research*, *53*(1), 64–80. doi:10.1016/0377-2217(91)90093-B

Michael, F. L., Noor, Z. Z., Zardari, N. H., & Meza, M. J. F. (2013). Analytical hierarchy process application in urban sustainability indicators prioritization. *Resources and Environment*, *3*, 1–5.

Michael, G., & Bell, H. (2007). Mixed routing strategies for hazardous materials: Decision-making under complete uncertainty. *International Journal of Sustainable Transportation*, *1*(2), 133–142. doi:10.1080/15568310601092013

Miche, Y., Sorjamaa, A., Bas, P., Simula, O., Jutten, C., & Lendasse, A. (2010). OP-ELM: Optimally pruned extreme learning machine.[PubMed]. *IEEE Transactions on Neural Networks*, *21*(1), 158–162. doi:10.1109/TNN.2009.2036259

Milazzo, M. F., Lisi, R., Maschio, G., Antonioni, G., Bonvicini, S., & Spadoni, G. (2002). Hazmat transport through, Messina town: From risk analysis suggestions for improving territorial safety. *Journal of Loss Prevention in the Process Industries*, *15*(5), 347–356. doi:10.1016/S0950-4230(02)00028-1

Milazzo, M. F., Lisi, R., Maschio, G., Antonioni, G., & Spadoni, G. (2010). A study of land transport of dangerous substances in Eastern Sicily. *Journal of Loss Prevention in the Process Industries*, *23*(3), 393–403. doi:10.1016/j.jlp.2010.01.007

Miller, E. (1987). The Therac-25 Experience.*Proceedings of Conference State Radiation Control Program Directors*. Washington, DC: Division of Engineering Safety, Office of Nuclear Regulatory Research, U.S. Nuclear Regulatory Commission.

Miller, H. J. (2003). What about People in Geographic Information Science. *Computers, Environment and Urban Systems*, *27*(5), 447–453. doi:10.1016/S0198-9715(03)00059-0

Miller-Hooks, E. D., & Mahmassani, H. S. (1998). Optimal routing of hazardous materials in stochastic, time-varying transportation networks. *Transportation Research Record, 1645*, 143–151. doi:10.3141/1645-18

Milligan, G. W. (1980). An examination of the effect of six types of error perturbation on fifteen clustering algorithms. *Psychometrika, 45*(3), 325–342. doi:10.1007/BF02293907

Mills, G. S., & Neuhauser, K. S. (1998). Urban risks of truck transport of radioactive material. *Risk Analysis, 18*(6), 781–785. doi:10.1111/j.1539-6924.1998.tb01120.x

Minor, E. S., & Lookingbill, T. R. (2010). A Multiscale Network Analysis of Protected-Area Connectivity for Mammals in the United States. *Conservation Biology, 24*(6), 1549–1558. doi:10.1111/j.1523-1739.2010.01558.x PMID:20666801

Mitchel, L., Gary, H., Jonathan, R., & Sean, W. (2008). Urban Population Distribution Models and Service Accessibility Estimation. *Computers, Environment and Urban Systems, 32*(1), 66–80. doi:10.1016/j.compenvurbsys.2007.06.001

Mitchell, M. (1998). *An introduction to genetic algorithms*. Cambridge, MA: MIT Press.

Mitchell, W. J., & Casalegno, F. (2008). *Connected sustainable cities*. Boston: MIT Mobile Experience Lab Publishing.

Mitra, G., & Valente, P. (2007). The evolution of Web-based optimisation from ASP to e-Services. *Decision Support Systems, 43*(4), 1096–1116. doi:10.1016/j.dss.2005.07.003

Mitroff, I. I., & Linstone, H. A. (1993). *The unbounded mind: Breaking the chains of traditional business thinking*. New York: Oxford Univ. Press.

Mobile technologies – The digital fabric of our lives. (2013). Vodafone Institute for Society and Communications, Retrieved from http://www.vodafoneinstitut.de/uploads/media/Studies_komplett_FINAL_von_250913.pdf

Mobility, Access and Transportation. (1997). Bureau of Transportation Statistics. Retrieved from http://www.bts.gov/transtu/tsar/tsar97/chap06.pdf

Mohammadi, M., Torabi, S. A., & Tavakkoli-Moghaddam, R. (2014). Sustainable hub location under mixed uncertainty. *Transportation Research Part E, Logistics and Transportation Review, 62*, 89–115. doi:10.1016/j.tre.2013.12.005

Moncayo-Martínez, L. A., & Zhang, D. Z. (2011). Multi-objective ant colony optimisation: A meta-heuristic approach to supply chain design. *International Journal of Production Economics, 131*(1), 407–420. doi:10.1016/j.ijpe.2010.11.026

Monprapussorn, S., Watts, D. J., & Banomyong, R. (2009). Sustainable hazardous materials route planning with environmental consideration. *Asian Journal on Energy and Environment, 10*(02), 122–132.

Montgomery, D. C., & Runger, G. C. (1999). *Applied statistics and probability for engineers*. New York, USA: John Wiley & Sons.

Moore, J. E., Sandquist, G. M., & Slaughter, D. M. (1995). A route-specific system for risk assessment of radioactive materials transportation accidents. *Nuclear Technology, 112*(1), 63–78.

Mounce, R., Hollier, G., Smith, M., Hodge, V. J., Jackson, T., & Austin, J. (2013). A metric for pattern-matching applications to traffic management. *Transportation Research Part C, Emerging Technologies, 29*, 148–155. doi:10.1016/j.trc.2012.04.019

Murad, A. A. (2004). Creating a GIS application for local health care planning in Saudi Arabia. *International Journal of Environmental Health Research, 14*(3), 185–199. doi:10.1080/0960312042000218606 PMID:15203450

Murat, Y. S., Cakici, Z., & Yaslan, G. (2014). Use of Fuzzy Logic Traffic Signal Control Approach as Dual Lane Ramp Metering Model for Freeways. *Advances in Intelligent Systems and Computing*, *223*, 339–349. doi:10.1007/978-3-319-00930-8_30

Muter, I., Cordeau, J.-F., & Laporte, G. A. (2014). Branch-and-price algorithm for the multidepot vehicle routing problem with interdepot routes. *Transportation Science*, *48*(3), 425–441. doi:10.1287/trsc.2013.0489

Nacu, A. (2013). National roads of Romania (vectorial data set) [Data file]. Retrieved from http://www.geo-spatial.org/download/re-eaua-rutiera-na-ionala-a-romaniei-set-de-date-vectorial

Nallaivarothayan, H., Ryan, D., Denman, S., Sridharan, S., Fookes, C., & Rakotonirainy, A. (2013). Detecting anomalous events at railway level crossings. *Proceedings of the Institution of Mechanical Engineers. Part F, Journal of Rail and Rapid Transit*, *227*(5), 539–553. doi:10.1177/0954409713501296

Nam, T., & Pardo, T. A. (2011). Conceptualizing smart city with dimensions of technology, people, and institutions. *Proceedings of the 12th Annual International Digital Government Research Conference: Digital Government Innovation in Challenging Times*. New York, NY: ACM. doi:10.1145/2037556.2037602

Nardini, L., Aparicio, L., Bandoni, A., & Tonelli, S. M. (2003). Regional risk associated with the transport of hazardous materials. *Latin American Applied Research*, *33*(3), 213–218.

Nembhard, D. A., & White, C. C. III. (1997). Applications of non-order-preserving path selection to hazmat routing. *Transportation Science*, *31*(3), 262–271. doi:10.1287/trsc.31.3.262

Neutens, T., Witlox, F., & Demaeyer, P. (2007). Individual accessibility and travel possibilities: A literature review on time geography. *European Journal of Transport and Infrastructure Research*, *7*(4), 335–352.

Nextstopdesign (n. d.). Retrieved from http://www.nextstopdesign.com/welcome

Nguyen-Tuong, D., Seeger, M., & Peters, J. (2009). Model learning with local Gaussian process regression. *Advanced Robotics*, *23*(15), 2015–2034. doi:10.1163/016918609X12529286896877

Nikolova, E., & Stier-Moses, N. E. (2014). A mean-risk model for the traffic assignment problem with stochastic travel times. *Operations Research*, *62*(2), 366–382. doi:10.1287/opre.2013.1246

Nishi, T., & Maeno, R. (2010). Petri net decomposition approach to optimization of route planning problems for AGV systems. *IEEE Transactions on Automation Science and Engineering*, *7*(3), 523–537. doi:10.1109/TASE.2010.2043096

NiSullivan, B., Tang, Y., & Marquis, C. (2014). Persistently learning: How small-world network imprints affect subsequent firm learning. *Strategic Organization*, *12*(3), 180–199. doi:10.1177/1476127014543772

Niă, M. R. (2008). Analyze model for the ecological footprint of new residential spaces in the Bucharest Metropolitan Area. *Forum Geografic,* (7), 8, 200-207.

Niă, M. R. (2011). *Dinamica rezidenialului în zona metropolitană a municipiului Bucureti i proiecia ei în starea mediului.* (PH.D. Thesis). University of Bucharest.

Nocera, S., Maino, F., & Cavallaro, F. (2012). A heuristic method for determining CO2 efficiency in transportation planning. *European Transport Research Review*, *4*(2), 91–106. doi:10.1007/s12544-012-0073-x

Nwagboso, C. O. (1997). Advanced vehicle and infrastructure systems. Computer application, control and automation. Chapter 1. Introduction to intelligent transport systems. UK: Wiley (John) & Sons.

O'Kelly, M. E. (1986). The location of interacting hub facilities. *Transportation Science*, *20*(2), 92–106. doi:10.1287/trsc.20.2.92

O'Kelly, M. E. (1987). A quadratic integer program for the location of interacting hub facilities. *European Journal of Operational Research*, *32*(3), 393–404. doi:10.1016/S0377-2217(87)80007-3

O'Kelly, M. E. (1992). Hub facility location with fixed costs. *Papers in Regional Science*, *71*, 292–306.

O'Kelly, M. E. (2014). Air freight hubs in the FedEx system: Analysis of fuel use. *Journal of Air Transport Management*, *36*, 1–12. doi:10.1016/j.jairtraman.2013.12.002

Ocalir, E. V., Ercoskun, O. Y., & Tur, R. (2010). An integrated model of GIS and fuzzy logic (FMOTS) for location decisions of taxicab stands. *Expert Systems with Applications*, *37*(7), 4892–4901. doi:10.1016/j.eswa.2009.12.026

Ogle, R. A., Morrison, D. R., & Viz, M. J. (2005). Emergency response to a noncollision hazmat release from a railcar. *Process Safety Progress*, *24*(2), 81–85. doi:10.1002/prs.10065

Oktal, H., & Ozger, A. (2013). Hub location in air cargo transportation: A case study. *Journal of Air Transport Management*, *27*, 1–4. doi:10.1016/j.jairtraman.2012.10.009

Olivares-Benitez, E., Ríos-Mercado, R. Z., & González-Velarde, J. L. (2013). A metaheuristic algorithm to solve the selection of transportation channels in supply chain design. *International Journal of Production Economics*, *145*(1), 161–172. doi:10.1016/j.ijpe.2013.01.017

Omurbek, N., Ustundag, S., & Helvacioglu, O. C. (2013). Use of analytical hierarchy process (AHP) in location decision: A study in Isparta Region. *Journal of Administrative Sciences*, *21*, 101–116.

Ontario Protected Areas Planning Manual. (2009). Ontario Ministry of Natural Resources. Peterborough: Queen's Printer for Ontario. Retrieved from http://www.ontarioparks.com/english/planning_pdf/papm/papm_approved.pdf

Ostresh, L. M. J. (1975). An efficient algorithm for solving the two center location – allocation problem. *Journal of Regional Science*, *15*(2), 209–216. doi:10.1111/j.1467-9787.1975.tb00921.x

O'Sullivan, D., Morrison, A., & Shearer, J. (2000). Using desktop GIS for the investigation of accessibility by public transport: An isochrone approach. *International Journal of Geographical Information Science*, *14*(1), 85–104. doi:10.1080/136588100240976

Osvald, A., & Stirn, L. Z. (2008). A vehicle routing algorithm for the distribution of fresh vegetables and similar perishable food. *Journal of Food Engineering*, *85*(2), 285–295. doi:10.1016/j.jfoodeng.2007.07.008

Ozdemirel, A., Gokgur, B., & Eliiyi, D. T. (2012). An assignment and routing problem with time windows and capacity restriction. 15th Meeting of the Euro Working Group on Transportation. *Procedia: Social and Behavioral Sciences*, *54*, 149–158. doi:10.1016/j.sbspro.2012.09.734

Ozus, E., Akın, D., & Çiftçi, M. (2012). Hierarchical cluster analysis of multicenter development and travel patterns in Istanbul. *Journal of Urban Planning and Development*, *138*(4), 303–318. doi:10.1061/(ASCE)UP.1943-5444.0000121

Pallottino, S., & Scutellà, M. G. (1997). Dual algorithms for the shortest path tree problem. *Networks*, *29*, 125–133. doi:10.1002/(SICI)1097-0037(199703)29:2<125::AID-NET7>3.0.CO;2-L

Paltrinieri, N., Landucci, G., Molag, M., Bonvicini, S., & Spadonia, G., & Cozzani, V. (2009). Risk reduction in road and rail LPG transportation by passive fire protection. *Journal of Hazardous Materials*, *167*(1-3), 332–344. doi:10.1016/j.jhazmat.2008.12.122 PMID:19188020

Pamuk, F. S., & Sepil, C. (2001). A solution to the hub center problem via a single-relocation algorithm with Tabu search. *IIE Transactions*, *33*(5), 399–411. doi:10.1080/07408170108936838

Pan, F., Zhang, L., Lu, J., Zhao, J. J., & Wang, F. (2013). A method for determining the number of traffic conflict points between vehicles at major-minor highway intersections. *Traffic Injury Prevention*, *14*(4), 424–433. doi:10.1080/15389588.2012.713148 PMID:23531266

Parentela, E. M., & Sathisan, S. K. (1995). *GIS-based allocation of emergency response units along a transportation route. In 1995*. San Diego: ESRI User Conference.

Pascual-Hortal, L., & Saura, S. (2006). Comparison and development of new graph-based landscape connectivity indices: Towards the priorization of habitat patches and corridors for conservation. *Landscape Ecology*, *21*(7), 959–967. doi:10.1007/s10980-006-0013-z

Patel, M. H., & Horowitz, A. J. (1994). Optimal routing of hazardous materials considering risk of spill. *Transportation Research Part A, Policy and Practice*, *28*(2), 119–132. doi:10.1016/0965-8564(94)90033-7

Pătroescu, M. Niă, M. R., Iojă, I. C., & Vânău, G. O. (2009). The ecological footprint-indicator for analysing the environmental impact of residential surfaces in metropolitan areas [Case study: Bucharest Metropolitan Area]. *Strategies, Concepts and Technologies for Planning the Urban Future Conference, CITIES 3.0*

Peres, Y., Sotnikov, D., Sudakov, B., & Zwick, U. (2013). All-pairs shortest paths in $O(n^2)$ time with high probability. *Journal of the ACM*, *60*(4), 1-25.

Pesic, D. J., Blagojevic, M. D., & Glisovic, S. M. (2011). The model of air pollution generated by fire chemical accident in an urban street canyon. *Transportation Research Part D, Transport and Environment*, *16*(4), 321–326. doi:10.1016/j.trd.2011.01.012

Pet-Armacost, J. J., Sepulveda, J., & Sakude, M. (1999). Monte Carlo sensitivity analysis of unknown parameters in hazardous materials transportation risk assessment. *Risk Analysis*, *19*(6), 1173–1184. doi:10.1111/j.1539-6924.1999.tb01136.x PMID:10765455

Petersen, M. S., Sessa, C., Enei, R., Ulied, A., Larrea, E., Obisco, O., et al. (2009). *Report on Transport Scenarios with a 20 and 40 year Horizon*. DG TREN. Retrieved from http://ec.europa.eu/transport/themes/strategies/studies/doc/future_of_transport/ 2009_02_transvisions_report.pdf

Peters, J., & Hall, G. B. (1999). Assessment of ambulance response performance using a geographic information system. *Social Science & Medicine*, *49*(11), 1551–1566. doi:10.1016/S0277-9536(99)00248-8 PMID:10515636

Philipson, L. L., & Napadensky, H. S. (1982). The methodologies of hazardous materials transportation risk assessment. *Journal of Hazardous Materials*, *6*(4), 361–382. doi:10.1016/0304-3894(82)85003-6

Piantanakulchai, M., & Saengkhao, N. (2003). Evaluation of alternatives in transportation planning using multi-stakeholders multi-objectives AHP modeling. *Proceedings of the Eastern Asia Society for Transportation Studies*, *4*, 1613–1628.

Pijawka, K. D., Foote, S., Soesilo, A., Scanlon, R. D., & Cantilli, E. J. (1985). *Improving transportation of hazardous materials through risk assessment and routing. National Research Council*. Washington, DC, USA: Transportation Research Board.

Pillonetto, G., Chiuso, A., & De Nicolao, G. (2011). Prediction error identification of linear systems: A nonparametric Gaussian regression approach. *Automatica*, *47*(2), 291–305. doi:10.1016/j.automatica.2010.11.004

Pimentel, D. (2006). *Impacts of organic farming on the efficiency of energy use in agriculture*. Itaca: University, Organic Center - Cornell.

Pine, J. C., & Marx, B. D. (1997). Utilizing state hazardous materials transportation data in hazardous analysis. *Journal of Hazardous Materials*, *54*(1–2), 113–122. doi:10.1016/S0304-3894(96)01871-7

Pinto, R. (2012). Stock rationing under service level constraints in a vertically integrated distribution system. *International Journal of Production Economics*, *136*(1), 231–240. doi:10.1016/j.ijpe.2011.11.023

Pinzon, L. (2013). *Beyond connectivity: The impacts of social media in urban development in Puerto Ayora, Ecuador* [DPU Working Paper No. 160]. London: The Bartlett Development Planning Unit.

Piquer-Rodríguez, M., Kuemmerle, T., Alcaraz-Segura, D., Zurita-Milla, R., & Cabello, J. (2012). Future land use effects on the connectivity of protected area networks in southeastern Spain. *Journal for Nature Conservation*, *20*(6), 326–336. doi:10.1016/j.jnc.2012.07.001

Pirie, G. H. (1979). Measuring accessibility: A review and proposal. *Environment & Planning A*, *11*(3), 299–312. doi:10.1068/a110299

Pirkul, H., & Schilling, D. A. (1998). An efficient procedure for designing single allocation hub and spoke systems. *Management Science*, *44*(12 Part 2), s235–s242. doi:10.1287/mnsc.44.12.S235

Pishvaee, M. S., Farahani, R. Z., & Dullaert, W. (2010). A memetic algorithm for bi-objective integrated forward/reverse logistics network design. *Computers & Operations Research*, *37*(6), 1100–1112. doi:10.1016/j.cor.2009.09.018

Planas, E., Pastor, E., Presutto, F., & Tixier, J. (2008). Results of the MITRA project: Monitoring and intervention for the transportation of dangerous goods. *Journal of Hazardous Materials*, *152*(2), 516–526. doi:10.1016/j.jhazmat.2007.07.032 PMID:17884284

Planken, L., de Weerdt, M., & Van der Krogt, R. (2012). Computing all-pairs shortest paths by leveraging low treewidth. *Journal of Artificial Intelligence Research*, *43*, 353–388.

Pogarcic, I., Francic, M., & Davidovic, V. (2008). Application of AHP method in traffic planning. Proceedings of the *16th International Symposium on Electronics in Transport 2008.*

Pokharel, S. (2008). A two objective model for decision making in a supply chain. *International Journal of Production Economics*, *111*(2), 378–388. doi:10.1016/j.ijpe.2007.01.006

Possingham, H., Ball, I., & Andelman, S. (2000). Mathematical methods for identifying representative reserve netorks. In S. Ferson & M. Burgman (Eds.), *Quantitative methods for conservation biology* (pp. 291–305). New York: Springer-Verlag. doi:10.1007/0-387-22648-6_17

Power, D. J. (Ed.) (2013). *Engineering Effective Decision Support Technologies: New Models and Applications*. Hershey, PA: IGI Global. doi:10.4018/978-1-4666-4002-3

Pradhananga, R., Taniguchi, E., & Yamada, T. (2010). Ant colony system based routing and scheduling for hazardous material transportation. *Procedia: Social and Behavioral Sciences*, *2*(3), 6097–6108. doi:10.1016/j.sbspro.2010.04.022

Pradhananga, R., Taniguchi, E., Yamada, T., & Qureshi, A. G. (2014). Bi-objective decision support system for routing and scheduling of hazardous materials. *Socio-Economic Planning Sciences*, *48*(2), 1–14. doi:10.1016/j.seps.2014.02.003

Pratikakis, N. E., Realff, M. J., & Lee, J. H. (2009). Controlled exploration of state space in off-line ADP and its application to stochastic shortest path problems. *Computers & Chemical Engineering*, *33*(12), 2111–2122. doi:10.1016/j.compchemeng.2009.06.012

Pressey, R. L., Watts, M., Ridges, M., & Barrett, T. (2005). *C-Plan conservation planning software. User Manual.* NSW Department of Environment and Conservation Retrieved from http://www.edg.org.au/software/cplan/C-Plan_Manual_2005.pdf

Primack, R., Pătroescu, M., Rozylowicz, L., & Iojă, I. C. (2008). *Fundamentele conservării diversităţii biologice*. Bucureşti: Editura AGIR.

Public Transport Victoria. (2014). *Route timetables. Route 445 - Werribee Station - Hoppers Crossing Station via Werribee Plaza*. Retrieved from http://ptv.vic.gov.au/route/view/8266

Puerto, J., Ramos, A. B., & Rodríguez-Chía, A. M. (2013). A specialized branch & bound & cut for single-allocation ordered median hub location problems. *Discrete Applied Mathematics*, *161*(16-17), 2624–2646. doi:10.1016/j.dam.2013.05.035

Purdy, G. (1993). Risk analysis of the transportation of dangerous goods by road and rail. *Journal of Hazardous Materials*, *33*(2), 229–259. doi:10.1016/0304-3894(93)85056-K

Qiao, Y., Keren, N., & Mannan, M. S. (2009). Utilization of accident databases and fuzzy sets to estimate frequency of hazmat transport accidents. *Journal of Hazardous Materials*, *167*(1-3), 374–382. doi:10.1016/j.jhazmat.2009.01.097 PMID:19250750

Qiu, Y., & Yagan, O. (2014). Robustness in clustering-based weighted inter-connected networks. *The European Physical Journal B*, *87*(4), 89. doi:10.1140/epjb/e2014-40466-1

Quiñonero Candela, J., & Rasmussen, C. E. (2005). A unifying view of sparse approximate Gaussian process regression. *Journal of Machine Learning Research*, *6*, 1939–1959.

Radke, J., & Mu, L. (2000). Spatial decompositions, modeling and mapping service regions to predict access to social programs. *Geographic Information Sciences*, *6*(2), 105–112.

Rajabalipour Cheshmehgaz, H., Desa, M. I., & Wibowo, A. (2013). A flexible three-level logistic network design considering cost and time criteria with a multi-objective evolutionary algorithm. *Journal of Intelligent Manufacturing*, *24*(2), 277–293. doi:10.1007/s10845-011-0584-7

Raj, P. K., & Pritchard, E. W. (2000). Hazardous materials transportation on US railroads: Application of risk analysis methods to decision making in development of regulations. *Transportation Research Record*, *1707*, 22–26. doi:10.3141/1707-03

Ramaa, A., Rangaswamy, T. M., & Subramanya, K. N. (2009). A Review of literature on performance measurement of supply chain network.*Proceedings of the 2009 Second International Conference on Emerging Trends in Engineering & Technology* (pp. 802–807). IEEE. doi:10.1109/ICETET.2009.18

Ramudhin, A., Chaabane, A., & Paquet, M. (2009). On the design of sustainable green supply chains. In *Proceedings of the 2009 International Conference on Computers & Industrial Engineering* (pp. 979–984). Troyes: IEEE. doi:10.1109/ICCIE.2009.5223686

Rasmussen, M. S., Lusby, R. M., Ryan, D., & Larsen, J. (2011). *Subsequence Generation for the Airline Crew Pairing Problem [Technical Report]*. DTU Management.

Ray, S. C. (2004). *Data Envelopment Analysis: Theory and Techniques for Economics and Operations Research*. UK: Cambridge University Press. doi:10.1017/CBO9780511606731

Razi, M. A., & Athappilly, K. (2005). A comparative predictive analysis of neural networks (NNs), nonlinear regression and classification and regression tree (CART) models. *Expert Systems with Applications*, *29*(1), 65–74. doi:10.1016/j.eswa.2005.01.006

Redekopp, M., Simmhan, Y., & Prasanna, V. K. (2013, May 20-24). Optimizations and Analysis of BSP Graph Processing Models on Public Clouds. Proceedings of *IEEE 27th International Parallel and Distributed Processing Symposium*. Boston.

Redlands Institute website. (n. d.). University of Redlands. Retrieved from http://www.spatial.redlands.edu/redlandsinstitute/giscience/sds/

Redmill, F., & Rajan, J. (1997). *Human factors in safety-critical systems*. Newton, MA: Butterworth-Heinemann.

Regan, M. A., Oxley, J. A., Godley, S. T., & Tingvall, C. (2001). *Intelligent transport systems: Safety and human factors issues*. Royal Automobile Club of Victoria.

Regattieri, A., Gamberi, M., & Manzini, R. (2007). Traceability of food products: General framework and experimental evidence. *Journal of Food Engineering*, *81*(2), 347–356. doi:10.1016/j.jfoodeng.2006.10.032

Reilly, A., Nozick, L., Xu, N., & Jones, D. (2012). Game theory-based identification of facility use restrictions for the movement of hazardous materials under terrorist threat. *Transportation Research Part E, Logistics and Transportation Review*, *48*(1), 115–131. doi:10.1016/j.tre.2011.06.002

Ren, C., Yuan, X., Wang, J., Zhang, X., & Li, J. (2012). Study on emergency response rank mode of flammable and explosive hazardous materials road transportation. *Procedia Engineering*, *45*, 830–835. doi:10.1016/j.proeng.2012.08.246

Reniers, G. L. L., & Dullaert, W. (2013). A method to assess multi-modal hazmat transport security vulnerabilities: Hazmat transport SVA. *Transport Policy*, *28*, 103–113. doi:10.1016/j.tranpol.2012.05.002

Reniers, G. L. L., Jongh, K.D., Gorrens, B., Lauwers, D., Leest, M.V., & Witlox, F. (2010). Transportation Risk Analysis tool for hazardous Substances (TRANS) – A user-friendly, semi-quantitative multi-mode hazmat transport route safety risk estimation methodology for Flanders. *Transportation Research Part D, Transport and Environment*, *15*(8), 489–496. doi:10.1016/j.trd.2010.07.001

ReVelle, C., Cohon, J., & Shobrys, D.ReVelle. (1991). Simultaneous siting and routing in the disposal of hazardous wastes. *Transportation Science*, *25*(2), 138–145. doi:10.1287/trsc.25.2.138

Ritsema van Eck, J. R., & de Jong, T. (1999). Accessibility analysis and spatial competition effects in the context of GIS-supported service location planning. *Computers, Environment and Urban Systems*, *23*(2), 75–89. doi:10.1016/S0198-9715(99)00016-2

Rizet, C., Browne, M., Cornelis, E., & Leonardi, J. (2012). Assessing carbon footprint and energy efficiency in competing supply chains: Review – Case studies and benchmarking. *Transportation Research Part D, Transport and Environment*, *17*(4), 293–300. doi:10.1016/j.trd.2012.01.002

Robin-Jouan, Y., Ehrlich, J., Guillaumin, B., Delarche, M., & Dutech, M. (2005, April). Transport-specific communication services: Safety-based or critical applications for mobiles and cooperation with infrastructure networks. *Annales des Télécommunications*, *60*(3-4), 405–440.

Rodrigue, J. P., Comtois, C., & Slack, B. (2013). *The geography of transport systems*. Routledge.

Rodríguez-Martín, I., Salazar-González, J., & Yaman, H. (2014). A branch-and-cut algorithm for the hub location and routing problem. *Computers & Operations Research*, *50*, 161–174. doi:10.1016/j.cor.2014.04.014

Rodríguez, V., Alvarez, M. J., & Barcos, L. (2007). Hub location under capacity constraints. *Transportation Research Part E, Logistics and Transportation Review*, *43*(5), 495–505. doi:10.1016/j.tre.2006.01.005

Roeller, D. (2013). Engaging citizens through social media in sustainable urban mobility planning in Aberdeen (UK). *Eltis.org*. Retrieved from http://www.eltis.org/discover/case-studies/engaging-citizens-through-social-media-sustainable-urban-mobility-planning

Romer, H., & Haastrup, P., & StyhrPetersen, H. J. (1995). Accidents duringmarine transport of dangerous goods – distribution of fatalities. *Journal of Loss Prevention in the Process Industries*, *8*(1), 29–34. doi:10.1016/0950-4230(95)90059-X

Ronay, E., & Egger, R. (2013). NFC smart city: Cities of the future a scenario technique application (pp. 565–577). In Z. Xiang & I. Tussyadiah (Eds.), (2014). Information and Communication Technologies in Tourism. Springer.

Ronza, A., Vilchez, J. A., & Casal, J. (2007). Using transportation accident databases to investigate ignition and explosion probabilities of flammable spills. *Journal of Hazardous Materials*, *146*(1-2), 106–123. doi:10.1016/j.jhazmat.2006.11.057 PMID:17240052

Roundabouts: an Informational Guide. (2000). United States Department of Transportation, Federal Highway Administration.

Ruan, Z., Hui, P., Lin, H., & Liu, Z. H. (2013). Risks of an epidemic in a two-layered railway-local area traveling network. *The European Physical Journal B*, *86*(1), 13. doi:10.1140/epjb/e2012-30292-x

Ruifang, M. O. U. (2010). Environmental risk assessment model on dangerous goods in logistics. *ICLEM*, *2010*, 2023–2029.

Rygnestad, H. (2000). *Integrating Environmental Economics and Policy Analyses in a Geographical Information System*. Statens Jordbrugs-og Fiskeriøkonomiske Institut.

Saat, M. R., Werth, C. J., Schaeffer, D., Yoon, H., & Barkan, C. P. L. (2014). Environmental risk analysis of hazardous material rail transportation. *Journal of Hazardous Materials*, *264*, 560–569. doi:10.1016/j.jhazmat.2013.10.051 PMID:24239259

Saaty, T. L. (1977). A scaling method for priorities in hierarchical structures. *Journal of Mathematical Psychology*, *15*(3), 57–68. doi:10.1016/0022-2496(77)90033-5

Saaty, T. L. (1980). *The Analytical Hierarchy Process*. New York: McGraw Hill.

Saaty, T. L. (1988). *Multicriteria decision making: The analytic hierarchy process*. Pittsburgh: RWS Publications.

Saaty, T. L. (1994). *Fundamentals of decision making and priority theory with the AHP*. PA, U.S.A: RWS Publications.

Saaty, T. L. (2001). The seven pillars of the analytic hierarchy process.*Proceedings of the 15th Int'l Conf on Multiple Criteria Decision Making (MCDM)* (pp.1-15). New York: Springer.

Saaty, T. L. (2006). Rank from comparisons and from ratings in the analytic hierarchy /network processes. *European Journal of Operational Research*, *168*(2), 557. doi:10.1016/j.ejor.2004.04.032

Saccomanno, F. F., & Shortreed, J. H. (1993). Hazmat transport risks – societal and individual perspectives. *Journal of Transportation Engineering*, *119*(2), 177–188. doi:10.1061/(ASCE)0733-947X(1993)119:2(177)

Saccomanno, F. F., Shortreed, J. H., Van Aerde, M., & Higgs, J. (1989). Comparison of risk measures for the transport of dangerous commodities by truck and rail. *Transportation Research Record*, *1245*, 1–13.

Saccomanno, F., & Chan, A. (1985). Economic evaluation of routing strategies for hazardous road shipments. *Transportation Research Record*, *1020*, 12–18.

Saco, D. (2002). *Cybering democracy. Public space and the Internet*. Minneapolis, London: University of Minnesota Press.

Saddoune, M., Desaulniers, G., & Soumis, F. (2013). Aircrew pairings with possible repetitions of the same flight number. *Computers & Operations Research*, *40*(3), 805–814. doi:10.1016/j.cor.2010.11.003

Sadegh, A., Radwan, A. E., & Rouphail, N. M. (1988). ARTWORK: A simulation model of urban arterial work zones. *Transportation Research Record*, *1163*, 1–3.

Saksena, J. P., & Kumar, S. (1966). The routing problem with 'K' specified nodes. *Operations Research, 14*(5), 909–913. doi:10.1287/opre.14.5.909

Salle, C. T. P., Guahyba, A. S., Wald, V. B., Silva, A. B., Salle, F. O., & Nascimento, V. P. (2003). Use of artificial neural networks to estimate production variables of broilers breeders in the production phase. *British Poultry Science, 44*(2), 211–217. doi:10.1080/0007166031000088361 PMID:12828206

Samanlioglu, F. (2013). A multi-objective mathematical model for the industrial hazardous waste location-routing problem. *European Journal of Operational Research, 226*(2), 332–340. doi:10.1016/j.ejor.2012.11.019

Samuel, C., Keren, N., Shelley, M. C., & Freeman, S. A. (2009). Frequency analysis of hazardous material transportation incidents as a function of distance from origin to incident location. *Journal of Loss Prevention in the Process Industries, 22*(6), 783–790. doi:10.1016/j.jlp.2009.08.013

Santos, D., de Sousa, A., Alvelos, F., & Pioro, M. (2013). Optimizing network load balancing: An hybridization approach of metaheuristics with column generation. *Telecommunication Systems, 52*(2), 959–968. doi:10.1007/s11235-011-9604-3

Sarvi, M., & Kuwahara, M. (2008). Using ITS to improve the capacity of freeway merging sections by transferring freight vehicles. *Intelligent Transportation Systems. IEEE Transactions on, 9*(4), 580–588.

Sasaki, M., Suzuki, A., & Drezner, Z. (1999). On the selection of hub airports for an airline hub-and-spoke system. *Computers & Operations Research, 26*(14), 1411–1422. doi:10.1016/S0305-0548(99)00043-X

Satunin, S., & Babkin, E. (2014). A multi-agent approach to Intelligent Transportation Systems modeling with combinatorial auctions. *Expert Systems with Applications, 41*(15), 6622–6633. doi:10.1016/j.eswa.2014.05.015

Saura, S., & Rubio, L. (2010). A common currency for the different ways in which patches and links can contribute to habitat availability and connectivity in the landscape. *Ecography, 33*(3), 523–537. doi:10.1111/j.1600-0587.2009.05760.x

Saura, S., & Torne, J. (2009). Conefor Sensinode 2.2: A software package for quantifying the importance of habitat patches for landscape connectivity. *Environmental Modelling & Software, 24*(1), 135–139. doi:10.1016/j.envsoft.2008.05.005

Saz-Salazar, S., & Rausell-Köster, P. (2008). A Double-Hurdle model of urban green areas valuation: Dealing with zero responses. *Landscape and Urban Planning, 84*(3-4), 241–251. doi:10.1016/j.landurbplan.2007.08.008

Scanlon, R. D., & Cantilli, E. J. (1985). Assessing the risk and safety in the transportation of hazardous materials. *Transportation Research Record*, (1020).

Schaefer, A. J., Johnson, E. L., Kleywegt, A. J., & Nemhauser, G. L. (2005). Airline crew scheduling under uncertainty. *Transportation Science, 39*(3), 340–348. doi:10.1287/trsc.1040.0091

Schepers, A. W., Thibault, J., & Lacroix, C. (2000). Comparison of simple neural networks and nonlinear regression models for descriptive modeling of Lactobacillus helveticus growth in pH-controlled batch cultures. *Enzyme and Microbial Technology, 26*(5-6), 431–445. doi:10.1016/S0141-0229(99)00183-0 PMID:10713218

Schietekat, J. M., & Booysen, M. J. (2013, July). Detection of reckless driving in the Sub-Saharan informal public transportation system using acceleration-sensing telematics. Proceedings of EUROCON, (pp. 597-601). IEEE. doi:10.1109/EUROCON.2013.6625042

Schittenkopf, C., Deco, G., & Brauer, W. (1997). Two strategies to avoid overfitting in feedforward networks. *Neural Networks, 10*(3), 505–516. doi:10.1016/S0893-6080(96)00086-X PMID:18255655

Schoenemann, T., & Cremers, D. (2010). A combinatorial solution for model-based image segmentation and real-time tracking. *IEEE Transactions on Pattern Analysis and Machine Intelligence*, *32*(7), 1153–1164. doi:10.1109/TPAMI.2009.79 PMID:20489221

Schuster, A. (2010). Internet courses for senior citizens in Munich, Germany. *Eltis.org*. Retrieved from [REMOVED HYPERLINK FIELD] http://www.eltis.org/discover/case-studies/internet-courses-senior-citizens-munich-germany#field_case_study_user_contact

Schuurman, N., Fiedler, R.S., Grzybowski, S.C.W., & Grund, D., (2006). Defining rational hospital catchments for non-urban areas based on travel-time. *International Journal of Health Geographics*, *5*(43), 1–11. doi:10.1186/1476-072X-5-43 PMID:16390549

Schweitzer, L. (2006). Environmental justice and hazmat transport: A spatial analysis in southern California. *Transportation Research Part D, Transport and Environment*, *11*(6), 408–421. doi:10.1016/j.trd.2006.08.003

Scott-Andersson, A., Tysklind, M., & Fangmark, I. (2007). A method to relate chemical accident properties and expert judgements in order to derive useful information for the development of environment-accident index. *Journal of Hazardous Materials*, *147*(1-2), 524–533. doi:10.1016/j.jhazmat.2007.01.052 PMID:17320285

Scott, J., Larson, A., Jefferies, F., & Veenendaal, B. (2006). Small-area estimates of general practice workforce shortage in rural and remote Western Australia. *The Australian Journal of Rural Health*, *14*(5), 209–213. doi:10.1111/j.1440-1584.2006.00811.x PMID:17032297

Seeger, M. (2004). Gaussian processes for machine learning.[PubMed]. *International Journal of Neural Systems*, *14*(2), 69–106. doi:10.1142/S0129065704001899

Sender, J., & Clausen, U. (2013). Heuristics for solving a capacitated multiple allocation hub location problem with application in German wagonload traffic. *Electronic Notes in Discrete Mathematics*, *41*, 13–20. doi:10.1016/j.endm.2013.05.070

Serre, D. (2002). *Matrices: Theory and Applications*. Springer-Verlag.

Seuring, S. (2013). A review of modeling approaches for sustainable supply chain management. *Decision Support Systems*, *54*(4), 1513–1520. doi:10.1016/j.dss.2012.05.053

Seuring, S., & Müller, M. (2008). From a literature review to a conceptual framework for sustainable supply chain management. *Journal of Cleaner Production*, *16*(15), 1699–1710. doi:10.1016/j.jclepro.2008.04.020

Sevenant, M., & Antrop, M. (2010). Transdisciplinary landscape planning: Does the public have aspirations? Experiences from a case study in Ghent (Flanders, Belgium). *Land Use Policy*, *27*(2), 373–386. doi:10.1016/j.landusepol.2009.05.005

Shackel, B., & Richardson, S. J. (1991). *Human factors for informatics usability*. Cambridge University Press.

Shappert, L. B., Brobst, W. A., Langhaar, J. W., & Sisler, J. A. (1973). Probability and consequences of transportation accidents involving radioactive-material shipments in the nuclear fuel cycle. *Nuclear Safety*, *14*(6), 597–604.

Shaw, S. L. (1993). Hub structures of major US passenger airline. *Journal of Transport Geography*, *1*(1), 47–58. doi:10.1016/0966-6923(93)90037-Z

Shebalov, S., & Klabjan, D. (2006). Robust airline crew pairing: Move-up crews. *Transportation Science*, *40*(3), 300–312. doi:10.1287/trsc.1050.0131

Shen, X., Xie, C., Liu, H., & Qui, Z. (2013). Model and algorithm for routing and scheduling problem in hazardous materials transportation network. *Journal of Networks*, *8*(5), 1027–1034. doi:10.4304/jnw.8.5.1027-1034

Shen, X., Yan, Y., Li, X., Xie, C., & Wang, L. (2014). Analysis on tank truck accidents involved in road hazardous materials transportation in China. *Traffic Injury Prevention*, *15*(7), 762–768. PMID:24380669

Sherali, H. D., Brizendine, L. D., Glickman, T. S., & Subramanian, S. (1997). Low-probability high consequence considerations in routing hazardous materials shipments. *Transportation Science*, *31*(3), 237–251. doi:10.1287/trsc.31.3.237

Shim, J. P., Warkentin, M., Courtney, J. F., Power, D. J., Sharda, R., & Carlsson, C. (2002). Past, present, and future of decision support technology. *Decision Support Systems*, *33*(2), 111–126. doi:10.1016/S0167-9236(01)00139-7

Shorten, C. V., Galloway, J., Krebs, J. G., & Fleming, R. S. (2002). A 12-year history of hazardous materials incidents in Chester County, Pennsylvania. *Journal of Hazardous Materials*, *89*(1), 29–40. doi:10.1016/S0304-3894(01)00302-8 PMID:11734344

Shu-Xi, W. (2012). The Improved Dijkstra's Shortest Path Algorithm and Its Application. *Procedia Engineering*, *29*, 1186–1190. doi:10.1016/j.proeng.2012.01.110

Sims, K. R. E. (2013). Do Protected Areas Reduce Forest Fragmentation? A Microlandscapes Approach. *Environmental and Resource Economics*, *58*(2), 303–333. doi:10.1007/s10640-013-9707-2

Sivakumar, R. A., & Batta, R. (1994). The variance-constrained shortest path problem. *Transportation Science*, *28*(4), 309–316. doi:10.1287/trsc.28.4.309

Sivakumar, R. A., Batta, R., & Karwan, M. H. (1993). A network-based model for transporting extremely hazardous materials. *Operations Research Letters*, *13*(2), 85–93. doi:10.1016/0167-6377(93)90034-E

Skorin-Kapov, D., & Skorin-Kapov, J. (1994). On Tabu search for the location of interacting hub facilities. *European Journal of Operational Research*, *73*(3), 502–509. doi:10.1016/0377-2217(94)90245-3

Skorin-Kapov, D., Skorin-Kapov, J., & O'Kelly, M. (1996). Tight linear programming relaxations of uncapacitated p-hub median problems.*Proceedings of the National Decision Science Conference*(*Vol. 94*, pp. 582–593). doi:10.1016/0377-2217(95)00100-X

Slack, B. (1990). Intermodal transportation in North America and the development of inland load centers. *The Professional Geographer*, *42*(1), 72–83. doi:10.1111/j.0033-0124.1990.00072.x

Slack, B. (2001). Intermodal transportation. In A. Brewer, K. J. Button, & D. A. Hensher (Eds.), *Handbook of Logistics and Supply Chain Management* (1st ed., p. 700). Pergamon.

Smith, B. C., & Johnson, E. L. (2006). Robust airline fleet assignment: Imposing station purity using station decomposition. *Transportation Science*, *40*(4), 497–516. doi:10.1287/trsc.1060.0153

Sohn, J., & Park, S. (1998). Efficient solution procedure and reduced size formulation for P-hub location problem. *European Journal of Operational Research*, *108*(1), 118–126. doi:10.1016/S0377-2217(97)00201-4

Solyal, O., & Sural, H. (2012). The one-warehouse multi-retailer problem: Reformulation, classification, and computational results. *Annals of Operations Research*, *196*(1), 517–541. doi:10.1007/s10479-011-1022-0

Soni, G., & Kodali, R. (2012). A critical review of empirical research methodology in supply chain management. *Journal of Manufacturing Technology Management*, *23*(6), 753–779. doi:10.1108/17410381211253326

Soykan, B., & Erol, S. (2014). A Branch-and-Price Algorithm for the Robust Airline Crew Pairing Problem. *The Journal of Defense Sciences*, *13*(1), 37–74.

Sözen, A., Alp, I., & Kilinc, C. (2012). The efficiency assesstment of the hydro-power plants in Turkey by using data envelopment analysis. *Renewable Energy*, *46*, 192–202. doi:10.1016/j.renene.2012.03.021

Sözen, A., Alp, I., & Ozdemir, A. (2010). Assessment of operational and environmental performance of thermal power plants in Turkey by using data envelopment analysis. *Energy Policy*, *38*(10), 6194–6203. doi:10.1016/j.enpol.2010.06.005

Spadoni, G., Leonelli, P., Verlicchi, P., & Fiore, R. (1995). A numerical procedure for assessing risks from road transport of dangerous substances. *Journal of Loss Prevention in the Process Industries*, *8*(4), 245–252. doi:10.1016/0950-4230(95)00029-Z

Spichkova, M. (2011). Architecture: Requirements + Decomposition + Refinement. *In Softwaretechnik-Trends, 32(4)*. Retrieved from http://pi.informatik.uni-siegen.de/gi/stt/31_4/index.html

Spichkova, M., Hölzl, F., & Trachtenherz, D. (2012). Verified System Development with the AutoFocus Tool Chain. In *Proceedings of the 2nd Workshop on Formal Methods in the Development of Software*. Electronic Proceedings in Theoretical Computer Science 86. Retrieved from http://arxiv.org/abs/1207.1841

Spichkova, M., Zhu, X., & Mou, D. (2013). *Do we really need to write documentation for a system? CASE tool add-ons: generator+editor for a precise documentation.* Proceedings of the International Conference on Model-Driven Engineering and Software Development. Retrieved from http://arxiv.org/abs/1404.7265

Spichkova, M. (2012). Human Factors of Formal Methods.*Proceedings of the IADIS Interfaces and Human Computer Interaction 2012 (IHCI 2012)*. Lisbon, Portugal: IADIS Press.

Spichkova, M. (2014). Design of formal languages and interfaces: "Formal" does not mean "unreadable". In K. Blashki & P. Isaias (Eds.), *Emerging Research and Trends in Interactivity and the Human-Computer Interface* (pp. 301–314). Hershey, PA: IGI Global; doi:10.4018/978-1-4666-4623-0.ch015

Spichkova, M., & Simic, M. (2015). Towards formal modelling of autonomous systems. In E. Damiani, R. J. Howlett, L. C. Jain, L. Gallo, & G. De Pietro (Eds.), *Intelligent Interactive Multimedia Systems and Services* (pp. 279–288). Springer International Publishing; doi:10.1007/978-3-319-19830-9

Spinak, A., Chiu, D., & Casalegno, M. (2008). *SII-Sustainability Innovation Inventory*, Connected Urban Development, Retrieved from http://www.connectedurbandevelopment.org/mit/sustainability_innovation_inventory

Spinak, A., & Casalegno, F. (2012). Sustainable and equitable urbanism: The role of ICT in ecological culture change and poverty alleviation. In O. Y. Ercoskun (Ed.), *Green and Ecological Technologies for Urban Planning: Creating Smart Cities* (pp. 37–57). Hershey: Information Science Publishing. doi:10.4018/978-1-61350-453-6.ch003

Spitadakis, V., & Fostieri, M. (2012). WISETRIP-International multimodal journey planning and delivery of personalized trip information. *Procedia: Social and Behavioral Sciences*, *48*, 1294–1303. doi:10.1016/j.sbspro.2012.06.1105

Stareveren, R. V. (2014) *How open data access will revolutionize Istanbul's public transit systems.* Sustainablecities-collective. Retrieved from http://sustainablecitiescollective.com/reneevanstaveren/290891/how-open-data-access-will-revolutionize-istanbul-s-public-transit-systems

Steenbrink, P. A. (1974). *Optimization of transport networks.* London: J. Wiley.

Stegle, O., Fallert, S. V., MacKay, D. J. C., & Brage, S. (2008). Gaussian process robust regression for noisy heart rate data.[PubMed]. *IEEE Transactions on Bio-Medical Engineering*, *55*(9), 2143–2151. doi:10.1109/TBME.2008.923118

Sternberg, H., & Andersson, M. (2014). Decentralized intelligence in freight transport—A critical review. *Computers in Industry*, *65*(2), 306–313. doi:10.1016/j.compind.2013.11.011

Storøy, J., Thakur, M., & Olsen, P. (2013). The TraceFood Framework – Principles and guidelines for implementing traceability in food value chains. *Journal of Food Engineering*, *115*(1), 41–48. doi:10.1016/j.jfoodeng.2012.09.018

Strohmann, T. R., & Grudic, G. Z. (2002). *A Formulation for minimax probability machine regression* (T. G. Dietterich, S. Becker, & Z. Ghahramani, Eds.). Advances in Neural Information Processing SystemsCambridge, MA: MIT Press.

Sugumaran, V., & Sugumaran, R. (2007). Web-based spatial decision support systems (WEBSDSS): Evolution, architecture, examples and challenges. *Communications of the Association for Information Systems*, *19*, 844–875.

Suh, K., Smith, T., & Linhoff, M. (2012). Leveraging socially networked mobile ICT platforms for the last-mile delivery problem. *Environmental Science & Technology*, *46*(17), 9481–9490. doi:10.1021/es301302k PMID:22877137

Sulijoadikusumo, G. S., & Nozick, L. K. (1998). Multiobjective routing and scheduling of hazardous materials shipments. *Transportation Research Record*, *1613*, 96–104. doi:10.3141/1613-13

Sung, C. S., & Jin, H. W. (2001). Dual-based approach for a hub network design problem under nonrestrictive policy. *European Journal of Operational Research*, *132*(1), 88–105. doi:10.1016/S0377-2217(00)00114-4

Sun, J., Bai, Y., Luo, J., & Dang, J. (2009). Modelling of a chaotic time series using a modified minimax probability machine regression. *The Chinese Journal of Physiology*, *47*(4), 491–501.

Sun, X., Lu, H., & Wu, J. (2013). Bus detection based on sparse representation for transit signal priority. *Neurocomputing*, *118*, 1–9. doi:10.1016/j.neucom.2013.02.006

Sun, Z. L., Au, K. F., & Choi, T. M. (2007). A neuro-fuzzy inference system through integration of fuzzy logic and extreme learning machines.[PubMed]. *IEEE Transactions on Systems, Man, and Cybernetics. Part B, Cybernetics*, *37*(5), 1321–1331. doi:10.1109/TSMCB.2007.901375

Svozil, D., Kvasnicka, V., & Pospichal, J. (1997). Introduction to multilayer feed forward neural networks. *Chemometrics and Intelligent Laboratory Systems*, *39*(1), 43–62. doi:10.1016/S0169-7439(97)00061-0

Swoveland, C. (1987). Risk analysis of regulatory options for the transport of dangerous commodities by rail. *Interfaces*, *17*(4), 90–107. doi:10.1287/inte.17.4.90

Sylvie, D. (2007). Gis-based accessibility analysis for network optimal location model. *Cybergeo, Systems, Modelisation Geostatistiques*, Article 407.

Szeto, W. Y. (2013). Routing and scheduling hazardous material shipments: Nash game approach. *Transportmetrica B. Transport Dynamics*, *1*(3), 237–260. doi:10.1080/21680566.2013.861330

Taleai, M., Sharifi, A., Sliuzas, R., & Mesgari, M. (2007). Evaluating the compatibility of multi-functional and intensive urban land uses. *International Journal of Applied Earth Observation and Geoinformation*, *9*(4), 375–391. doi:10.1016/j.jag.2006.12.002

Tam, B., Ryan, D., & Ehrgott, M. (2014). Multi-objective Approaches to the Unit Crewing Problem in Airline Crew Scheduling. *Journal of Multi-Criteria Decision Analysis*, *21*(5-6), 257–277. doi:10.1002/mcda.1517

Tan, P., Steinbach, M., & Kumar, V. (2006). Introduction to data mining. Addison-Wesley.

Taniguchi, E., Thompson, R. G., & Yamada, T. (2014). Recent trends and innovations in modelling city logistics. *Procedia: Social and Behavioral Sciences*, *125*, 4–14. doi:10.1016/j.sbspro.2014.01.1451

Taş, D., Gendreau, M., Dellaert, N., van Woensel, T., & de Kok, A. G. (2014). Vehicle routing with soft time windows and stochastic travel times: A column generation and branch-and-price solution approach. *European Journal of Operational Research*, *236*(3), 789–799. doi:10.1016/j.ejor.2013.05.024

Taylor, M. A. P., Zito, R., Smith, N., & D'Este, G. (2005). Modelling the impacts of transport policies to reduce greenhouse gas emissions from urban freight transport in Sydney. *Journal of the Eastern Asia Society for Transportation Studies*, *6*, 3135–3150.

Tena-Chollet, F., Tixier, J., Dusserre, G., & Mangin, J.-F. (2013). Development of a spatial risk assessment tool for the transportation of hydrocarbons: Methodology and implementation in a geographical information system. *Environmental Modelling & Software*, *46*, 61–74. doi:10.1016/j.envsoft.2013.02.010

Tezcan, O. (2007). Analytic hierarchy process method and application in area selection of ready mixed concrete plant. *National Journal of Ready Mixed Concrete, Nov.-Dec.*, 58-62.

The Eight-Development Plan of Turkey, Transportation, Maritime Transport (in Turkish). (2001). State Planning Organization. Ankara: Republic of Turkey Prime Ministry State Planning Organization.

The Velib blog. (n. d.). P'tit Vélib'. Retrieved from http://blog.velib.paris.fr/en/ptit-velib/

Theodoridis, S., & Koutroubas, K. (1999). *Pattern Recognition*. London: Academic Press.

Thomadsen, T., & Larsen, J. (2007). A hub location problem with fully interconnected backbone and access networks. *Computers & Operations Research*, *34*(8), 2520–2531. doi:10.1016/j.cor.2005.09.018

Toh, R. S., & Higgins, R. C. (1985). The impact of hub and spoke network centralization and route monopoly on domestic airline profitability. *Transportation Journal*, *24*, 16–27.

TomTom Report. (2013). *TomTom European Congestion Index*. Retrieved from http://www.tomtom.com/lib/doc/trafficindex/2013-1101%20TomTomTrafficIndex2013Q2EUR-km.pdf

Topcuoglu, H., Corut, F., Ermis, M., & Yilmaz, G. (2005). Solving the uncapacitated hub location using genetic algorithms. *Computers & Operations Research*, *32*(4), 967–984. doi:10.1016/j.cor.2003.09.008

Topuz, E., Talinli, I., & Aydin, E. (2011). Integration of environmental and human health risk assessment for industries using hazardous materials: A quantitative multi criteria approach for environmental decision makers. *Environment International*, *37*(2), 393–403. doi:10.1016/j.envint.2010.10.013 PMID:21111481

Toumazis, I., & Kwon, C. (2013). Routing hazardous materials on time-dependent networks using conditional value-at-risk. *Transportation Research Part C, Emerging Technologies*, *37*, 73–92. doi:10.1016/j.trc.2013.09.006

Transportation Strategy Report (in Turkish) (pp. 23–30). (2004Ministry of Transportation. Ankara.

Trépanier, M., & Leroux, M., & MarcellisWarin, N. (2009). Cross-analysis of hazmat road accidents using multiple databases. *Accident; Analysis and Prevention*, *41*(6), 1192–1198. doi:10.1016/j.aap.2008.05.010 PMID:19819367

Triantaphyllou, E., & Mann, S. H. (1995). Using the analytic hierarchy process for decision making in engineering applications: Some challenges. *International Journal of Industrial Engineering: Applications and Practice*, *2*, 35–44.

Tsai, C. P., & Lee, T. L. (1999). Back-propagation neural network in tidal-level forecasting. *Journal of Waterway, Port, Coastal, and Ocean Engineering*, *125*(4), 195–202. doi:10.1061/(ASCE)0733-950X(1999)125:4(195)

Tsai, C. P., Shen, J. N., & Kerh, T. (1999). Wave forecasting using neural network model. In B. Kumar & B. H. V. Topping (Eds.), *Artificial Intelligence Applications in Civil and Structural Engineering* (pp. 125–130). Edinburgh, UK: Civil-Comp. Press. doi:10.4203/ccp.62.5.3

Tsatcha, D., Saux, E., & Claramunt, C. A. (2014). Bidirectional path-finding algorithm and data structure for maritime routing. *International Journal of Geographical Information Science*, *28*(7), 1355–1377. doi:10.1080/13658816.2014.887087

Tsompanoglou, S., & Photis, Y. N. (2013). Measuring urban concentration: A spatial cluster typology based on public and private sector service patterns. *World Review of Science. Technology and Sustainable Development*, *10*(4), 185–202.

Tunc, A. (2003). *Traffic Engineering and Applications*. Ankara, Turkey.

U.S. Department of Transportation & U.S. Department of Commerce. (2010). *Hazardous materials transportation 2007 commodity flow survey*.

UK Trade and Investment (2003). Retrieved from https://www.uktradeinvest.gov.uk/ukti/appmanager/ukti/splash

Umit, H., & Kara, B. Y. (2003). Büyük ölçekli bir tehlikeli madde taşımacılığının coğrafi bilgi sistemleri tabanlı analizi: Shell Türkiye örneği[in Turkish]. *Yöneylem Araştırma Dergisi*, *15*, 35–47.

UN Habitat. (2013). *Planning and design for sustainable urban mobility*. New York: Routledge.

Underwood, J. N., Wilson, S. K., Ludgerus, L., & Evans, R. D. (2013). Integrating connectivity science and spatial conservation management of coral reefs in north-west Australia. *Journal for Nature Conservation*, *21*(3), 163–172. doi:10.1016/j.jnc.2012.12.001

United Nations. (2013). *European agreement concerning the international carriage of dangerous goods by road*. Geneva: Author.

Urban, D., & Keitt, T. (2001). Landscape connectivity: A graph-theoretic perspective. *Ecology*, *82*(5), 1205–1218. Retrieved from http://amadeus.ecs.umass.edu/mie373/landscapes_msts.pdf doi:10.1890/0012-9658(2001)082[1205:LCAGTP]2.0.CO;2

USAID (U.S. Agency for International Development). (2003). Retrieved from http://www.usaid.gov/iraq/

USDOT (U.S. Department of Transportation). (2008). *Emergency response guidebook*. Washington, DC, USA: PHMSA.

Uses of social media in public transportation . (2012). Transit Cooperative Research Program. Retrieved from http://onlinepubs.trb.org/onlinepubs/tcrp/tcrp_syn_99.pdf

Vahidnia, M. H., Alesheikh, A. A., & Alimohammadi, A. (2009). Hospital site selection using fuzzy AHP and its derivatives. *Journal of Environmental Management*, *90*(10), 3048–3056. doi:10.1016/j.jenvman.2009.04.010 PMID:19477577

Vaidya, O. S., & Kumar, S. (2006). Analytic hierarchy process: An overview of applications. *European Journal of Operational Research*, *169*(1), 1–29. doi:10.1016/j.ejor.2004.04.028

Van Aerde, M., Stewart, A., & Saccomanno, F. (1988). Estimating the impacts of L.P.G. spills during transportation accidents. *Journal of Hazardous Materials*, *20*, 375–392. doi:10.1016/0304-3894(88)87023-7

Van Raemdonck, K., Macharis, C., & Mairesse, O. (2013). Risk analysis system for the transport of hazardous materials. *Journal of Safety Research*, *45*, 55–63. doi:10.1016/j.jsr.2013.01.002 PMID:23708476

Van Vliet, D. (1978). Improved shortest path algorithms for transport networks. *Transportation Research*, *12*(1), 7–20. doi:10.1016/0041-1647(78)90102-8

Vanaerde, M., Shortreed, J., Stewart, A. M., & Matthews, M. (1989). Assessing the risks associated with the transport of dangerous goods by truck and rail using the riskmod model. *Canadian Journal of Civil Engineering*, *16*(3), 326–334. doi:10.1139/l89-058

Vargas, L. G. (1990). An overview of the analytic hierarchy process and its applications. *European Journal of Operational Research*, *48*(1), 2–8. doi:10.1016/0377-2217(90)90056-H

Vasconcelos, A. D., Nassi, C. D., & Lopes, L. A. S. (2011). The uncapacitated hub location problem in networks under decentralized management. *Computers & Operations Research*, *38*(12), 1656–1666. doi:10.1016/j.cor.2011.03.004

Vasudevan, S., Ramos, F., Nettleton, E., & Durrant Whyte, H. (2009). Gaussian process modeling of large-scalwe terrain. *Journal of Field Robotics*, *26*(10), 812–840. doi:10.1002/rob.20309

Verma, M. (2009). A cost and expected consequence approach to planning and managing railroad transportation of hazardous materials. *Transportation Research Part D, Transport and Environment*, *14*(5), 300–308. doi:10.1016/j.trd.2009.03.002

Verma, M. (2011). Railroad transportation of dangerous goods: A conditional exposure approach to minimize transport risk. *Transportation Research Part C, Emerging Technologies*, *19*(5), 790–802. doi:10.1016/j.trc.2010.07.003

Verma, M., & Verter, V. (2007). Railroad transportation of dangerous goods: Population exposure to airborne toxins. *Computers & Operations Research*, *34*(5), 1287–1303. doi:10.1016/j.cor.2005.06.013

Verma, M., & Verter, V. (2010). A lead-time based approach for planning rail–truck intermodal transportation of dangerous goods. *European Journal of Operational Research*, *202*(3), 696–706. doi:10.1016/j.ejor.2009.06.005

Verma, M., Verter, V., & Zufferey, N. (2012). A bi-objective model for planning and managing rail-truck intermodal transportation of hazardous materials. *Transportation Research Part E, Logistics and Transportation Review*, *48*(1), 132–149. doi:10.1016/j.tre.2011.06.001

Verter, V., & Kara, B. Y. (2001). A GIS-based framework for hazardous materials transport risk assessment. *Risk Analysis*, *21*(6), 1109–1120. doi:10.1111/0272-4332.216179 PMID:11824686

Verter, V., & Kara, B. Y. (2008). A path-based approach for Hazmat transport network design. *Management Science*, *54*(1), 29–40. doi:10.1287/mnsc.1070.0763

Vilchez, J. A., Sevilla, S., Montiel, H., & Casal, J. (1995). Historical analysis of accidents inchemical plants and in the transportation of hazardous materials. *Journal of Loss Prevention in the Process Industries*, *8*(2), 87–96. doi:10.1016/0950-4230(95)00006-M

Vliet, J. V., White, R., & Dragicevic, S. (2009). Modeling urban growth using a variable grid cellular automaton. *Computers, Environment and Urban Systems*, *33*(1), 35–43. doi:10.1016/j.compenvurbsys.2008.06.006

Volpert, W. (1991). Work Design for Human Development. In C. Floyd et al. (Eds.), *Software Development and Reality Construction*. Berlin, Germany: Springer.

Vuchic, V. R. (1999). *Transportation for livable cities*. NJ, USA: Rutgers Center for Urban Policy Research.

Wagner, B. (2007). An exact solution procedure for a cluster hub location problem. *European Journal of Operational Research*, *178*(2), 391–401. doi:10.1016/j.ejor.2006.02.011

Wagner, B. (2008). A note on location of hubs in a competitive environment. *European Journal of Operational Research*, *184*(1), 57–62. doi:10.1016/j.ejor.2006.10.057

Walczak, S., & Sincich, T. (1999). A comparative analysis of regression and neural networks for university admissions. *Information Sciences*, *119*(1-2), 1–20. doi:10.1016/S0020-0255(99)00057-2

Wang, C. (2008). Optimization of hub-and-spoke two-stage logistics network in regional port cluster. *Systems Engineering – Theory &Practice, 28*(9), 152–158.

Wang, F., & Sun, J. (2012). *Distance metric learning in data mining.SDM Conference (Tutorial).*

Wang, J., Kang, Y., Kwon, C., & Batta, R. (2012). Dual toll pricing for hazardous materials transport with linear delay. *Networks and Spatial Economics*, *12*(1), 147–165. doi:10.1007/s11067-011-9156-9

Wang, Y. H., Li, Y., Yang, S. L., & Yang, L. (2005). Classification of substrates and inhibitors of P-glycoprotein using unsupervised machine learning approach.[PubMed]. *Journal of Chemical Information and Modeling*, *45*(3), 750–757. doi:10.1021/ci050041k

Wang, Y., Cao, F., & Yuan, Y. (2011). A study on effectiveness of extreme learning machine. *Neurocomputing*, *74*(16), 2483–2490. doi:10.1016/j.neucom.2010.11.030

Warshall, S. (1962). A theorem on Boolean matrices. *Journal of the ACM*, *9*(1), 11–12. doi:10.1145/321105.321107

Warsing, D. P., Souza, G. C., & Greis, N. P. (2001). Determining the value of dedicated multimodal cargo facilities in a multi-region distribution network. *European Journal of Operational Research*, *133*(1), 81–93. doi:10.1016/S0377-2217(00)00185-5

Watts, M., Stewart, R., Dan, S., & Kircher, L. (2011). Using the Zonae Cogito Decision Support System. The Ecology Centre, University Of Queensland. Retrieved from http://www.uq.edu.au/marxan/docs/ZonaeCogitoManual_8March2011.pdf

Watts, M. E., Ball, I. R., Stewart, R. S., Klein, C. J., Wilson, K., & Steinback, C. et al. (2009). Marxan with Zones: Software for optimal conservation based land- and sea-use zoning. *Environmental Modelling & Software*, *24*(12), 1513–1521. doi:10.1016/j.envsoft.2009.06.005

Weide, O., Ryan, D., & Ehrgott, M. (2010). An iterative approach to robust and integrated aircraft routing and crew scheduling. *Computers & Operations Research*, *37*(5), 833–844. doi:10.1016/j.cor.2009.03.024

Welles, W. L., Wilburn, R. E., Ehrlich, J. K., & Kamara, J. M. (2009). New York hazardous substances emergency events surveillance (HSEES) data support emergency response, promote safety and protect public health. *Journal of Loss Prevention in the Process Industries*, *22*(6), 728–734. doi:10.1016/j.jlp.2008.07.013

Wessberg, N., Molarius, R., Seppala, J., Koskela, S., & Pennanen, J. (2008). Environmental risk analysis for accidental emissions. *Journal of Chemical Health and Safety*, *15*(1), 24–31. doi:10.1016/j.jchas.2007.07.001

Wickens, C. D., & Hollands, J. G. (2000). *Engineering psychology and human performance.* Upper Saddle River, NJ: Prentice Hall.

Widodo, K. H., Nagasawa, H., Morizawa, K., & Ota, M. (2006). A periodical flowering–harvesting model for delivering agricultural fresh products. *European Journal of Operational Research*, *170*(1), 24–43. doi:10.1016/j.ejor.2004.05.024

Wijeratne, A. B., Turnquist, M. A., & Mirchandani, P. B. (1993). Multiobjective routing of hazardous materials in stochastic networks. *European Journal of Operational Research*, *65*(1), 33–43. doi:10.1016/0377-2217(93)90142-A

Wikborg, U., & Lee, T. E. (2013). Noncyclic scheduling for timed discrete-event systems with application to single-armed cluster tools using pareto-optimal optimization. *IEEE Transactions on Automation Science and Engineering*, *10*(3), 699–710. doi:10.1109/TASE.2012.2217128

Williams, C. K. I. (1998). Prediction with Gaussian processes: From linear regression and beyond. In M. I. Jordan (Ed.), *Learning and inference in graphical models* (pp. 599–621). Kluwer Academic Press; doi:10.1007/978-94-011-5014-9_23

Wiszniewska, A. (2013). Public Transportation Ticket on a Mobile Phone in Warsaw (Poland). *Eltis.org*. Retrieved from [REMOVED HYPERLINK FIELD]http://www.eltis.org/discover/case-studies/public-transportation-ticket-mobile-phone-warsaw-poland

World Commission on Environment and Development. (1987). *Our Common Future*. Oxford, United Kingdom: Oxford University Press.

Wu, Y. H., & Miller, H. J. (2002). Computational tools for measuring space-time accessibility within transportation networks with dynamic flow. *Journal of Transportation and Statistics*, *4*(1), 1–14.

Xie, Y., Lu, W., Wang, W., & Quadrifoglio, L. (2012). A multimodal location and routing model for hazardous materials transportation. *Journal of Hazardous Materials, 227– 228*, 135– 141.

Xifeng, T., Ji, Z., & Peng, X. (2013). A multi-objective optimization model for sustainable logistics facility location. *Transportation Research Part D, Transport and Environment*, *22*, 45–48. doi:10.1016/j.trd.2013.03.003

Xiong, G., & Wang, Y. (2014). Best routes selection in multimodal networks using multi-objective genetic algorithm. *Journal of Combinatorial Optimization*, *28*(3), 655–673. doi:10.1007/s10878-012-9574-8

Xu, J., Liu, Q., & Wang, R. (2008). A class of multi-objective supply chain networks optimal model under random fuzzy environment and its application to the industry of Chinese liquor. *Information Sciences*, *178*(8), 2022–2043. doi:10.1016/j.ins.2007.11.025

Yaman, H. (2008). Star P-hub median problem with modular arc capacities. *Computers & Operations Research*, *35*(9), 3009–3019. doi:10.1016/j.cor.2007.01.014

Yaman, H., Kara, B. Y., & Tansel, B. C. (2007). The latest arrival hub location problem for cargo delivery systems with stopovers. *Transportation Research Part B: Methodological*, *41*(8), 906–919. doi:10.1016/j.trb.2007.03.003

Yang, D. H., Goerge, R., & Mullner, R.(2006). Comparing GIS-Based Methods of Measuring Spatial Accessibility to Health Services. *Journal of Medical Systems*, *30*(1), 23–32. doi:10.1007/s10916-006-7400-5 PMID:16548411

Yang, J., Li, F., Zhou, L., Zhang, L., Huang, L., & Bi, J. (2010). A survey on hazardous materials accidents during road transport in China from 2000 to 2008. *Journal of Hazardous Materials*, *184*(1-3), 647–653. doi:10.1016/j.jhazmat.2010.08.085 PMID:20863620

Yan, Z., & Yuan, C. W. (2003). Ant colony optimization for navigating complex labyrinths. In G. Y. Wang, Q. Liu, Y. Y. Yao, & A. Skowron (Eds.), *Rough Sets, Fuzzy Sets, Data Mining, and Granular Computing, Lecture Notes in Artificial Intelligence, Volume:2639* (pp. 445–448). Berlin: Springer Verlag. doi:10.1007/3-540-39205-X_74

Yates, J., Wang, X. H., & Chen, N. N. (2014). Assessing the effectiveness of k-shortest path sets in problems of network interdiction. *Optimization and Engineering*, *15*(3), 721–749. doi:10.1007/s11081-013-9220-z

Yen, J. W., & Birge, J. R. (2006). A stochastic programming approach to the airline crew scheduling problem. *Transportation Science*, *40*(1), 3–14. doi:10.1287/trsc.1050.0138

Yu, B., Zhu, H., Cai, W., Ma, N., Kuang, Q., & Yao, B. (2013). Two-phase optimization approach to transit hub location – the case of Dalian. *Journal of Transport Geography*, *33*, 62–71. doi:10.1016/j.jtrangeo.2013.09.008

Yu, R., Leung, P., & Bienfang, P. (2006). Predicting shrimp growth: Artificial neural network versus nonlinear regression models. *Aquacultural Engineering*, *34*(1), 26–32. doi:10.1016/j.aquaeng.2005.03.003

ZareMehrjerdi, Y. Z. (2009). Excellent supply chain management. *Assembly Automation*, *29*(1), 52–60. doi:10.1108/01445150910929866

Zhang J., & Zhao L. (2007). Risk analysis of dangerous chemicals transportation. *Systems Engineering - Theory & Practice, 27*(12), 117-122.

Zhang, G. P., Patuwo, B. E., & Hu, M. Y. (2001). A simulation study of artificial neural networks for nonlinear time-series forecasting. *Computers & Operations Research*, *28*(4), 381–396. doi:10.1016/S0305-0548(99)00123-9

Zhang, J. J., Hodgson, J., & Erkut, E. (2000). Using GIS to assess the risks of hazardous materials transport in networks. *European Journal of Operational Research*, *121*(2), 316–329. doi:10.1016/S0377-2217(99)00220-9

Zhang, X. G., Zhang, Y. J., Zhang, Z. L., Mahadevan, S., Adamatzky, A., & Deng, Y. (2014). Rapid physarum algorithm for shortest path problem. *Applied Soft Computing*, *23*, 19–26. doi:10.1016/j.asoc.2014.05.032

Zhao, L., Wang, X., & Qian, Y. (2012). Analysis of factors that influence hazardous material transportation accidents based on Bayesian networks: A case study in China. *Safety Science*, *50*(4), 1049–1055. doi:10.1016/j.ssci.2011.12.003

Zhao, P., Lu, B., & de Roo, G. (2011). The impact of urban growth on commuting patterns in a restructuring city: Evidence from Beijing. *Papers in Regional Science*, *90*(4), 735–754. doi:10.1111/j.1435-5957.2010.00343.x

Zhou, P., Chen, X., Wu, Y., & Shang, Z. (2010). Gaussian process: An alternative approach for QSAM modeling of peptides.[PubMed]. *Amino Acids*, *38*(1), 199–212. doi:10.1007/s00726-008-0228-1

Zhou, P., Tian, F., Chen, X., & Shang, Z. (2008). Modeling and prediction of binding affinities between the human amphiphysin SH3 domain and its peptide ligands using genetic algorithm-Gaussian processes.[PubMed]. *Biopolymers*, *90*(6), 792–802. doi:10.1002/bip.21091

Zhu, Q. Y., Qin, A. K., Suganthan, P. N., & Huang, G. B. (2005). Evolutionary extreme learning machine. *Pattern Recognition*, *38*(10), 1759–1763. doi:10.1016/j.patcog.2005.03.028

Zhu, X., & Dale, A. P. (2000). Identifying Opportunities for Decision Support Systems in Support of Regional Resource Use Planning: An Approach through Soft Systems Methodology. *Environmental Management*, *26*(4), 371–384. doi:10.1007/s002670010094 PMID:10954800

Zografos, K. G., & Androutsopoulos, K. N. (2002). Heuristic algorithms for solving hazardous materials logistical problems. *Transportation Research Record*, *1783*, 158–166. doi:10.3141/1783-20

Zografos, K. G., & Androutsopoulos, K. N. (2008). A decision support system for integrated hazardous materials routing and emergency response decisions. *Transportation Research Part C, Emerging Technologies*, *16*(6), 684–703. doi:10.1016/j.trc.2008.01.004

Zografos, K. G., & Davis, C. (1989). Multi-objective programming approach for routing hazardous materials. *Journal of Transportation Engineering*, *115*(6), 661–673. doi:10.1061/(ASCE)0733-947X(1989)115:6(661)

Zografos, K. G., & Samara, S. (1989). Combined location–routing model for hazardous waste transportation and disposal. *Transportation Research Record*, *1245*, 52–59.

Zografos, K. G., Vasilakis, G. M., & Giannouli, I. M. (2000). A methodological framework for developing a DSS for hazardous material emergency response operations. *Journal of Hazardous Materials*, *71*, 503–521. doi:10.1016/S0304-3894(99)00096-5 PMID:10677678

Zuazola, G I.J., Moreno, A., Landaluce, H., Angulo, I., Perallos, A., Hernández-Jayo, U., Sainz, N., Ziai, M.A., Batchelor, J.C., Elmirghani, J.M.H. (2013). Telematics system for the intelligent transport and distribution of medicines. *Intelligent Transport Systems, 7*(1), 131-137.

Zunxiong, L., Xin, X., & Zhang, D. (2006). Predict chaotic time series using minimax probability machine regression. *Information Technology Journal*, *5*(3), 529–533. doi:10.3923/itj.2006.529.533

About the Contributors

Ebru V. Ocalir-Akunal is associate professor at the City and Regional Planning Department of the Gazi University, Ankara, Turkey. She graduated from the City and Regional Planning Department of Gazi University in 1993. She completed her master's studies in the Regional Planning Department of the METU in 1996. She got her Ph. D. degree from the Institute for Transport Planning and Traffic Engineering of Vienna University of Technology in 2003. She carried out post-doctoral studies in the Research Centre of Transport Planning and Traffic Engineering of Vienna University of Technology in 2010. Her main topics of research are the development of spatial structures and mobility and decision support systems.

***

Cengiz Akcam is working as Civil Engineer at General Highway Directorate of Denizli Brach, Turkey.

Darcin Akin has undergraduate (graduated as valedictorian) and master's degrees in Civil Engineering from Dokuz Eylul University in 1987 and 1992, respectively. He received his doctoral degree from Michigan State University (MSU), USA in 2000 in Civil Engineering. After graduation from MSU, he joined the Department of City and Regional Planning at Gebze Institute of Technology (GIT), Kocaeli. He taught graduate courses at GIT in transportation engineering and planning and supervised graduate thesis in several aspect of urban planning. In 2012, He joined the Department of Urban and Regional Planning at Yildiz Technical University, Istanbul. He also served as a technical advisor during the preparation of Istanbul Master Transport Plan from 2005 to 2010. He also led the project of Trip Generation Handbook for the city of Istanbul. His research interests include transportation engineering and planning. He has published and presented papers at national and international journals and conferences, including ASCE journals, TRB Meetings and the TRB Journal. He also talked several conferences organized by the Ministry of Transportation, Maritime Affairs and Communications, and the Metropolitan Municipalities of Istanbul and Kocaeli. Dr. Akin is currently a full-professor and the chairman of the Department of Civil Engineering at Gediz University in Izmir, Turkey.

Serdar Alasalvar graduated from Department of Civil Engineering in Pamukkale University, Denizli, Turkey in 2000. Upon his graduation he worked at several consulting firms and developed several civil engineering projects. He served at Air Force Logistics Command as a Third Lieutenant. He has been working as a civil engineer at Istanbul Metropolitan Municipality since 2008. He received his master's degree from Bahcesehir University, Istanbul, Turkey with a thesis titled Modelling Urban Growth and Expansion Using Interdistrict Travel Data: Case Study of Sakarya Province, Turkey, under the supervision of Dr. Darcin Akin.

Turan Arslan received his BS degree in Civil Engineering from the Yildiz Technical University, Istanbul, Turkey, in 1993 and the MSc and PhD degrees in the Civil Engineering from Illinois Institute of Technology, Chicago, USA, in 1997 and 2003, respectively. He joined the staff of Mustafa Kemal University, Hatay, Turkey in 2004 as an Assistant Professor. He became an Associate Professor in 2011. Since June 2012, he has joined Uludag University and been with the Department of Civil Engineering and Head of Transportation Division in the Uludag University. His research interests focus on transportation modeling, travel behavior, and decision support systems.

Can Elmar Balas 1. Family name: Balas 2. First names: Can Elmar 3. Date of birth: 1968 4. Nationality: Turkish 5. Civil status: Married 6. Education: Institution (Date from - Date to) Degree(s) or Diploma(s) obtained: Middle East Technical University (METU), Ankara 1986 – 1990 B.Sc. in Civil Engineering Middle East Technical University (METU), Ankara 1990 – 1992 M.Sc. in Coastal Engineering Middle East Technical University (METU), Ankara 1992 – 1998 Ph.D. in Coastal Engineering 8. Membership of professional bodies: American Society of Civil Engineers (ASCE) Mediterranean Coastal Environment (MEDCOAST) International Maritime Organization (IMO) Formal Safety Assessment (FSA).

Marco Bortolini is a Post-Doctoral fellow at the Department of Industrial Engineering of the University of Bologna, Italy. In 2013, he received his PhD degree from the University of Padova, Italy, Industrial Engineering doctoral school. His main research activities include the study of effective strategies and tools to design industrial plants and logistic systems, the study of optimization and heuristic models applied to integrated supply chain management, the feasibility analysis of production plants and renewable energy systems. In these fields, he takes part in several research projects in partnership with other universities, scientific centers and private companies. He is author of 60 scientific contributions for international journals, book chapters and conference sessions.

Ziya Cakici is working at Pamukkale University as a Research Assistant since 2012.

Fatih Çipil obtained his Ph.D. in Department of Environmental and Technical Research on Accidents from the Gazi University. His research field lies in the area of environmental and safety traffic. He is working on applications of efficiency analysis, renewable energy technologies, data envelopment analysis and artificial intelligence applications.

Aykut Süreyya Duyguvar is Industrial Engineer graduated from Istanbul Technical University in 2004. He is productivity analysis officer in military and serving with the rank of captain. He is now Ph.D.Candidate in Supply Chain Department in Turkish Military Academy Defence Science Institute.

Sebnem Duzgun has been performing research on spatial information technologies, spatial, spatio-temporal data analyses and mining in Earth science and engineering as well as interdisciplinary studies. She is the author of two books, five book chapters and more than 200 scientific papers.

Ozge Yalciner Ercoskun is an associate professor in the City and Regional Planning Department of the Gazi University, Ankara, Turkey. She graduated from the City and Regional Planning Department of the Istanbul Technical University in 1998. She completed her master's studies in the Geodetic and Geographic Information Technologies Department of the METU in 2002. She got her Ph.D. degree from the City and Regional Planning Department of the Gazi University in 2007. She has attended several national and international congresses; summer schools and workshops related to ecological urban planning and geographic information systems. She has written 3 books and more than 50 papers on sustainable urban design and ecological and smart urban planning, geographic information systems and information technologies. She has worked as a researcher in many national and institutional projects. She has awards about sustainability and urban growth, sustainable tourism.

Serpil Erol is a professor in the Department of Industrial Engineering at Gazi University, She received her Ph.D. from the Ankara University in 1987. Her current research interests are in systems analysis and design, production planning, inventory theory, supply chain management, logistics management, flexible manufacturing systems, distribution and transportation. Professor Erol has over 90 publications of several journals. She attended and presented papers at the national and international conferences.

Kivanc Ertugay is a GIS specialist urban planner. His research interests are; Accessibility measures, accessibility modeling, Geographical Information Systems (GIS), Transportation Planning, Service/Catchment Area, Location/Allocation, Supply/Demand related analysis, and Spatial Data Analysis.

Mauro Gamberi is an Associate Professor of industrial plants and logistics at the Department of Management and Engineering, University of Padova, Italy. His key research topics deal with the design and optimization of complex production systems, manufacturing processes and assembly lines, inbound and outbound networks and warehousing systems. He, further, focuses on the mechanical plant design, validation and field-test developing collaborations with private companies and public research centers.

Margaret Hamilton, BSc (Hons), PhD, University of Wollongong, Australia, is an Associate Professor in the School of Computer Science and Information Technology at RMIT University in Melbourne Australia. Her current research centres on Mobility, Human Computer Interaction and Computer Science Education. Smart transportation has been her passion, and her recent research directions have involved smart bus applications as well as integration of real-time, multimodal public transport systems to provide passengers with up to the minute accurate optimised travel options. She currently supervises 5 PhD students and researches into the smart sensing and cloud services research area. Her teaching currently involves Introduction to Information Technology, Professional Computing Practice (involving ethics, employability skills, legal obligations), Advanced Professional Development, and has included Web Programming, Programming in C++, Computing Theory. She is currently holds two A* research grants from the Australian Government, and is working on developing sustainability resources throughout RMIT University.

Jagan J. is currently working as the Junior Research Fellow (JRF) for the Sponsored Research project by BRNS in the Centre for Disaster Mitigation and Management (CDMM), VIT University, Vellore. His project area covers the Cone penetration tests and liquefaction susceptibility of soils. He is also pursuing the M.S. (By Research) in Geotechnical Engineering in the same department. His current research is in unsaturated soil by using the data mining techniques. He also touches the area like slope stability, site characterization, pile foundation, etc. He had completed his Bachelor of Engineering (B.E.) in Civil Engineering in Adhiyamaan College of Engineering, Hosur.

Serhan Karabulut graduated from the Land Forces Academy, Department of Mechanical Engineering in 2003. He completed his master's studies in Gazi University, Traffic Planning and Implementation in 2007. He got his Ph. D. degree from Gazi University, Department of Environmental and Technical Research of Accidents in 2014. His research area is the dangerous goods transportation. Karabulut is currently serving as an officer in the Turkish Land Forces, Transport Command.

Yetis Sazi Murat is a full time Professor at Department of Civil Engineering, Pamukkale University, Denizli, Turkey. He has been working on various topics of transportation science (traffic engineering, transportation planning, public transportation etc.) since 1994. Use of Artificial Intelligence Methods in Transportation Science is one of the main research interests of him. He has many research papers on these subjects.

Mihaita Iulian Niculae is a research assistant at the Centre for Environmental Research and Impact Studies at the University of Bucharest. He studies landscape ecology, the dynamics of the countryside and conservation planning. Niculae has attended courses in environmental economics and management at the Luigi Bocconi Comercial University, Milan. His work has been presented in national and international scientific conferences and been published in several number of articles in international databases indexed journals.Since 2006, he has been an active participant in research projects related to biodiversity conservation and integrated waste management.

Mihai Razvan Nita Researcher at the Centre for Environmental Research and Impact Studies and Lecturer at the Faculty of Geography, University of Bucharest. Phd in Geography since 2011, main research topics include green infrastructures, environmental conflicts and environmental impact assessments.

Francesco Pilati is a Ph.D. student in Mechatronics and Product Innovation Engineering at the Department of Management and Engineering, University of Padova, Italy. His key research topics include innovative approaches for production system design. He implements these approaches in several fields, such as warehousing, assembly lines and plant facility design. His research, moreover, focuses on the development of methods and techniques to consider, simultaneously, different industrial plant aspects as economic profitability, environmental sustainability and technical performance.

Alberto Regattieri is a Full Professor at the Department of Industrial Engineering of the University of Bologna, Italy. He teaches undergraduate and graduate courses in logistics, operations and industrial plant design. His current research interests include logistics, the optimal design of manufacturing and service systems, the production planning and control and the development of effective maintenance strategies for industrial plants. He manages several research projects on these fields granted by public institutions and private companies.

Sangeeta Roy is an Undergraduate in Civil Engineering. She completed her B.Tech (Civil) in School of Mechanical and Building Science, VIT University, Vellore. She has vast knowledge in the field of geotechnical engineering and also in soft computing. She also has good exposure on pile foundation, soil slope stability, etc. She also developed her knowledge on the integration of soft computing and transportation engineering. Her area of interest covers the soil engineering with the combination of data mining techniques.

Pijush Samui is a professor at Centre for Disaster Mitigation and Management in VIT University, Vellore, India. He obtained his B.E. at Bengal Engineering and Science University; M.Sc. at Indian Institute of Science; Ph.D. at Indian Institute of Science. He worked as a postdoctoral fellow at University of Pittsburgh (USA) and Tampere University of Technology (Finland). He is the recipient of CIMO fellowship from Finland. Dr. Samui worked as a guest editor in Disaster Advances journal. He also serves as an editorial board member in several international journals. Dr. Samui is editor of International Journal of Geomatics and Geosciences. He is the reviewer of several journal papers. Dr. Samui is a Fellow of the International Congress of Disaster Management and Earth Science India. He was the recipient of Shamsher Prakash Research Award for the year of 2011.

Bülent Soykan is a Ph.D. candidate at Old Dominion University, having received his M.S. in Technology Management from Defense Sciences Institute, Turkish Military Academy. His research interests include combinatorial optimization, airline crew scheduling, and simulation. He has published research in The Journal of Defense Sciences.

Adnan Sozen is Professor in the Department of Energy Systems Engineering at the Faculty of Technology of Gazi University. He obtained his M.Sc. and Ph.D. in Mechanical engineering from the Gazi University. His research field lies in the area of environmental and energy. He is working on applications of efficiency analysis, renewable energy technologies, solar heating/cooling systems, energy systems, environmental efficiency, data envelopment analysis and artificial intelligence applications. Professor SÖZEN has published more than 120 papers in reputable academic journals (75 of them are covered by SCI articles) and conferences.

Maria Spichkova received her Bachelor Degree in Computer Science at Moscow Power Engineering Institute (Technical University, Russia) in 2000 and a Master's Degree in Computer Science at Dresden University of Technology (Germany) in 2003, respectively. She received a PhD in Computer Science in 2007 at Technical University Munich (Germany), where she worked in a researcher and lecturer role between 2003 and 2013. At the Technical University Munich, Maria Spichkova was involved both as a participant and as project leader in a large number of researched industrial projects in collaboration with AVL, Microsoft, DENSO Automotive, Robert Bosch GmbH, BMW Car IT, etc. She also supervised Bachelor and Master's theses, taught a number of lecture courses (e.g., Applied Logic in Engineering, Modelling of distributed systems, Fundamentals of system and program development) as well as seminar courses. Currently, she is a Lecturer at the School of Computer Science and IT, RMIT University, Australia. She conducts research related to specification, modelling, testing, and verification of safety-critical and distributed systems, as well as to human factor related areas.

Gabriel Ovidiu Vanau is a geographer whose research interests focus on land use and environmental planning, urban-rural interaction and linkages, protected areas governance in relation to economic development. Gabriel Ovidiu Vanau holds a doctorate in geography from the University of Bucharest. His doctoral thesis examined the fringe of Bucharest as an urban-rural interface. He works as a research assistant at the Centre for Environmental Research and Impact Studies and as a teaching assistant at the Faculty of Geography, both within the University of Bucharest. He is strongly committed to integrating education and outreach into research projects, particularly by exposing his students to the excitement of scientific research.

Zafer Yilmaz 1996- Bachelor degree from Turkish Military Academy (Systems Engineering) 2003- MS degree from Middle East Technical University (Informatics Institute Modelling and Simulation Department) 2015- PhD. degree in Supply and Logistic Management Department of Turkish Military Academy Defense Sciences Institute.

Index

A

B

C

D

E

F

G

H

I

L

M

N

O

P

R

S

T

U

W

www.ingramcontent.com/pod-product-compliance
Lightning Source LLC
LaVergne TN
LVHW060353300926
849162LV00006B/1256